*Engineering Materials*

Springer

*Berlin*
*Heidelberg*
*New York*
*Barcelona*
*Hong Kong*
*London*
*Milan*
*Paris*
*Singapore*
*Tokyo*

Valentin G. Gavriljuk · Hans Berns

# High Nitrogen Steels

## Structure, Properties, Manufacture, Applications

With 199 Figures

Springer

Prof. Dr. Valentin G. Gavriljuk
Institute of Metal Physics
36 Vernadsky Prospect
252142 Kiev
Ukraine

Prof. Dr. -Ing. Hans Berns
Lehrstuhl Werkstofftechnik
Ruhr Universität
44780 Bochum
Germany

Cataloging-in-Publication Data applied for

Die Deutsche Bibliothek - CIP-Einheitsaufnahme
**Gavriljuk, Valentin G.:**
High nitrogen steel: structure, properties, manufacture, applications
Valentin G. Gavriljuk; Hans Berns.
Berlin; Heidelberg; NewYork; Barcelona; Hong Kong; London; Milan;
Paris; Singapur; Tokyo: Springer, 1999
(Engineering materials)
ISBN 3-540-66411-4

ISBN 3-540-66411-4 Springer-Verlag Berlin Heidelberg New York


Printed in Germany



Typesetting: Camera-ready copies from authors
Cover-Design: MEDIO, Berlin
Printed on acid-free paper SPIN 10730039 62/3020 5 4 3 2 1 0

to BELA and ISOLDE

# Preface

During the second half of this century the nitrogen concentration of steel moved in both directions: down in constructional grades by oxygen blowing to prevent brittle fracture upon age hardening at ambient temperature and up in stainless grades to improve strength, corrosion resistance and austenite stability. The gradual recognition of the beneficial effects of this element on the properties of high alloy steels led to a widespread development of high nitrogen steels (HNS) documented by numerous applications and the proceedings of five HNS conferences at Lille (France) 1988, Aachen (Germany) 1990, Kiev (Ukraine) 1993, Kyoto (Japan) 1996 and Helsinki/Stockholm (Finnland/Sweden) 1998.

The word "high" in HNS is not clearly defined but accounts e.g. for 0.1 mass% of nitrogen in creep resistant steels, 0.9 mass% in stainless grades or 2 mass% in tool steels. "High" best refers to "intentionally raised" by appropriate alloying or by pressure and powder metallurgy. For convenience steel grades are designated by their approximate content of alloying elements in mass% omitting unintentional minor additons of Si, Mn, C, P, S asf. As an example Cr22Ni5Mo3N0.2 stands for a stainless duplex grade.

The present book starts by comparing the effects of nitrogen and carbon on the atomic structure and interaction within the crystal lattice. Next the constitution of HNS is investigated leading to specific microstructural features. Based on this background the mechanisms behind the mechanical, chemical and tribological properties of HNS are derived. This concludes the fundamentals treated in chapters one to three. Chapter four comes down to shop practice of manufacturing HNS which, in respect to e.g. melting, hot working, and welding, may be quite different from respective carbon grades. In chapter five, different HNS for special applications are presented like e.g. hardenable steels for stainless bearings, high strength austenitic steels for non-magnetic retaining rings, superaustenitic stainless sheet for the chemical industry or solution nitrided impellers for pumps. Chapter six is an attempt to summarize key aspects of HNS.

The book is meant for material scientist working in the field of high alloy steels and also for engineers engaged in materials technology, material selection and design. The work covers a wide scope from the atomic structure to the application of HNS and was written by a metal physicist and an engineer.

The authors would like to express their gratitude for the assistance of numerous persons in the institutes at Kiev and Bochum who carried out research on HNS or helped to prepare this book for print. Thanks are especially extended to Dr. Bela Shanina, Dr. Alexander Tarasenko, Dipl.-Ing. Andreas Packeisen and Dipl.-Ing. Martin Schellewald for the layout of the manuscript and figures. The authors also acknowledge the financial support of their research by the National Academy of

Science and the Science and Technology Center in Ukraine, by the Federal Ministry of Research and Education and the Research Foundation of Germany, by the State of Northrhine-Westphalia and by the European Union.

V.G. Gavriljuk

H. Berns

# Contents

# 1 Structure

High nitrogen steels belong to iron-based multicomponent solid solutions which are constructed by substitutional (iron, chromium, nickel, manganese etc.) as well as interstitial (nitrogen, carbon) atoms and, in addition, can be strengthened by the particles of nitrides. The mechanical and electrochemical properties of solid solutions are mainly determined by the behaviour of solutes and dislocation properties.

The distribution of solute atoms is far from that typical for ideal solid solutions where solute atoms do not interact with each other. Even in the absence of nitrogen, the atoms of alloying elements reveal a clear tendency to some preference in relation to their nearest atomic neighbourhood. It is well known, for example, that nickel atoms prefer to have the iron ones as nearest neighbours in iron-based austenite, which is described in terms of short range atomic ordering, whereas the iron-manganese alloys are characterised by the preference of Mn-Mn interatomic bonds, which results in the short-range decomposition of solid solution, i.e. the clustering of manganese atoms. A similar behaviour is also observed for chromium and molybdenum atoms in the iron matrix. Nitrogen, being dissolved in austenite, affects this atomic correlation, and this is one of the important preconditions for nitrogen influence in steel.

Another pronounced feature of nitrogen steels concerns the dislocation structure and properties of individual dislocations. During plastic deformation of austenitic steels, nitrogen promotes the planar slip and changes the distribution of dislocations from a tangled structure to so-called pile-up's. The strong interaction between nitrogen atoms and dislocations in austenite is not generally expected as interstitial atoms induce the cubic distortions in fcc lattice and only interstitial-substitutional atomic complexes cause small deviations from the cubic symmetry as compared to those in bcc crystals where interstitials induce remarkable tetragonal distortions and reveal a strong interaction with dislocations. Nevertheless, the value of binding energy between nitrogen atoms and dislocations in austenitic steels exceeds that for carbon atoms and it is close to that between interstitials (carbon, nitrogen) and dislocations in ferrite. Both of these factors, peculiarities of dislocation structure and interstitials-dislocation interaction, are important for the mechanical properties of nitrogen steels.

A proper interpretation of nitrogen effects in steel should be based on the knowledge of electronic properties because the electronic structure determines atomic interactions as well as properties of crystal lattice imperfections. The electron exchange between solute atoms of different types affects the atomic distribution in solid solutions, including tendencies to form atomic clusters, dislocation atmospheres, grain boundary segregations as well as precipitates. For this reason, the electronic structure is also responsible for the thermodynamical stability of different phases, which is important for phase transformations and electrochemical properties. Finally, the electronic properties control the splitting of individual dislocations (the value of stacking fault energy) and the type of dislocation assemblies, which affects the formation of structure during plastic deformation and, therefore, mechanical and tribological properties.

The aim of this chapter is to describe high nitrogen steels from the viewpoint of physical metallurgy, in order to obtain some fundamental basis for the explanation of beneficial nitrogen effects. The state of nitrogen atoms in austenite and martensite and the constitution of multicomponent nitrogen-containing solid solutions and precipitates will be discussed in detail with a particular attention to the electron exchange, atomic correlation and properties of dislocations.

## 1.1 State of nitrogen atoms in austenite

By the term "state" we mean a number of parameters or properties which characterise the electronic constitution of nitrogen atoms dissolved in austenite, their distribution in the fcc iron-based lattice and the interaction with metallic atoms and crystal lattice imperfections. The crystal lattice of metals is formed by ionic cores (atomic nuclei and closed electronic shells around them) and free electrons. Interactions between atoms constituting the crystal are caused by free electrons and the overlapping unfilled electronic shells. Roughly, the first ones are responsible for the metallic component of interatomic bonding, the second ones form covalent bonds.

That is why we are going to start the discussion of nitrogen state in austenite from the electron exchange betwen nitrogen and metallic atoms (Sect. 1.1.1). The results of theoretical and experimental studies will be presented.

Next, it will be shown in Sect. 1.1.2 that the atomic distribution, i.e. the short range atomic order in fcc iron-based alloys correlates with the electronic structure and, in particular, with the type of interatomic bonding: metallic or covalent. As a reference, we use the behaviour of carbon in austenitic steels and discuss the distribution of nitrogen atoms in austenite in the comparison to that of carbon ones.

Alloying elements (substitutional solutes) affect the distribution of nitrogen atoms. The nature of interstitial-substitutional interaction, formation of interstitial-substitutional complexes (*i-s*) and their crystalline symmetry are described in Sect. 1.1.3.

The interaction between nitrogen atoms and crystal lattice defects is a special case of atomic distribution which depends on the short range atomic order in the perfect solid solution as well as on the structure of defects, mainly on the dislocation splitting. Thus, we will also describe the interaction of interstitial atoms with vacancies, dislocations, grain boundaries and discuss some peculiarities of the substructure of austenitic nitrogen steels (Sects 1.1.4 and 1.1.5).

### 1.1.1 Electronic structure

Nitrogen and carbon atoms occupy octahedral interstitial sites in the crystal lattice of austenite. Figure 1.1 shows different types of sites (octa- and tetrahedral) in the fcc and bcc lattice. Their sizes in the fcc and bcc iron (the sizes of the largest spheres that could be placed in these holes) are given in Table 1.1. A preference to occupy octahedral sites is clear for the fcc lattice where the octahedral site has a higher volume, whereas in the bcc iron the tetrahedral site provides a larger space for an interstitial atom.

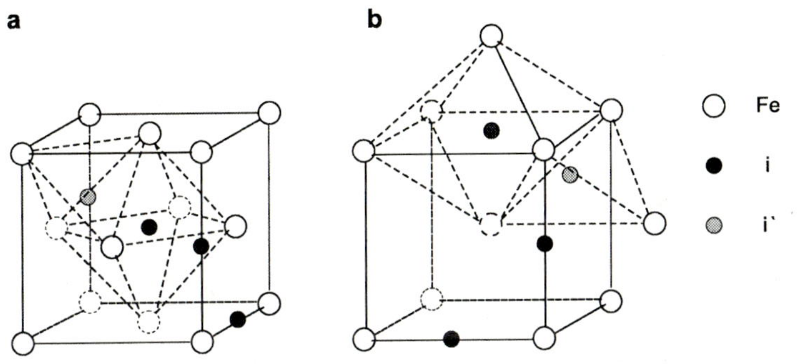


**Fig. 1.1.** Octa- (i) and tetrahedral (i`) interstitial sites: **a** in fcc and **b** in bcc lattices

**Table 1.1.** Size (nm) of octa- ($r_6$) and tetrahedral ($r_4$) holes in fcc and bcc crystal lattices of iron

| Lattice | $r_6$ | $r_4$ |
|---|---|---|
| fcc | 0.053 | 0.029 |
| bcc | 0.019 | 0.036 |

However, an interstitial atom in the octahedral site of bcc lattice (see Fig. 1.1b) shifts two iron atoms in <001> direction by about 0.05 nm for each atom (see, for example, Krivoglas and Tikhonova 1960 and Moss 1967), which increases the effective volume of the octahedral site and causes tetragonal distortions in bcc interstitial solid solutions, the latter being, in particular, important for the interaction between interstitials and dislocations and, consequently, for solid solution strengthening. As was already stated above, distortions caused by interstitial impurities in the fcc lattice have a cubic symmetry and contribute much less to the strengthening.

The atomic radii of nitrogen and carbon depend on the coordination number CN, i.e. the number of nearest neighbour atoms. According to Pauling (1960), they are equal to $r_C = 0.0771$ nm and $r_N = 0.0740$ nm for CN = 4 (so-called tetrahedral covalent radii for the diamond-like lattice). A correction for CN = 6, which describes the case of a simple cubic lattice, leads to to $r_C = 0.0890$ nm and $r_N = 0.0854$ nm (Ledbetter and Austin 1987). It is noteworthy that, for both ways of representation, the atomic radii of nitrogen is smaller than that of carbon. It follows also from the comparison of the atomic radii with the size of the holes (see Table 1.1) that the dissolution of interstitials is accompanied by significant distortions, and the size factor markedly affects the solubility of nitrogen and carbon in $\gamma$- and $\alpha$-iron.

Numerous measurements of the lattice parameters of nitrogen and carbon austenites were performed at different contents of interstitials. Some of the data available are presented in Table 1.2 as a percentage increase of the lattice parameter per the atomic per cent of nitrogen or carbon. It is surprising that nitrogen atoms, having a smaller radius as compared to carbon, cause a higher dilation being dissolved in the fcc lattice. As will be shown further, such a behaviour originates from the peculiarities of electron exchange in austenitic nitrogen and carbon steels.

**Table 1.2.** Relative change of the lattice parameter caused by nitrogen and carbon in iron austenites, $100\Delta a/a_0 \cdot c_{\text{at.\%}}$ (in per cent)

| Alloy | Nitrogen | Carbon | Reference |
|---|---|---|---|
| Fe | 0.224 | 0.210 | Liu Cheng et al. (1990) |
| Fe | - | 0.199 | Ridley et al. (1969) |
| Fe | - | 0.190 | Ridley and Stuart (1970) |
| Fe | 0.218 | - | Paranjpe et al. (1950) |
| Fe-18Cr-10Ni | 0.240 | 0.218 | Ledbetter and Austin (1987) |
| Fe-18Cr-10Ni | 0.234-0.279 | 0.125-0.167 | Irvine et al. (1961) |
| Fe-25Cr-18Ni | 0.214 | - | Kikuchi et al. (1976) |
| Fe-25Cr-28Ni-2Mo | 0.211 | - | Kikuchi et al. (1976) |

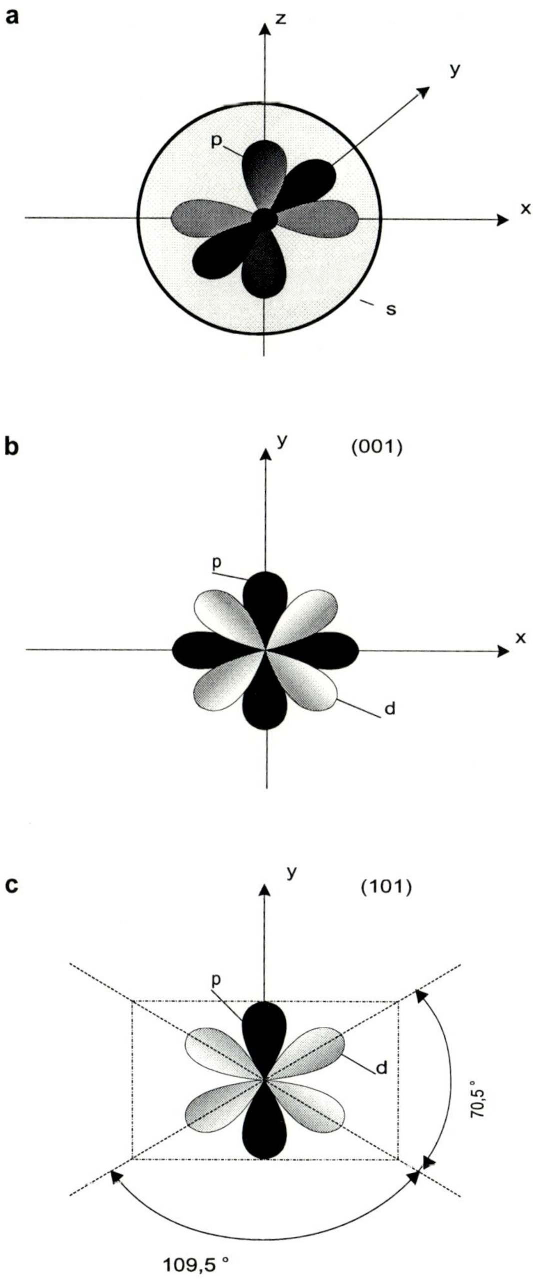


**Fig. 1.2.** A space symmetry of *s*-, *p*- and *d*-electrons in the cubic lattice of transition metals. **a** - a sphere of *s*-electrons and leaves of *p*-electrons in three dimensions; **b** - plane (001): $p_x$- and $p_y$-electrons (black leaves), $d_{xy}$–electrons (bright leaves); **c** - plane (101): $p_y$-electrons (black leaves), $d_{xyz}$–electrons (bright leaves)

Steels belong to iron-based alloys and the most part of alloying elements are of the same 3*d* period of transition metals as iron (Ni, Co, Cu, Mn, Cr, V, Ti). It is essential for the electronic structure of atoms that it is built by electrons having a different space symmetry and, respectively, a different space distribution in the crystal lattice, which depends on the value of so-called orbital quantum number of electrons. While *n*, the main quantum number, determines the number of the electronic shell (number of the period in the periodic table of elements), *l*, the orbital quantum number, determines the orbital electron moment and can have 2*n*-1 values ($l = 0$ is denoted by *s*, $l = 1$ by *p*, $l = 2$ by *d*, …). Space localisations of electrons in atoms of transition metals are shown in Fig. 1.2. The spherical symmetry is typical for *s*- electrons, whereas *p*- and *d*-electrons have a leaf symmetry.

The electronic structure of an iron atom is described as $1s^2 2s^2 2p^6 3s^2 3p^6 3d^6 4s^2$, where 1,2,… denote the value of *n*, while *s*, *p*, *d* correspond to different values of *l* and the superscripts show the number of electrons in each state. The unfilled external shell has 3*d*- and 4*s*-electrons. In the metallic crystal lattice, as it was obtained from the data of Mössbauer spectroscopy (Walker et al. 1961), the iron atoms have the structure ….$3d^7 4s^1$.

External *s*-electrons in metals become free. They form so-called "conduction electrons" which give the main contribution to metallic conductivity. Free electrons cause in the crystal lattice the metallic bonds which are delocalised in the interatomic space and, for this reason, provide high ductility and fracture toughness. If the metallic bonds prevail, the mutual shift of atoms in the crystal lattice (e.g. the shift caused by the dislocation movement in the slip plane) does not cause a significant decrease in their bonding. External *d*-electrons form interatomic bonds in some definite directions (see Fig. 1.2b), although they are partly delocalised and contribute to electro- and thermoconductivity. The localised *d*-electrons provide covalent bonds in metals which are responsible for the brittleness. The *p*-shell in iron is closed and does not contribute to atomic interactions.

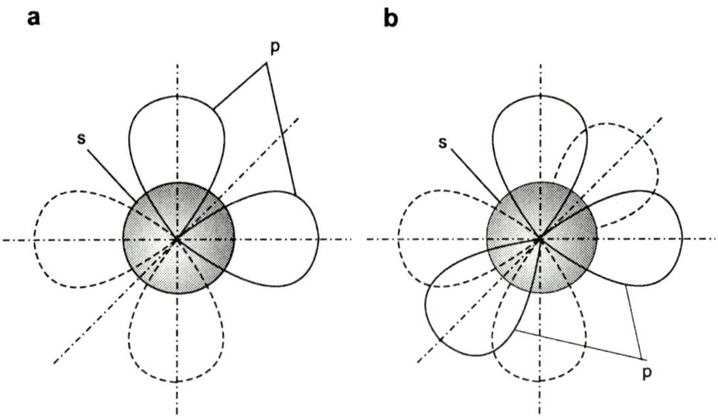


**Fig. 1.3.** Electron structure of free carbon **a** and nitrogen **b** atoms

The electronic structures of carbon and nitrogen atoms are $1s^2 2s^2 2p^2$ and $1s^2 2s^2 2p^3$, respectively (see Fig. 1.3). One excessive *p*-electron tightens the 2p-electron shell of the nitrogen atom and decreases its radius.

Thus, the electron exchange between iron and nitrogen(carbon) atoms in the fcc iron lattice amounts to an interaction between the iron 4*s*-, 3*d*- electrons and *s*, *p*-electrons of the interstitials, resulting in their complicated hybridisation.

### 1.1.1.1 Interatomic electron exchange

Theoretical calculations of the electron band structure for pure γ-iron and for γ-iron alloyed by nitrogen(carbon) allow to forecast a redistribution of the electron density in austenitic steels due to nitrogen or carbon.

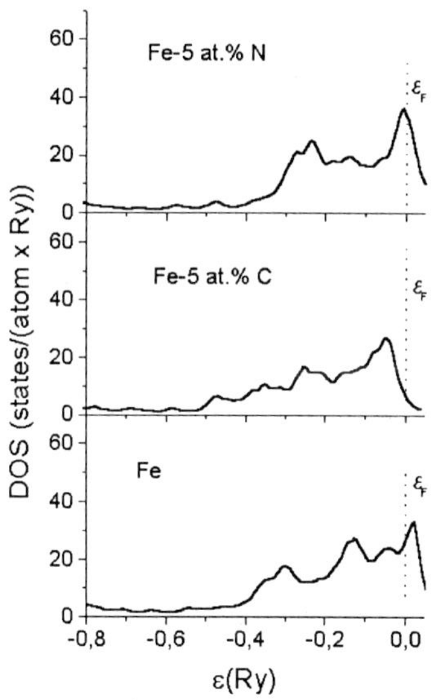


**Fig. 1.4.** Density of electron states (DOS) for fcc iron and fcc alloys Fe-5 at.% C and Fe-5 at.% N

The electron energy bands were calculated by Shanina (see Gavriljuk et al. 1999) using the *ab initio* method of norm-conserving transferable pseudopotential developed by Bachelet et al. (1982) and Greenside and Schlüter (1983). The obtained values of atomic potentials and corresponding wave functions for nitrogen, carbon and iron atoms were used in order to calculate the energy bands for pure fcc iron and for Fe-5 at.% N and Fe-5 at.% C alloys (Fig. 1.4). The value of Fermi energy in the pure fcc iron, $\varepsilon_F = 0.3835$ a.u. (1 a.u. is equal to 27.2 eV), is the same as was obtained in *ab initio* multi-electron calculations performed by Wood (1962) who used the method of augmented plane waves.

As follows from Fig. 1.4, carbon shifts the maximum of the electron state density below the Fermi level, whereas nitrogen contributes to the electronic states just at the Fermi level. Thus, one can expect that nitrogen atoms in austenite increase the concentration of free electrons, i.e. enhance the metallic component of atomic interactions. Carbon atoms are expected to contribute to the localised electrons, i.e. to enhance covalent bonds. The calculations of the distribution of *s*-electrons in the vicinity of iron and interstitial atoms show that the interaction of both nitrogen and carbon with the host atoms in the austenitic lattice results in a decrease of the contribution from iron atoms to the concentration of *s*-electrons, the nitrogen effect being stronger (see Fig. 1.5). At the same time, nitrogen gives a significant contribution to the density of *s*-electrons centered at N atoms, whereas the contribution from carbon atoms to *s*-electrons is negligible.

These results of theoretical calculations enable one to explain some experimental observations available. First of all, according to the data of electrotransfer (Seith 1955, Nakajima and Hirano 1978), the nitrogen atoms in austenite carry a negative electric charge, whereas carbon is charged positively. Based on these results, one comes usually to the conclusion that nitrogen atoms are acceptors and carbon ones are donors of electrons. In fact, the situation is more complicated and there is no clear donor-acceptor interaction. During the electrotransfer under applied voltage, the nitrogen atom in fcc iron can move towards the anode together with a cloude of free electrons in its vicinity. A carbon atom behaves as a positively charged ion because it gives its electrons to the 3*d*-band of iron and makes no essential contribution to the conduction electron band.

**Table 1.3.** Parameters of Mössbauer spectra of Fe-N and Fe-C austenites

| Alloy (mass %) | Isomeric shift* (mm/s) | Quadruple interaction* (mm/s) | Reference |
|---|---|---|---|
| Fe-1.86C | 0.04 | 0.63 | DeCristofaro and Kaplow (1977) |
| Fe-1.88C | 0.043 | 0.643 | Genin and Flinn (1968) |
| Fe-2.7N | 0.145 | 0.25 | Foct (1973) |
| Fe-2.34N | 0.21 | 0.29 | DeCristofaro and Kaplow (1977) |

* the values are given in mm/s and, related to the light velocity, $3{\cdot}10^{11}$ mm/s, and the energy of $\gamma$ quanta, 14.4 keV, they correspond to the shifts in the nuclear energy levels of $\sim 10^{-9}$ eV

The theoretically forecast increase in the concentration of free electrons in austenite due to nitrogen is also consistent with the above data of lattice dilatation of nitrogen austenite (see Table 1.2). A higher lattice dilation of nitrogen austenite as compared to the carbon one is natural if a weaker metallic component of interatomic bonding is enhanced by nitrogen. An important experimental evidence of the space distribution of the electron charge in the vicinity of iron atoms in Fe-N and Fe-C austenites can be obtained using Mössbauer spectroscopy (see Table 1.3). Because of their spherical space symmetry, only *s*-electrons have some density at the iron nuclei. If solute atoms are located in the neighbourhood of iron atoms, a shift of the nuclear levels (isomeric shift) occurs which is proportional to the change of the charge density of *s*-electrons at the iron nuclei. The positive value of the isomeric shift coresponds to a decrease in the charge density of *s*-electrons at the iron nucleus and it is much higher in the nitrogen austenite than in the carbon one. Just such a behaviour is expected, as, according to the above mentioned results (see Fig. 1.5), the electron exchange between interstitial and iron atoms results in a decrease of the density of s-electrons in the vicinity of iron atoms and this effect is stronger in iron-nitrogen alloys than in iron-carbon ones.

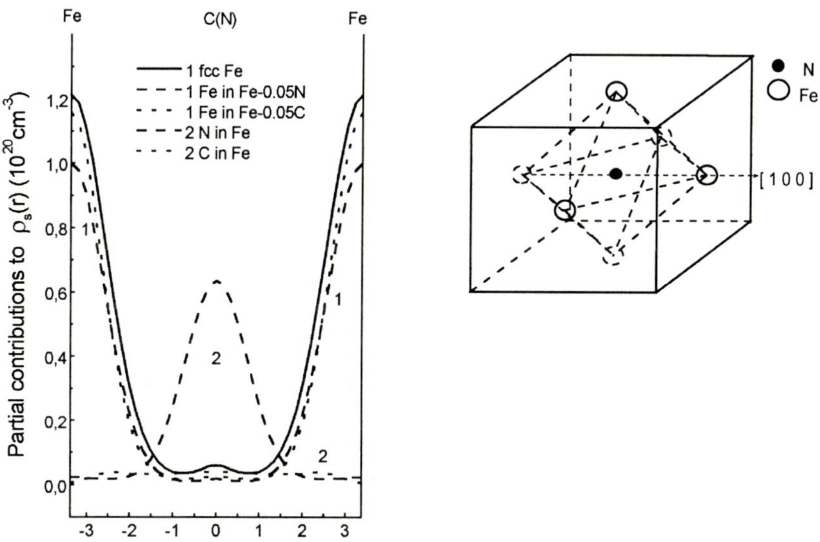


**Fig. 1.5.** Space distribution of partial density of free electrons in direction [100] in the fcc iron cell containing carbon(nitrogen) atom in the octahedral site (shown on the right): (1) contributions to *s*-electron density from Fe atoms: solid lines – in pure Fe, dashed lines – in Fe-5 at.%N, dotted lines – in Fe-5 at.%C; (2) contributions to *s*-electron density from nitrogen (dashed lines) and carbon (dotted lines) atoms in the fcc alloys Fe-5 at.%N and Fe-5 at.%C, respectively

Other experimental evidence of different distributions of *s*-electrons in nitrogen and carbon austenites is based on so-called quadruple interaction which characterises the gradient of the electric field around the iron nuclei. It is obvious that with increasing density of free electrons, the electric field in the fcc crystal lattice acquires a smaller deviation from the cubic symmetry and the value of its gradient will decrease, which, in fact, takes place if carbon in iron austenite is replaced by nitrogen (Table 1.3).

### *1.1.1.2*
### *Experimental studies of electronic structure*

Experimental data of the electron exchange between the solute and host atoms in austenitic steels were obtained by means of electron spin resonance, ESR (see Baran et al. 1992, Gavriljuk et al. 1993, Shanina et al. 1994, 1995, 1998). ESR enables one to obtain a valuable information about the concentration of free and localised electrons in paramagnetic alloys. Typical austenitic steels are paramagnetic and, therefore, this experimental method can be successfully used.

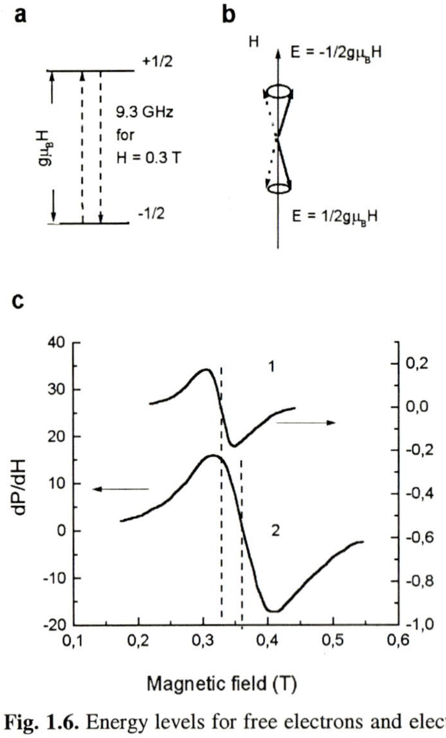


**Fig. 1.6.** Energy levels for free electrons and electron spin resonance. **a** - the splitting of energy levels corresponding to two spin orientations of electrons under magnetic field, the microwave field (GHz) has to be applied in order to transfer electrons to the upper level, **b** - two spin orientations for a free electron under applied constant magnetic field *H*, *g* is a factor of spectroscopic splitting (characterises the extent of splitting of energy levels), $\mu_B$ is the Bor magneton; **c** - signals of ESR for two austenitic steels Cr20Ni16Mn6 (0.14C) (1) and Cr21Ni16Mn7 (0.13N) (2) as written in coordinates: "the derivative of absorbed energy *P* on the magnetic field *H* - magnetic field *H*"

***Interstitial solid solutions.*** An essence of ESR is shown schematically in Fig. 1.6. A static magnetic field, being applied to the paramagnetic material, removes the magnetic degeneration of the electron energy levels and the magnetic moments of electrons (spins) become oriented along the vector of magnetic field or in the opposite direction (Fig. 1.6b), which corresponds to two different values of their energy (Fig. 1.6a). The distribution of electrons between these two states (two energy levels) is controlled by Boltzmann statistics. If a microwave field of an appropriate power is applied to the sample, some part of electrons is transferred from the low level to the upper one and their spins change the direction. The energy spent for this transfer is proportional to the concentration of free electrons, and the ESR signal (shown in Fig. 1.6c as a derivative of the absorbed energy on magnetic field) is recorded as a function of the value of magnetic field. As follows from Fig. 1.6c, the ESR signal of the nitrogen-containing austenitic steel is two orders higher as compared to the carbon one at comparable contents of interstitials. It is the direct experimental evidence of an increase in the concentration of free electrons in iron-based austenite due to nitrogen, i.e. of the enhanced metallic component of interatomic bonding in nitrogen austenite. This result helps to understand the nature of high fracture toughness of austenitic nitrogen steels and many other features of their mechanical behaviour (see Chap. 3).

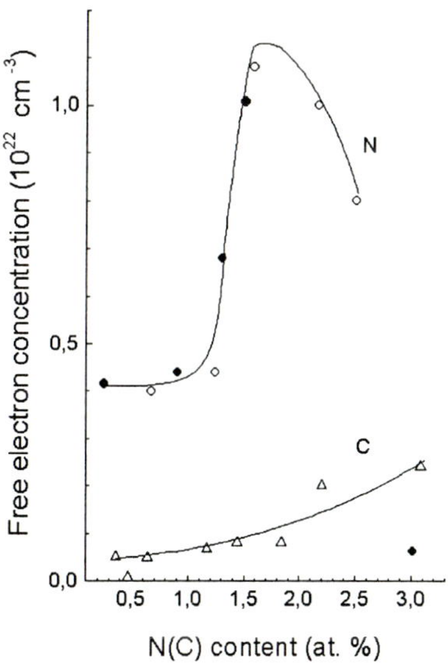


**Fig. 1.7.** Effect of nitrogen (open circles) and carbon (open triangles) on the concentration of free electrons in austenitic Cr18Ni16Mn10 steel. Data for Cr20Ni16Mn6 steel are shown by filled circles being diminished 40 times

The concentration of free electrons in austenitic nitrogen and carbon steels is shown in Fig. 1.7 as a function of the content of interstitial elements. An increase in the carbon content does not markedly change the electron concentration, whereas its complicated behaviour is observed in nitrogen steel. Its increase occurs in a definite range of nitrogen content between 0.7 and 2 at.% (approximately 0.15 to 0.5 mass %) of nitrogen. If to accept that the metallic interatomic bonding is decisive for construction steels where the fracture toughness is of particular significance, one can suppose that the above mentioned concentration interval is optimal for the alloying of austenitic construction steels by nitrogen. This suggestion is supported by studies of structure and mechanical properties of high-nitrogen steels. Foct and Akdut (1993) observed a cleavage-like fracture of austenite in duplex steel containing 0.62 % of N (in fact, the nitrogen content in austenite was higher because of nitrogen redistribution between ferrite and austenite). Müllner et al. (1994) have studied brittle fracture of austenitic nitrogen steels containing about 1 mass % of N.

Such an embrittlement is consistent with a decreasing concentration of free electrons in austenitic steels when the nitrogen content exceeds 0.6 mass % (Fig. 1.7). One can also claim that the concentration dependence shown in this Figure has some general character for CrNiMn steels: the data for the other steel having less manganese satisfy the same concentration dependence being diminished 40 times. This means that manganese deteriorates the metallic interatomic bonding, which in turns is consistent with the promotion of brittle fracture of austenitic steels due to manganese.

***Substitutional solid solutions.*** The data of electron exchange in N(C)-free iron-based substitutional solid solutions are needed for a reliable evaluation of the role of interatomic bonding on the structure and properties of nitrogen steels. These studies were conducted on a number of interstitial-free stainless steels alloyed by a varying content of Cr, Ni, Mn, Mo, Cu etc. (see Shanina et al. 1998).

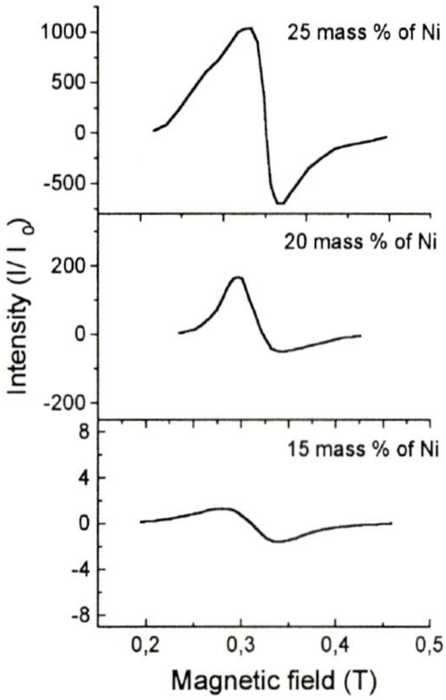


**Fig. 1.8.** Effect of nickel on the electron spin resonance in austenitic Cr15Ni(x) steel

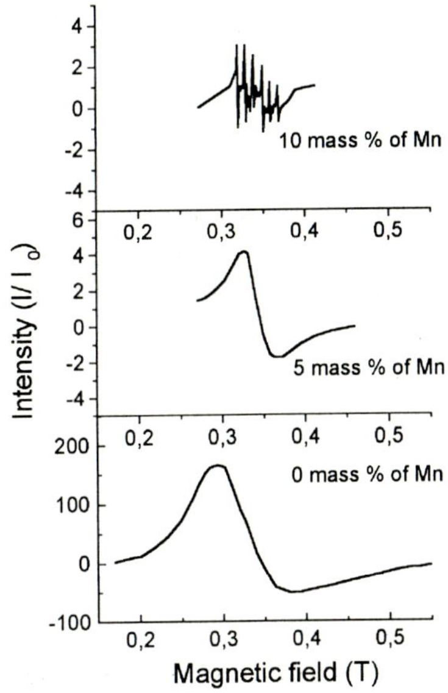


**Fig. 1.9.** Effect of manganese on the electron spin resonance in austenitic Cr15Ni20Mn(x) steel

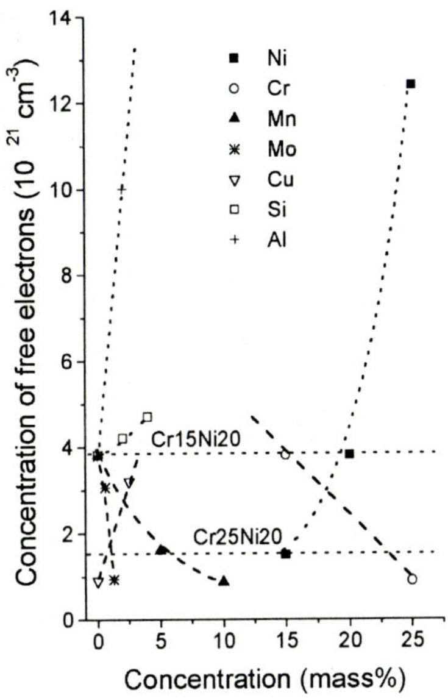


**Fig. 1.10.** Effect of alloying elements on the concentration of free electrons in austenitic CrNi steels. Dashed lines show the levels of the free electron concentrations corresponding to steels Cr15Ni20 and Cr25Ni20

As examples, the ESR signals are shown in Figs 1.8 and 1.9 for Cr15Ni(x)Mn(x) steels. The intensity of the ESR signal is enhanced by 3 orders if the nickel content is increased from 15 to 25 mass % in austenitic Cr15Ni(x) steel. This is an evidence of an increasing concentration of free electrons, i.e. enhanced metallic bonding in austenite due to nickel. If manganese is added to austenitic CrNi steels, the ESR signal is markedly decreased and, at 10 % of Mn, instead of ESR, a small signal of electron paramagnetic resonance of Mn atoms is observed due to *d*-electrons localised at Mn atoms. As was also mentioned above (see Fig. 1.7), manganese decreases the concentration of free electrons in austenitic CrNiMnN steel. The summarised data of the influence of different elements on the concentration of free electrons in austenitic steels are given in Fig. 1.10. It follows from these data that nickel, copper, silicon, aluminum assist the metallic interatomic bonding, whereas manganese, chromium, molybdenum decrease the concentration of free electrons and, therefore, enhance covalent interatomic bonding.

Si and Al are the elements containing no *d*-electrons. In relation to the d-elements it is noteworthy that Ni and Cu occupy the places to the right of Fe in the periodic table while Mn, Cr, Mo are located to the left.

On account of this circumstance, the following qualitative explanation can be given for the observed change in the electronic structure of austenitic steels due to chemical composition. If iron is alloyed by elements having a smaller amount of *d*-electrons (Mn, Cr etc.), holes are created in the 3*d*-band of iron. These empty places have to be filled by 4*s*-electrons of the iron and of alloying elements because their energy is higher than that of *d*-electrons, which leads to a decrease in the density of free electrons and results in the promotion of covalent interatomic bonding. If nickel, cobalt or copper are added to iron, their excessive *d*-electrons become free and contribute to the conduction band.

Thus, a comparison of the electron exchange between iron atoms and interstitial or substitutional solutes in austenite leads to the conclusion that, in this relation, the Fe-N alloy is similar to iron alloys with substitutional elements located to the right of iron in the periodic table. The effect of carbon in iron austenite is close to that of the elements to the left of iron.

### 1.1.2 Distribution of nitrogen and carbon atoms

Austenitic steels are based on the γ-iron solid solution. By definition, solid solutions are phases the composition of which can be changed continuously without violating their homogeneity. In fact, however, they are not homogeneous on submicroscopical scale because the distribution of solute atoms is determined by their interactions with the host atoms and other solutes. These interactions affect the homogeneity of solid solutions, solubility of nitrogen and, in particular, control the temperature and sequence of precipitation from the solid solution. In a solid solution as a statistical assembly, the interactions between atoms are expressed by thermodynamical parameters, the knowledge of which is important for the studies of nitrogen steels.

Before discussing the nitrogen distribution in austenite, we shall briefly describe the thermodynamical parameters and relations which are needed for such a discussion. A more detailed information can be found in a number of the specialised books (see e.g. Lupis 1983, Elliot et al. 1963, Wagner 1962).

#### *1.1.2.1 Thermodynamical characteristics of solid solutions*

***Enthalpy***. According to the first law of thermodynamics, energy remains constant during any process, which is written as

$$dE = \delta Q + \delta W, \tag{1.1}$$

where $E$, $Q$ and $W$ are energy, heat and work, respectively. The term "work" includes all nonthermal processes. It usually implies the expenditure of energy against an external pressure: $W = -\int P_{ext} dV$, where $dV$ is a change in a system's volume. For constant pressure one introduces a special function, called enthalpy:

$$H = E + PV. \quad (1.2)$$

On account of (1.1)

$$dH = d(E + PV) = \delta Q + \delta W + VdP + PdV = \delta Q + VdP,$$

and, therefore,

$$(dH)_P = (\delta Q)_P,$$

i.e. enthalpy is the absorption or release of heat by a system under constant pressure.

***Entropy*** is a state function which characterises the degree of irreversibility and disordering:

$dS = \delta Q_{rev}/T$ for reversible

and

$dS > \delta Q_{irrev}/T$ for irreversible processes.

$\delta Q$ is zero for isolated systems and, therefore, their entropy always has a tendency to increase during irreversible processes, which constitutes one of the formulations of the second law of thermodynamics.

For diluted solid solutions, the most important one is the so-called "configurational" entropy which determines a number of ways to construct a solution from the constituting atoms according to the Boltzmann formula $S = k_B lnW$, where $k_B$ is the Boltzmann constant and $W$ is the statistical sum of states.

The following functions of state are used for different external conditions.

***Helmholz free energy***, $F = E - TS$, characterises the state of a system at constant volume and temperature.

***Gibbs free energy***, $G = E + PV - TS = H - TS$, is used to describe processes at constant pressure and temperature, which is a more common case. The condition of a system's equilibrium is $(\delta G)_{T,P} \geq 0$, i.e. any reversible change leads to an increase in Gibbs energy while, at equilibrium, Gibbs energy is at a minimum. Spontaneous, i.e. irreversible processes can occur when the Gibbs energy is decreased. Thus, the Gibbs free energy is an equivalent of the potential energy in mechanics.

***Standard states.*** In thermodynamical calculations and measurements, one deals with the relative quantities, e.g. with a change in energy or enthalpy. Therefore, it is useful to choose some reference, or standard states. A state of a material at the pressure of 1 bar and with a most stable structure at a given temperature is usually chosen as its standard state. For example, the bcc α-Fe and graphite are the standard states of iron and carbon, respectively. The thermodynamical functions of standard states are marked with the sign °. In thermodynamical tables, the standard enthalpies at different temperatures are often presented as $H_T^\circ - H_{298}^\circ$.

***Chemical potential and activity.*** The phases in alloys belong to so-called open systems which allow mass transfer through their boundaries. For this reason, it is necessary to know some additional parameters (e.g. the chemical composition) as compared to the closed systems where energy is completely determined by pressure and temperature. Entropy $S$, volume $V$, and the number of moles $n_i$ of the components are chosen as independent variables for the thermodynamical description of alloys.

The thermodynamical properties of a substance in a solution differ from those for its pure state, this is why so-called partial molar quantities are being used to characterise the behaviour of components in solutions. The partial molar quantities depend on the chemical composition and do not depend on the amount of solution. For any component they are obtained as derivatives of some quantities characterising the solution as a whole on the number of moles of this component. For example, the partial molar volume of $i$-th component:

$$V_i = \left(\frac{\partial V}{\partial n_i}\right)_{P,T,n_{j(j \neq i)}},$$

where $V$ is the volume of the whole solution, $n_j$ is the number of moles of the $j$-th component.

One of the most important molar quantities in the thermodynamics of solutions is the partial molar Gibbs free energy:

$$G_i = \left(\frac{\partial G}{\partial n_i}\right)_{P,T,n_{j(j \neq i)}} = \left(\frac{\partial E}{\partial n_i}\right)_{S,V,n_{j(j \neq i)}} = \mu_i . \tag{1.3}$$

The chemical potential $\mu_i$ of the $i$-th component characterises a change of energy when a small amount of this component is added to the system without changing the number and amount of other components. It is a measure of the component's tendency to leave the system.

This circumstance can be illustrated by the example of a nitrogen-containing duplex (ferritic-austenitic) steel. A typical steel of this class contains chromium (22-25 mass %), nickel (3-5 %) and nitrogen (~0.2 %). The contents of nitrogen in ferrite and austenite are different and controlled by the equality of chemical potentials of nitrogen. Nitrogen is mainly dissolved in austenite. At the same time, chromium belongs to the ferrite-forming elements and its concentration in the ferrite is higher than in the austenite of duplex steel. As nitrogen prefers chromium in the nearest neighbourhood rather than iron, it means that the presence of chromium in ferrite decreases the chemical potential of nitrogen and its content in ferrite will be increased. For the same reason, the high nitrogen content in austenite decreases the chemical potential of chromium and this results in its increasing concentration in austenite in the presence of nitrogen. As a result, nitrogen duplex steel has an increased nitrogen content in ferrite and increased chromium content in austenite as compared to nitrogen-free steel. It follows from the above mentioned that, if the nitrogen-free duplex steel undergoes solid state gaseous nitrogen saturation at high temperature, the chromium contents in austenite and ferrite will be changed. The migration of nitrogen from austenite into ferrite and, *vice versa*, the counteracting chromium migration from ferrite to austenite have to occur until $\mu_N^{\alpha} = \mu_N^{\gamma}$.

It is remarkable that, in this case, migration of nitrogen can occur from the phase where its content is low (ferrite) into the phase with a higher content (austenite), i.e. it is the inequality of chemical potentials and not that of concentrations that determines the direction of diffusion.

The same tendency can be observed if two pieces of steel having different chemical compositions are placed into a sealed capsule at a sufficiently high temperature: diffusion of nitrogen through the gas phase will occur from the steel where its content is lower if the other steel contains more elements (Cr, Mn, Mo, V etc.) decreasing the nitrogen chemical potential. Let us illustrate this case by the quantitative computation for two austenitic steels Cr14Ni18 and Cr18Ni14 alloyed by 0.2 and 0.3 mass % of nitrogen, respectively, and heated at 1200 °C for a time sufficient for thermodynamical equilibrium. According to the calculation (see Fig. 1.11), after the equilibrium state is achieved ($\mu_N^{\mathrm{Cr14Ni18}} = \mu_N^{\mathrm{Cr18Ni14}} = 1.61 \cdot 10^5$ J), the steel Cr14Ni18 contains 0.162 mass % of N whereas in the steel Cr18Ni14 the content of nitrogen is increased to 0.338 mass %. Thus, in the described Gedanken experiment, nitrogen really migrates from the steel of its low content into that of higher nitrogen.

It is important to note (see also Eq. (1.3)) that the Gibbs free energy of a system (solution) is equal to the sum of the chemical potentials of its components:

$$G = \sum_i n_i \mu_i . \tag{1.4}$$

According to definition (1.3) and relation (1.4), the chemical potential of the *i-th* component is equal to its Gibbs energy per 1 mol. At constant temperature

$$(d\mu_i)_T = (dG_i)_T = (v_i dP)_T$$

and

$$(d\mu_i/dP)_T = v_i ,$$

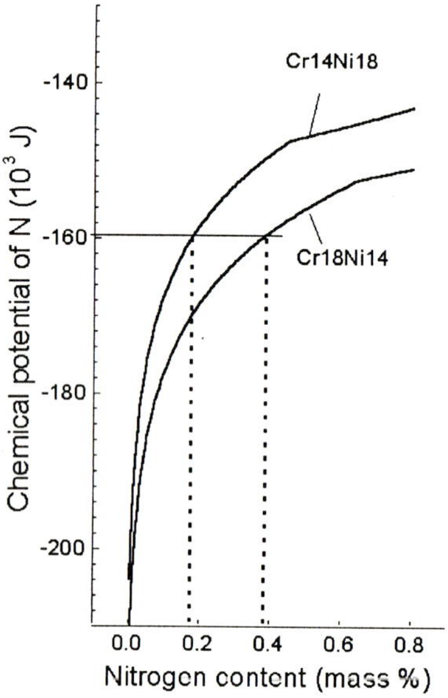


**Fig. 1.11.** Chemical potentials of nitrogen in steels Cr18Ni14(0.3N) and Cr14Ni18(0.2N) at 1200°C. By dotted lines, the nitrogen concentrations in both steels are indicated in the equilibrium state, if two steels contact each other in the sealed capsule

i.e. the change in the chemical potential of the $i$-th component due to pressure is equal to its molar volume. However, in the solid state, the molar volume is small as compared to the gaseous state and the effect of pressure on the chemical potential of solid phases is insignificant. This is why *activity* is introduced instead of the partial pressure. Using the equation of states $PV_i = RT$ one obtains:

$$(d\mu_i)_T = RT(d\ln P)_T .$$

*Activity* $a_i$ is determined by the same equation:

$$(d\mu_i)_{T,P} = RT(d\ln a_i)_{T,P} \quad (1.5)$$

and $\mu_i = \mu_i^*(T,P) + RT \ln a_i$; $\mu_i^*$ is the chemical potential of the $i$-th component in a reference state which is the same as the standard state $\mu_i^\circ$ but for variable pressure.

Activity characterises a change in the chemical potential due to the change in the composition of solutions at constant temperature and pressure. As a reference state for the $i$-th component in a given phase, one chooses the pure component $i$ with the same structure and, therefore, $a_i = 1$ for the reference state, i.e. at the concentration $x_i = 1$. Activity $a_i \to 0$ at $x_i \to 0$ and for small concentrations

$$a_i = x_i . \quad (1.6)$$

Solutions which satisfy relation (1.6) are denoted as ideal. This relation is true for diluted solutions where one can neglect the interaction between the solute atoms. The real solutions deviate from this relation and the deviation is accounted for by the activity coefficient $\gamma_i$ which is derived from the relation

$$a_i = \gamma_i x_i.$$

***The Raoult and Henri laws*** describe the concentration dependence of activities and refer to the solvent and solutes in solutions, respectively. The behaviour of the solvent $j$ is determined by Raoult's law:

$$(\partial a_j / \partial c_j)_{c_j \to 1} = 1,$$

if its concentration $c_j \to 1$ and relation (1.6) is fulfilled.

According to Henri's law, the activity of solutes in a diluted solution is proportional to their concentration. Henri's law is fulfilled if, at infinitely small concentration $x_i$, the activity coefficient $\gamma_i^\infty$ has a finite value:

$$\gamma_i^\infty = (\gamma_i)_{x_i \to 0} = (\partial a_i / \partial x_i)_{x_i \to 0} \neq 0.$$

Raoult's law is automatically true if Henri's law occurs, but Henri's law does not follow from Raoult's law.

In practice, metallurgists use a "harder" formulation of Henri's law: *at small concentrations of the solutes, the activity coefficient* $\gamma_i^\infty$ *has not only a finite, but a constant value:*

$$(\partial \ln \gamma_i / \partial x_i)_{x_i \to 0} \neq 0,$$

which means a linear relation between the activity of the solutes and their concentration and this is denoted as Henri's law of the first order.

It is important to note (see the discussion of this topic by Lupis 1983), that Henri's law of the first order contains a hidden requirement of short range atomic interactions: the interaction potential has to decrease strongly with an increase in the distance between atoms. The statistical modelling (Hill 1960) shows that derivatives from $\ln\gamma_i$ on concentrations include some integrals which converge only if the interatomic potentials decrease abruptly with an increase in the distances between atoms, which is typical of metallic solutions. That is why Henri's law of the first order is correct for metals.

### 1.1.2.2
### Thermodynamical models of Fe-C(N) austenite

In an ideal solid solution, where the interaction between interstitials is absent, the relative partial enthalpies of solutes are equal to zero and the relative partial entropies originate from the statistical distribution of solute atoms in the lattice (configurational entropies). This statement is written as follows (McLellan 1972):

$$\Delta\overline{H}_i^{ideal} = \overline{H}_i^{ideal} - \overline{H}_i^o = 0,$$

$$\Delta\overline{S}_i^{ideal} = \overline{S}_i^{ideal} - \overline{S}_i^o = -k\,ln\frac{c_i}{1-2c_i} = -k\,ln\frac{\theta}{1-\theta},$$

where
$H_i^{ideal}$ and $H_i^0$ are the partial molar enthalpies,
$S_i^{ideal}$ and $S_i^0$ are the partial molar entropies in an indefinitely diluted solid solution and in a pure dissolved component $i$, respectively,
$c_i$ is the atomic concentration,
$\theta$ is the fraction of interstitial atoms $n_i/N$
$n_i$ is the number of interstitial atoms,
$N$ is the number of host atoms occupying the lattice sites.

The partial molar free Gibbs energy (chemical potential) for interstitial atoms in an ideal solid solution is

$$\mu^{ideal} = \mu^o + kTln\frac{\theta}{1-\theta},$$

where $\mu^o = \overline{H}_i^o - T\overline{S}_i^o$.

The chemical potential of interstitial atoms in a real solid solution (see also (1.5))

$$\mu = \mu^o + kTlna_i$$

and the interaction between interstitial atoms is accounted for by the change of thermodynamical activity with the concentration of interstitials.

The distribution of carbon and nitrogen atoms and the interaction between them in the iron austenite are usually studied based on the measurements of concentration and temperature dependences of thermodynamical activities.

***Models of distribution of carbon.*** Darken and Smith (1946) were the first to study thermodynamical properties of Fe-C alloys. They have supposed that the carbon atom in austenite has another carbon atom or none as a nearest neighbour. The results of measurements of thermodynamical activity were consistent with the theoretical calculations at the condition that the fraction of C-C pairs was equal to one third of their amount for an ideal solid solution, i.e. for the random distribution of carbon atoms. In the following, a number of models were proposed and they can be conventionally assembled in two groups.

In the models of "hard repulsion", the deviation of the concentration dependence of carbon activity from that of an ideal solid solution is attributed to the "non-ideal" dependence of configurational entropy on the carbon content, whereas the partial enthalpy of carbon dissolution is considered to be independent of the composition. These models lead to the conclusion of hard mutual repulsion of carbon atoms, which results in a blocking of neighbouring octahedral holes for the occupation by other carbon atoms. This idea was varied from a simple blocking (see e.g. Speiser and Spretnak 1955) to an overlapping of blocked interstitial sites (McLellan et al. 1967, Spraque and McLellan 1968) and to a dependence of the degree of overlapping on the carbon content (Gallacher et al. 1969, Lee 1974).

In the models of "soft repulsion", the entropy is supposed to be "ideal", and the deviations from Henri's law are considered as originating from an increase in the solution heat of carbon with its increasing content. At this approach, the existence of paired carbon atoms beside single ones is possible. The theoretical models of soft repulsion suppose the existence of interaction between interstitial atoms being in the value of $kT$ order. They contain some fitting parameter (the interaction energy $w_{i\text{-}i}$) which is determined from a comparison with experimental data. Interactions are taken into account only between nearest neighbouring interstitial atoms and nearest interstitial-iron neighbours, that is why these models are identified as "first order models".

Another distinction between soft repulsion models and those of the hard repulsion is that they deal with "regular" solutions which are close to the ideal ones, although, in fact, they are diluted real solid solutions. For their thermodynamical description, one uses the non-configurational (vibrational a.o.) partial molar entropy $S_i^v$ along with the configurational one, so that the expression for the chemical potential is written as (McLellan 1972)

$$\mu_i = \overline{H}_i^\infty - TS_i^v + kT\ln\frac{\theta}{1-\theta} \tag{1.7}$$

where $$\overline{H}_i^\infty = lim_{\theta\to 0}\left(\frac{\partial H}{\partial n_i}\right),$$

$$\overline{S}_i^v = lim_{\theta\to 0}\left(\frac{\partial S}{\partial n_i}\right) + k\,ln\frac{\theta}{1-\theta}, \qquad \frac{\theta}{1-\theta} = a_{config},$$

and the activity of the solute element in the relation to its standard state is:

$$a_i = \frac{\theta}{1-\theta} exp\left(\frac{\overline{H}_i^{\infty} - \overline{H}_i^{o}}{kT}\right) exp\left\{-\left(\frac{\overline{S}_i^{v} - \overline{S}_i^{o}}{k}\right)\right\} =$$

$$a_{config}\, exp\left(\frac{\Delta G}{kT}\right) \tag{1.8}$$

where the exponential part is the non-configurational contribution to the activity. For low concentrations or high temperatures, when the interaction energies are negligible, the activity approaches its configurational part. It is shown that the diluted solutions of carbon (McLellan 1964) and nitrogen (McLellan 1965, Swartz 1969, Dunn and McLellan 1971) in γ iron are quasi-regular and the relation (1.7) is true for them. A possibility to apply the theory of quasi-regular solution for the studies of non-diluted (θ~0.1) carbon and nitrogen solutions in fcc iron was also motivated by McLellan et al. (McLellan and Chraska 1971, McLellan and Farraro 1980).

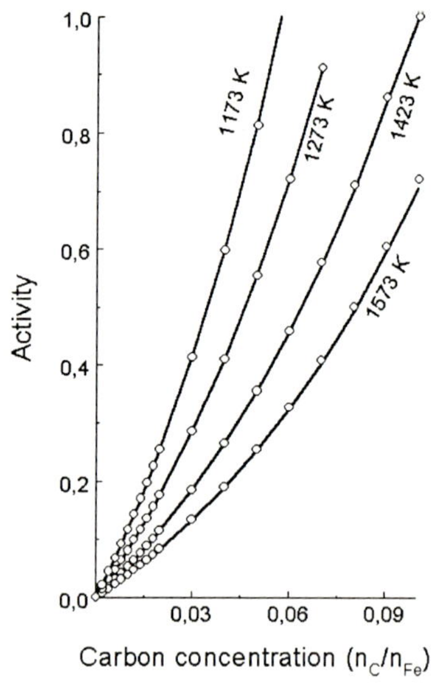


**Fig. 1.12.** The carbon activity in fcc iron according to Ban-ya et al. (1970)

Two most widely used approximations of the soft repulsion models were developed by McLellan and Dunn (1969, 1970) (MD model of "quasi-chemical first order approximation") and Aaronson, Domian and Pound (1966) (ADP model). Shiflet et al. (1978) have shown that both models are true for $w_{i\text{-}i}/kT \leq 1$ and coincide with a good accuracy.

The most reliable measurements of carbon activity in iron austenite were performed by Ban-ya et al. (1969, 1970) (see Fig. 1.12). For both theories (MD and

ADP), a comparison with these data gives the value of C-C repulsion interaction energy in the first coordination sphere $w_{C\text{-}C} \approx 0.08\text{-}0.09$ eV (7-8 kJ/mole).

***Models of distribution of nitrogen.*** The number of thermodynamical theories of nitrogen in fcc iron is limited compared to those for carbon because of a shortage of experimental data on nitrogen activity and some problems concerning their interpretation. The nitrogen activity data obtained by Atkinson and Bodsworth (1970) are mainly used (see Fig. 1.13). The authors have used the dissociation of ammonia, $NH_3$, to provide the high nitrogen activity for solid state saturation of iron samples with nitrogen.

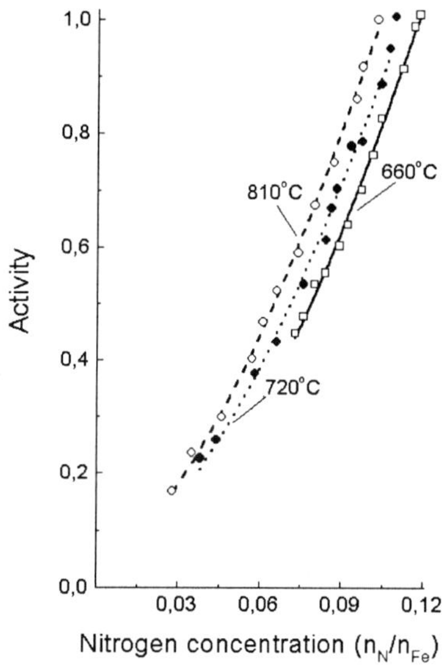


**Fig. 1.13.** The nitrogen activity in fcc iron according to Atkinson and Bodsworth (1970)

McLellan and Alex (1970) have applied the "quasi-chemical first order approximation" to the treatment of these data and obtained the value of N-N interaction energy in the first coordination sphere $w_{N\text{-}N} \approx 0.036\text{-}0.040$ eV, which is twice less than the above mentioned data for carbon. It is remarkable that Bodsworth (1970) interpreted his own data based on the hard repulsion model and stated that McLellan's approach leads to a change in sign of the nitrogen solution enthalpy with increasing nitrogen content, which is hardly acceptable.

The reliability of the above mentioned experimental data obtained by Atkinson and Bodsworth is a controversial issue. Frisk (1991) calculated the concentration dependence of nitrogen activity in the fcc Fe-N alloy, based on the experimental data of $\alpha\text{-}\gamma$ phase boundaries and nitrogen solubility in both phases, and showed a discrepancy between the calculated data and data by Atkinson and Bodsworth. Earlier, such a discrepancy for the data at high temperatures was discussed by

Hillert and Jarl (1975) who supposed that the experimental data were affected by some problems arising from the $NH_3$ gas dissociation.

It would be natural to avoid this problem using the saturation of γ-iron by nitrogen at high temperatures in the equilibrium with $N_2$ gas at $p(N_2) = 1$ bar. However, the nitrogen solubility in γ-iron at such a pressure is low (~0.1 at. %) and insufficient for studies of N-N interaction (see e.g. Schenck et al. 1963, Zitter and Habel 1973).

Bashchenko et al. (1985) obtained the nitrogen activity from the data of nitrogen solubility in γ-iron in equilibrium with gaseous nitrogen at high pressures. However, at the obtained nitrogen content (1.0-1.1 mass %, i.e. 3.8-4.2 at. %), the deviations in the concentration behaviour of configurational nitrogen activity from the ideal one are again not sufficient for the tests of the theoretical models.

Thus, one can state that reliable experimental data of thermodynamical activity of nitrogen in fcc binary Fe-N alloys are absent in a wide concentration range (0 - 10 at. % of N) and the evaluation of an N-N interaction energy based on activity data is not sufficiently correct.

In general, while analysing the task of a correct choice between the models of "hard" and "soft" repulsion for both carbon and nitrogen distributions in iron austenites, one comes to the following conclusions: (1) the account of *i-i* interaction in the first coordination sphere only (first order models) seems to be incorrect because of the strong local distortions caused by carbon and nitrogen atoms in γ-iron; (2) the thermodynamical measurements themselves are insufficient for this purpose and experimental data obtained from some independent measurements are necessary.

A good possibility in this relation is provided by the use of Mössbauer spectroscopy as a sensitive method for studies of local neighbourhood of iron atoms in solid solutions and, based on these experimental data, by Monte Carlo simulation of solid solutions, which allows one to obtain the values of *i-i* interaction energies in different coordination spheres and describe the distribution of interstitial atoms in γ-iron. The realisation of this programme will be presented in the following sections.

### 1.1.2.3
### Monte Carlo method for thermodynamical simulation of solid solutions

This method and its applications in the physics of condensed matter are described elsewhere (see Binder 1987, 1992). It allows one to calculate the coordinates of constituents constructing any thermodynamical system (for instance, atoms occupying the sites of a crystal lattice in a solid solution) if their interaction parameters (interaction energies) are known. As coordinates, one uses the multidimensional vector $\mathbf{R}_j$ of the phase coordinates of the system which describes all possible characteristics of atoms (their coordinates in our case). If $E(\mathbf{R}_j)$ is the energy of the system in some $\mathbf{R}_j$ state, the probability (density of probabilities) to find the system in this state is

$$P(\mathbf{R}_j) = \frac{1}{Z} exp\left[ -\frac{E(\mathbf{R}_j)}{k_B T} \right] \quad ,$$

where $Z$ is the statistical sum of the system ("grand partition function"), $k_B$ is the Boltzmann constant, T is the temperature in Kelvin units, and, for any parameter $A(\mathbf{R}_j)$ characterising a given state of the system (e.g. a parameter of short range atomic order), its thermodynamical average value is

$$\langle A \rangle = \frac{1}{N} \sum_{j=1}^{N} A(\mathbf{R}_j) \quad ,$$

where $N$ is a number of occasional calculations of this parameter in given states being arbitrarily chosen.

Initially, the constituents (atoms) are supposed to be randomly distributed in the system. The probability of transitions $P_{tr}$ ($\mathbf{R}_j \rightarrow \mathbf{R}_i$) which lead the system to equilibrium is determined as

$$P_{tr} = \frac{exp\left[ -\frac{\delta E}{k_B T} \right]}{1 + exp\left[ -\frac{\delta E}{k_B T} \right]}$$

where $\delta E = E(\mathbf{R}_i) - E(\mathbf{R}_j)$. The lattice gas model is used for interstitial solid solutions and the energy of system is determined as the sum of pair interactions between interstitial atoms (Bugaev and Chepulski 1995):

$$E = \frac{1}{2} \sum_{i \neq j} w(\mathbf{r}_i - \mathbf{r}_j) C(\mathbf{r}_i) C(\mathbf{r}_j)$$

where summation is performed on all interstitial sites $\mathbf{r}_i$, $w(\mathbf{r}_i - \mathbf{r}_j)$ is the energy of pair interactions between interstitial atoms in positions $\mathbf{r}_i$ and $\mathbf{r}_j$, $C(\mathbf{r}_i)$ is equal to 0 if the interstitial atom occupies this site or 1 if not.

In practice, the modelling process looks as follows. A finite fcc lattice (e.g. 24×24×24 lattice parameters$^3$) is chosen with periodical boundary conditions. The interstitial sites of this finite fcc lattice are randomly filled by the interstitials to a necessary concentration. Then, the following steps are realised:

(1) an occupied interstitial site is chosen randomly;
(2) a next interstitial site is chosen randomly;
(3) if this interstitial site is occupied, the procedure is repeated from step (1);

(4) if not, the interstitial atom can jump into this site with some probability;
(5) the procedure is repeated from step (1) until the system reaches thermodynamical equilibrium (energy of system and short range order parameters fluctuate near some average values);
(6) After the system reaches its thermodynamical equilibrium, the statistical averaging of determined parameters is performed.

Thus, if one knows the values of interaction energies between interstitial atoms in some solid solution, one can obtain all the parameters of its equilibrium atomic distribution (parameters of short and long range atomic orders, the fractions of host atoms having different neighbourhoods of interstitials etc.).

The combination of Mössbauer spectroscopy and Monte Carlo simulation enables one to solve an inverse task: to determine possible values of interaction energies leading to those atomic configurations which are obtained from Mössbauer studies of Fe-C and Fe-N austenites.

### 1.1.2.4 C-C interaction and atomic configurations of carbon in Fe-C austenite

Mössbauer spectroscopy was applied to studies of binary Fe-C and Fe-N austenites because of its high resolution in the distinction between different nearest neighbourhoods of iron atoms (see e.g. Gielen and Kaplow 1967, Genin and Flinn 1968, DeCristofaro and Kaplow 1977, Foct et al. 1977 etc.).

Mössbauer spectrum of a binary Fe-C austenite (Fig. 1.14) consists of two components belonging to iron atoms having a pure iron neighbourhood ($Fe_0$) and carbon atoms along with the iron ones as nearest neighbours ($Fe_C$). As it was shown by Gavriljuk and Nadutov (1983) (see also Bugaev et al. 1983), two kinds of iron atoms, i.e. those having one ($Fe_1$) and two carbon atoms as neighbours in the first coordination sphere ($Fe_{2\text{-}90}$) contribute to the same $Fe_C$ component of the Fe-C spectrum. The validity of this interpretation was confirmed in the thermodynamical studies by Oda et al. (1994), Sozinov et al. (1997) and Balanyuk et al. (1998).

The values of C-C interaction energies in the first ($w_1$) and second ($w_2$) coordination spheres of the interstitial sublattice in Fe-C austenite, as obtained by Monte Carlo modelling of atomic configurations satisfying the experimental Mössbauer data of the abundance of different iron sites, are shown in Fig. 1.15. It follows from this analysis that the repulsion between carbon atoms occupying neighbouring interstitial sites (90° pairs) is weak whereas carbon atoms as second nearest neighbours (180° pairs) repulse each other strongly.

The long range atomic order in Fe-C austenites and, in particular, the existence of ordered structures like $Fe_8C$ is discussed elsewhere (see e.g. Bauer et al. 1988, 1990, Schön et al. 1993). The possibility of such an ordering was evaluated using Monte Carlo modelling (Sozinov et al. 1997). It was shown (see Fig. 1.16) that the $Fe_8C$ structure can be formed only if a strong repulsion exists in the first three coordination spheres. For the $Fe_4C$ structure, one needs a strong repulsion in the first, a weak one in the second and a strong one in the third coordination sphere.

The conditions for both structures are not consistent with the values of C-C interaction energies satisfying the experimental data.

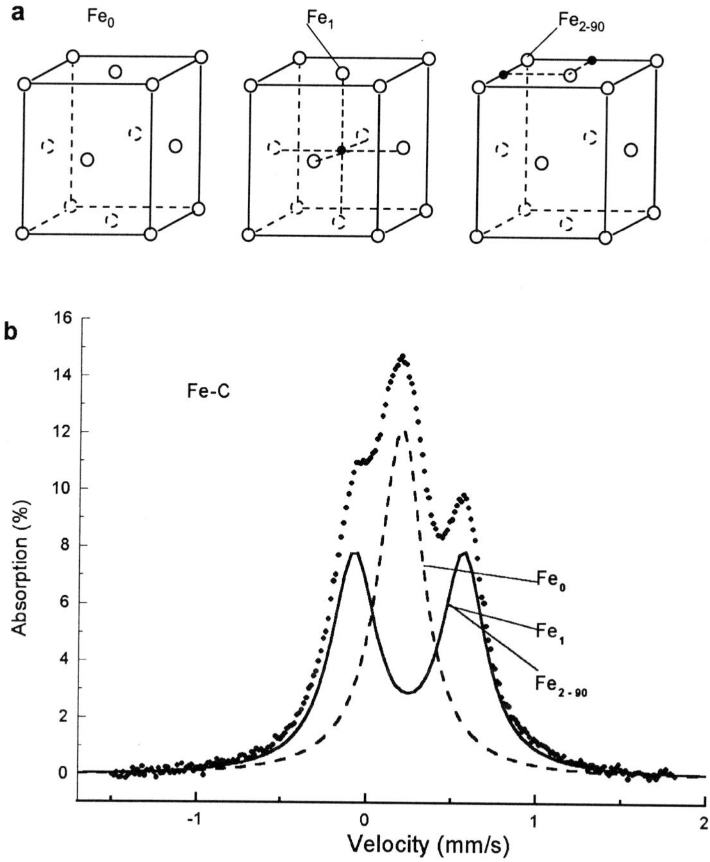


**Fig. 1.14.** Mössbauer spectrum of Fe-C austenite **b** and atomic configurations corresponding to its components **a**. A single carbon atom and a carbon atom pair in the first coordination sphere (90° pairs) are shown

The values of C-C interaction energies in different coordination spheres can be also simulated using the experimental data of carbon activity (Balanyuk et al. 1998 I). According to relation (1.8), the carbon activity in an Fe-C alloy

$$a_C = a_{config} \exp(\Delta G/k_B T)$$

consists of two contributions: the configuration activity $a_{config}$ which depends directly on the carbon distribution and the non-configurational part $\exp(\Delta G/k_B T)$, where $\Delta G$ is the difference between the thermodynamical potentials of a carbon atom in its standard state, e.g. in graphite, and in an infinitely diluted Fe-C solid solution. The non-configurational contribution cannot be calculated as the Fe-C interaction is unknown.

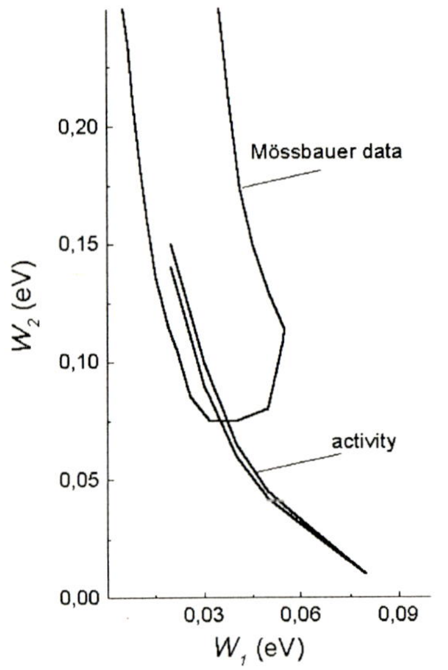


**Fig. 1.15.** The values of pair C-C interaction energies in two coordination spheres calculated by Monte Carlo modelling based on the experimental Mössbauer data (Oda et al. 1994) and the data of thermodynamical activity measured by Ban-ya et al. (1970)

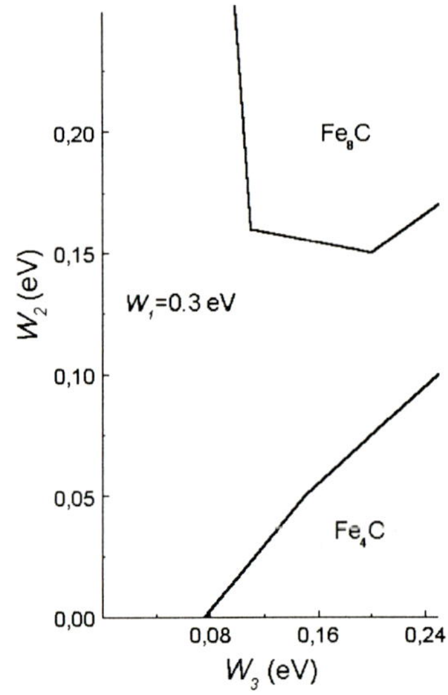


**Fig. 1.16.** The areas of existence of ordered $Fe_4C$ and $Fe_8C$ structures at different values of C-C interaction energies in the second ($w_2$) and third ($w_3$) coordination spheres and constant $w_1$ = 0.3 eV in the first coordination sphere

Nevertheless, taking into account that $a_{\mathrm{config}} \sim \theta$ for infinitely diluted solutions, one can evaluate the non-configurational contribution to the carbon activity from the concentration dependence of $a_C$ at small $\theta$. Such an assessment was performed by Dunn and McLellan (1970). Using these data, Sozinov et al. (1997) have calculated the concentration dependence of carbon activity for different values of C-C interaction energies. In contrast to the above discussed "first order models", they took into consideration the interaction betwen carbon atoms in several coordination spheres. The results of calculations are presented in Fig. 1.17. It is clear that only the values of C-C interaction energies, which are consistent with a soft repulsion between carbon atoms in the first coordination sphere and a hard repulsion in the second one, satisfy the experimental data of carbon activity. High values of repulsion energy in the first coordination sphere, i.e. between nearest C-C neighbours, are not consistent with the measured data of carbon activity (see also Fig. 1.15 where the values of C-C interactions in the first and second coordination spheres satisfying the data of carbon activity measured by Ban-ya et al. are shown).

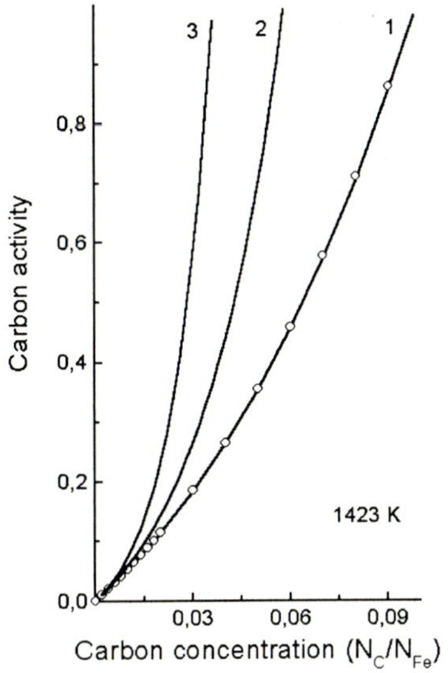


**Fig. 1.17.** The carbon activity in fcc Fe-C alloy at T = 1423 K. Open circles are the experimental data from Ban-ya et al. (1969, 1970). The solid lines are the computation data obtained using the non-configurational contribution as determined by Dunn and McLellan (1970) and the following values of C-C interaction energies: (1) $w_1 = 0.036$ eV, $w_2 = 0.075$ eV (the data obtained by Oda et al. 1994); (2) $w_1 = 0.3$ eV, $w_2 = 0.2$ eV; (3) $w_1 = 0.3$ eV, $w_2 = 0.2$ eV, $w_3 = 0.2$ eV

Thus, one can state that the computer modelling of experimental data available leads to the conclusion that carbon atoms can occupy neighbouring interstitial sites. At the same time, the strong C-C repulsion between carbon atoms occupying interstitial sites as second nearest neighbours prevents the formation of 180° carbon pairs and, therefore, the structure $Fe_4C$ in Fe-C austenite is not realised. The $Fe_8C$ structure which requires hard repulsion in the first, second and third spheres does also not exist. Let us note preliminarily that such a distribution of carbon atoms corresponds to the strengthening of the covalent interatomic bonding due to carbon in the iron austenite.

### *1.1.2.5*
### *N-N interaction and atomic configurations of nitrogen in Fe-N austenite*

The experimental data about nitrogen distribution in binary Fe-N austenite were provided by Mössbauer spectroscopy (see Foct 1973, DeCristofaro and Kaplow 1977, Gavriljuk et al. 1990 I etc.). The Mössbauer spectrum of Fe-N austenite (see Fig. 1.18) consists of three components which originate from iron atoms having no nitrogen atoms as nearest neighbours ($Fe_0$), one nitrogen atom and, possibly, two in 90° configurations in nearest interstitial sites ($Fe_1$ and $Fe_{2\text{-}90}$) and two nitrogen atoms in 180° dumb-bell-like configurations as nearest neighbours ($Fe_{2\text{-}180}$).

The first experimental evidence of the existence of 180° N-N pairs was obtained by Foct (1973). Thus, the experimental data reveal a difference in the carbon and nitrogen disributions in the $\gamma$ iron, as 180° interstitial pairs, i.e. an interstitial atom having another interstitial atom as a next nearest neighbour, do not exist in Fe-C austenite because of high C-C repulsion in the second coordination sphere of the interstitial sublattice. The range of interaction energies between nitrogen atoms occupying the neighbouring interstitial sites ($w_1$) and the sites in the second coordination sphere ($w_2$), which are consistent with experimental Mössbauer data of the abundance of iron sites having no, one and two nitrogen atoms as nearest neighbours, is shown in Fig. 1.19.

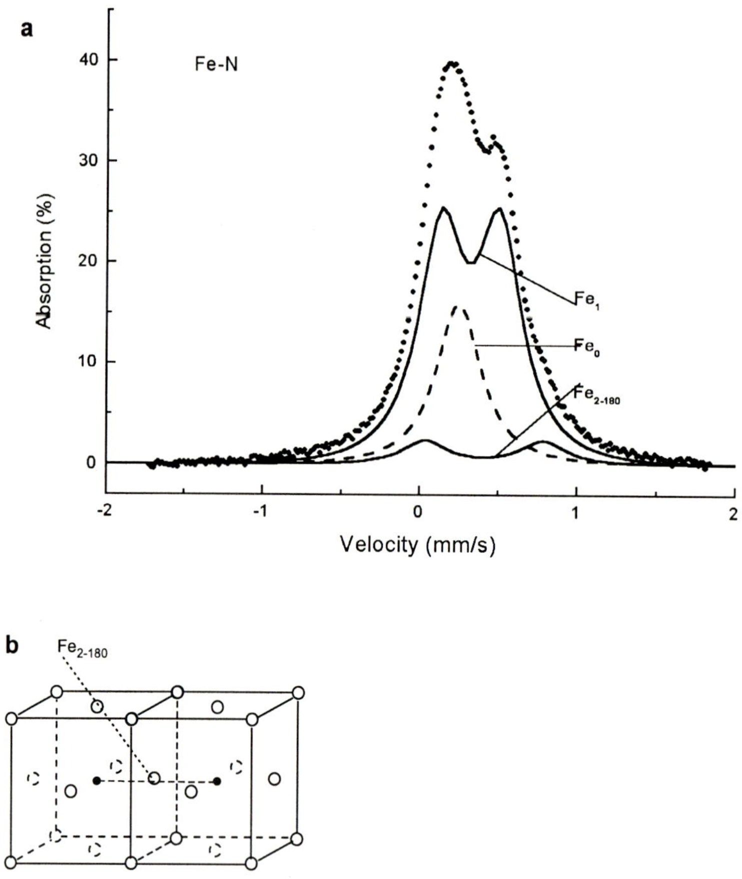


**Fig. 1.18.** The Mössbauer spectrum of the binary Fe-N austenite **a** and corresponding atomic cofigurations **b**. See configurations of atoms $Fe_0$ and $Fe_1$ in Fig. 1.14**a**. Nitrogen atoms in 180° configurations ($Fe_{2\text{-}180}$) are, in fact, the structural elements of $Fe_4N$ phase

It is seen that the computer modelling satisfies the experimental data only for a strong repulsion between nitrogen atoms in the first coordination sphere and a rather weak one in the second coordination sphere. Thus, one can state that, in contrast to carbon, there is a smaller probability for nitrogen atoms to occupy the nearest interstitial sites and a more favourable thermodynamical conditions for occupying interstitial sites in the second coordination sphere.

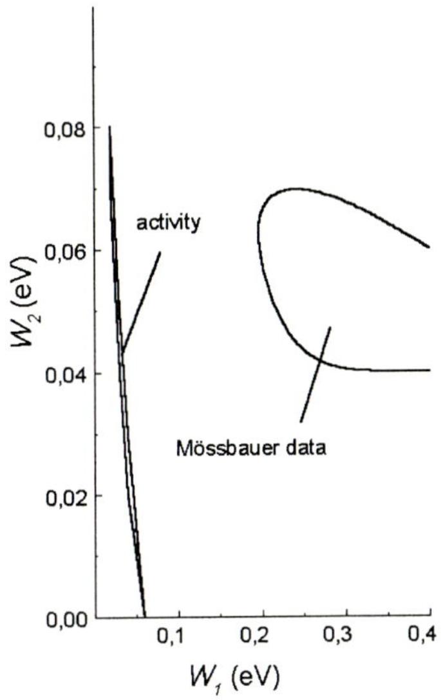


**Fig. 1.19.** The values of N-N interaction energies in two first coordination spheres satisfying the Mössbauer data of nitrogen distribution in Fe-N austenite, as obtained by means of the computer modelling (Sozinov et al. 1999). The range of interaction energies satisfying the data obtained by Atkinson and Bodsworth (1970) for the nitrogen acivity is also shown

Above we have discussed the difficulties which are met while measuring the thermodynamical activity of nitrogen in the γ iron. A computer simulation of a solid solution of nitrogen in the γ iron satisfying the experimental data of the nitrogen activity can be performed exactly in the same way as it was done for Mössbauer data of the nitrogen distribution. The results of this computer modelling are also shown in Fig. 1.19. It is clear that the data of nitrogen activity available are not consistent with the experimental data of nitrogen distribution in Fe-N austenite.

Therefore, it is expedient, based on the data of N-N interaction energies satisfying the experimental data of nitrogen distribution, to calculate the concentration dependence of the chemical potential of nitrogen and its thermodynamical activity in the γ iron. Such calculations were performed by Sozinov et al. (1999), and the results are given in Fig. 1.20 along with the experimental data of Atkinson and Bodsworth, the calculations for the case of an ideal solid solution ($w_1 = 0$, $w_2 = 0$) and for the case of the "quasi-chemical first order approximation" ($w_1 \approx 0.04$ eV, $w_2 = 0$ according to McLellan and Alex 1970).

It is seen that the deviation from the behaviour of an ideal solid solution is much higher than it follows from the experimental data of nitrogen activity and from the thermodynamical calculations performed taking into account the N-N interaction in the first coordination sphere only.

***Short range atomic ordering.*** A higher probability of the 180° N-N pairs as compared to the 90° ones, as it follows from the computer simulation of the nitrogen γ solid solution, means a short range ordering of nitrogen atoms in Fe-N austenite. In fact, 180° N-N configurations are the structural elements of $Fe_4N$ $\gamma'$ phase, and it would be attractive to interpret the nitrogen austenite as a $\gamma'$ phase having a number of empty sites of nitrogen atoms because of an insufficient nitrogen content. Therefore, it is important to discuss an opportunity for the formation of ordered structures like $Fe_4N_{1-x}$ or $Fe_8N_{1-x}$ ($x = n/N$ is the relative nitrogen concentration) in the nitrogen austenite.

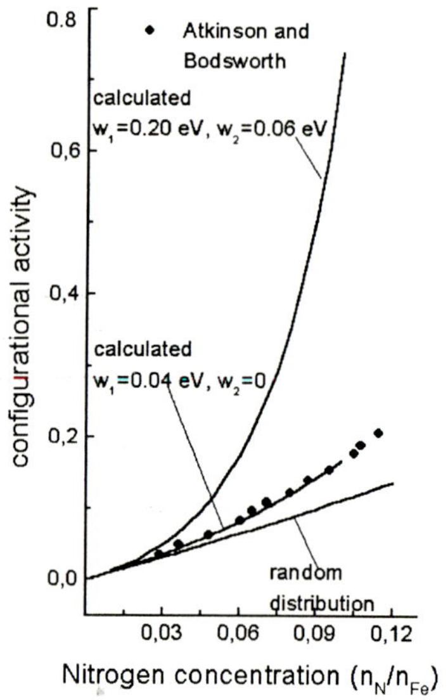


**Fig. 1.20.** The configurational activity of nitrogen in the fcc alloy Fe-N at 1223 K as calculated using the N-N interaction parameters obtained from the analysis of Mössbauer spectra: ($w_1 = 0.20$ eV, $w_2 = 0.06$ eV). The experimental points are taken from Atkinson and Bodsworth (1970). The results of calculation for random nitrogen distribution ($w_1 = 0$, $w_2 = 0$) and for the "quasi-chemical first order approximation" accounting for the N-N interaction in the first coordination sphere only ($w_1 \approx 0.04$ eV, $w_2 = 0$ according to McLellan and Alex 1970) are also shown

The only experimental evidence for a long range order in Fe-N austenite was presented by Suyazov et al. (1976) by means of transmission electron microscopy. The computer simulation of nitrogen austenite provides one with the data of N-N interaction parameters which can be consistent with the existence of ordered structures. The results of computer simulation are shown in Fig. 1.21. One can conclude from these data that the existence of an $Fe_8N$ structure requires the repulsion in all three spheres and the needed value of repulsion energy $w_2$ in the second sphere exceeds the values obtained from the computer simulation based on the experimental data (see Fig. 1.19). For the $Fe_4N$ structure, there exists some area of small $w_2$ values, which are consistent with the combination of positive large $w_1$ and small $w_2$. Thus, at the condition of repulsion in the third coordination sphere ($w_3 > 0$), the $Fe_4N_{1-x}$ structure in austenite can exist.

The most important conclusion from the above consideration is that, in contrast to carbon, the distribution of nitrogen atoms in iron austenite reveals at least features of short range atomic ordering. It is remarkable that the tendency of nitrogen atoms to ordering in austenite corresponds to an increase in the concentration of free electrons due to nitrogen, i.e. to an enhanced metallic component of interatomic bonding.

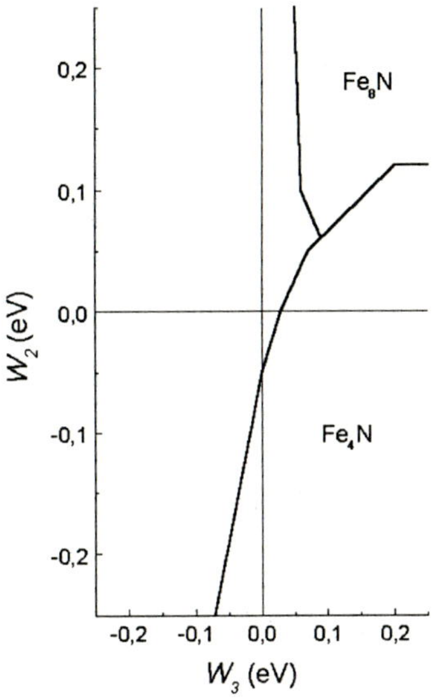


**Fig. 1.21.** The ranges for the existence of ordered structures in the binary Fe-9.3 at.%N austenite (the relative nitrogen concentration $n_N/N_{Fe} = 0.1025$) at the varying values of N-N interaction energies in the second ($w_2$) and third ($w_3$) coordination spheres and the constant value of $w_1 = 0.2$ eV; T = 873 K

A comparison of the data of nitrogen and carbon distributions in fcc iron with the peculiarities of the electronic structure of nitrogen and carbon austenites suggests a decisive role of the electron exchange between solute and host atoms in the short range atomic order.

### 1.1.2.6 Correlation between atomic interaction and short range atomic order

The data discussed in the first two sections enable one to trace some important correlations between the electronic structure and the atomic distribution in iron-based austenites. We shall characterise the short range atomic order (SRO) according to Cowley (1950). The Cowley parameter $\alpha_i$ of SRO is determined by the ratio between the fractions of interatomic bonds formed by atoms of the same or different kinds:

$$\alpha_i = 1 - P_i = (1 - N^{AB})/Nc_B(1 - c_B)Z = 1 - \frac{N^{AB}_{real}}{N^{AB}_{random}} ,$$

where $i$ denotes the $i$-th coordination sphere, $P_i$ is the probability to find an A atom as $i$-th neighbour (i.e. in the $i$-th coordination sphere) in relation to a B atom located in the origin of coordinates, $N^{AB}$ is the actual number of AB pairs in the solid solution, $N$ is the total number of atoms in the solid solution, $c_B$ is the concentration of B atoms, $Z$ is the coordination number (12 for the first coordination sphere in an fcc lattice).

Thus, $N(1 - c_B)$ is the number of $A$ atoms in solid solution. For a random distribution, each $A$ atom has $Zc_B$ of $B$ neighbours, that is why $Nc_B(1 - c_B)Z$ is the number of $AB$ pairs in an ideal solid solution. If short range ordering occurs in the $AB$ alloy, $AB$ pairs are preferential as compared to the $AA$ or $BB$ ones. Therefore, the numerator in the above fraction is higher than the denominator and $\alpha$ is negative. If short range decomposition (clustering) is the case, $AA$ and $BB$ pairs are preferred and $\alpha$ is positive.

It seems reasonable to seek a possible correlation between an effect of alloying elements on the interatomic bonding in austenitic steels and the type of atomic distribution, based on studies of the influence of alloying elements on the electron exchange and on the data available of SRO in binary and multicomponent iron alloys.

Some attempts were undertaken to find a correlation between the electronic structure and long range atomic order (Küntzler 1973). The formation of superstructures like FeCo, $Cu_3Au$, $Ni_3Fe$, $Ti_3Al$ etc. was likely to correspond to a decrease in the state density at the Fermi surface, as measured by the low temperature electronic specific heat, which was attributed to the formation of new Brillouin zones touching the Fermi surface and lowering the electronic energy near the Fermi surface. In general, the formation of superstructures is expected to be accompanied by a decrease in the state density because it means a decrease in the energy of the crystal lattice. In the case of SRO, the situation seems to be more complicated as both kinds of SRO, short range ordering and clustering, lead to a decrease in the energy of a solid solution.

Friedel (1974) has shown that the valence electrons of solute atoms can cause such a change of the screened ("clothed" according to Friedel) pseudopotential in the area of "Friedel's oscillations" that the sign of interaction of a host atom with the charge fluctuations at nearest solute atoms is changed. A recent study by Abrikosov et al. (1998) devoted to a detailed analysis of three most popular techniques for the total energy calculations of random solid solutions (the supercell method, the Connolly-Williams method and the coherent potential approximation) shows how complicated this problem is even in the absence of interatomic correlations.

***Experimental data of short range atomic order.*** A number of studies were devoted to the SRO in iron-based alloys. It is known that nickel in iron-based fcc alloys promotes short range atomic ordering as Fe-Ni bonds are preferred in the comparison to Fe-Fe and Ni-Ni ones (see e.g. Menshikov and Yurchikov 1984, Menshikov et al. 1971, Roth et al. 1978). A peculiarity of short range atomic ordering in fcc Fe-Ni alloys with a nickel content of less than 50% is that it is accompanied by a decomposition of the solid solution and the formation of nickel-rich submicrovolumes close to the superstructure FeNi. It is of particular interest that short range order was not found in austenitic solid solution Fe-25.8Ni-0.88C treated at 1453 K (Butler et al. 1991). Thus, it seems that carbon prevents short range ordering in Fe-Ni alloys.

Fe-Cr alloys are prone to short range decomposition, i.e. to the clustering of chromium atoms (Solomon and Levinson 1978). Although these data refer to bcc iron alloys, it is natural to think that Cr-Cr and Fe-Cr interactions are not changed significantly when the bcc lattice is replaced by the fcc one. This is confirmed by the experimental observations of chromium atom clusters in austenitic steels made by means of the ion field microscopy (Litvinov et al. 1986).

Using electron irradiation in order to accelerate the migration of solute atoms, Rotman et al. (1989) observed a short range decomposition of fcc Fe-Cr-Ni solid solutions resulting in the formation of (Cr,Fe)-rich and Ni-rich microareas in high nickel alloys, while chromium and nickel fluctuations coincided in low nickel alloys.

Manganese atoms in the binary fcc Fe-Mn alloys are found to form clusters (see e.g. Amigood and Litvinov 1983). The clusters of molybdenum are also found in iron-based alloys (Ericsson and Cohen 1971). Copper atoms are prone to precipitate from iron and even to form small clusters of pure copper.

Silicon and aluminum are well known as elements the distributions of which in the iron is characterised by the SRO and even the long range atomic order, e.g. by the formation of ordered structures like $Fe_3Si$ (Warlimont 1968, Vlasova 1972) and $Fe_3Al$ (Houska and Averbach 1962, Vlasova 1963), respectively, with the parameter $\alpha_1 = -1/3$.

To characterise the extention of the SRO phenomenon, it may be added that even in liquid metallic alloys the distribution of alloy components is not random and a chemical short range order is observed far above the melting point by means of high-temperature X-ray diffraction (Il'insky et al. 1995).

A tendency to form clusters in multicomponent iron-based solid solutions was studied using the data of contributions to magnetic susceptibility from different electron subsystems: free electrons, isolated localised *d*-electrons (isolated atoms of alloying *d*-elements) and superparamagnetic clusters, i.e. clusters of *d*-atoms (see Shanina et al. 1995, 1998). This analysis was based on different temperature dependences of the magnetic susceptibilities caused by the above electron subsystems.

Let us denote the magnetic susceptibility of free electrons (conduction electrons) as $\chi_{s0}$. This is so-called Pauli susceptibility and it does not depend on the temperature. The magnetic susceptibility of isolated localised *d*-electrons $\chi_{d0}$ changes with the temperature according to the Curie-Weiss law:

$$\chi_{d0} = C_1/(T - \theta_p) ,$$

where $C_1$ and $\theta_p$ are the Curie constant and paramagnetic Curie temperature, respectively.

The subsystems of free *s*-electrons and of isolated localised *d*-electrons can exchange their electrons and the contributions of their interaction to the magnetic susceptibility are described as

$$\chi_s = \chi_{s0}(1 + \alpha_1\chi_{d0}) \quad \text{and} \quad \chi_d = \chi_{d0}(1 + \alpha_1\chi_{s0}) ,$$

where $\alpha_1$ is the parameter of exchange interaction between the free electrons and isolated localised *d*-electrons (isolated atoms of *d*-elements). It is seen that the ratio of magnetic susceptibilities of *s*- and *d*-electron subsystems $\chi_r^{-1} = \chi_s/\chi_d$ has to be a linear function of temperature.

Along with isolated *d*-electrons, superparamagnetic clusters exist in the alloys of 3*d*-metals. The magnetic cluster is a group of *d*-atoms with polarised magnetic moments (spins) as a result of their interatomic interaction. The magnetic macromoment of the cluster *M* is proportional to the number of atoms included in the cluster. The clusters of *d*-atoms contribute to the whole susceptibility and change the relative parts of *s*- and *d*-contributions. The changed value of $\chi_{s0}$ is written as

$$\chi'_{s0} = \chi_{s0} / (1 - \alpha_2\chi_{d2}(T))$$

where $\alpha_2$ is the parameter of exchange interaction between the free electrons and clusters, $\chi_{d2}$ is the magnetic susceptibility of the superparamagnetic cluster system which obeys the Langevin law (see Vonsovsky 1971)

$$\chi_{d2} = C_2 \cdot L(\theta/T), \ \ C = \chi_{d2} \text{ at } T = 1,$$

$$L(\theta/T) = coth(\theta/T) - T/\theta, \ \ \theta = MH/k_B B,$$

$\theta$ is the Curie temperature of clusters (proportional to the number of atoms in the cluster), *H* is an external magnetic field, *B* is a constant.

A temperature dependence of the relative magnetic susceptibility $\chi_r^{-1} = \chi_s/\chi_d$ is described by the formula

$$\chi_r^{-1}(T) = \alpha_1\chi_{s0} + \chi_{s0}\chi_{d1}^{-1} + \alpha_2\chi_{d2}\chi_{s0}\chi_{d1}^{-1} . \tag{1.9}$$

It follows from the above analysis that the temperature dependence of the relative magnetic susceptibility $\chi_r^{-1} = \chi_s/\chi_d$ has to be non-monotonous if the the tendency to clustering prevails. As examples, the temperature dependences of $\chi_r^{-1}$ are shown in Fig. 1.22 for interstitials-free austenitic Cr15Ni25, Cr15Ni20 and Cr15Ni20Mo1.3 steels. It is seen that the clustering effect is enhanced if the nickel content is decreased in CrNi steel or if this steel is additionally alloyed by molybdenum.

A tendency to form clusters can be characterised by the parameter

$$K_c = \alpha_2 C_2 \cdot \theta / \alpha_1 C_1 . \tag{1.10}$$

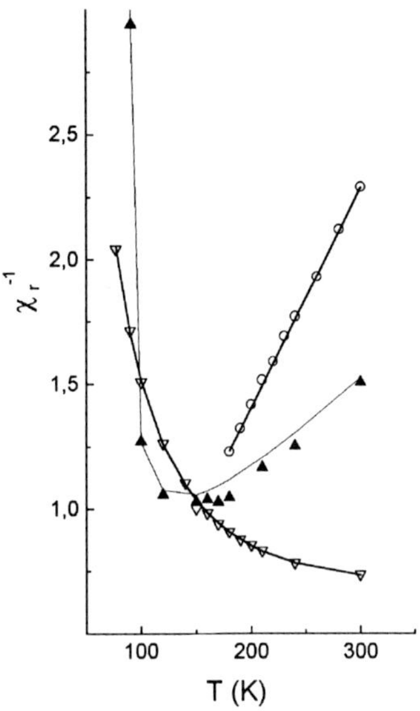


**Fig. 1.22.** Temperature dependence of the relative magnetic susceptibility $\chi_r^{-1} = \chi_s/\chi_d$ for interstitial-free austenitic steels Cr15Ni25 (circles), Cr15Ni20 (open triangles) and Cr15Ni20Mo1.3 (closed triangles), according to Shanina et al. (1998). For conveniency, the data for Cr15Ni25 steel are increased by 10 times. Decrease in the nickel content and alloying by molybdenum cause a non-monotonous temperature behaviour of $\chi_r^{-1}$, which is a sign of the tendency to clustering

As follows from the above mentioned, $\alpha_2 C_2$ is proportional to the number of clusters in the sample and $\theta$ is proportional to the number of atoms in the cluster, therefore, their product characterises the number of *d*-electrons included into the cluster system. The denominator in the above fraction, $\alpha_1 C_1$, is proportional to the number of localised *d*-electrons belonging to single *d*-atoms. Therefore, $K_c$ describes the relation between the values of the interaction energies of conduction electrons in mean fields of the cluster system and of isolated localised *d*-electrons. The more the value of $K_c$, the higher the effect of clustering is. Some limitations in using Eq. (1.10) can be encountered if the contributions of an alloying element to isolated *d*-electrons (denominator) and clusters (numerator) are comparable.

The data relating interatomic bonding and a type of SRO are presented in Table 1.4. One can see that enhanced metallic interatomic bonds promote short range atomic ordering whereas the covalent bonds assist clustering.

A simple qualitative explanation of this correlation can be proposed for the alloys constructed from the elements of 3*d*-group. As was discussed in Sect. 1.1.1, if the iron is alloyed by elements occupying the places left to the iron in the periodic table, i.e. those having a smaller number of *d*-electrons as compared to iron, and, as a result, *s*-electrons of iron and of the alloying element have to be spent for the filling of empty places in the 3*d*-band of alloy, the contribution of *d*-electrons to interatomic bonding grows, i.e. the covalent bonds prevail in the atomic interactions. If the latter is the case, the interatomic bonds have to be stronger between atoms having a higher number of *d*-electrons, as the overlapping integral of the electron wave functions between these atoms is larger. For the alloys constructed by the iron and elements left to the iron (Mn, Cr, V etc.), it means that iron atoms have to prefer the iron ones as nearest neighbours whereas atoms of alloying elements (substitutional solutes) will be pushed out from the solid solution and be forced to form clusters.

**Table 1.4.** The values of the concentration of free electrons $n_0$, the clustering coefficient $K_c$ and correlation between the types of interatomic bonding and SRO in austenitic steels

| Alloy | $n_0$ $10^{20}cm^{-3}$ | $K_c$ | Varying element | Enhanced bonds | Assisted SRO |
|---|---|---|---|---|---|
| Cr15Ni15 | 0.14 | 378 | - | - | - |
| Cr15Ni20 | 0.37 | 5,3 | Ni | metallic | ordering |
| Cr25Ni20 | 0.10 | 254 | Cr | covalent | clustering |
| Cr15Ni20Mn5 | 0.16 | 42 | Mn | covalent | clustering |
| Cr15Ni20Mn10 | 0.09 | - | Mn | covalent | clustering |
| Cr15Ni20Mo1.3 | 0.09 | 84 | Mo | covalent | clustering |
| Cr25Ni20Cu2.5 | 0.31 | 71 | Cu | metallic | ordering |
| Cr15Ni20Si2 | 90 | 0 | Si | metalic | ordering |
| Cr15Ni20Al2 | 45 | 0 | Al | metallic | ordering |
| Cr18Ni16Mn10(0.4N) | 2.6 | 19 | N | metallic | ordering |
| Cr18Ni16Mn10(0.3C) | 0.19 | 4.9 | C | covalent | clustering |

On the contrary, if the elements occupying the places right to the iron in the periodic table (Co, Ni, Cu) are added to the iron, their exessive *d*-electrons have to be hybridised with the free electrons of iron and acquire their symmetry, i.e. to contribute to the conduction band of the alloy and strengthen the metallic interatomic bonds. It follows from Table 1.4 that Cr, Mn, Mo assist the clustering whereas Ni and Cu promote the ordering in austenitic CrNi steels. It is noteworthy that the copper is known to form clusters while dissolved in the iron. Nevertheless, in the multicomponent iron-based solid solution, an increase in the concentration of free electrons, provided due to the copper, promotes, as the whole, a more homogeneous distribution of *d*-atoms in austenite (compare the values of $K_c$ for Cr25Ni20 and Cr25Ni20Cu2.5 steels in Table 1.4), even if the copper atoms themselves are prone to form clusters.

A special opportunity to test the correlation revealed between interatomic bonding and the SRO is provided due to the alloying of steel by Si and Al both of which do not belong to the 3*d*-group and, at the same time, are known to cause short range ordering and to form ordered structures with iron. Fig. 1.23 shows the electron spin resonance spectra in austenitic Cr15Ni20 steel doped by 2 and 4 mass % of Si or Al. The intensity of ESR signal is extremely increased due to both elements. It means an enhanced metallic interatomic bonding due to Si and Al (see the values of concentrations of free electrons in Table 1.4 and Fig. 1.10). In consistence with the above quoted studies of short range ordering in Fe-Si and Fe-Al alloys, the coefficient $K_c$ of clustering is equal to zero for Si(Al)-doped steels

Cr15Ni20 (Table 1.4), i.e. the temperature dependences of *g*-factor and magnetic susceptibility reveal no sign of clustering in these steels.

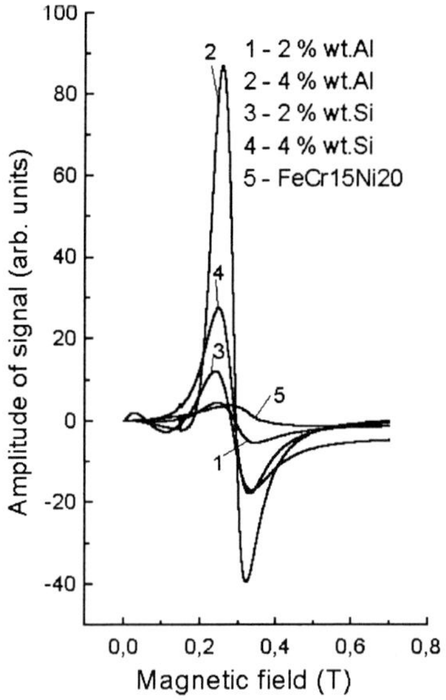


**Fig. 1.23.** Electron spin resonance spectra of austenitic steel Cr15Ni20 alloyed by 2 and 4 mass % of Si or Al

Thus, one can state that an effect of the type of interatomic bonding on the SRO has some general character: the metallic bonds enhance the short range atomic ordering and covalent bonds cause the clustering. Let us now analyse a more complicated situation when interstital atoms, nitrogen and carbon, are added to austenitic steel. As shown in Sect 1.1.1, alloying with nitrogen increases the concentration of free electrons in the $\gamma$ iron and austenitic steels, and, therefore, the metallic bonds are enhanced due to nitrogen, whereas carbon contributes its electrons to the 3*d*-band mainly, so that the covalent bonds become preferable. The temperature dependence of relative magnetic susceptibility $\chi_r^{-1} = \chi_s/\chi_d$ for Cr18Ni16Mn10 steel alloyed by 0.31 and 0.4 mass % of nitrogen or 0.15 and 0.43 mass % of carbon is shown in Fig. 1.24. The data for carbon austenite demonstrate clearly a significant effect of clustering, whereas, in the nitrogen austenite, the contributions from free electrons and isolated localised *d*-electrons prevail. The values of the concentration of free electrons and other characteristics of the electron subsystems for these two steels are presented in Table 1.5. It follows from the comparison that the concentration of free electrons and, respectively, their contribution to magnetic susceptibility are one order higher in the nitrogen steel than in the carbon one. While increasing the concentration of free electrons, nitrogen contributes insignificantly to the systems of isolated localised d-electrons (see the values of $\chi_{d1}$ and $\alpha_1\chi_{d1}$) and superparamagnetic clusters (see $\alpha_2\chi_{d2}$ and $\theta$).

The tendency to clustering in the austenitic carbon steel is clear according to the values of $\alpha_2\chi_{d2}$ and $\theta$ and the temperature dependence of the relative magnetic susceptibility. At the same time, because of a strong contribution of carbon to *d*-electrons ($\alpha_1\chi_{d1}$ in the denominator of formula 1.10), the value of $K_c$ is smaller than that for the nitrogen steel. Thus, there is some limitation in the application of formula (1.10) for interstitial solid solutions. The number $N_a$ of atoms in a cluster can be approximately evaluated if some average value of the magnetic moment for each *d*-atom, e.g. ~$2.5\mu_B$, is suggested.

Then, using the relation $\Theta = MH/k_B$, where $M = g\mu_B N_a$, and the value of the external magnetic field $H = 0.3$ T used in the experiment, we obtain $\Theta \approx N_a$, which corresponds to an average size of clusters in austenitic carbon steels of about 1 nm.

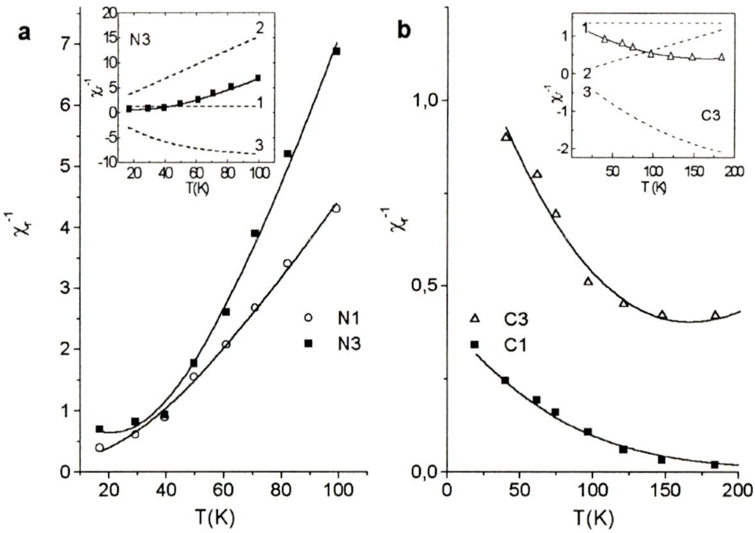


**Fig. 1.24.** Temperature dependence of the magnetic susceptibility of austenitic Cr18Ni16Mn10 steels alloyed with 0.31(N1) and 0.4(N3) mass % of nitrogen **a** or 0.15(C1) and 0.43(C3) mass % of carbon **b**. The fitting of experimental data to formula (1.9) is shown by the solid lines. The inserts show contributions to the magnetic susceptibility from free electrons (1), localised *d*-electrons (2) and superparamagnetic clusters (3) as obtained from the fitting. Deviation from the linear relation is caused by the superparamagnetic clusters of substitutional atoms

One can conclude, based on the above discussion, that nitrogen can affect the distribution of alloying elements in austenitic steels through the interstitial-substitutional interactions and through the change in the electronic structure, i.e. due to promotion of the metallic interatomic bonds. The latter seems to be a more significant factor of nitrogen alloying.

The experimental observations of nitrogen influence on the distribution of alloying elements in austenitic steels are performed using the method of small angle neutron scattering, SANS (see Gavriljuk et al. 1997). This method allows one to obtain an information concerning the mass (chemical) inhomogeneity of solid solutions on the scale of several to tens of nm.

**Table 1.5.** The values of parameters of electron subsystems in Cr18Ni16Mn10 steel alloyed by 0.54 mass % of nitrogen or 0.4 mass % of carbon. $\chi_{s0}$ and $\chi_{d1}$ are magnetic susceptibilities of *s*- and *d*-electron subsystems; $\alpha_1\chi_{d1}$ and $\alpha_2\chi_{d2}$ are the local magnetic fields of the isolated magnetic moments and the clusters, respectively, due to their exchange interaction with free electrons; $\theta$ is Curie temperature of clusters

| N(C) content, mass % | $10^{20}n_0$ ($cm^{-3}$) | $10^7\chi_{s0}$ | $10^5\chi_{d1}$ | $\alpha_1\chi_{d1}$ ($T$=1K) | $\alpha_2\chi_{d2}$ ($T$=1K) | $\theta$ (K) | $K_c$ |
|---|---|---|---|---|---|---|---|
| 0.54N | 2.4 | 10.0 | 5.4 | 15.7 | -1.14 | 140 | 10.2 |
| 0.4C | 0.19 | 0.8 | 80.0 | $6.0\times10^3$ | -33 | 715 | 4.0 |

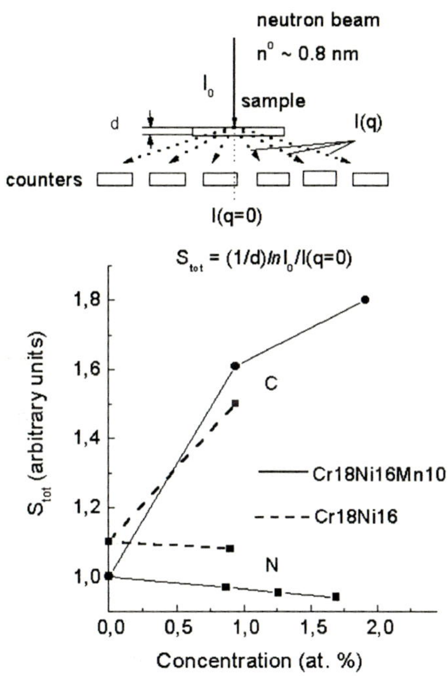


**Fig. 1.25.** A scheme of small angle neutron scattering measurements and an effect of nitrogen and carbon on the total cross section of neutron scattering $S_{tot}$ in austenitic Cr18Ni16Mn10 and Cr18Ni16 steels. The wavelength of neutrons is about 0.8 nm; $S_{tot}$ is measured as a weakening of the intensity $I_0$ of the primary neutron beam

The experimental technique is schematically shown in Fig. 1.25 together with the measured data for austenitic CrNiMn and CrNi steels alloyed by nitrogen or carbon. An incident neutron beam is scattered on the inhomogeneities of mass distribution in the sample and the weakening of its intensity is described as the total cross section of scattering $S_{tot} = lnI_0/I$, where $I_0$ and $I$ are the intensities of the incident and passed neutron beams, respectively. As the nuclei of different elements differ in their neutron scatterings (the scattering lengths for chromium $b_{Cr} = 0.352\cdot10^{-12}$ cm, manganese $b_{Mn} = -0.36\cdot10^{-12}$ cm, iron $b_{Fe} = 0.96\cdot10^{-12}$ cm, nickel $b_{Ni} = 1.03\cdot10^{-12}$ cm, $b_C = 0.66\cdot10^{-12}$ cm, $b_N = 0.94\cdot10^{-12}$ cm, see in detail Gurevich and Tarasov 1965), atomic clusters have to enhance the neutron scattering. As follows from Fig 1.25, the alloying of austenitic CrNi and CrNiMn steels by nitrogen decreases the total cross section of neutron scattering, whereas carbon causes a striking increase of $S_{tot}$. It has to be emphasised that this effect is not due to scatterings of neutrons on nitrogen or carbon atoms themselves, because, if it would be the case, the effect has to be the opposite one as the scattering cross section of neutrons on nitrogen atoms is about twice and the absorption cross section more than 500 times as large as those on carbon ones. Moreover, the fractions of nitrogen and carbon atoms in solid solutions are too small in order to cause such a large scattering. So, it is an effect of nitrogen and carbon on the distribution of substitutional solutes, i.e on the short range atomic order in austenitic steels.

As discussed earlier, Fe-Cr and Fe-Mn alloys are prone to short range decomposition as Cr-Cr and Mn-Mn bonds are stronger than those of Fe-Cr and Fe-Mn. Fe-Ni alloys reveal short range atomic ordering (Fe-Ni bonds are preferred), although, as the superstructure FeNi has an equiatomic composition, in Fe-Ni alloys containing less than 50 % of nickel the ordering is accompanied by the decomposition of a solid solution, resulting in the formation of nickel-rich microvolumes. Thus, in absence of interstitials, austenitic CrNi and CrNiMn steels are prone to the short range decomposition, i.e. clustering. The alloying with nitrogen diminishes this tendency whereas carbon enhances it.

## 1.1.3 Interaction of nitrogen atoms with substitutional solutes

Nitrogen atoms in austenitic steels prefer to be neighbours of solute atoms belonging to elements located left to the iron in the periodical table as these decrease the chemical potential of nitrogen in iron. A number of experimental studies confirm the tendency to form *s-i* (substitutional-interstitial) atomic complexes and some examples are presented below.

By means of X-ray and electron diffraction, Jack (1989) and Driver et al. (1979) observed interstitial-substitutional clusters after nitriding at 400-800 °C of fcc iron alloys with 34 at.% Ni and small additions of Ti, Nb, V or Mo. Clustering occurred periodically in <100> direction and was interpreted as a modulated structure of which the wavelength increases with temperature being extrapolated to infinity at about 780 °C.

Cr-N complexes in austenitic steel Cr15Ni15 were studied by Oda et al. (1990) using X-ray absorption spectroscopy (X-ray Absorption Near Edge Structure, XANES, and Extended X-ray Absorption Fine Structure, EXAFS). The authors observed that silicon in this steel enhances a gathering of nitrogen atoms around chromium atoms. Plastic deformation was shown to break *s-i* complexes, and, based on this result, low cycle fatigue softening, which takes commonly place in Cr-austenitic Ni and Cr-Mn steels doped by nitrogen, was attributed to the breakdown of *s-i* complexes.

Mo-N complexes in the austenitic phase of duplex CrNiMoN steels were studied by means of atom probe field ion microscopy (see Wahlberg et al. 1989). A special feature in these observations was an uniform distribution of Mo-N complexes in the material and the absense of large clusters or precipitates. The authors observed also Cr-N complexes in steels with low molybdenum content and, on account for the comparable interaction energies for Mo-N and Cr-N in iron austenite, they suggested the existence of Mo-N-Cr and Cr-N complexes which were difficult to be identified due to a different field evaporation behaviour of such complexes.

We shall analyse the thermodynamical basis for the existence of complexes *s-i* in nitrogen austenite, which is important for the comparison of different substitutional solutes in relation to their interaction with nitrogen and carbon atoms. The solid solution strengthening of austenitic nitrogen(carbon) steels is determined by the symmetry of these complexes and the strength of distortions induced by them in the fcc crystal lattice.

### 1.1.3.1 Estimation of interaction energies s-i

A physical reason for the formation of *s-i* complexes in solid solutions of Fe-Me-N(C) lies in the difference between interaction energies of Fe-N(C) and Me-N(C). This *s-i* interaction can be evaluated using "Fe-base database" of the ThermoCalc program created by the Swedish Royal Institute of Technology (see e.g. Sundman 1995).

A number of thermodynamical data of Fe base liquid and solid systems containing substitutional and interstitial solute elements were obtained up to now (see e.g. Dinsdale 1991, Hertzman and Sundman 1982, Frisk 1991, Hertzman and Jarl 1987, Ono 1986 etc.). Using this data and some analytical methods (see Lupis 1983, Jarl 1978, Ko and McLellan 1983 and references there), one can estimate the values of interaction energies between interstitial and substitutional atoms. However, there is a large scatter in the experimental data available, which creates difficulties in their analysis. Fortunately, in some cases, such an estimate can be performed due to the development of a self-consistent thermodynamic database which is used in the CALPHAD approach (CALCulation of PHAse Diagrams) for the calculation of multicomponent phase diagrams.

Traditionally, the interaction energies *s-i* in the *Fe-s-i* systems are estimated from experimental data of the influence of substitutional alloying elements on the value of the chemical potential, activity or coefficient of activity of interstitial atoms.

Let us suppose that the binary system *Fe-i* and the ternary *Fe-s-i* one are in contact with some external system containing *i* atoms with a chemical potential $\mu$ and that an exchange of *i* atoms between these systems is possible. In this case, $\mu = \mu_i = \mu_i^s$, and the ratio between atomic concentrations of interstitial atoms in solid solutions *Fe-i* and *Fe-s-i* (see for details Sozinov and Gavriljuk 1999) can be described as

$$\frac{\theta_i^s}{\theta_i} = 1 + z_1\left[exp\left(-\frac{\Delta E_1}{kT}\right) - 1\right]\theta_s \quad , \tag{1.11}$$

where
$\theta_i$ is the part of interstitial sites occupied by $i$ atoms in the *Fe-i* solution,
$\theta_i^s$ is the same for the *Fe-s-i* solution,
$z_1$ is the coordination number in the first coordination sphere of the interstitial sublattice (the number of nearest neighbouring interstitial sites around an interstitial site in the origin of coordinates, $z_1 = 12$ in the fcc lattice),
$\Delta E_1$ is a change in the energy of system (solid solution) when one "virtual" atom $i$ is put into an interstitial site near the substitutional atom $s$.
$\Delta E_1$ has the meaning of an interaction energy $s$-$i$ in the first coordination sphere.

Equation (1.11) constitutes a base for the calculation of the interaction energy $s$-$i$. It can be written in a more simple form

$$\frac{\theta_i^s}{\theta_i} = 1 + \gamma\theta_s \tag{1.12}$$

where $\gamma$ is determined from Eq. (1.11).

In this consideration, we take into account only the interaction between nearest neighbouring $s$ and $i$ atoms. It is clear from (1.11) that, for a repulsive interaction $s$-$i$ ($\Delta E_1 > 0$), the value of $\gamma$ cannot be smaller than $-z_1$.

In case of an attractive character of the interaction $s$-$i$ ($\Delta E_1 < 0$), the value of $\gamma$ will be positive.

In order to find $\gamma$, the calculation of chemical potentials of C and N have to be performed for some small values of their concentrations ($\theta_i = 10^{-5}$) in the binary fcc alloys Fe-C and Fe-N. It is natural that, due to the different "standard states" accepted in thermodynamics for C (graphite) and N (gas $N_2$ at P = 1 bar), the calculated values of chemical potentials are different. However, if to fix these values of the chemical potential and vary the concentration of the substitutional atoms $\theta_s$ (e.g. from 0 to $10^{-3} \div 10^{-4}$) in Fe-Me-C and Fe-Me-N alloys, one can obtain the dependence $\theta_i^s$ on $\theta_s$ at a constant chemical potential of interstitial atoms. Of course, the two-phase region has to be avoided. It turns out that in all cases this dependence has a very weak deviation from linearity and can be approximated by Eq. (1.12).

The values of $\gamma$ obtained in such a manner for different alloying elements are presented in Table 1.6 along with the interaction energies $\Delta E_1$ calculated using Eq. (1.11).

The values of $\gamma$ for Si-C and Si-N interactions are smaller than their possible minimum magnitudes ($-z_1$) for the model in which only the interaction at the nearest distances is taken into account. This means that, in fact, interaction has a long-range character, which is possible not only for Si, but for other alloying elements interacting with interstitial atoms.

It is seen that nitrogen has a stronger interaction with substitutional elements (except niobium) than carbon. All substitutional elements presented in Table 1.6, except Ni and Si, reveal an attractive atomic interaction with nitrogen and carbon.

**Table 1.6.** Values of $\gamma$ at 1423 K and interaction energies $\Delta E_1$ for Fe-Me-C and Fe-Me-N alloys

| Me | C | | N | |
|---|---|---|---|---|
| | $\gamma$ | $\Delta E_1$, eV | $\gamma$ | $\Delta E_1$, eV |
| Nb | 23.7 | -0.196 | 17.5 | -0.167 |
| V | 16 | -0.159 | 35 | -0.236 |
| Cr | 8.4 | -0.107 | 23 | -0.193 |
| Mo | 7.9 | -0.103 | 9.1 | -0.113 |
| W | 5.8 | -0.083 | 18.1 | -0.171 |
| Mn | 3.1 | -0.051 | 7.8 | -0.102 |
| Ni | -3.6 | 0.112 | -4 | 0.135 |
| Si | -8.1 | * | -9.7 | * |

*- couldn't be estimated from Eq. (1.11) for the model in which only the interaction at the nearest distances is taken into account

### 1.1.3.2
### Crystalline symmetry of s-i defects

To clarify the role of *s-i* complexes in strengthening of austenitic steels, some knowledge is needed on their crystalline symmetry and local distortions created by them in the fcc crystal lattice of austenite. Such an opportunity is provided by the study of inelastic relaxation caused by non-cubic defects in an fcc austenitic matrix. The theory of inelastic relaxation in crystalline solids is described in detail elsewhere (see e.g. Novick and Berry 1972).

Internal friction results from elementary jumps of nitrogen atoms from one interstitial site into another one in order to provide a more favourable orientation of non-cubic defects under an applied stress as a response to the external cyclic mechanical loading. Interstitial atoms in the fcc crystal lattice are defects of the cubic symmetry and, therefore, they cannot cause any mechanical after-effect when an external loading of changing amplitude is applied. The *s-i* complexes only induce a local deviation from the cubic symmetry (see Fig. 1.26) and, under an applied stress, the energy of distortions induced by these defects depends on their crystallographic orientation and on the type of the applied stress field.

Rozin and Finkelstein (1953) have discovered a relaxation peak of internal friction at 300 °C at a frequency of mechanical vibrations of ~1 Hz in austenitic steel

Cr25Ni20 containing carbon. They explained the damping as arising from the migration of carbon atoms in the fcc lattice but did not propose any specific mechanism of relaxation. It was found in later studies that, indeed, the activation enthalpy of relaxation is equal to that for the migration of interstitials, and different hypotheses were proposed for the elementary mechanism of relaxation. In the following, we use the abbreviation "RF" to denote this kind of relaxation.

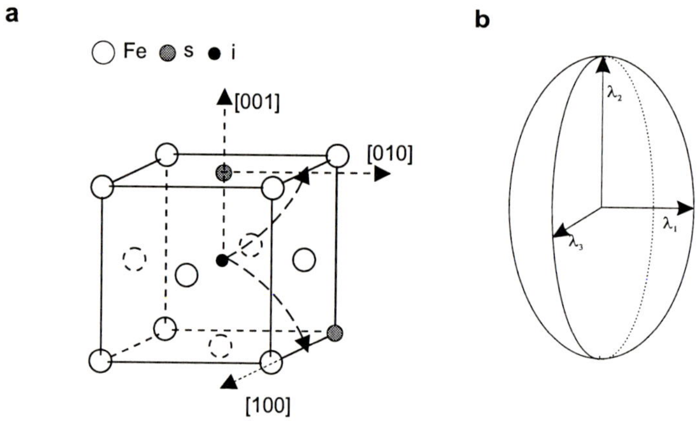


**Fig. 1.26.** An example of *s-i* defect created in fcc lattice **a**. The defect has tetragonal symmetry which corresponds to the strain ellipsoide **b** with $\lambda_1=\lambda_3\neq\lambda_2$ where $\lambda_i$ are the main components of the strain tensor. If a contracting stress acts along [001], the interstitial atom can jump into other site in order to decrease the elastic energy, and this relaxation process is accompanied by the absorption of energy (internal friction)

Ke and Tsien (1956) studied the RF relaxation in austenitic carbon-containing manganese and chromium-manganese-nickel steels. The authors attributed the carbon-induced damping peak to the reorientation, under external stress, of atomic pairs formed by carbon atoms with substitutional atoms or vacancies which, in both cases, create non-cubic distortions in the fcc lattice and, thus, cause a Snoek-like relaxation. The latter takes place in the bcc lattice where, even in absense of substitutional solutes, single interstitials occupying octahedral holes induce tetragonal distortions (Fig. 1.1b).

Wu and Wang (1958) examined fcc FeNiC alloys and proposed a relaxation mechanism based on the reorientation of a carbon atom pair, of which one-half is located in a vacancy of the fcc lattice. Ulitchny and Gibala (1973) tested the effects of the solution treatment temperature, electron irradiation and cold work on damping in FeNiC austenites and concluded that the relaxation is caused by carbon atom-vacancy pairs.

Nitrogen-containing iron-based alloys were firstly studied by Verner et al. (1961, 1965). A reorientation of nitrogen atom pairs located within the third to the fifth coordination spheres was proposed as a source of relaxation.

Thus, there is an uncertainty in the interpretation of relaxation peaks caused by nitrogen and carbon, although there is no doubt that damping is due to the migration of interstitial atoms. In accordance with the theory of inelastic relaxation, two critical tests can be made in order to clarify the nature of defects responsible for relaxation. The first one is based on the measurement of the dependence of the relaxation strength (the intensity of relaxation peak) upon the content of interstitials. The concentration dependence has to be linear if single interstitial atoms are involved in the relaxation and square for interstitial pairs. The second one is concerned with the crystallographic symmetry of non-cubic defects in the cubic lattice. Each type of the defect has a unique orientation dependence of the relaxation strength, which can be studied using single crystals.

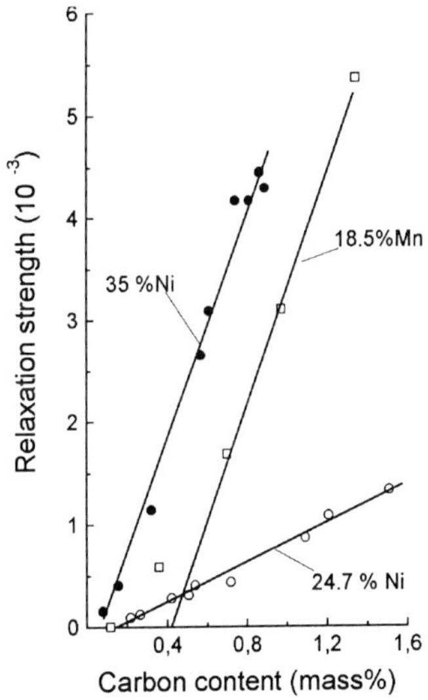


**Fig. 1.27.** The concentration dependence of the Rozin-Finkelstein relaxation strength in austenitic carbon steels: 18.5 mass % of Mn (Ke and Tsien 1956), 35 % Ni (Wu and Wang 1958) and 24.7 % Ni (Gavriljuk 1987 I)

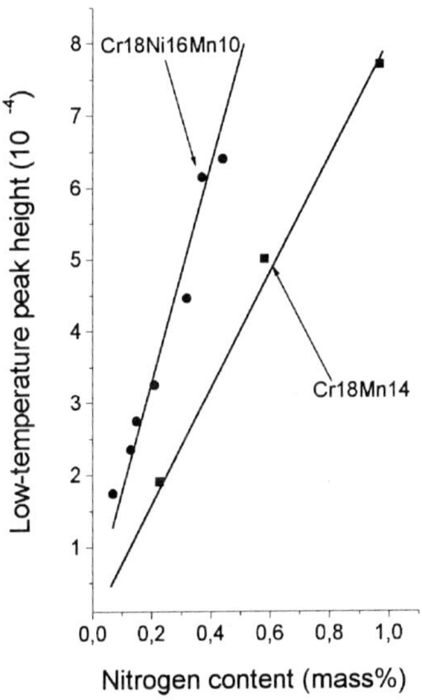


**Fig. 1.28.** The concentration dependence of the Rozin-Finkelstein relaxation strength in austenitic nitrogen steel Cr18Ni16Mn10 (Gavriljuk et al. 1998). The data for Cr18Mn14 steel is also shown according to Banov et al. (1978)

The relaxation strength versus the interstitials content is shown in Figs 1.27 and 1.28 for different austenitic steels. In both carbon- and nitrogen-alloyed steels, a linear concentration dependence of the relaxation strength is revealed. A special feature of carbon steels is that the concentration dependence cannot be extrapolated to the origin of coordinates and the RF peak exists at carbon contents above about 0.1 to 0.3 mass %, whereas in nitrogen steels the linear concentration dependence starts from the origin of coordinates.

One assumes (see e.g. Ke and Tsien 1956) that the vacancies in the austenitic crystal lattice absorb carbon atoms and that the RF relaxation can be observed after a saturation of existing vacancies by carbon. However, in view of significant amount of carbon (0.45 to 1.35 at. %) to be absorbed by vacancies, the necessary density of vacancies seems to be not realistic.

The different behaviour of chromium and manganese carbon steels as compared to the high-nickel carbon and nitrogen steels can be attributed to a marked difference in the formation of grain-boundary segregations by the above elements. As will be shown in Sect. 1.1.4.3, nitrogen in austenitic steels is not noticeably prone to form a grain-boundary segregation whereas carbon reveals a strong affinity to grain boundaries. As a result of the segregation phenomena, an apparent square-like concentration dependence of the relaxation strength is sometimes observed in the range of small carbon contents in austenite, whereas, at high concentration of carbon and in the whole range of nitrogen contents, the linear behaviour occurs only. It is also remarkable that, with increasing nickel content, the concentration dependence of RF relaxation strength is extrapolated nearer to the zero carbon content. As was revealed by Petrov and Trophimova (1998), nickel pushes carbon out of the grain boundaries.

Therefore, single carbon and nitrogen atoms, not their pairs, contribute to the relaxation. It means that *s-i* complexes are responsible for the damping in carbon and austenitic nitrogen steels.

A type of *s-i* complexes was clarified by studies of the orientation dependence of the relaxation strength (Gavriljuk et al. 1998). According to Novick and Berry (1972), the relaxation strength caused by non-cubic defects in a cubic crystal is described by the equation

$$\delta G^{-1} = \delta S_{44} + 4(\delta(S_{11} - S_{12}) - 0.5\cdot\delta S_{44})\cdot\Gamma,$$

where $\delta G$ is the relaxation of the shear modulus (the relaxation strength), $S_{ij}$ are the components of the tensor of elastic compliances, $\Gamma = \gamma_1^2\gamma_2^2 + \gamma_2^2\gamma_3^2 + \gamma_3^2\gamma_1^2$ is an orientation factor, $\gamma_i$ are the directing cosines of the angles between the crystal axes and the main axis of the dipole which is formed by elastic distortions caused by a non-cubic defect in a cubic lattice (see Fig. 1.26).

In the cubic lattice, the orientation dependence of the relaxation strength is determined by the symmetry of the non-cubic defect: for the tetragonal <100> defect $\delta G^{-1}_{<100>} = 0$ and $\delta G^{-1}_{<111>} = (4/3)\delta(S_{11} - S_{12})$, while for the orthorhombic <110> defect $\delta G^{-1}_{<100>} = \delta S_{44}$ and $\delta G^{-1}_{<111>} = (4/3)\delta(S_{11} - S_{12}) + (1/3)\delta S_{44}$.

The internal friction spectra of single crystals (Fig. 1.29) provide a clear evidence of the orientation dependence of the relaxation strength, and, moreover, the type of orientation dependence is being changed with increasing percentage of substitutional solutes. The relaxation is absent for <100> orientation in the steel Cr18Ni10, while it does not vanish for both of the orientations in steel Cr18Ni8Mn10. In accordance with the relations given above, it means that Cr18Ni10(0.2N) austenite is characterised by tetragonal type defects, while the

Cr18Ni8Mn10(0.5N) alloy presents non-cubic defects the symmetry of which is not higher than the orthorhombic one (the tetragonal *s-i* dipole and the orthorhombic $s_1,s_2$-*i* one are shown in the upper left corners of Figs. 1.1.29a,b). A replacement of "*s-i*" defects having tetragonal symmetry by "$s_1,s_2$-*i*" ones, the symmetry of which is lower, is consistent with an increase in the content of alloying elements leading to an increasing probability for an interstitial nitrogen atom to have two and more substitutional solute atoms as nearest neighbours.

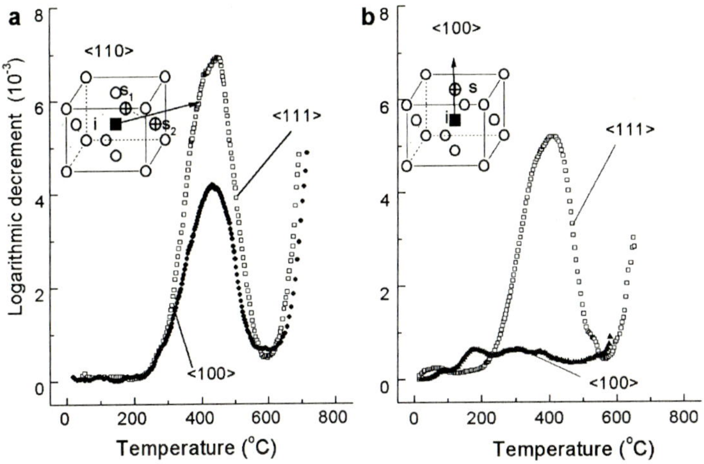


**Fig. 1.29.** The orientation dependence of the Rozin-Finkelstein relaxation in single crystals of Cr18Ni8Mn10N0.5 **a** and Cr18Ni10N0.2 **b**. Orthorhombic <110> (**a**) and tetragonal <100> (**b**) *s-i* defects are shown in the upper left corners

Thus, one can state that single nitrogen atoms create *s-i* complexes with atoms of alloying elements in austenitic steels and the crystallographic symmetry of *s-i* complexes in the as-quenched state is determined by the fraction of substitutional solutes. It has also to be noted that the strength of the inelastic relaxation caused by interstitial atoms is a measure of distortions induced by *s-i* complexes in fcc austenite lattice. We will use these data while discussing the interaction between nitrogen atoms and dislocations in the following section.

## 1.1.4 Interaction of nitrogen atoms with crystal lattice imperfections

The knowledge of interactions between nitrogen atoms and vacancies, dislocations and grain boundaries is of practical significance as the controlling strengthening mechanisms in austenitic nitrogen steels include these interactions to a more or less extent.

### 1.1.4.1. Vacancies

The interaction of nitrogen and carbon atoms with vacancies was studied in detail for bcc α-iron using internal friction and positron annihilation (see e.g. Weller and Diehl 1976, Vehanen et al. 1982). It was shown that nitrogen atoms in ferrite form complexes with monovacancies $V_1$-N which acquire a decreased mobility and decompose at 220 °C with a dissociation enthalpy of 1.4 eV. The dissociation enthalpy consists of the $V_1$-N binding enthalpy and the migration enthalpy of the nitrogen atom as a more mobile component in these complexes. Thus, the binding enthalpy of nitrogen atoms to monovacancies can be evaluated as ~0.7 eV. The $V_1$-C complexes in ferrite are more stable and their dissociation temperature and enthalpy are 290 °C and 1.6 eV, respectively. Therefore, the mobility of decorated vacancies, i.e. of complexes of monovacancies with interstitials $V_1$-$i$ is decreased because, for their movement, an expenditure of energy for the dissociation of complexes is needed. Decorated bivacancies $V_2$-$i$ can easily move, however, their fraction in the thermodynamical equilibrium is insignificant.

The interaction of nitrogen atoms with vacancies in austenite is smaller than in ferrite because of the smaller distortions of the fcc crystal lattice by interstitials. Perhaps, for this reason, the swelling of austenitic steels under irradiation is not much affected by nitrogen (see Igata et al. 1981, Gavriljuk et al. 1991). At the same time, this interaction has some non-trivial aspect concerned with an increase of vacancy concentration due to interstitial atoms. The dilation of the crystal lattice by interstitials decreases the enthalpies of formation and migration of undecorated vacancies, which has to result in an enhanced selfdiffusion.

The thermodynamics of interstitial-vacancy interactions in solid solutions was analysed by McLellan (1988). He has shown that the free enthalpy of formation of monovacancies varies with the interstitial concentration $\theta$ as

$$H_{1v}^{f}(\theta) = H_{1v}^{f}(0) - BI\theta_i V_{1v}^{f} / V_v^0$$

where
$B$ is the bulk modulus,
$V_v^0$ is the molar volume of fcc Fe,
$V_{1v}^f$ is the formation volume of a monovacancy, i.e. $V_{1v}^f = (\partial H_{1v}^f/\partial P)_T$,
$I$ is a measure of lattice distortions caused by interstitials.

The numerical evaluation made by Mclellan gave a decrease in $H_{1v}^f$ of about 4 kJ·mol$^{-1}$, i.e. of 10 % change at 1273 K and $\theta = 0.07$, which corresponds to a fivefold increase in the vacancy concentration. This result was used for an explanation of the enhanced diffusivity of Fe atoms in the iron austenite with increasing carbon concentration.

The above consideration sheds a light on the remarkable data of as-quenched defects in nitrogen-containing stainless austenitic steels obtained by Kikuchi et al. (1974, 1978, 1985).

They have shown that the ageing of steel Cr25Ni28N0.5 at temperatures between 650 and 850 °C leads to the formation of secondary defects which are Frank sessile and double faulted loops, beside the perfect prismatic loops. No secondary defects were observed in the steel of the same basic composition, but without nitrogen. These observations give an experimental evidence for the exessive (superabundant) vacancies induced by nitrogen in austenitic steels.

The same structural features were obtained in phosphorus-containing austenitic stainless steels as phosphorus, being substitutionally dissolved in iron, causes nevertheless significant distortions in iron-base solid solutions. Moreover, the dislocation loops were observed to disappear when an additional ageing is performed at temperatures above 850 °C and to appear again after a third ageing below 850 °C. The addition of phosphorus is also found to accelerate the precipitation of $Cr_2N$ nitrides during ageing, i.e. it assists the migration of chromium atoms in austenite.

It seems that the excessive vacancies induced by interstitials in solid solutions are a general case as the same effect was observed to be caused by hydrogen in different metals (see e.g. Watanabe et al. 1996, Gavriljuk et al. 1996).

### 1.1.4.2 Dislocations

It is naturally to expect that the interaction of nitrogen atoms with dislocations plays an important role in the strengthening of austenitic steels. A comparative study of the pinning of dislocations in austenitic steels by nitrogen and carbon atoms was performed by Gavriljuk et al. (1987, 1989).

The values of enthalpy of binding interstitials to dislocations were obtained using the internal friction technique which allows one to measure the energy losses caused by the nucleation and movement of dislocations (see Fig. 1.30). In the absence of relaxation processes, the internal friction background is mainly due to the vibrations of dislocation segments. If some relaxation process occurs (like the strain-induced migration of interstitial atoms), its contribution to internal friction remains unchanged with increasing strain amplitude at constant temperature, because the Snoek-like relaxation processes are amplitude-independent.

In the annealed state, the dislocations are pinned by interstitial atoms. For annealed or aged materials, the starting stress of dislocation sources is lower than that of unpinning dislocations decorated by interstitials. The value of stress required for the unpinning of dislocations from interstitials is about of $G/10 \div G/50$ (see e.g. Hirth and Lothe 1970), which is much higher than the ultimate stress. That is why the plastic strain at the yield strength of annealed or aged materials is caused by the nucleation of new dislocations, not by pinned ones breaking away, and the subsequent work hardening is due to the intersection of the forest of immobile pinned dislocations by the new gliding ones. Therefore, if some stress lower than that sufficient for the start of dislocation sources is applied, the sample behaves elastically and the internal friction is not changed by the strain amplitude $\varepsilon$ (see short initial parts of the curves in Fig. 1.30 at 550 and 650 °C). If the critical shear stress is achieved and new dislocations are emitted, an amplitude-dependent internal friction occurs because the higher the applied stress, the larger is the area swept by gliding dislocations and, respectively, the greater is the inelasticity.

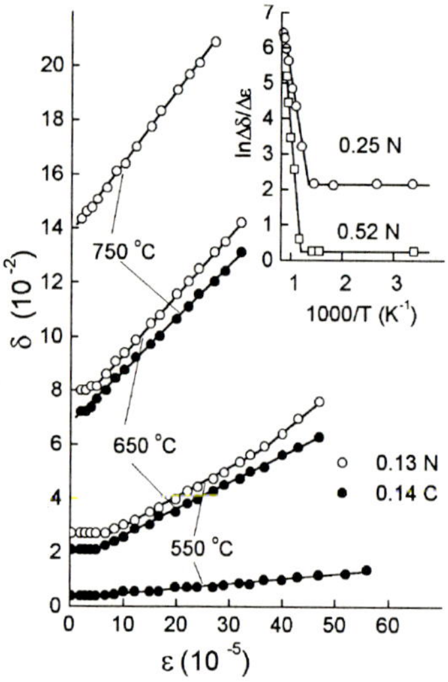


**Fig. 1.30.** The logarithmic decrement $\delta = Q^{-1}/\pi$, where $Q^{-1}$ is internal friction, *versus* the strain amplitude for austenitic steel Cr20Ni16Mn6 doped by nitrogen (0.13 mass %) or carbon (0.14 mass %) at different temperatures. A treatment of the experimental data in the Arrhenius coordinates is shown as insert

At a constant frequency of induced mechanical vibrations, the slope of the amplitude-dependent part of $\delta(\varepsilon)$ curves characterises the mobility of dislocations (the value of area swept by dislocations). Below some critical temperature, this slope does not depend on the temperature (see the insert in Fig. 1.30). It means that deformation is still provided by the slip of emitted unpinned dislocations, while the previously existing dislocations, as before, are pinned by interstitials. Above this critical temperature, about 400-500 °C for austenitic steels depending on the degree of pinning, the pinning becomes sufficiently weak to enable the dislocation to break away from interstitials and the interstitial atoms become sufficiently mobile, so that dislocations move dragging their interstitial clouds. At these temperatures, the interaction of new dislocations with interstitial atoms in the solid solution results also in a dragging of interstitials by dislocations. The higher the temperature, the weaker is the interstitial-dislocation interaction and the more exhausted are the interstitial clouds round the dislocations. As a result, the larger is the slope of the amplitude-dependent part of the $\delta(\varepsilon)$ curves. Treating these experimental data in Arrhenius coordinates, one can obtain the values of enthalpy of interstitial-dislocation binding (see Table 1.7).

It follows from the slope of the $\delta(\varepsilon)$ curves in Fig. 1.30 that, in the whole temperature range, nitrogen atoms lock dislocations more effectively than carbon ones. This conclusion is also supported by the data of Table 1.7. An important experimental observation is that the effective energy of binding between nitrogen atoms and dislocations is increased with the nitrogen content in austenite, whereas such a dependence is not significant for carbon. Similar values of the binding enthalpy and their concentration dependence were obtained by Kozlov et al. (1998) in the studies of the steel Cr18Ni15Mn10N using TEM technique for measurements of the bending angles of dislocation lines on obstacles in solid solutions as proposed by Fleischer and Hibbard (1963) (see Table 1.8).

**Table 1.7.** Enthalpy $H_b$ of the binding of nitrogen and carbon atoms with dislocations in austenitic steel Cr20Ni16Mn6

| N(C) content (mass %) | 0.14C | 0.4C | 0.13N | 0.25N | 0.36N | 0.52N |
|---|---|---|---|---|---|---|
| $H_b \pm 0.05$ eV | 0.54 | 0.66 | 0.65 | 0.69 | 0.88 | 1.25 |

Thus, one can state that nitrogen atoms in austenite possess a higher energy of binding with dislocations than carbon atoms and have to serve as stronger obstacles for the dislocation slip.

**Table 1.8.** Enthalpy $H_b$ of the binding of nitrogen atoms with dislocations in aus tenitic steel Cr18Ni15Mn10N (after Kozlov et al.. 1998)

| N content (mass %) | 0.065 | 0.17 | 0.26 | 0.36 | 0.56 | 0.65 |
|---|---|---|---|---|---|---|
| $H_b$ (eV) | 0.6 | 0.62 | 0.68 | 0.75 | 1.11 | 1.18 |

The data of studies on the inelastic relaxation caused by *s-i* complexes (Rozin-Finkelstein relaxation) and on the electron structure of austenitic steels alloyed by nitrogen and carbon, as described in the previous sections, allow one to explain a mechanism of interstitial-dislocation interaction in austenitic steels and to clarify the remarkable difference between nitrogen and carbon atoms in their binding to dislocations.

Three kinds of contributions to the interaction between interstitials and dislocations are mainly discussed: elastic, chemical and electrostatic.

It is generally admitted that a main contribution to the interaction between dislocations and impurity atoms is provided by elastic distortions. The first analysis was made by Cottrell and Bilby (1949) who calculated the interaction between the hydrostatic components of stress fields created by an edge dislocation and interstitials, which results in the formation of interstitial clouds in the vicinity of the nuclei of edge dislocations. Later on, Cochardt et al. (1955) have shown that non-cubic defects, like carbon atoms in bcc iron, interact also with the shear stress fields of screw dislocations and the enthalpy of this binding is comparable to that of edge dislocations. Further treatment of this problem was proposed by Suzuki (1979).

In the fcc lattice, *s-i* complexes can interact only with screw dislocation components because of the cubic symmetry of distortions caused by single *i* atoms.

As the experimental data are satisfactorily treated in Arrhenius coordinates (see Fig. 1.30), one can state that the formation of Cottrell clouds, obeying the exponential temperature dependence, is really the controlling mechanism for interstitials-dislocation interaction in austenitic steels. The chemical interaction, i.e. the formation of Suzuki's clouds at stacking faults, obeys a linear temperature dependence.

As shown in the previous paragraph, the Rozin-Finkelstein relaxation can serve as an appropriate tool for the evaluation of local distortions caused by defects in an fcc lattice. A correlation between lattice distortions and RF relaxation strength can be illustrated by the data of high-nickel carbon steels in Fig. 1.27. The slope of the concentration dependence of the RF relaxation strength is much higher for Fe-35Ni-C steel than for Fe-24.7Ni-C one. The data obtained by Wu and Wang (1958) show that with a further increase in the Ni concentration up to 50 mass %, the RF relaxation strength is also increased. A reason for such an effect of nickel lies in the volume magnetostriction. High-nickel iron austenites belong to the invar alloys and reveal strong magnetic properties. In accordance with the idea of Weiss (1963) about two magnetic states of the iron atoms in invar alloys, ferromagnetic and antiferromagnetic, the ferromagnetic interaction results in a large atomic volume and, as a consequence, in a remarkable increase in the lattice parameter of Fe-Ni invar alloys. For this reason, the Fe-C complexes in Fe-Ni-C austenites cause larger distortions with increasing nickel content in the range of invar concentrations and provide an increasing contribution to the RF relaxation strength, which is confirmed by the data in Fig. 1.27.

One can imagine that a difference between the interactions of nitrogen and carbon atoms with dislocations in austenitic steels can be attributed to a difference in the strain fields, created by s-N and s-C complexes in the iron fcc lattice. Really, according to the measurements of the lattice parameters (see Table 1.1), the nitrogen atoms induce higher distortions in the fcc austenite than the carbon ones. Thus, one could expect that s-N complexes cause higher local distortions in austenite as compared to the s-C ones.

However, the comparison of the data of the concentration dependence of the RF relaxation strength in austenitic nitrogen and carbon steels (Fig 1.31) leads to the conclusion that there is no significant difference between the asymmetry of the strain ellipsoides $\Delta\lambda$ (see Fig. 1.26) created by s-C and s-N complexes in austenitic steels. This means that not only the lattice dilation controls the interaction of defects with screw dislocations. A deviation from the cubic symmetry that determines the value of $\Delta\lambda$ is also important and it is not always consistent with the volume dilation.

Mössbauer data concerning the non-cubic symmetry of the electric fields around iron atoms in Fe-C and Fe-N austenites can be used to shed a light on this topic (see Table 1.3). The electric field gradient in the cubic iron lattice is created by a non-symmetrical distribution of the electron charge caused by the iron-nitrogen(carbon) electron exchange, i.e. it characterises directly the local distortions of the crystal lattice induced by interstitials. Therefore, in accordance with the data presented in Table 1.3, the distribution of electrons in the Fe-C complex is less symmetrical than in Fe-N one, and, respectively, a deviation from the cubic symmetry of the local distortion is higher in the carbon austenite in spite of a larger lattice dilation induced by nitrogen atoms.

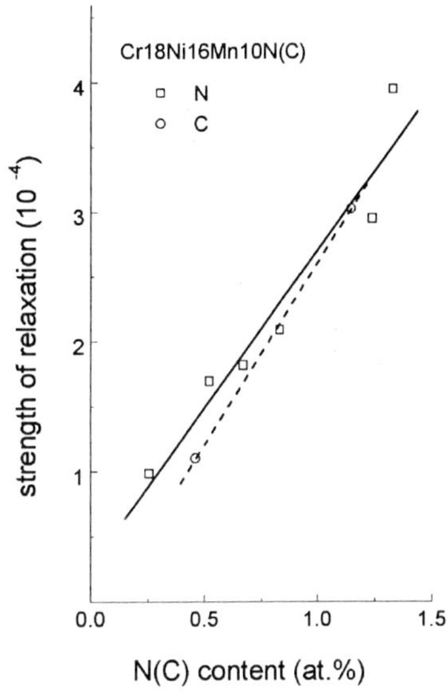


**Fig. 1.31.** The relaxation strength caused by s-N and s-C atomic complexes in austenitic steel Cr18Ni16Mn10 as a function of nitrogen or carbon content. The relaxation strength was determined as an area under the internal friction peaks in the coordinates $Q^{-1} = f(1000/T)$, T in K

Thus, one can state that the elastic contribution to the interaction of interstitial atoms with dislocations in austenitic steels is not a unique factor determining the pinning of dislocations.

On account of theoretical calculations and experimental studies of the electron exchange in Fe-C and Fe-N austenites and alloyed austenitic steels (see Sect. 1.1.1), it seems reasonable that an additional dislocation pinning can be caused by the electrostatic interaction between nitrogen(carbon) atoms and dislocations (see Cottrell 1953). A possible role of this kind of interstitial-dislocation interaction in metals was discussed by Kornyushin (1970). Indeed, nitrogen atoms in austenite are negatively charged whereas carbon atoms carry a positive electric charge (Seith 1955). The dislocation nuclei have a shortage of electrons and, therefore, are positively charged. Therefore, an electrostatic attraction between nitrogen atoms and dislocations has to exist in addition to the elastic one, whereas, in the carbon austenite, the elastic attraction, as compared to nitrogen, has to be counterbalanced by the electrostatic repulsion. These small additions of opposite sign to the prevailing elastic interaction can possibly explain the observed higher pinning of dislocations in austenitic steels by nitrogen atoms as compared to carbon ones.

### 1.1.4.3 Grain boundaries

There are few data of grain-boundary segregations caused by nitrogen. Briant (1987) has used Auger spectroscopy to measure nitrogen concentration in steel Cr18Ni9 after intergranular fracture and observed some nitrogen content at grain boundaries in the samples aged at 600-700 °C for 5-100 hours. A remarkable feature of these observations was that nitrogen segregation did not depend on the temperature and duration of ageing.

Cosegregation of nitrogen and chromium and nitrogen and molybdenum was studied by Grabke et al. (1989) in Fe-Cr-N and by Eltester and Uebing (1996) in Fe-Mo-N alpha-iron single crystals at elevated temperatures. Two-dimensional surface compounds CrN and MoN were observed in approximately two atomic layers, but the formation of three-dimensional compounds was ruled out.

A comparative study of nitrogen and carbon grain boundary segregations was performed in ferritic steels (see Rudy and Huggins 1966). No significant nitrogen segregation was observed in contrast to high enrichment in carbon at grain boundaries. Earlier, Lagerberg and Josefsson (1955) have also shown that carbon assists the creation of veins in ferrite sub-boundaries produced during hot deformation while nitrogen does nor cause such an effect, which is obviously concerned with the difference in the segregation effects of carbon and nitrogen.

Thus, one can conclude that nitrogen has at least lower tendency to form grain boundary segregations in steel as compared to carbon.

Nitrogen partitioning between a solid solution, grain boundaries and precipitates was studied by Petrov et al. (1999) in high-alloy austenitic steels using electron energy loss spectroscopy (EELS).

No detectable difference between the nitrogen contents of the matrix and grain boundaries was found in the steel Cr21Ni6Mn9N1 containing 1.05 mass % of nitrogen and solution treated at 1180 °C as well as in that annealed at 800 °C for 200 hours (Fig. 1.32).

The average values of the nitrogen content from 10 measurements performed on the solution treated steel (Fig. 1.32b) are $8.4 \cdot 10^{13}$ and $7.2 \cdot 10^{13}$ atoms/cm$^2$ for the matrix and grain boundary, respectively. The measured concentration of nitrogen in the nitride $Cr_2N$, $5.19 \cdot 10^{14}$ atoms/cm$^2$, was consistent with the expected atomic fraction of nitrogen.

A rather large scatter of experimental data is observed for the matrix as well as for the grain boundaries, which can be attributed to some inhomogeneities in the distribution of solute atoms.

A weak affinity of nitrogen to grain boundaries is particularly interesting because the other important interstitial element in steel, carbon, is prone to grain-boundary segregation (Petrov 1993). The segregation behaviour of these elements can be analysed based on two main factors determining the affinity of solute atoms to free surfaces: atomic size and electronic structure.

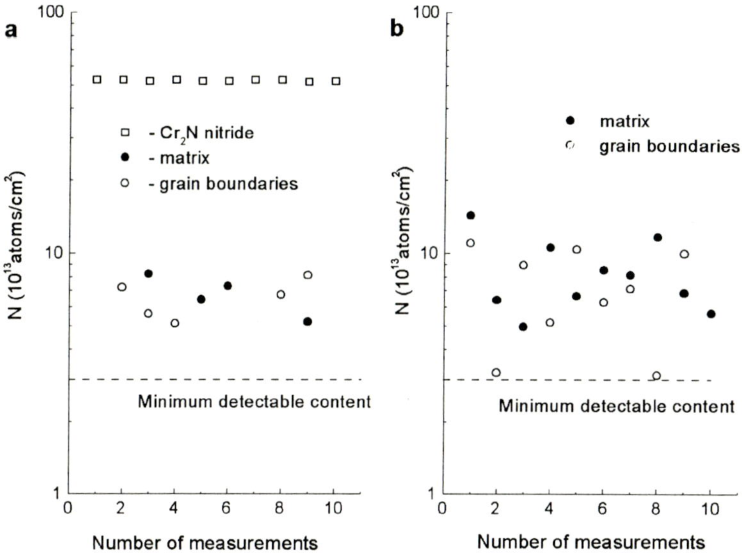


**Fig. 1.32.** Nitrogen content in the $Cr_2N$ nitride, the austenitic matrix and at the grain boundaries of steel Cr21Ni6Mn9N1 after annealing at 800 °C for 200 h **a**; the same for the matrix and grain boundaries after solution treatment at 1180 °C **b**. The minimum detectable nitrogen content equal to $3 \cdot 10^{13}$ atoms/$cm^2$ is shown by the dashed lines (after Petrov et al. 1999)

The significance of the electronic factor can be illustrated by an increased affinity of nickel to grain boundaries in austenitic steels (Petrov and Trophimova 1998) in spite of a small difference in the atomic size of nickel and iron in the fcc lattice of $\gamma$ iron. An increase in the concentration of free electrons caused by nickel in austenite (see Fig. 1.8), i.e. an increase in the electron energy of the crystal lattice, stimulates the migration of nickel atoms to the grain boundaries where free electrons can be easy accomodated. Moreover, according to Petrov and Trophimova (1998), nickel atoms push the carbon ones out of grain boundaries, i.e. the electronic factor prevails over the size factor in this case.

As discussed above (Table 1.2), nitrogen causes a bit higher dilation of the austenitic crystal lattice as compared to carbon and could be expected to have a higher affinity to grain boundaries. Therefore, the different behaviour of nitrogen and carbon in relation to grain boundaries is obviously not due to the size factor. A difference in their electron exchange with matrix atoms has to be accounted for. Like nickel, nitrogen increases the concentration of free electrons in the fcc iron, whereas carbon decreases it (Sect. 1.1.1). Thus, nitrogen in austenitic steels represents a particular case, where both factors, increased elastic distortions and increase in the electron energy, are involved in the game. One can suppose that, in this case, free electrons rearrange around the nitrogen atoms in order to decrease the energy of elastic distortions. This suggestion is based on the following results.

*Ab initio* calculations of the electronic structure (see Fig. 1.5) show a significant increase in the concentration of *s*-electrons in the vicinity of nitrogen atoms located in the octahedral sites of fcc iron, which is consistent with the data of electrotransfer. Free electrons have a spherical space symmetry and, being delocalised, can decrease elastic distortions in the crystal lattice.

As was already shown above (Table 1.3), Mössbauer studies give an experimental evidence of such an effect: a smaller electric field gradient, i.e. a more symmetrical distribution of electron density is observed in the Fe-N austenite as compared to that in the Fe-C one.

The weak tendency of nitrogen atoms to form grain boundary segregation can be used to explan prominent properties of nitrogen steels, such as high resistance to intercrystalline corrosion and good toughness which result from a significant delay of grain boundary precipitation of chromium nitrides as compared to the precipitation of chromium carbides (see in detail Chap. 3).

### 1.1.5 Dislocations and twins in nitrogen austenite

The substructure formed during heat treatment or cold work of austenitic nitrogen steels affects strongly their properties such as an increased yield strength and fracture toughness, improved fatigue life, wear resistance, creep resistance, stress relaxation, enhanced cold work hardening etc.

Planar dislocation slip and pronounced twinning are the features of deformed nitrogen austenites. A preference of the deformation mode is mainly determined by properties of the individual dislocations, in particular by their splitting affecting the mobility, their interaction with obstacles, cross slip, climb etc. On the other hand, the dislocation splitting is controlled by an important thermodynamical parameter, the stacking fault energy (SFE).

#### *1.1.5.1 Stacking fault energy*

Dislocations in austenitic steels are usually split, i.e. they consist of stacking faults limited by two partial dislocations. Stacking faults are errors in a normal sequence of atomic planes (ABCABC... of the fcc or ABAB... of the hcp lattices) and, in general, the value of SFE characterises the thermodynamical stability of the crystal structure. A decrease in the value of SFE is accompanied by an increasing splitting of dislocations, i.e. by the growth of stacking faults.

There is a scattering in the data of the nitrogen effect on the SFE of austenitic steels measured on steels of different chemical compositions. Before discussing the experimental data available, one should remark that, in particular in TEM studies, correct values of stacking fault energy can be obtained if the splitting is measured on fresh dislocations induced by slight deformation before preparing the foils. In such a way, one can avoid an effect of segregation of impurity atoms at dislocations changing the extent of splitting.

According to transmission electron microscopy (TEM) measurements performed by Fawley et al. (1968), the variation of nitrogen content within 0.005-0.05 mass % in the steel Cr20Ni20 had a slightly decreasing effect on the SFE (from 400 to 380 $mJ/m^2$).

By means of TEM, Swann (1963) showed that the addition of 0.12 mass % of nitrogen to the steel Cr18Ni13 produced a slight lowering of SFE from 160 to 130 $mJ/m^2$.

Dulieu and Nutting (1964) obtained nearly the same result in TEM studies of the steel Cr18Ni10. An increase in the nitrogen content from 0.02 to 0.25 mass % resulted in decreasing the SFE from 100-120 to 90 $mJ/m^2$. According to the same authors, nickel increases the SFE of austenite whereas chromium decreases it.

Stoltz and Vander Sande (1980) found a decrease of the SFE due to nitrogen in austenitic CrNiMn steels. A peculiarity of these observations was that an increase of the nitrogen content above 0.24 mass % did not affect the SFE.

A comprehensive study of SFE in austenitic CrNi, CrNiMn and NiMn steels was performed by Schramm and Reed (1975) who used an X-ray diffraction profile analysis for this purpose. They obtained that nitrogen, as well as chromium, silicon and manganese, decrease the SFE whereas carbon and nickel increase it.

Thus, the above mentioned authors have found nitrogen to decrease the SFE of austenitic steels.

In some contrast to these results, Taillard and Foct (1989) did not observe any significant influence of nitrogen on the SFE in their TEM study of steel 316L doped by 0.08 or 0.25 mass % of nitrogen.

Moreover, the addition of 0.05 to 0.23 mass % of nitrogen to the austenitic chromium-manganese steel Cr13Mn19 led to an increase of SFE from 320 to 400 $mJ/m^2$, according to TEM results by Petrov (1978).

A non-monotonous influence of nitrogen was revealed by Fujikura et al. (1975) who tested the concentration dependence of SFE in X-ray diffraction studies (see Fig. 1.33). The probability of stacking faults in the steel Cr18Ni10Mn8N was decreased at small additions of nitrogen (0.2-0.3 mass %) and increased at higher nitrogen contents. For comparison, the concentration dependence of SFE in austenitic carbon steel, according to Petrov (1978), is shown in the same Figure. A decrease of SFE was observed for a carbon content of about 0.3 mass %, which was attributed to splitting caused by the segregation of carbon atoms at dislocations. However, an opposite effect in nitrogen austenite as shown in Fig. 1.33, i.e. the narrowing of dislocations resulting in a decrease of the probability of stacking faults in the range of small nitrogen concentrations, cannot be caused by segregations. Therefore, some other reason for a non-monotonous concentration dependence of SFE in austenitic nitrogen(carbon) steels may exist.

Some grounds for this suggestion are provided by the data of the correlation between SFE and electronic properties.

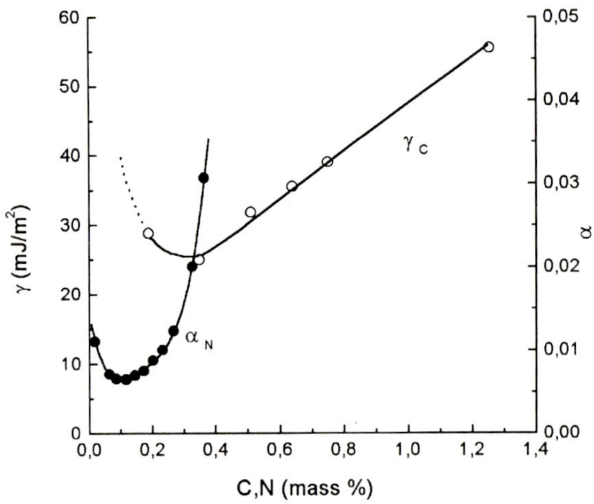


**Fig. 1.33.** Effects of carbon on the stacking fault energy γ in Fe-12Mn austenite (according to Petrov 1978) and nitrogen on the probability α of stacking faults in Cr18Ni10Mn8 austenite (according to Fujikura et al. 1975)

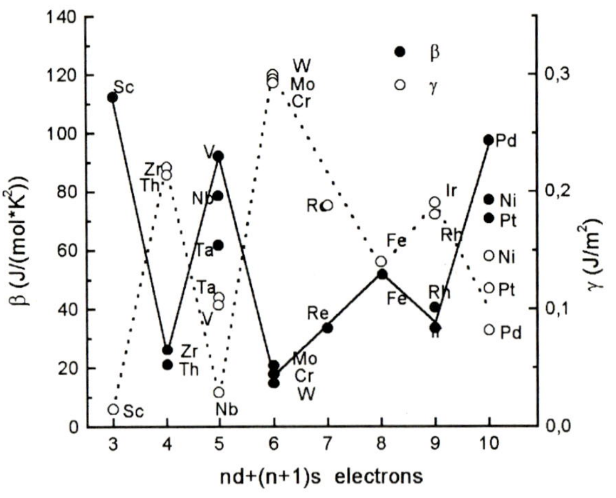


**Fig. 1.34.** Stacking fault energy γ and the electron heat capacity coefficient β versus the number of external nd + (n+1)s electrons in transition metals (after Noskova et al. 1965)

Noskova et al. (1965) have collected data of SFE and the electron heat capacity in pure metals and alloys, the latter characterising the electronic structure. It was shown that, in transition metals, the SFE changes non-monotonously with the number of $s$+$d$ electrons revealing a series of maxima and minima (Fig. 1.34). The same behaviour, but of opposite sign, occurs for the coefficient $\beta$ of electron heat capacity which is proportional to the state density $D_F$ at the Fermi surface, $\beta = (2/3)k_B^2\pi D_F$, where $k_B$ is the Boltzmann constant (Fig. 1.35).

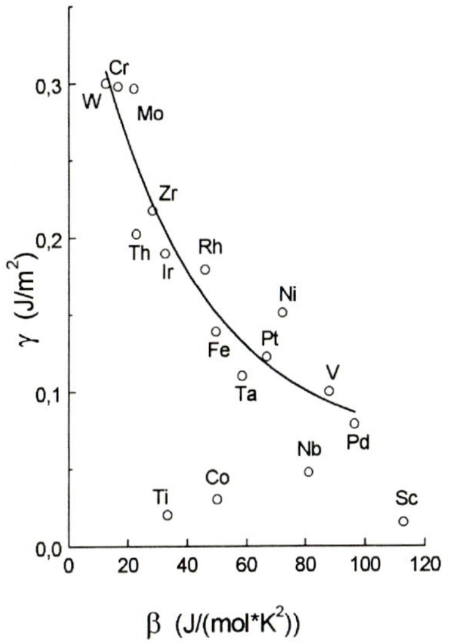


**Fig. 1.35.** Correlation between stacking fault energy γ and the coefficient of electron heat capacity β from Fig. 1.34 (after Noskova et al. 1965)

Thus, an inverse dependence of the SFE on the state density at the Fermi surface is revealed in a number of experiments, although some deviations from this correlation take place.

This result means that the value of SFE is mainly determined by the fraction of free electrons in an alloy. The higher the latter, the lower is SFE and, respectively, the more split are dislocations. Based on these findings, one can assume that a non-monotonous concentration dependence of the SFE in austenitic nitrogen steels as well as a scattering of the data of the nitrogen effect measured in steels of different basic compositions may arise from the non-monotonous behaviour of the state density at the Fermi surface with increasing number of $s$+$d$ electrons (see Fig. 1.34).

Some experimental evidence of the non-monotonous nitrogen effect on SFE was obtained by us in TEM studies of the steel Cr18Ni16Mn10 alloyed with different nitrogen content (Fig. 1.36). The stacking fault energy in this steel increases up to 0.4 mass % of nitrogen and decreases at a higher nitrogen content. The data of the state density at the Fermi surface taken from Fig. 1.7 are also shown in Fig.

1.36. One can see that the non-monotonous concentration behaviour occurs for both the γ and $D_F$ values in spite of some shift of their maxima. Thus, a qualitative correlation between stacking fault energy and the state density at the Fermi surface in nitrogen CrNiMn austenitic seels appears, although, in this case, it is not the inverse proportionality as for pure metals and substitutional alloys (compare with Fig. 1.35).

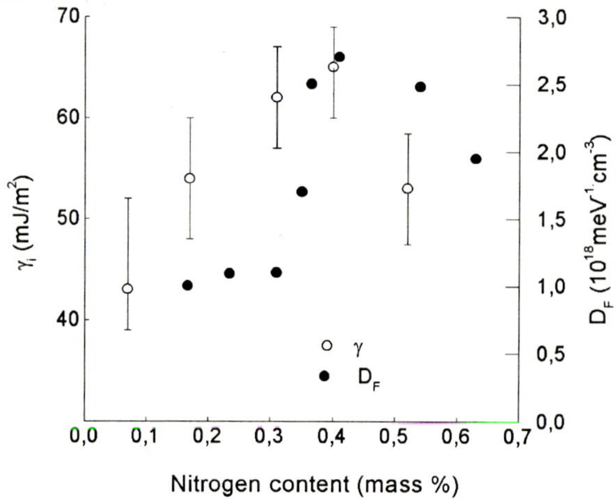


**Fig. 1.36.** Effect of nitrogen on the SFE, γ, and the state density at the Fermi surface, $D_F$, in he steel Cr18Ni16Mn10

At the same time, the effect of nitrogen on the SFE in austenitic nickel-free CrMn steels is quite different. According to TEM measurements, in austenitic Cr15Mn17 steels alloyed by 0.22, 0.45 and 0.8 mass % of nitrogen, the SFE is equal to 25, 17 and 51 mJ/m$^2$, respectively, i.e. opposite to the concentration dependence of SFE in CrNiMn steels.

Thus, the effect of nitrogen on the stacking fault energy depends on the basic chemical composition of austenitic steels. As the electron density at the Fermi surface plays an important role for the SFE, one can suggest that its non-monotonous concentration behaviour occurs if the Fermi sphere touches the boundaries of the Brilouin zone in the reciprocal space, which results in a change of shape of the Fermi surface. Such a behaviour is expected e.g. in high-nickel steels as nickel significantly increases the concentration of free electrons.

### *1.1.5.2*
### *Dislocation structure and twins*

Through the stacking fault energy, nitrogen affects the dislocation structure and, in particular, the deformation twinning as a mode of plastic deformation, which can play a decisive role in deformation and fracture, especially at low temperatures.

A change in the dislocation structure of the steel Cr18Ni16Mn10 with increasing nitrogen content is characterised by a decrease in the splitting of dislocations. Some typical examples are shown in Fig. 1.37. Extended dislocations and pile-ups prevail at low nitrogen contents, whereas single narrow dislocations are often seen in the high nitrogen steel.

After cold work (Fig. 1.38) twinning occurs along with dislocation slip in low nitrogen austenite. The density of twins is decreased with increasing nitrogen content. Nevertheless, planar slip remains a remarkable feature of the cold worked nitrogen austenite (Fig. 1.38 b-d).

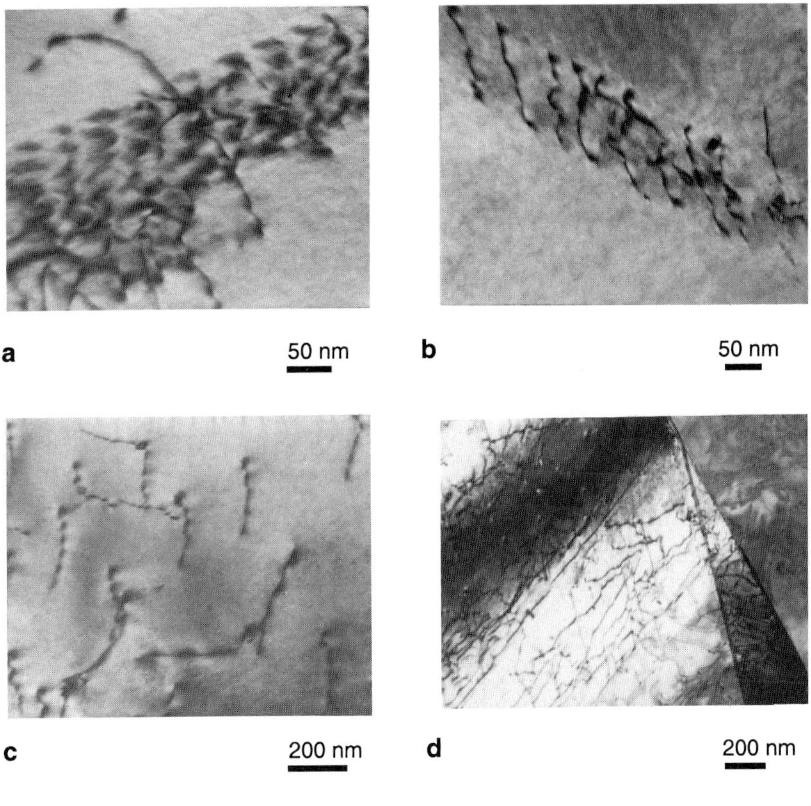


**Fig. 1.37.** Dislocation structure of the steel Cr18Ni16Mn10 after solution treatment at 1150 °C, 30 min, followed by water quenching: **a** 0.17 mass % N, **b** 0.3 mass % N, **c** 0.4 mass % N, **d** 0.54 mass % N

Quite a different dislocation structure is formed in austenitic nickel-free CrMn nitrogen steels (see Fig. 1.39). In accordance with the lower SFE (see Sect. 1.5.2.1), extended stacking faults are observed in the solution treated steel

Cr15Mn17 steel (Fig. 1.39 a,c). The decrease in the stacking fault energy due to nitrogen leads to a pronounced twinning during deformation (Fig. 1.39 b,d).

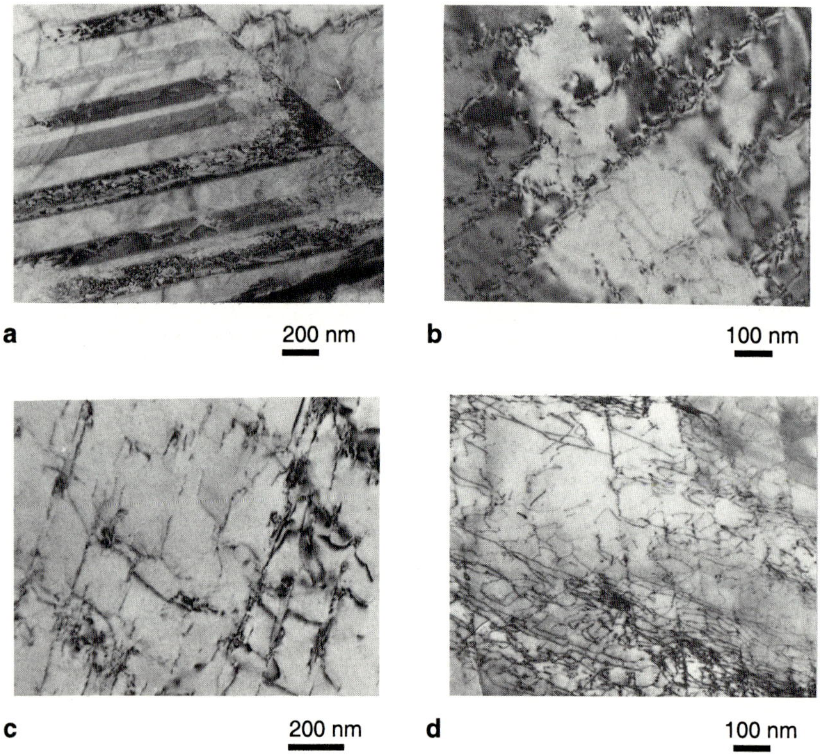


**Fig. 1.38.** Dislocation structure of the steel Cr18Ni16Mn10 after cold rolling to 12 % of decrease in the thickness: **a** 0.17 mass % N, **b** 0.3 mass % N, **c** 0.4 mass % N, **d** 0.54 mass % N

Comparing this result with the above data for CrNiMn steels, one can conclude that mechanical twinning during plastic deformation of austenitic nitrogen steels depends clearly on the effect of nitrogen on the stacking fault energy.

A peculiarity of the dislocation structure of cold worked austenitic nitrogen steels is the planar slip (see examples in Figs 1.38 c,d). The higher the nitrogen content, the more pronounced is the planar slip.

It is a common place to attribute the slip planarity to a low SFE as the split dislocations possess a lower ability to cross slip. However, as follows from Figs. 1.37b,c and 1.36, an increase of SFE does not lead to less pronounced planar slip. So, there exists a more general reason for nitrogen to affect the dislocation structure, which does not depend on the effect of nitrogen on the SFE. It is attributed to short range atomic order in austenitic nitrogen steels (see Grujicic et al. 1989, Sassen et al. 1989 etc.).

It was shown earlier (Thomas 1963, see also Gerold and Karnthaler 1989) that short range order in the alloys assists planar slip because each successive dislocation needs a lower stress to pass over the slip plane as compared to the first one which passes through the ordered crystal and consumes an additional energy to shift the atoms from the low energy sites (see also Sect. 3.1.1.1 about the role of the type of short range atomic order in the planar slip and strengthening).

Thus, two main features determine the substructure of nitrogen austenite: planar slip and twinning. They will be considered again in detail while discussing the cold work hardening of austenitic nitrogen steels (see Sect. 3.1.3).

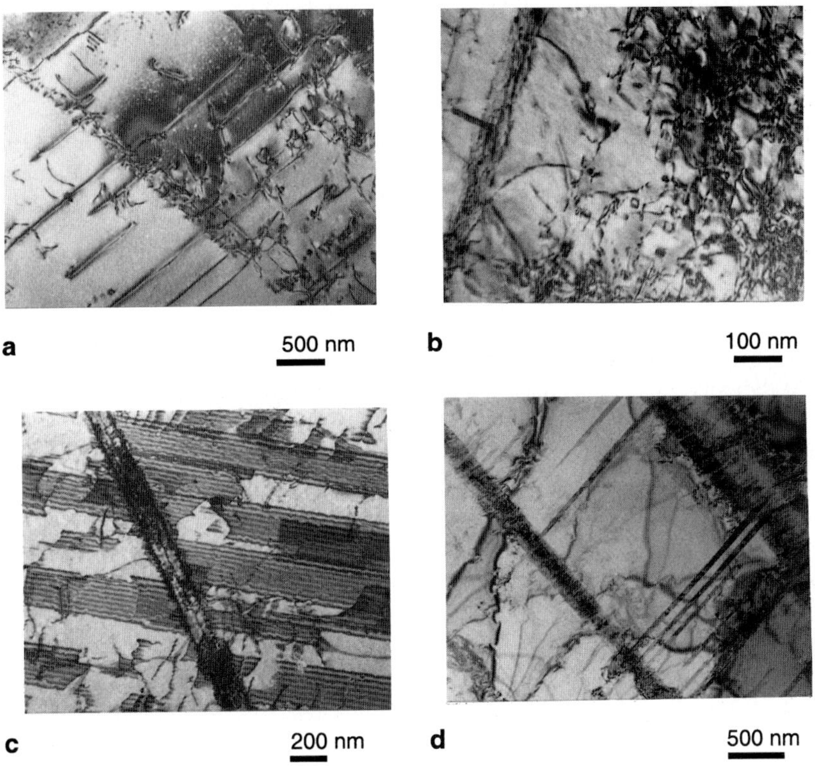


**Fig. 1.39.** Structure of the steel Cr15Mn17 alloyed with 0.22 **a,b** and 0,45 **c,d** mass % N after solution treatment **a,c** and cold work of 12 % **b,d**. The values of SFE are 25.5 $mJ/m^2$ for the steel with 0.22 % N and 17.3 $mJ/m^2$ for that with 0.45 % N

## 1.2 Distribution of nitrogen atoms in martensite

As the martensitic transformation is a diffusionless one, i.e. it is accompanied by atomic displacements less than the interatomic distance, the atomic distribution in ferrous martensites is inherited from the parent austenitic phase. However, the quenching of steels at room temperature does not retain the initial atomic configurations of nitrogen and carbon atoms as they existed in the austenite. The matter is that, in accordance with the enthalpy of migration of nitrogen and carbon atoms in α-iron (about 0.8 eV), they are mobile at room temperature. That is why the crystal structure of nitrogen martensite as well as of carbon martensite at room temperature is far from the freshly formed one and has to be characterised as an aged structure. A really virgin iron-based martensite can be observed if the start temperature of martensitic transformation is below -50 °C.

## 1.2.1
## Freshly formed Fe-N martensite

Unusual peculiarities of the crystallographic structure of virgin martensites were revealed by measurements of X-Ray diffraction. Lysak and Vovk (1965, 1971) were first to show that Fe-C and Fe-Mn-C martensites obtained after quenching in liquid nitrogen have a decreased tetragonality as compared to that predicted by a well known formula

$$c/a = 1 + 0.046c_C,$$

where $c_C$ is the carbon content in mass %.

At the same time, an increased tetragonality was observed in freshly formed Fe-Ni-C (Lysak and Danilchenko 1971), Fe-Al-C (Watanabe and Wayman 1971) and carbon-free Fe-Ni (Bondar et al. 1988, Ullakko and Gavriljuk 1992) martensites. After ageing at room temperature, the c/a ratio approached its normal value determined by the carbon content in steel (see the above mentioned equation).

It is remarkable that, in contrast to carbon martensites, freshly formed Fe-N martensite has an abnormally high tetragonality (Ullakko and Gavriljuk 1992, Ullakko et al. 1994).

A change of tetragonality during heating of freshly formed martensites from liquid helium to room temperature is shown in Fig. 1.40. Abnormally high tetragonality of 33,5Ni, 25Ni-0.7C and 2.39N martensites disappears during heating in the temperature range of 100 to 180 K. The alloy with 20 % of Ni is placed at some intermediate position between two groups of martensites. It has a higher tetragonality, than is expected according to the carbon content, but its behaviour during heating is similar to that of low-tetragonal 1.76C and 3Mn-1.6C martensites.

Several mechanisms were proposed to explain abnormal tetragonality. Lysak and Nikolin (1966) have attributed the low tetragonality to some distribution of carbon atoms between octa- and tetrahedral interstitial sites as a result of the γ→ε transition supposed to be an intermediate step in the martensitic γ→α′ transition. Both high and low tetragonalities were interpreted by Roitburd and Khachaturyan (1970) in terms of the distribution of carbon atoms on the x-, y-, z-sublattices of octahedral interstitial sites in the bcc lattice.

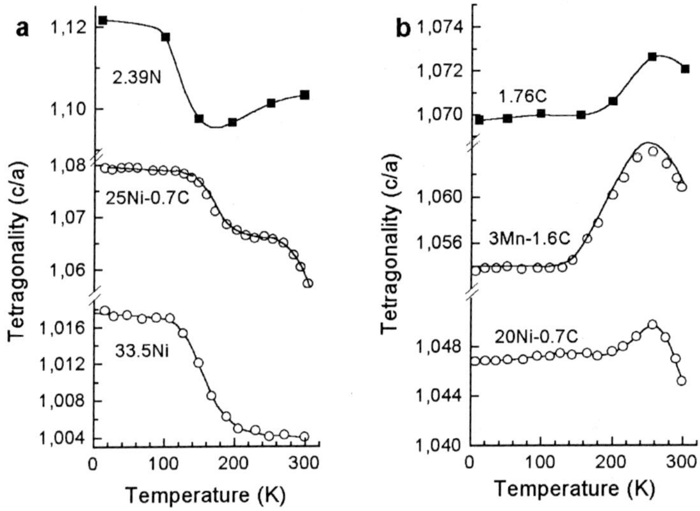


**Fig. 1.40.** Change of the c/a ratio during heating of freshly formed martensites from liquid helium to 300 K. **a** – Fe-2.39N, Fe-25Ni-0.7C, Fe-33.5Ni martensites with abnormally high tetragonality, **b** High-tetragonal Fe-20Ni-0.7C and low-tetragonal Fe-3Mn-1.6C and Fe-1.76C martensites

These hypotheses were tested using Mössbauer spectroscopy which was proven to be sensitive to a change in the nearest neighbourhood of iron atoms. The suggested transfer of carbon atoms from tetrahedral to octahedral sites or the redistribition of interstitial atoms between x-, y- and z-sublattices of the octahedral sites has to be accompanied by a change in the values of the hyperfine fields and electric field gradients at the iron nuclei, which causes shifts of the lines in Mössbauer spectra and a redistribution of their intensities. As shown by Gridnev et al. (1977), nothing is changed in Mössbauer spectra at temperatures where the abnormal tetragonality disappears, i.e. there is no redistribution of carbon atoms in this temperature range. At the same time, according to Mössbauer data, the only process concerned with a redistribution of interstitial atoms in the freshly formed martensite starts during heating above 220 K and it is clustering leading to the disappearance of single carbon atoms in the solid solution. In other words, anomalies of tetragonality of freshly formed martensites are not concerned with any special metastable distribution of interstitials. A high tetragonality of the interstitial-free Fe-33.5Ni martensite (see Fig. 1.40) is consistent with this conclusion.

An unambigous evidence that high tetragonality in Fe-Ni-C as well as in carbon-free Fe-Ni martensites is due to the coherency of $\gamma$-$\alpha'$ interfaces in the as-quenched state was obtained by Ullakko and Gavriljuk (1992) by means of X-Ray and internal friction techniques. A coherent bond between the initial austenitic matrix and the freshly formed martensite prevents so-called Bain's strain which has to decrease the *c/a* ratio from the ideal one corresponding to the bct element in the fcc lattice to that in the martensite lattice (see Bain's scheme of $\gamma \rightarrow \alpha'$ transition in iron in Fig. 1.41). The coherency of two lattices prevents the final compression along the $c$ axis and some increased tetragonality of martensite is retained at low temperatures if dislocations are not sufficiently mobile in order to provide the relaxation of coherent stresses due to microplastic deformation. A slight deformation of the freshly formed martensite at low temperatures or an increase of the temperature were shown to lead to the breaking of the coherency and to the disappearance of the abnormally high tetragonality (see for details in Ullakko and Gavriljuk 1992).

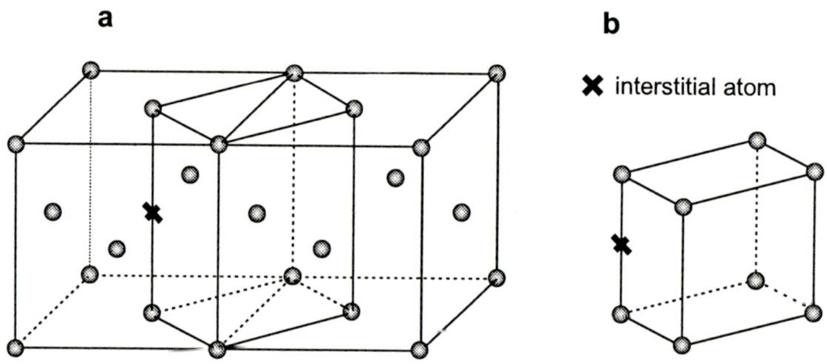


**Fig. 1.41.** An element of bct structure in the fcc lattice **a** and Bain's strain leading to the martensitic bct structure **b**. The strains are $\varepsilon_1 = \varepsilon_2 = a\sqrt{2}/a_0$ and $\varepsilon_3 = c/a_0$ where $a_0$ is the lattice parameter of fcc austenite, $c$ and $a$ are parameters of bct martensite. The positions of interstitial atoms are shown by ×

As to the abnormally low tetragonality in Fe-C and Fe-Mn-C virgin martensites, the most probable mechanism was proposed by Mirzayev et al. (1982). They have calculated the effect of thin $\{110\}_\alpha$ twins on the distribution of intensity within the X-Ray reflections resulting in a shift of their gravity center and, consequently, in a decrease of the distance between the reflections of the martensite dublet in the X-ray diffraction patterns. The following heating leads to an increase in the thickness of twins, the recovery of the normal profile of X-ray reflections, and the disappearance of this apparently low tetragonality.

For further consideration, it is worth noting that we encounter again opposite effects of nitrogen and carbon in solid solutions (see Fig. 1.40). The similarity in the effects of nitrogen and nickel is also remarkable. Earlier it was shown (see Sects. 1.1.1.2 and 1.1.2.6) that both elements increase the concentration of free electrons and assist short range atomic ordering in austenitic steels. Now one can see that both nickel and nitrogen assist the coherency between the lattices of freshly formed martensite and parent austenite, which results in an increased c/a ratio in the former. Therefore, some correlation between short range ordering, coherency at interfaces and an increased tetragonality of virgin martensites can be supposed.

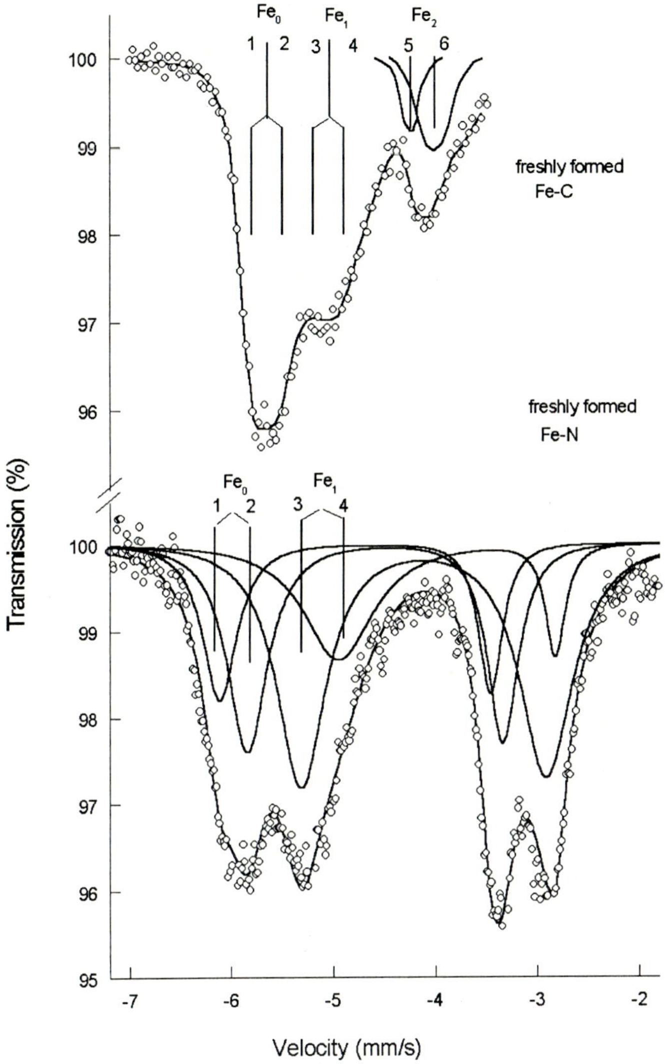


**Fig. 1.42.** Mössbauer spectra of virgin martensites in the alloys Fe-9at.%C and Fe-9.3at.%N as obtained by quenching in liquid nitrogen. Measurements are performed at T = 80 K. Only parts of spectra are shown: the outer line (nuclear transition -1/2→-3/2) in the Fe-C spectrum and two lines (transitions -1/2→-3/2 and -1/2→-1/2) in the Fe-N spectrum. Each line consists of several components $Fe_i$ belonging to iron atoms having $i$ intersitial atoms as nearest neighbours ($i$ = 0, 1, ...). In turn, these sublines have the fine structure caused by the different orientations of the electric field gradient $z$ and the axis of easy magnetisation [001] for atoms $Fe_1$ and $Fe_2$ (see Fig. 1.43) and by the presence of interstitial atoms as second nearest neighbours for atoms $Fe_0$. It is clearly seen that $Fe_0$, $Fe_1$ and $Fe_2$ components are present in the Fe-C spectrum, whereas only $Fe_0$ and $Fe_1$ components exist in the Fe-N spectrum

This suggestion is also supported by the experimental data on aluminum steels (Watanabe and Wayman 1971). Fe-Al alloys are characterised by short range atomic ordering and Fe-Al-C martensite has an abnormally high tetragonality.

On the other hand, carbon and carbon+manganese, being prone to clustering in solid solution, provide an abnormally low tetragonality in the as-quenched martensite.

Let us now discuss the distribution of nitrogen atoms in freshly formed martensite. Based on the different N-N and C-C interactions and atomic distributions in austenite, it is natural to expect a different constitution of nitrogen and carbon bct solid solutions formed during quenching. This difference is clearly demonstrated by the data of Mössbauer studies (Fig. 1.42).

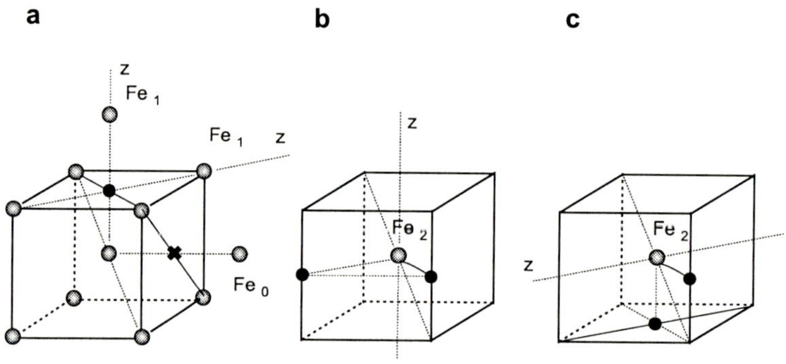


**Fig. 1.43.** Atomic configurations in martensites corresponding to iron atoms having no ($Fe_0$), one ($Fe_1$) interstitial atom as nearest neighbour **a**, and two ($Fe_2$) interstitial atoms in nearest neighbourhood **b**, **c**. Splitting of the lines in Mössbauer spectra (Fig. 1.42) is due to different Fe-$i$ configurations and due to different relative orientations of the axis $z$ of the strain tensor (electric field gradient) and of the axis of easy magnetisation [001]

According to Mössbauer data, iron atoms having no ($Fe_0$), one ($Fe_1$) and two ($Fe_2$) interstitial atoms as nearest neighbours (see atomic configurations in Fig. 1.43) are present in the spectrum of Fe-C virgin martensite. The spectrum of Fe-N martensite gives a clear evidence that paired nitrogen atoms in the neighbouring interstitial sites ($Fe_2$ atoms correspondingly) do not exist in Fe-N virgin martensite (see also Rochegude and Foct 1986 where Mössbauer spectra in binary Fe-N solid solutions are analysed in detail). Thus, the experiment confirms an essential difference in the distributions of nitrogen and carbon atoms in freshly formed martensites as they are inherited from the parent austenite. As will be discussed in the following, these peculiarities of the atomic distribution are retained during ageing and result in a different tempering behaviour of nitrogen and carbon martensites.

## 1.2.2
## Early stages of ageing and tempering of nitrogen martensite

A priority in the studies of ageing of freshly formed nitrogen martensite belongs to Suyazov et al. (1976) who, based on TEM studies of martensite formed by cooling in liquid nitrogen, have concluded that ageing at room temperature (RT) leads to short range atomic ordering of nitrogen atoms, whereas carbon atoms form clusters.

A difference in the redistribution of nitrogen and carbon atoms during ageing below room temperature can again be clarified using Mössbauer spectroscopy, as was demonstrated for freshly formed martensites in the preceding section. A change in the nearest neighbourhood of iron atoms in freshly formed martensite with increasing temperature is clearly seen in Mössbauer spectra (Fig. 1.44).

The component $Fe_1$ in the spectra of Fe-C martensite of the iron atoms having single carbon atoms as nearest neighbours disappears starting at -50 °C while the intensity of the component $Fe_2$ belonging to the iron atoms with paired carbon atoms is increased. Therefore, clustering of carbon starts at low temperatures and it is completed at room temperature.

In contrast, no remarkable change occurs in Fe-N martensite during low temperature ageing in spite of a higher mobility of nitrogen atoms in α-iron as compared to carbon atoms. Therefore, the opposite tendencies concerning the distributions of nitrogen and carbon atoms in the iron lattice take also place in martensites.

A reason for such a behaviour is that the ageing of nitrogen martensite develops towards the ordered α″-structure $Fe_{16}N_2$ where, exactly as in the freshly formed Fe-N martensite, there is no iron atom having two or more nitrogen atoms as nearest neighbours (see Chap. 2, Fig. 2.5). The iron nitride $Fe_{16}N_2$ precipitates as an intermediate phase during tempering of Fe-N martensite. It is isostructural with the parent iron-nitrogen martensite and differs only in the ordered distribution of nitrogen atoms. Short range ordering of nitrogen atoms during ageing occurs without any change in the nearest neighbourhood of iron atoms, and the fraction of the iron atoms having nitrogen atoms as next to nearest neighbours (the second coordination sphere on the interstitial sublattice) can only be changed.

Fundamental studies of the room temperature ageing and of early stages of tempering of nitrogen martensite were performed by Mittemeijer with collaborators (see e.g. a review of these studies given by Böttger and Mittemeijer 1998 and Böttger et al. 1998). Transmission electron microscopy, dilatometry and differential thermal analysis were used in their complex studies.

The morphology of low- and high-nitrogen martensites was shown to be similar to that of carbon martensites (Van Gent et al. 1985). The habit plane $\{232\}_\gamma$ in the nitrogen lath martensite was found to be close to $\{575\}_\gamma$ habit plane in the carbon lath martensite. In the nitrogen plate martensite, the habit plane $\{259\}_\gamma$ was observed, which is consistent with the data available for the carbon plate martensite.

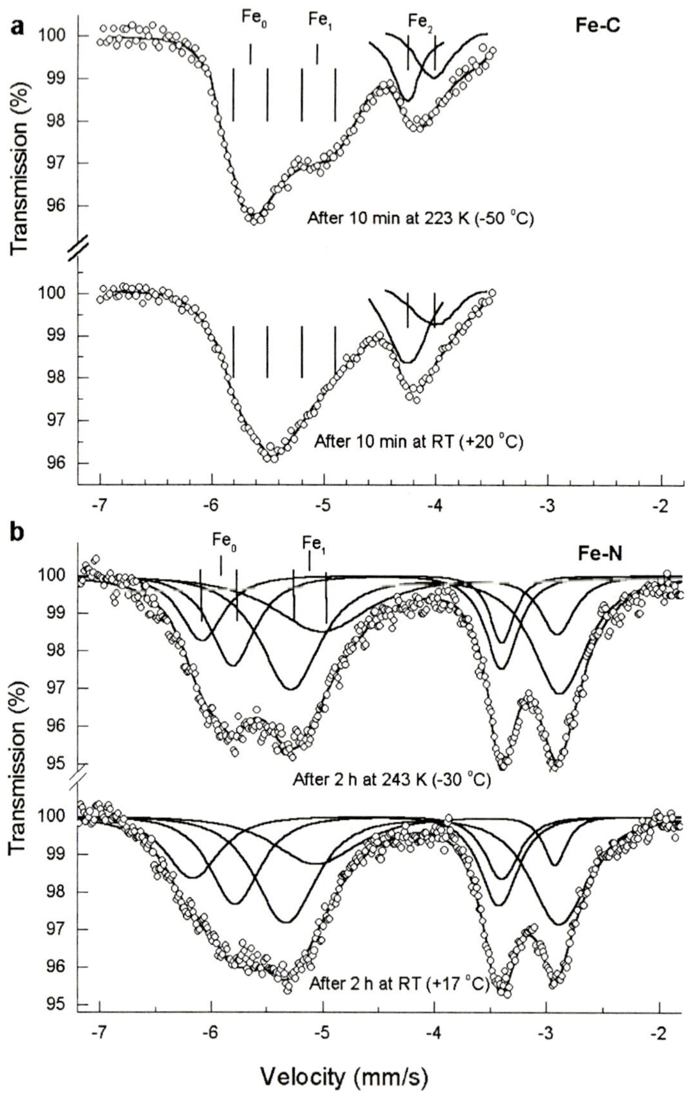


**Fig. 1.44.** Mössbauer spectra of Fe-C **a** and Fe-N **b** martensites measured at 80 K after low temperature ageing. The component $Fe_1$ belonging to the iron atoms having one carbon atom as nearest neighbour (see Fig. 1.43) disappears due to ageing of Fe-C martensite while the component $Fe_2$ is increased. The component $Fe_1$ is not changed and the component $Fe_2$ does not appear during ageing of Fe-N martensite, i.e. no clustering of nitrogen atoms occurs

The crystallographic orientation of nitrogen martensite in relation to the parent austenite was found to be intermediate between those for Kurdjumov-Sachs and Nishiyama-Wassermann relationships but nearer for that described by Nishiyama-Wassermann, i.e. $(111)_\gamma \| (011)_{\alpha'}$ and $[112]_\gamma \| [011]_{\alpha'}$. The plate nitrogen martensite is twinned along $\{112\}_{\alpha'}$, but a fine $\{110\}_{\alpha'}$ twinning is found within the $\{112\}_{\alpha'}$ twins, exactly as it is the case in the carbon martensite.

Also no evidence of clustering as a preprecipitation stage was obtained during ageing at room temperature. The coherent $\alpha''$- $Fe_{16}N_2$ platelets with the habit plane $\{001\}_{\alpha'}$ were observed to precipitate at room temperature. The stresses at the coherent interfaces are supposed to be responsible for an increase in the hardness of aged nitrogen martensite.

A comparison with the aged carbon martensite reveals that the intermediate carbide phases ($\varepsilon$ or $\eta$), which are also precipitated coherently with the parent martensite, have a more complicated structure than $\alpha''$-nitride (Jack 1951, Murphy and Whiteman 1970, Nagakura et al. 1983). Their precipitation needs a significant redistribution of carbon atoms and some change in the distribution of iron atoms, the latter being accelerated by pipe diffusion. Their crystallographic structures have a larger incompatibility with that of the parent carbon martensite as compared to those between $\alpha''$-nitride and parent nitrogen martensite. Larger distortions of the crystalline lattice at the coherent interfaces lead to higher stresses. That is why the hardness of aged carbon martensites is always higher than that of nitrogen martensites at equal contents of interstitials.

Ageing of nitrogen martensite at 130 and 170 °C was shown to cause a break of coherency, which is accompanied with a decrease in hardness and a release of heat (Van Gent et al. 1985). After prolonged ageing at room temperature (25 to 40 hours), incoherent $\alpha''$ precipitates were also found using neutron diffraction (Böttger et al. 1990).

Thus, the first stage of tempering of nitrogen martensite can be described as a decomposition of $\alpha'$-martensite into $\alpha''$- $Fe_{16}N_2$ and $\alpha$-ferrite:

$$\alpha' \rightarrow \alpha'' + \alpha.$$

The growth of the $\alpha''$- $Fe_{16}N_2$ precipitates is controlled by the diffusion of nitrogen (the diffusion distance of nitrogen atoms does not exceed 50 nm for one day of ageing at room temperature, as evaluated by Böttger et al. 1990). The local depletion of the parent martensite by nitrogen during ageing and low temperature tempering leads to an increase in the misfit between the crystal lattices of $\alpha''$-$Fe_{16}N_2$ and the matrix. As shown by Van Genderen et al. (1997), this misfit can be diminished at the expense of some increase in the volume of $\alpha''$, which is achieved by introducing the structural vacancies of nitrogen atoms in the nitride lattice, i.e. by the formation of $Fe_{16}N_{2-x}$. Calculations of the Gibbs free energy for the perfect and nitrogen-deficient $\alpha''$-phase in the $\alpha'$- or $\alpha$-matrix of iron have confirmed that the perfect $Fe_{16}N_2$ nitride is favoured in a $\alpha'$-martensite matrix, whereas the nitrogen-deficient $Fe_{16}N_{2-x}$ is preferential in a $\alpha$-ferrite matrix.

In qualitative accordance with these results, Tanaka et al. (1997) have found a new nitride α‴-$Fe_{16}N$ formed coherently with α″- $Fe_{16}N_2$ during tempering for 7 days at 160 °C. This rarely observed new low temperature nitride can be obtained from α″- $Fe_{16}N_2$ by removing nitrogen atoms from each second (110) plane.

Above 220 °C the low temperature nitride α″- $Fe_{16}N_2$ transforms into γ′-nitride $Fe_4N$ (see e.g. Liu 1990). This process is accompanied by a decrease in the hardness and a negative volume effect. The value of the activation energy (~196 kJ/mole) lies in between that for the pipe (134 kJ/mole) and volume (251 kJ/mole) diffusion of iron, which allows to suggest that both processes are rate controlling.

Both α″- $Fe_{16}N_2$ and ε/η carbide are metastable transition precipitates in tempered martensites. Transformation of α″ nitride into stable γ′-$Fe_4N$ nitride occurs between 450 and 650 K while conversion of ε/η carbide into cementite takes place above 570 K.

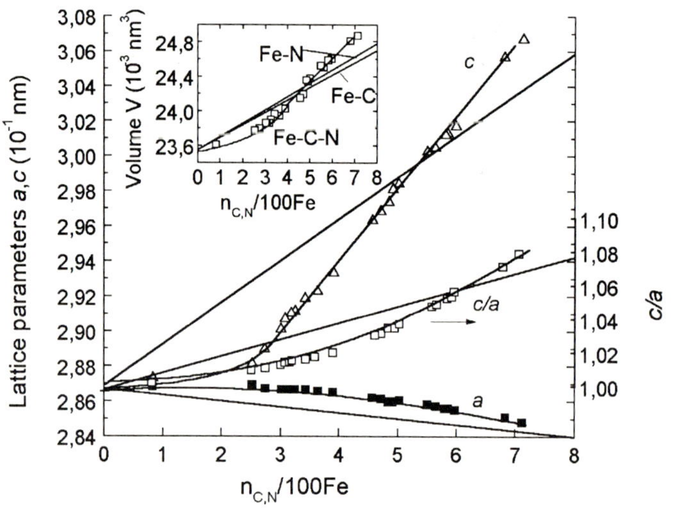


**Fig. 1.45.** Variation of the martensite lattice parameters and of the martensite unite cell volume (insert) for Fe-N, Fe-C and Fe-C-N alloys with the atomic concentration of interstitials ($n_{C,N}$/100 Fe atoms) after Chen and Tang (1990). Data for Fe-N and Fe-C are represented by straight lines

A peculiarity of ageing and tempering of ternary iron-nitrogen-carbon martensite is that the transition nitride α″ and carbide ε/η develop separately and simultaneously between 370 and 450 K. This important conclusion was made based on X-ray diffraction studies using synchrotron radiation of high incident intensity (Van Genderen et al. 1992) as well as the combination of dilatometric and calorimetric experiments (Liu et al. 1992). No low temperature carbonitrides were observed. Therefore, a preliminary redistribution of carbon and nitrogen atoms has to occur before precipitation. According to the atom probe analysis (Böttger et al. 1996), interstitials of one type (nitrogen or carbon) prevail in the enriched regions. As will be shown in the following (see Sect. 2.5), such a non-trivial behaviour occurs also during tempering of iron-nitrogen-carbon martensites alloyed by chromium.

In addition, it is worth noting that measurements of the lattice parameters of iron-nitrogen carbon martensites performed by Chen and Tang (1990) at room temperature (i.e. in the aged state) have revealed a non-linear dependence of the $c$ and $a$ parameters and of the martensite unit cell volume on the interstitial atom concentration, in contrast to the linear concentration relations for the iron-nitrogen and iron-carbon martensites (Fig. 1.45).

This result provides an evidence for some interaction between carbon and nitrogen atoms in the aged martensite. It was shown in this chapter that the distribution of nitrogen atoms is characterised by short range ordering while carbon atoms have a tendency to clustering. The distribution of interstitials in the combined C+N solid solution has some special features. We will see in the following (see Chap. 2), that alloying of steels by nitrogen+carbon leads to an increased stability of austenite to martensitic transformation and of martensite to precipitation of nitrides/carbides during tempering.

# 2 Constitution

The constitution of steel is characterised by the number of constituting phases, their type, shape, size and chemical composition, and by the atomic distribution within the phases.

Variations in the phase compositions and in the concentration of solutes create a unique opportunity to control the structure and properties, which is of special significance for the design of nitrogen steels. That is why the knowledge of nitrogen solubility in different phases and of phase diagrams for multicomponent nitrogen-containing iron-based systems is so important. The data available are presented and discussed in detail in this Chapter.

A brief description of some thermodynamical parameters for the characterisation of nitrogen solubility in multicomponent alloys is given in the first section. Next, an analysis of the binary Fe-N diagram and a short description of the iron nitrides is proposed. The constructing of multicomponent phase diagrams is demonstrated in the third section with a reference to the ThermoCalc programme. Two most important systems, Fe-Cr-Ni-N and Fe-Cr-Mn-N, and complex MeN phases are analysed based on calculations and experimental data. Precipitates from two most important matrices for the design of nitrogen steels, fcc and bcc solid solutions, are described in the two last sections.

## 2.1. Thermodynamics of multicomponent solutions

In multicomponent metallic solutions, the solutes $i$ affect their solubility and this influence is accounted for by so-called interaction parameters. They are derived from the expansion of the activity coefficient $\gamma_i$ of the $i$-th component with respect to small concentrations:

$$ln\gamma_i = ln\gamma_i^\infty + \left(\frac{\partial\, ln\gamma_i}{\partial x_i}\right)_{x_i\to 0} \cdot x_i + \frac{1}{2}\left(\frac{\partial^2 ln\gamma_i}{\partial x_i^2}\right)_{x_i\to 0} \cdot x_i^2 + \dots \qquad (2.1)$$

or

$$ln\gamma_i = \sum_{k=0}^{\infty} J_k^{\,i} x_i^{\,k},$$

where

$$J_k^{\,i} = \frac{1}{k}\left(\frac{\partial^k \ln\gamma_i}{\partial x_i^{\,k}}\right)_{x_i\to 0}.$$

Coefficient $J_k^{\,i}$ is the interaction parameter of the $k$-th order.

The interaction parameter of zero order is $\ln\gamma_i^{\infty}$, i.e., it represents the value of the activity coefficient for an infinitely diluted solution when $x_i \to 0$.

The first order interaction parameter $\partial\ln\gamma_i/\partial x_i$ is denoted as $\varepsilon_i^i$. It was introduced by Chipman (1955) and Wagner (1962) in order to characterise an influence of the concentration of the $i$-th component on its activity coefficient. In other words, it determines the slope of function (2.1) at $x_i \to 0$.

The second order interaction parameter

$$0.5\cdot(\partial^2 \ln\gamma_i/\partial x_i^2)_{x_i} = r_i$$

was proposed by Lupis and Elliott (1967) for higher concentrations of solutes. It characterises the curvature of function (2.1) at $x_i \to 0$. The interaction parameters are introduced for an infinite dilution and their application to finite concentrations has some limitations.

In industry, one uses concentrations in mass % instead moles. For this case, the activity coefficient is denoted as $f_i \cong \gamma_i/\gamma_i^{\infty}$, and the activity is the product of the concentration and the activity coefficient, $a_i = c_i\cdot f_i$. The interaction parameters are calculated in exactly the same way as for the molar concentrations:

$$e_i^i = (\partial \lg f_i/\partial \%i)_{\%i\to 0}$$ for the first order,

$$r_i^i = 0.5\cdot[\partial^2 \lg f_i/(\partial \%i)^2]_{\%i\to 0}.$$ for the second order.

The practical significance of taking the interaction parameters into account is clearly seen while studying the effect of chromium, manganese and nickel on the solubility of nitrogen in iron. The concentration dependence of $\lg f_N^X$ is shown in Fig. 2.1 for X = Cr, Mn, Ni as

$$\lg f_N^X = e_N^N + e_N^X\cdot\%X$$ (solid curves)

and (2.2)

$$\lg f_N^X = e_N^N + e_N^X\cdot\%X + r_N^X\cdot(\%X)^2$$ (dashed curves).

In this example, we use the values of interaction parameters for different substitutional elements given by Feichtinger and Zheng (1991). We neglect the effect of nitrogen concentration in solution $e_N^N\cdot\%N$, which would change the concentration-free member of Eqs. (2.2). It follows from Fig. 2.1 that, really, the interaction parameter of the first order, $e_N^X$, determines the slope of $\lg f_N^X$ at X → 0, while the interaction parameter of the second order, $r_N^X$, characterises its curvature. This behaviour is marked in Fig. 2.1 for 15 % Cr.

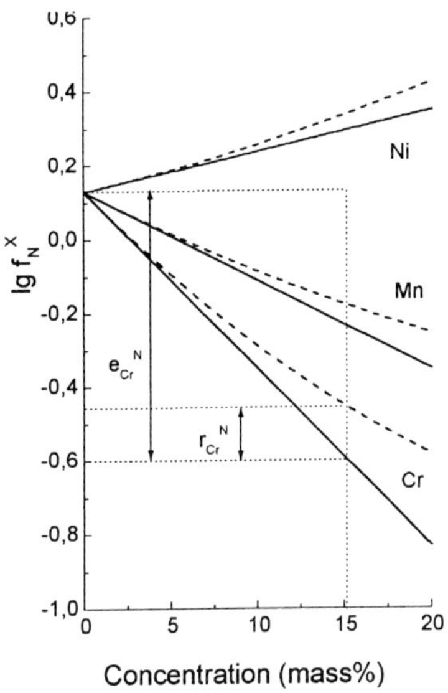


**Fig. 2.1.** An illustration of the interaction parameters characterising effects of Cr, Mn and Ni upon the coefficient of nitrogen activity in iron at 1600 °C. The interaction parameter of the first order, $e_N^X$, characterises the slope of $\lg f_N^X$ curve at X→0, whereas that of the second order, $r_N^X$, is responsible for its curvature. The values of interaction parameters given by Feichtinger and Zheng (1991) are used

Substitutional solutes affect the solubility of nitrogen to a different degree and in different directions. Chromium changes the nitrogen solubility stronger than manganese and nickel, and, as a result, it is necessary to account for $r_N^{Cr}$ already starting from 2.5% of the chromium concentration, whereas, in the case of alloying by manganese and nickel, the effect of $r_N^X$ is notable at a content above 5%. As was shown by Grigorenko and Pomarin (1989), the dependence of the logarithm of the nitrogen concentration in iron upon the manganese concentration can be described by the equation of second order for the whole range of manganese contents, whereas it is not sufficient for the iron-chromium alloys where it is necessary to account for the interaction parameter of the third order for chromium concentrations above 40%.

## 2.1.1 Solubility of nitrogen in liquid iron and steel

The usage of interaction parameters for the calculation of nitrogen solubility in melts can be demonstrated by two examples for iron and an iron-based solution at 1600 °C.

The reaction of dissolution of nitrogen in iron is described as $1/2N_2 = N_{Fe}$. The value of standard Gibbs free energy for this reaction $\Delta G° = 48400\ J$ at 1600 °C

(Elliot et al. 1963) and, from the relation $\lg K = -\Delta G°/RT = -1.35$, one obtains the value of the reaction constant, $K = 0.045$.

Since $K = a_N/p_{N2}^{1/2} = \%N \cdot f_N/p_{N2}^{1/2} = 0.045$ and, for a diluted solution, $a_N \approx c_N$ and $f_N = 1$, the solubility of nitrogen is equal to ~0.045 % at $p_{N2} = 1$ bar. This value can be corrected using the relation:

$$\lg f_N = e_N^N \cdot \%N + r_N^N \cdot (\%N)^2 + ...$$

For a concentration range of about 0.045%, the second term in the above equation can be neglected as compared to the first one. The value of $e_N^N$ is about of 0.1 and $e_N^N \cdot \%N < 0.01$, i.e. the correction for $f_N$ is less than 1% of nitrogen solubility. Therefore,

$$\%N \cong 0.045 p_{N_2}^{1/2},$$

i.e. the solubility of nitrogen in the iron is proportional to the square root of pressure. This relation is known as the Sieverts' law and it is a consequence of Henri's law of first order (see Sect. 1.1.2.1). The measurements of nitrogen solubility at different pressures give a value of $K \approx 0.043$ (see e.g. Schenck et al. 1962).

Let us calculate, as a second example, the solubility of nitrogen at 1600 °C in a steel containing (mass %) 18Cr, 18Mn and 2Mo:

$$\lg f_N^X = e_N^N \cdot \%N + e_N^{Cr} \cdot \%Cr + r_N^{Cr} \cdot (\%Cr)^2 + e_N^{Mn} \cdot \%Mn +$$

$$r_N^{Mn} \cdot (\%Mn)^2 + e_N^{Mo} \cdot \%Mo.$$

Using interaction parameters of Satir-Kolorz et al. (1990), see Table 2.1, one obtains $\lg f_N^X = -1.1059$ and $\%N = K/f_N^X = 0.57$.

It is important for practical use that CrMn compositions provide a sufficiently high nitrogen solubility at the normal pressure of 1 bar.

### 2.1.1.1
### *Effect of alloying elements*

In practice, it is useful to describe the influence of different solutes on the nitrogen solubility by an equivalent effect of some reference element. A concept of the equivalent factor is based on a similar concentration behaviour of the logarithmic specific coefficients of activity (see e.g. Fig. 2.1), which allows to recalculate them to one "effective" curve for a chosen reference element:

$$\lg f_N^{X^{ref}} = e_N^{X_i} \cdot (c_N^{X_i} \cdot \%X_i) + r_N^{X_i} \cdot (c_N^{X_i} \cdot \%X_i)^2 + .....$$

For the multicomponent solutions, the equivalent concentration

$$[\%X^{ref}]_{eq} = \Sigma c_N^{X_i} \cdot (\%X_i)$$

and the equivalent factor $c_N^{X_i}$ determines the content of a reference element $X^{ref}$ which provides the same nitrogen activity as $X_i$.

On these terms, the nitrogen activity in steel at 1600 °C can be expressed by the carbon-equivalent concentration (Schürmann and Kunze 1967):

$$\lg f_N^X = 0.123(\%C)_{eq} + 0.003(\%C)_{eq}^2$$

for the concentration range $0 \leq (\%C)_{eq} \leq 6$ and

$$\lg f_N^X = 0.123(\%C)_{eq} + 0.00225(\%C)_{eq}^2$$

for the concentration range $-12 \leq (\%C)_{eq} \leq 0$.

The correction for an arbitrary temperature of the melt is given by the relation

$$\lg f_{N,T}^X = (-0.75 + 3280/T) \cdot \lg f_{N,1873}^X.$$

A similar approach relating to the chromium-equivalent concentration was proposed by Satir-Kolorz et al. (1991) resulting in the following equation for the nitrogen activity coefficient in iron at 1600 °C:

$$\lg f_N = e_N^N \cdot \%N - 0.048 \cdot (\%Cr)_{eq} + 3.5 \cdot 10^{-4} \cdot (\%Cr)_{eq}^2.$$

The temperature correction can be the same as for carbon-equivalent factors, based on an assumption that the effect of temperature on nitrogen solubility should be the same.

Shown in Table 2.1 are the data of interaction parameters and equivalent factors for some important alloying elements in iron at 1600 °C or their temperature dependences, as obtained by different researchers. A possibility to predict the extention of the austenitic field is important, in particular for martensitic steels where the solution treatment has usually to be performed in the austenitic state. That is why the content of alloying elements $X_\gamma$ providing the solubility of e.g. 0.5 mass % of nitrogen at 1100 °C in the $\gamma$ phase is also presented in Table 2.1. It should be noted that thermodynamical parameters can not be calculated *ab initio*. The researchers obtain different values for interaction parameters (compare e.g. the data of Feichtinger and Zheng 1991, Grigorenko and Pomarin 1990, Feichtinger 1993, Satir-Kolorz et al. 1990, Feichtinger et al. 1990, Schürmann and Kunze 1967) and their use for the calculation of nitrogen solubility in steel allows a semiquantitative evaluation only. It is also necessary to account for the cross-effects of different substitutional solutes, (Cr-Mn, Cr-Ni etc.), which makes evaluations even more complicated.

**Table 2.1.** First and second order interaction parameters $e_N^X$ (Satir-Kolorz et al. 1990) and $r_N^X$ (Grigorenko and Pomarin 1989) and equivalent factors of carbon $c_{eq}^C$ (Schürmann and Kunze 1967) and chromium $c_{eq}^{Cr}$ (Satir-Kolorz et al. 1990) for the dissolution of nitrogen in iron at 1600 °C. At 1100 °C and a nitrogen content of e.g. 0.5 mass % the phase field of austenite extends to a content of X = $X_\gamma$ above which nitrides or δ-ferrite (in brackets) appear

| X | $e_N^{X_i}$ | $r_N^{X_i}$ | $c_{eq}^C$ | $c_{eq}^{Cr}$ | $X_\gamma$ (mass %) |
|---|---|---|---|---|---|
| Ti | -0.930 | - | -7.88 | 19.40 | $1.73\cdot10^{-7}$ |
| V | -0.098 | (6.2/T)-0.00005 | -0.85 | 2.05 | $3.43\cdot10^{-3}$ |
| Nb | -0.050 | (5.14/T)-0.0018 | - | 1.05 | $5.57\cdot10^{-4}$ |
| Cr | -0.048 | (3.3/T)-0.001 | -0.382 | 1.00 | 1.6(18.5) |
| Mn | -0.021 | (0.80/T)-0.00038 | -0.163 | 0.50 | |
| Mo | -0.013 | 0.000079 | -0.094 | 0.27 | (12.6) |
| W | -0.002 | - | 0.013 | 0.04 | (13.4) |
| Cu | 0.006 | - | 0.0463 | -0.12 | |
| Co | 0.010 | - | 0.077 | -0.20 | |
| Ni | 0.011 | 0.000035 | 0.0813 | -0.22 | |
| Al | 0.040 | - | - | -0.85 | |
| Si | 0.043 | - | 0.444 | -0.90 | |
| B | 0.083 | - | - | -1.73 | |
| C | 0.118 | (-28/T)+0.035 | 1.00 | -2.46 | |
| N | 0.13 | - | - | -2.70 | |

### 2.1.1.2 Effect of temperature

Figure 2.2 shows the temperature dependence of nitrogen solubility in different phases of pure iron. For development of nitrogen steels the solubility of nitrogen in the liquid phase is critical. It increases with temperature in the same manner as in bcc solid iron phases. It reflects a tendency in liquid metals to retain the short range atomic order typical for their solid phases with a less closely packed lattice (see e.g. Il'insky et al. 1995). Alloying affects the temperature dependence of the nitrogen solubility in the liquid iron significantly.

The influence of chromium is drastic and non-monotonous: an increase in the chromium content makes the slope of the [%N] = f(T) curve even negative starting from 2.3-3.0% of Cr (Torkhov et al.1971, Bezobrazov et al. 1960).

Nickel causes an insignificant effect and, moreover, the data available are contradicting: (*i*) an increase in nitrogen solubility with temperature was observed by Pomarin et al. (1972) and Schenck et al. (1959), (*ii*) no temperature influence was reported for nickel at contents above 2% by Kashyap and Parlee (1958) and (iii) a decrease in nitrogen solubility with temperature was obtained by Wentrup and Reif (1949) in Fe-Ni-N alloys where nickel was added up to 20%.

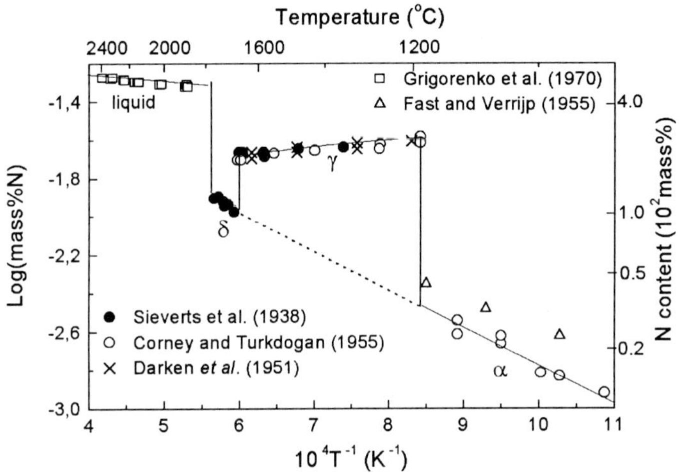


**Fig. 2.2.** Solubility of nitrogen in pure iron at $p_{N_2} = 1$ bar

In Fe-Mn-N alloys, the tendency is the same, but weaker, as for Fe-Cr-N: an inversion of the temperature dependence was observed at 12% of Mn (Schürmann and Kättlitz 1981).

The addition of molybdenum within 3-10% makes the nitrogen solubility in liquid iron independent of temperature (Kashyap and Parlee 1958, Schenck et al. 1958), while higher molybdenum contents invert the temperature dependence like chromium and manganese (Schürmann and Kättlitz 1981).

### 2.1.1.3
### Effect of pressure

Sieverts' law $[\%N]_{Fe} = p_{N2}^{1/2}$ for the dissolution of nitrogen in the pure liquid iron was confirmed in numerous studies (see e.g. Sieverts and Zapf 1935, Fedotov and Samarin 1958). It is not confirmed for multicomponent iron alloys. Chromium causes deviations from the linear relation at concentrations above 10% (Grigorenko and Pomarin 1989). A simple empirical formula to account for the chromium content and pressure was proposed by Bezobrazov et al. (1960) for 1600 °C and $p$ from 1 to 50 bar:

$$[\%N]_{Fe-Cr} = [\%N]_{Fe} \cdot c_{Fe} \cdot p_{N_2}^{1/2} + [\%N]_{Cr} \cdot c_{Cr}^2 \cdot p_{N_2}^{\alpha} \quad , \tag{2.3}$$

where $\alpha = 0.4 - 0.26c_{Cr}$, $c_i$ is the mass fraction of the $i$-th element ($c_i = [\%i]/100$). The limits of the applicability of Sieverts' law for the nitrogen solubility in Fe-Cr alloys at different pressures, as calculated by Besobrazov et al. according to Eq. (2.3), and the experimental data obtained by different authors (see Torkhov et al. 1971) are shown in Fig. 2.3. One can see that the deviations from the Sieverts' law become significant with an increase in the chromium and nitrogen content and that formula (2.3) is consistent with the experimental data.

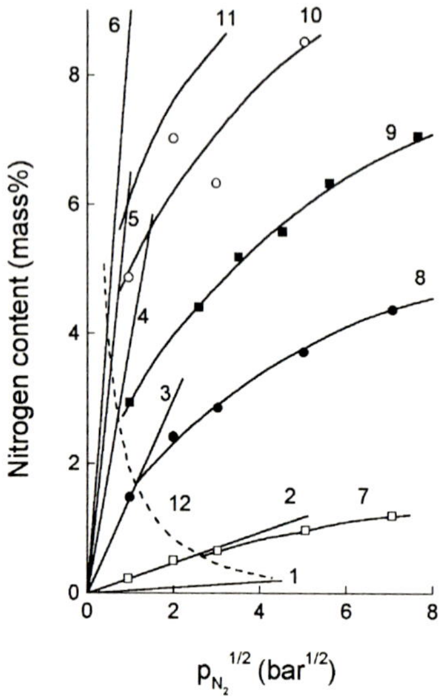


**Fig. 2.3.** Solubility of nitrogen in liquid Fe-Cr alloys at 1600 °C and different pressures after Bezobrazov et al. (1960). The experimental data are obtained by Torkhov et al. (1971). 1 – nitrogen solubility in pure iron according to Frehser and Kubisch (1963); 2 to 6 – a calculated solubility for 20, 50, 70, 90, 100 % of Cr, respectively, according to Torkhov et al. (1971); 7 to 11 – the calculations using Eq. 2.3. The dashed line 12 marks a limit for the validity of the Sieverts' law

The data for liquid Fe-Mn alloys are limited. At pressures below 1 bar, Sieverts' law is valid up to 100% of Mn (Beer 1961). The same result was obtained for pressures of 1 to 3 bar in Fe-Mn alloys with manganese contents up to 4% (Schenck et al. 1962). Deviations from Sieverts' law are expected at higher pressures or manganese contents.

Liquid Fe-Mo alloys with molybdenum contents up to 13% were studied at 1600 °C and pressures below 1 bar (Schenck et al. 1959), and Sieverts' law was confirmed.

Multicomponent alloys reveal the same tendency for the effect of pressure on the nitrogen solubility. In steel containing (mass %) 17.6Cr, 7.8Ni, 1.9Mo and 1.9Al, the Sieverts' law was found to be valid at temperatures between 1565 and 1760 °C for pressures below 1 bar (Wada and Pehlke 1977*). In FeCrxMn0.5 alloys, the deviations from Sieverts' law were essentially the same as those of FeCr alloys (Rashev and Ivanov 1979).

## 2.2 Binary Fe-N diagramme

As follows from Fig. 2.2, the solubility of nitrogen in α- and δ-iron in equilibrium with gaseous nitrogen at $p_{N2}$ = 1 bar is not significant and tends to increase with temperature, while γ-iron dissolves more nitrogen, but the effect of temperature is opposite to that in bcc phases. This difference in the temperature behaviour of nitrogen solubility is unusual as compared to that of carbon, where the solubility increases with temperature in both the bcc and the fcc iron lattices. As a first approximation, an explanation based on different heats of solution for carbon and nitrogen in iron can be proposed.

According to Chipman (1972), the change in the free energy accompanying the solution of graphite in γ-iron is:

$$\Delta G^\circ = \Delta H^\circ - T\Delta S^\circ = 10520 - 4.21T \text{ (in cal)}$$

with the heat of solution 10520 cal (or 44045 J).

For the solution of carbon in α- and δ-iron, the free energy is:

$$\Delta G^\circ = 25400 - 11.4T\,,$$

and the heat of solution is 25400 cal (or 106345 J). Both values of $\Delta H^\circ$ are positive, i.e. the reactions of carbon solution in both the fcc and the bcc iron phases are accompanied by the absorption of heat.

The situation is quite different for the solution of nitrogen in iron in equilibrium with gaseous nitrogen at $p_{N2} = 1$ bar (see Fromm and Gebhardt 1976, original data are given in cal)

$$\tfrac{1}{2}N_2 = [N]_{\alpha\text{-Fe}} \qquad \Delta G^\circ = 8347 + 9.9T,$$

$$\tfrac{1}{2}N_2 = [N]_{\gamma\text{-Fe}} \qquad \Delta G^\circ = -1920 + 15.2T,$$

$$\tfrac{1}{2}N_2 = [N]_{\delta\text{-Fe}} \qquad \Delta G^\circ = 6083 + 11.8T,$$

i.e. the reaction of nitrogen dissolution is endothermal in the bcc iron and exothermal in the fcc iron with the values of the solution heat (in J) 34947, −8038 and 25468 in α, γ and δ phases, respectively.

These data enable one to conclude that in equilibrium with gas at $p_{N2} = 1$ bar the solubility of nitrogen in γ-iron is expected to be decreased with increasing temperature in accordance with Le Chatelier's principle which postulates that any process taking a system out of the equilibrium (an increase in temperature during heating) is prevented. Vice versa, the solubility of nitrogen in α- and δ-iron and that of carbon in all the iron phases (in equilibrium with graphite) is increased with increasing temperature because the system stays in equilibrium that way.

While discussing the solubility of nitrogen in the solid phases of iron, one has to distinguish between the cases where iron is in equilibrium with gaseous nitrogen or with an iron nitride. The latter is achieved when the nitriding power is increased using an increasing pressure of gaseous nitrogen $N_2$ or the dissociation of ammonia $NH_3$. The difference in the nitriding powers of gaseous nitrogen and of dissociated ammonia is illustrated by a significant difference in the enthalpies of their dissociation (see, Fromm and Gebhardt 1976): $\Delta H^\circ = 86400$ cal or 361740 J for $\tfrac{1}{2}N_2 = N$ and $\Delta H^\circ = 12050$ cal or 50450 J for $NH_3^{gas} = \tfrac{1}{2}N_2 + {}^3/_2H_2$. It is clear from this comparison that, at the same temperature, high pressures of gaseous nitrogen are needed to produce nitrogen contents in iron compatible with ones produced by dissociated ammonia.

The phase diagram Fe-N is shown in Fig. 2.4 according to Hansen and Anderko (1958).

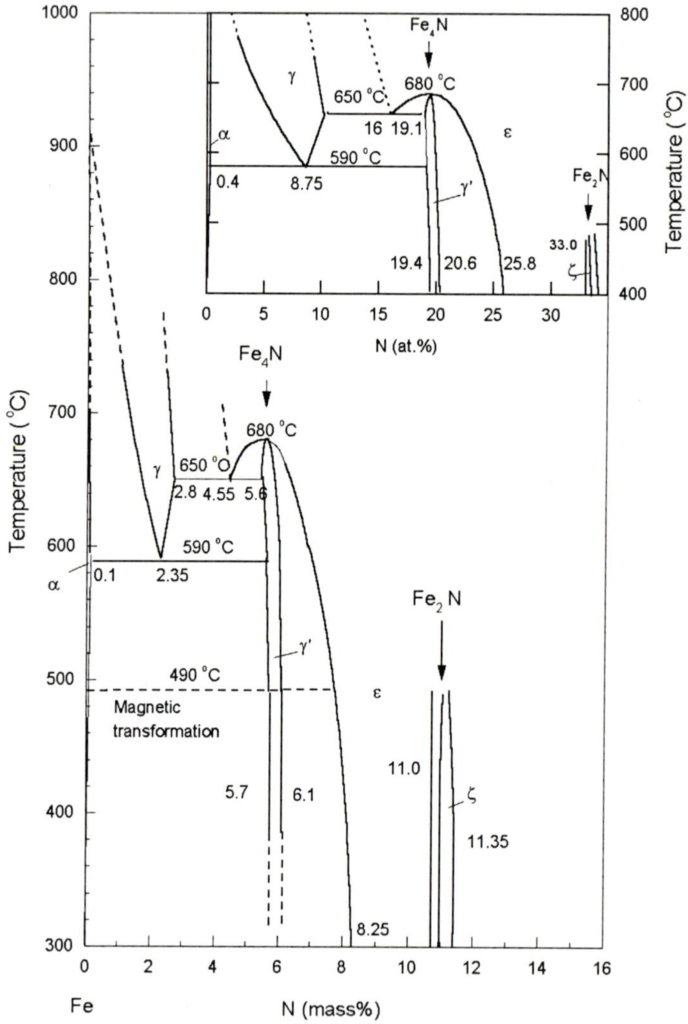


**Fig. 2.4.** Phase diagram Fe-N (after Hansen and Anderko 1958)

Unlike the usual way of presenting phase diagrams, it is constituted not for $p_{N2} = 1$ bar, but for very high partial pressures of nitrogen provided by the dissociation of ammonia. In fact, it is a projection of various equilibria in the temperature-pressure-concentration diagram onto a temperature-concentration plane (see also the discussion in Trans. Met. Soc. of AIME (1950) vol. 188, p. 1354). Any heat treatment of iron in gaseous nitrogen at $p_{N2}$ = 1 bar, previously nitrided at higher partial nitrogen pressures, will result in the nitrogen content according to Fig. 2.2 and not to Fig. 2.4.

As follows from the phase diagram, four iron nitride phases of increasing nitrogen content can be obtained: $Fe_8N$ ($\alpha''$), $Fe_4N$ ($\gamma'$), $Fe_3N$ ($\varepsilon$) and $Fe_2N$ ($\zeta$).

***The* $\alpha''$ *phase*** is an intermediate phase between $\alpha$-iron and $\gamma'$. It was discovered by Dijkstra (1949) in his studies of internal friction in $\alpha$-iron aged at temperatures below 300 °C after saturation by nitrogen at 575 °C. Later on Jack (1951) found the same phase after low-temperature ageing of nitrogen martensite and determined the dimensions of its unit cell:

$$a = 0.572 \text{ nm}, \; c = 0.629 \text{ nm}, \; c/a = 1.1.$$

The $\alpha''$-nitride has a body-centered tetragonal lattice (see Fig. 2.5a), however, it can also be presented as an ordered solid solution of nitrogen in the $\gamma$-iron where nitrogen atoms occupy one of the possible sublattices of the octahedral sites in fcc lattice and each second nitrogen atom site in this sublattice is vacant (see Fig. 2.5b).

a

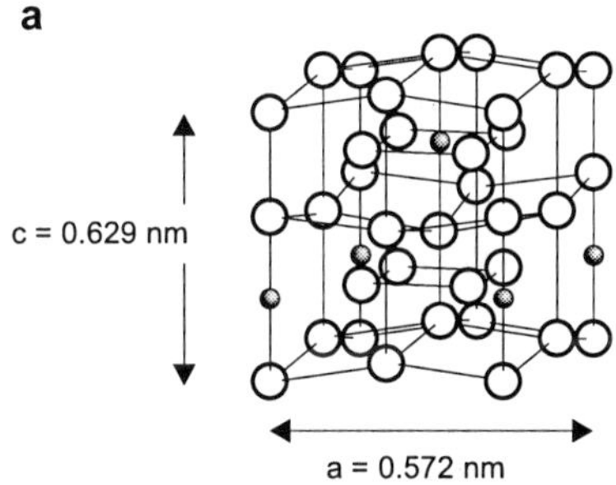


b

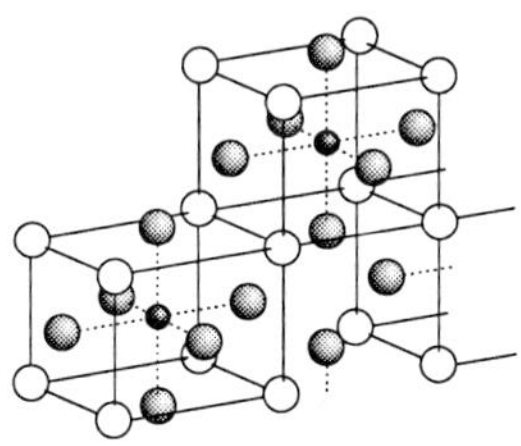

**Fig. 2.5.** Two presentations of $\alpha''$-iron nitride structure according to Jack (1950): **a** - as bct lattice, **b** - as fcc lattice where each second cell is occupied by nitrogen atoms

It can be understood from the presentation in Fig. 2.5b why $\alpha''$-nitride $Fe_8N$ is an intermediate phase between pure $\alpha$-iron and $\gamma'$-nitride $Fe_4N$. In this relation, it is also noteworthy, that the volume per metal atom in $\alpha''$-nitride (0.0129 $nm^3$) has a mid-value between those for bcc $\alpha$-iron (0.012 $nm^3$) and fcc $\gamma'$-nitride (0.0137 $nm^3$). The open circles in Fig. 2.5b belong to the iron atoms having no nitrogen in the nearest neighbourhood. The large distance between these atoms as compared to that in pure iron results in an extremely large magnetic moment (about $M \approx 3\mu_B$ compared to $M \approx 2\mu_B$ per atom in pure α-iron), which makes α″-nitride a promising material for magnetic applications.

Fast and Verrijp (1955) studied the nitrogen solubility in α-iron in equilibrium with nitrides and gaseous nitrogen using the internal friction technique. For the equilibrium with α″-nitride $Fe_8N$, it was found

$$c_{Fe_8N} = 3.3\times10^2 e^{-9900/RT} \text{mass \% N}$$

with the solution enthalpy ΔH = 9900 cal/g-atom N (41450 J/mol).

***The γ′ phase ($Fe_4N$)*** has an fcc lattice and exists across a narrow range of compositions (see Fe-N diagram, Fig. 2.4). The lattice parameter changes from 0.378 nm at 5.29% of N to 0.379 nm at 5.71% of N (in mass %, see Paranjpe et al. 1950). The crystal structure can be easily understood from Fig. 2.5b. Unlike α″-nitride, each elementary cell in the γ′-nitride is occupied by a nitrogen atom and there are two types of iron atoms: having no and two nitrogen atoms as nearest neighbours. The former occupy the corners of the cubic lattice, the latter are located in the centres of the faces and form N-Fe-N dumb-bell configurations that are clearly seen in Mössbauer spectra of Fe-N alloys (see Sect. 1.1.2.5).

**Table 2.2.** Content of nitrogen (mass %) in α-iron in equilibrium with α″, γ′ and gaseous nitrogen at $p_{N2}$ = 1 bar, according to Fast and Verrijp (1955)

| T (°C) | In equilibrium with α″ | In equilibrium with γ′ | In equilibrium with $N_2$ at 1 bar |
|---|---|---|---|
| 20 | $1.4\cdot10^{-5}$ | | |
| 100 | $5.2\cdot10^{-4}$ | | |
| 200 | $8.8\cdot10^{-3}$ | | |
| 300 | $5.5\cdot10^{-2}$ | $8.4\cdot10^{-3}$ | |
| 400 | $2.0\cdot10^{-1}$ | $2.5\cdot10^{-2}$ | |
| 500 | | $5.5\cdot10^{-2}$ | $9.0\cdot10^{-4}$ |
| 585 | | $9.5\cdot10^{-2}$ | $1.4\cdot10^{-3}$ |
| 700 | | | $2.4\cdot10^{-3}$ |
| 800 | | | $3.3\cdot10^{-3}$ |
| 900 | | | $4.5\cdot10^{-3}$ |

The solubility of nitrogen in the α-iron in equilibrium with γ′-nitride was obtained by Fast and Verrijp (1955) as

$$c_{Fe_4N} = 12.3\times10^2 e^{-8300/RT} \text{mass \% N}$$

with the solution enthalpy for γ′-nitride in α-iron is ΔH = 8300 cal/g-atom N (34750 J/mol). For comparison, the enthalpy of nitrogen dissolution in α-iron in equilibrium with gaseous nitrogen at $p_{N2}$ = 1 bar, according to the same authors, is ΔH = 7200 cal/g-atom N (30145 J/mol). It is less than the above mentioned data obtained from the thermodynamical measurements (Fromm and Gebhardt 1976). Nevertheless, the internal friction method allows to compare the nitrogen solubility in ferrite in equilibrium with nitrides $Fe_8N$, $Fe_4N$ and gaseous nitrogen at the pressure of 1 bar (Table 2.2). It follows from these data that the precipitation from ferrite supersaturated with nitrogen occurs at low temperatures in the following sequence: $\alpha'_{supersat} \rightarrow \alpha'' \rightarrow \gamma'$, and nitriding by gaseous nitrogen at 1 bar provides smaller nitrogen contents in ferrite as compared to those in equilibrium with nitrides.

***The ε phase*** covers a wide concentration range of homogeneity in the binary Fe-N diagram (Fig. 2.4). It has a hexagonal close-packed structure where some of the octahedral interstitial sites are occupied by nitrogen atoms (Fig. 2.6).

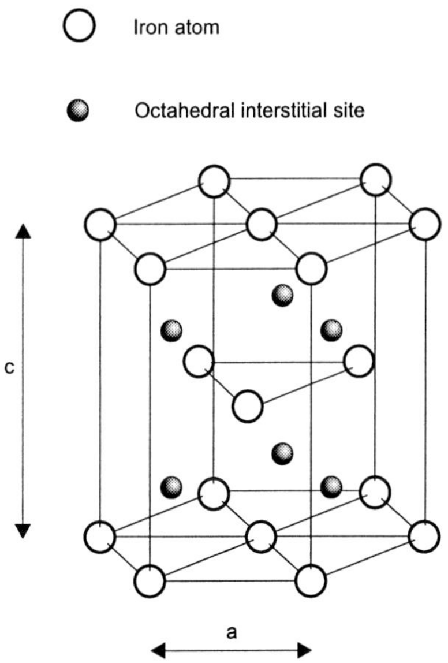


Fig. 2.6. Structure of ε-nitride

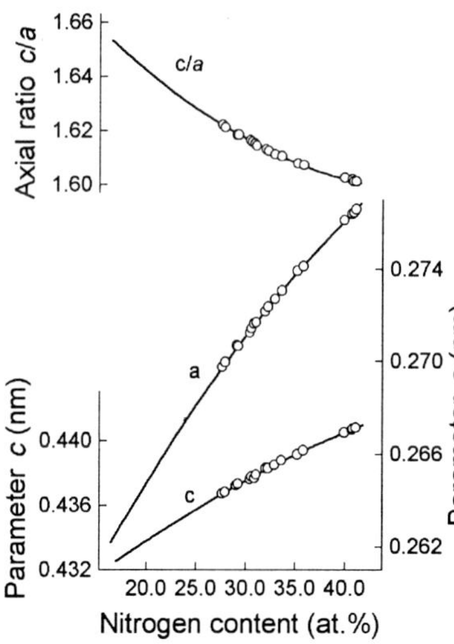


**Fig. 2.7.** Lattice parameters of ε-$Fe_3N$ nitride (after Paranjpe et al. 1950)

The layers of interstitial sites are orthogonal to the c-axis and each layer is equidistant between two planes of the iron atoms, so that each iron atom has six interstitial sites in the nearest neighbourhood. The lattice parameters and the *c/a* ratio, as determined by the X-ray method (Paranjpe et al. 1950), are presented in Fig. 2.7 as functions of the nitrogen content. There is some uncertainty in relation to the measurements performed at nitrogen contents below 7.3 mass % because of the high rate of eutectoid reaction for the decomposition of the ε phase during quenching from the high-temperature ε field and, as a result, of the difficulties to obtain samples with a homogeneous ε phase.

There are three kinds of iron atoms in ε-nitride in the whole concentration range of its homogeneity: with one ($Fe_1$), two ($Fe_2$) and three ($Fe_3$) nitrogen atoms as nearest neighbours. The long range atomic ordering occurs during annealing of ε-nitride at comparatively low temperatures (~550 °C), which leads to an increase in the fraction of $Fe_2$ atoms, and it is accompanied by an increase in the magnetisation. In the ordered superstructure $Fe_3N$ each iron atom is adjacent to two nitrogen atoms only. Magnetic properties of the disordered ε phase reveal a non-monotonous concentration dependence with a maximum corresponding to the stoichiometric $Fe_3N$ composition (DeCristofaro and Kaplow 1977, see Fig. 2.8).

The nature of stabilisation of the ε phase by nitrogen at a concentration of more than 10 at. % in iron is not simple, and it is obviously related to peculiarities of the distribution of nitrogen atoms in the hcp lattice. It cannot be explained as easily as for the substitution of the bcc crystal structure by an fcc one due to inreasing nitrogen content, which is due to the lower energy of nitrogen in the fcc iron as compared to the bcc one ($\Delta E \sim 0.3$ eV/atom). The number of iron atoms in the nearest neighbourhood of a nitrogen atom in the first and second coordination spheres and the Fe-N distances are the same for fcc and hcp iron structures. That is why it is doubtful that Fe-N interaction energies are significantly different in both phases.

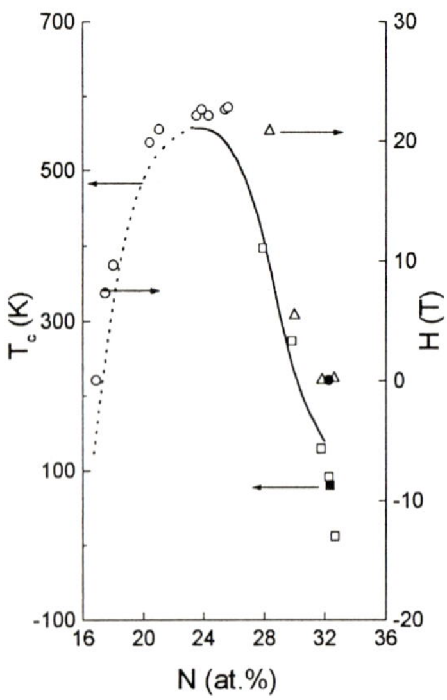


**Fig. 2.8.** Concentration dependence of hyperfine field that is proportional to the saturation magnetisation (open circles – DeCristofaro and Kaplow 1977, solid circles – Chabanel et al. 1968, open triangles – Mekata et al. 1972) and of Curie temperature (open squares – Mekata et al. 1972, solid squares – Chabanel et al. 1968, solid line – Bridelle 1955, dotted line – DeCristofaro and Kaplow 1977) for ε-$Fe_3N$ nitride

A possible reason for the preference of an hcp lattice in Fe-N alloys with increasing nitrogen content can be concerned with a difference in the values of N-N interaction. As was shown above (see Sect. 1.1.2.5), a strong repulsion exists between nitrogen atoms as nearest neighbours in fcc iron. In the hcp iron, a strong N-N repulsion takes place only along the hexagonal axis as the nearest-neighbour distance in the c-direction is $\sqrt{2/3}$ of the closest distance in the hexagonal plane. That is why the number of interstitial sites blocked for the occupation by nitrogen atoms is smaller in the hcp structure, which has to result in some entropy preference as the nitrogen content increases. Some energy preference is also expected because the interaction between nitrogen atoms contributes to the enthalpy of nitrogen dissolution.

***The ζ phase***, as well as ε, was originaly found by Hägg (1929). It has an orthorhombic structure with dimensions of the unit cell $a = 0.275$ nm, $b = 0.482$ nm and $c = 0.443$ nm. It exists in the narrow concentration range between 11.1 and 11.35 mass % of N (see Fig. 2.4). No evidence of ε+ζ field was observed. As will be shown in Sect. 2.5, the ζ nitride precipitates as an intermediate phase during the tempering of 15%Cr nitrogen martensite.

## 2.3 Multicomponent phase diagrams

As a rule, nitrogen steels have to meet some special requirements (stability in agressive environments, at low or high temperatures, under a complicated stress state etc.), which determines their complicated chemical composition. That is why it is so important to know the phase equilibrium and the limits of nitrogen solubility in multicomponent solid solutions.

### 2.3.1 The calculation of phase diagrams

The construction of phase diagrams for multicomponent systems needs a number of experimental data, and the best way is to combine measurements and computations. One of the most popular procedures for this purpose is known as the CALPHAD approach (CALculation of PHAse Diagrams). According to this technique, the equilibria and thermodynamic properties of multicomponent iron-based liquid and solid phases are calculated using a self-consistent database for the corresponding lower-order systems. First, the properties of each phase are studied using mathematical methods as well as available thermochemical data. Next, the thermodynamic equilibria between the phases can be calculated based on the condition of minimum Gibbs energy. In this section we will briefly describe this procedure following thermodynamical calculations performed by Hertzman (1987), Hertzman and Jarl (1987), Frisk and Hillert (1989), Frisk (1991), Frisk and Qiu (1994) etc.

One starts from binary systems, where the mixing behaviour of two elements in each phase is determined in terms of a mathematical expression based on the thermodynamic properties of pure elements (the heat of solution, heat capacity etc.). Such expressions contain certain adjustable (fitting) parameters to describe the temperature and pressure dependences of phase equilibrium. Experimental data on the solubility of pure elements in each phase are used to determine the adjustable parameters.

In ternary systems, ternary mixing effects are introduced, but the necessary experimental information is relatively reduced if the data for binary interactions are known. The interaction between four and more atoms is less significant and the need for new experimental data in the analysis of higher-order systems decreases drastically. Thus, it is relatively easy to conctruct the phase diagram of a multicomponent system if thermodynamic experimental data and calculations for lower-order systems are available.

Substitutional solutions represent the simplest case because all the atoms occupy the same kind of sites. The molar Gibbs energy for a certain α phase in a multicomponent substitutional solution is given by the expression

$$G_m^\alpha = \sum x_i \, {}^oG_i^\alpha + RT \sum x_i \ln x_i + {}^E G_m^\alpha, \tag{2.4}$$

where $G_i^\alpha$ is the Gibbs energy for the $i$-th component in the α phase and $x_i$ are atomic concentrations. The first summation represents the sum of Gibbs energies of elements. The second summation reflects the mixing entropy for random distribution. The exess Gibbs energy ${}^E G_m$ accounts for the deviation from an ideal solution. It contains contributions from non-random distribution and interaction energies between the solute atoms. In most cases, a simple model is used for the representation of ${}^E G_m$:

$$^E G_m = \sum_i \sum_j x_i x_j L_{i,j} + \sum_i \sum_j \sum_k x_i x_j x_k L_{i,j,k} + ..., \tag{2.5}$$

where parameters $L$ characterise the interaction energies between the elements ($i$, $j$, $k$....) constituting the system. As a rule, they cannot be calculated and are selected on the basis of some experimental data and the minimisation procedure for the Gibbs energy of the system, i.e. they are the above mentioned adjustable parameters. Let us present, as an example, the values of $L_{Cr,Fe}{}^{bcc}$ = 22000 - 11.375T, $L_{Fe,Ni}{}^{bcc}$ = 580 - 5.327T (J/mol) which were used by Hertzman and Sundman (1985) for the studies of the Fe-Cr-Ni system.

If interstitials are added to the substitutional solution, one considers two sublattices: for the substitutional and interstitial elements, respectively. As most interstitial sites are vacant, the vacancies *Va* are treated as an additional element. Thus, in Eq. (2.4), one should consider separately the summations over all substitutional elements and the interstitial ones plus the vacancies. For the *M-N* system, one could use, for example, a model of the solution extending from the pure metal

$M_IVa_I$ to the compound $M_IN_I$ and the following expression for the molar Gibbs energy

$$G_m = \sum_i \sum_j y_i y_j {}^oG_{ij} + RT[a\sum y_i \ln y_i + c\sum y_j \ln y_j] + {}^EG_m, \qquad (2.6)$$

where $y$ denotes the site fraction of the $i$-th or $j$-th elements, $i$ belongs to all the substitutional elements and $j$ to all the interstitial ones and vacancies, respectively; $a$ and $c$ are the number of sites in each sublattice per formula unit.

We will illustrate the above approach based on the studies performed for the Fe-Cr-Ni-N and Fe-Cr-Mn-N systems containing the fcc and bcc substitutional and interstitial solutions and nitrides. Both systems are of practical significance for stainless steels.

## 2.3.2 The Fe-Cr-Ni-N phase diagram

The experimental data needed for the Fe-Cr-Ni-N system were obtained for the systems of pure Fe (Fernández Guillermet and Gustafsson 1985), Cr (Andersson 1985), Ni (Dinsdale 1991) and Fe-Cr (Hertzman and Sundman 1982, Andersson and Sundman 1987), Fe-Ni and Cr-Ni (Hertzman and Sundman 1985), Fe-N (Agren 1979, Frisk 1991), Cr-N (Frisk 1991, Jarl 1977), Ni-N (Fernández Guillermet and Frisk 1991) systems.

The following ternary systems were studied: Fe-Cr-Ni by Hertzman and Sundman (1985), Hillert and Qui (1990); Fe-Ni-N by Frisk and Qiu (1994), Jarl (1978); Fe-Cr-N by Hertzman and Jarl (1987), Frisk (1990); Cr-Ni-N by Ono (1986). The concentration range for the bcc and hcp phases in the Fe-Cr-N system was established by Hertzman and Jarl (1987), Fernández Guillermet and Gustafsson (1985) and Andersson and Sundman (1987).

The $\alpha$ and $\gamma$ phases in the ternary Fe-Cr-Ni system represent substitutional solid solutions and, in accordance with Eq. (2.4), their Gibbs energy is written as

$$G_m^\alpha = x_{Fe}\,{}^oG_{Fe}^\alpha + x_{Cr}\,{}^oG_{Cr}^\alpha + x_{Ni}\,{}^oG_{Ni}^\alpha + RT(x_{Fe}\ln x_{Fe} + x_{Cr}\ln x_{Cr} + x_{Ni}\ln x_{Ni}) + x_{Fe}x_{Cr}L_{FeCr} + x_{Fe}x_{Ni}L_{FeNi} + x_{Cr}x_{Ni}L_{CrNi} + x_{Fe}x_{Cr}x_{Ni}L_{FeCrNi},$$

where, as before, $\alpha$ denotes a phase ($\alpha$, $\gamma$ or liquid) and other symbols have the same meaning as in (2.4). The expressions for the $\alpha$ and $\gamma$ solid phases contain an additional term accounting for magnetism. The data for pure metals and for binary systems can be found in the original studies given above. A more complicated expression is used for the $\sigma$ phase based on a three-sublattice model as proposed by Andersson et al. (1986) for the formula $(Fe,Ni)_8Cr_4(Cr,Fe,Ni)_{18}$ of the $\sigma$ phase. It is supposed that the $\sigma$ phase does not dissolve nitrogen (see also Sect. 2.4).

One of the results of the thermodynamic assessment of the Fe-Cr-Ni system performed by Hillert and Qiu (1990) is given in Fig. 2.9 as a section of the phase diagram at 74 mass % of Fe, which corresponds to the common compositions of stainless Cr-Ni steels.

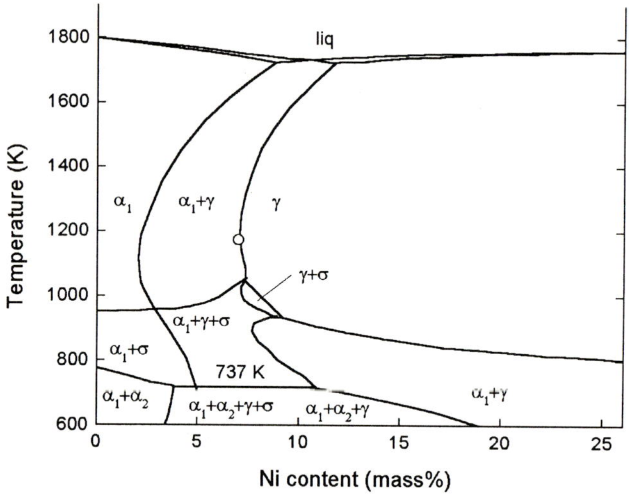


**Fig. 2.9.** Calculated vertical section of the Fe-Cr-Ni diagram at 74 mass % of Fe (after Hillert and Qiu 1990). The experimental point at 1173 K belongs to the boundary $\alpha$+$\gamma$ taken from the experimental Fe-Cr-Ni diagram (see Fig. 2.25)

The concentration-temperature range of the $\gamma$ phase is most important for practical use. The calculated boundary of the stable $\gamma$ phase is consistent with the experimental data available as confirmed by the example of the comparison with the section of the experimental phase diagram Fe-Cr-Ni at 1173 K (see Fig. 2.25). An interesting prediction is made for the precipitation of the $\sigma$ phase directly from $\gamma$ and its subsequent transformation to $\alpha$ during cooling. The $\alpha_2$ solid solution, having an increased Cr content as compared to $\alpha_1$, appears at lower temperatures as a result of this transformation.

In accordance with Eq. (2.6), the bcc, fcc and hcp phases in the Fe-Ni-Cr-N system are described using a two-sublattice model where Fe, Cr and Ni atoms occupy the first sublattice and N and Va the second one, and the expression for the Gibbs energy is written (see Frisk 1991) as

$$G_m = y_{Fe} y_N {}^0G_{Fe:N} + y_{Fe} y_{Va} {}^0G_{Fe:Va} + y_{Cr} y_N {}^0G_{Cr:N} + y_{Cr} y_{Va} {}^0G_{Cr:Va}$$
$$+ y_{Ni} y_N {}^0G_{Ni:N} + y_{Ni} y_{Va} {}^0G_{Ni:Va} + aRT(y_{Fe} \ln y_{Fe} + y_{Cr} \ln y_{Cr} + y_{Ni} \ln y_{Ni})$$
$$+ cRT(y_N \ln y_N + y_{Va} \ln y_{Va}) + {}^EG_m + G_m^{mg},$$

where the colon separates the elements of different sublattices, $^{o}G_{M:Va}$ relates to the pure metals M = Fe, Cr, Ni in a non-magnetic state, $G_m^{mg}$ represents the contribution of the magnetic ordering to interatomic interaction (Hillert and Jarl 1978), $^{o}G_{M:N}$ is the Gibbs energy for a hypothetical state $M_1N_1$.

The exess Gibbs energy $^{E}G_m$ contains, according to Eq. (2.6), the interaction parameters $L_{Fe,Cr:N}$, $L_{Fe,Cr:Va}$, $L_{Fe,Ni:N}$, ..., $L_{Fe:N,Va}$, $L_{Cr:N,Va}$, ..., $L_{Fe,Cr,Ni:Va}$, $L_{Fe,Cr:N}$, $L_{Fe,Cr:N,Va}$, $L_{Cr,Ni:N,Va}$ which are being fixed based on the experimental assessments or varied during the procedure of minimising the Gibbs energy.

Certain limitations in the calculations follow from the well known experimental data for the binary alloys.

***The fcc phase*** in the binary Fe-N system has to be limited by the composition $Fe_1N_{0.25}$, not $Fe_1N_1$, corresponding to the $\gamma'$ phase as follows from the Fe-N phase diagram (see Fig. 2.4). This limitation originates from the strong repulsion between nitrogen atoms in the first coordination sphere of the fcc lattice, which was considered in detail in Sect. 1.1.2.5 while discussing the distribution of nitrogen in austenite. The solubilities of chromium and nickel in the $\gamma'$ phase are neglected. In the Cr-N system, the fcc phase does not exist and the Cr-N interaction in the fcc phase is evaluated from the experimental studies of fcc Fe-Cr-N alloys. In the Ni-N system, the fcc phase is the most stable, although the $\varepsilon$ phase exists for the range of nitrogen contents from 0.15 to 0.25 mass % below 1700 and 500 K, respectively (see Fernandez Guillermet and Frisk 1991).

***The bcc solution*** of nitrogen in iron, chromium and nickel, exists only at low nitrogen contents because of the small size of the octahedral holes in the bcc crystal lattice and, as a result, of the easy formation of nitrides. No deviation from Henri's law is observed and, therefore, the interaction between nitrogen atoms is usually neglected.

***The ε phase*** with the ideal composition $FeN_{0.5}$ has, as stated above, an hcp crystal lattice. Therefore, the hcp state is introduced as a reference one for each substitutional element, even if the hcp lattice cannot be formed in pure metals. The formula $M_1N_{0.5}$, instead of $M_1N_1$, is again explained by the repulsion of nitrogen atoms in the nearest neighbourhood and by the close arrangement of the neighbouring interstitial sites in the hcp lattice (see also corresponding remarks in Sect. 2.2). The solid solution based on this composition in Fe-Cr-N and Ni-Cr-N systems has two elements in each sublattice and can be represented as $(Fe,Cr)_1(Va,N)_{0.5}$ or $(Ni,Cr)_1(Va,N)_{0.5}$. A thermodynamic analysis shows a tendency for a miscibility gap, i.e. for the separation into two compounds (see Frisk and Hillert 1989). For the Fe-Cr-N system, it results in $Fe_1Va_{0.5}$ +$Cr_1N_{0.5}$ decomposition.

***The cubic* CrN *nitride*** has a narrow concentration-temperature range. It is known to dissolve a certain amount of iron and is represented by the formula $(Cr,Fe)_1N_1$.

The solubility of nickel in it is usually neglected. The $CrN$-$Cr_2N$ equilibrium was studied by Mills (1970) and Kotlar et al. (1973).

***The π nitride*** with a β-manganese structure has been discovered and studied in detail by Kikuchi and colleagues (see Kikuchi et al. 1973, 1981, Kajihara et al. 1993). It is transformed from the di-chromium nitride $Cr_2N$, the latter being, in fact, an intermediate precipitate, not the final one. The ideal π nitride has the composition $Cr_3Ni_2N$. For the Fe-Cr-Ni-N system it was described as $(Cr)_{12}(Fe_{0.5}Ni_{0.4}Cr_{0.1})_8N_4$ (see Kikuchi et al. 1989) or $(Cr)_{12.8}(Fe,Ni)_{7.2}N_4$ (Frisk 1991).

Such thermodynamical assessments have enabled one to construct the Fe-Cr-Ni-N phase diagram (Frisk and Hillert 1989, Hertzman 1987, Frisk 1991), and some important results of the calculations are shown in Figs. 2.10 and 2.11 after Frisk and Hillert (1989) and Frisk (1991).

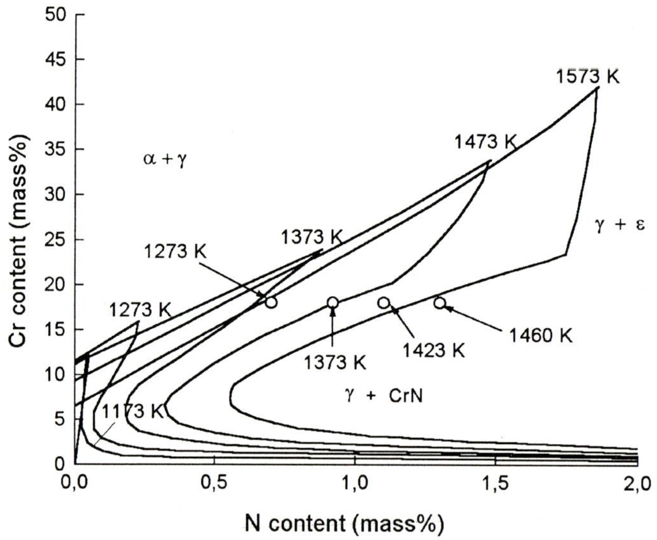


**Fig. 2.10.** Extension of the γ single-phase field in Fe-Cr system at various temperatures (after Frisk and Hillert 1989). The experimental data for 18 % of Cr are shown according to Ustinovshikov et al. (1996)

It is seen that the higher the temperature, the more nitrogen and chromium can be dissolved in the γ iron (Fig. 2.10). However, the agreement with the experimental data (Ustinovshikov et al. 1996) is rather poor. In the alloy Fe-18 mass % Cr, austenite dissolves more nitrogen and the γ single-phase region exists at lower temperatures as predicted by the calculations.

The extension of the γ phase in the Fe-Cr-Ni-N system at 1273 K is shown in Fig. 2.11 for different contents of nickel. One can conclude that the solubility of nitrogen in the γ phase decreases with an increase in the nickel content. The results of experimental measurements of the fcc/fcc + hcp boundaries performed by Hertzman (1987) for different nickel contents are shown to be consistent with the calculations.

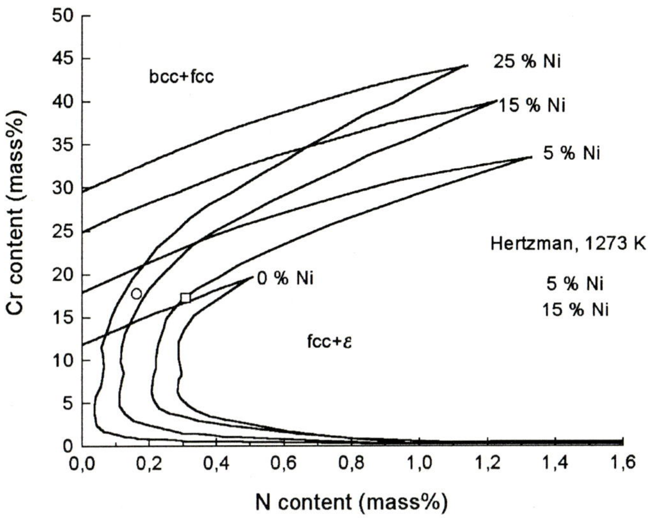


**Fig. 2.11.** Extension of γ single-phase field in Fe-Cr-N at 1273 K and at various Ni contents (after Frisk 1991). Experimental data by Hertzman (1987) are shown

***Experimental data on the nitrogen solubility in* Fe-Cr-Ni *austenite.*** The solubility of nitrogen in liquid and solid Fe-Cr-Ni alloys has been studied by Feichtinger et al. (1989) and Kikuchi et al. (1989). An important conclusion is drawn from the studies of the temperature dependence of the nitrogen solubility in different phases in equilibrium with gaseous nitrogen at 1 bar (see Fig. 2.12).

The addition of chromium to Fe-Ni-N alloys causes not only an increase in the solubility of nitrogen in all phases, but, moreover, makes the nitrogen solubility in solid phases higher than in the liquid one, and, therefore, the solidification of these steels is not accompanied by an outgassing of nitrogen, which would result in porosity of the ingots. Experimental data by Wada and Pehlke (1977) for the nitrogen solubility in the liquid phase of different Fe-Cr-Ni compositions are shown in the same Figure. A clear consistency exists betweeen these experimental data and thermodynamical calculations performed by Frisk (1991).

The nitrogen solubility in the solid state is particularly important because beneficial properties of nitrogen-containing stainless austenitic steels are mainly due to nitrogen in solid solution, not in nitrides. As nitrogen belongs to gaseous elements, it is necessary to distinguish between nitrogen solubilities in equilibrium with gaseous nitrogen or with nitrides.

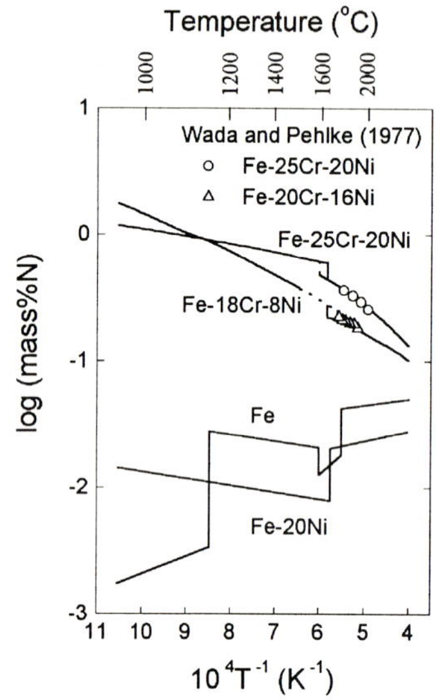


**Fig. 2.12.** Temperature dependence of the nitrogen solubility at 1 bar pressure of gaseous nitrogen in iron and different austenitic steels (after Kikuchi et al. 1989 and Frisk 1991). The experimental data by Wada and Pehlke (1977) for Cr25Ni20 and Cr20Ni16 steels are also shown

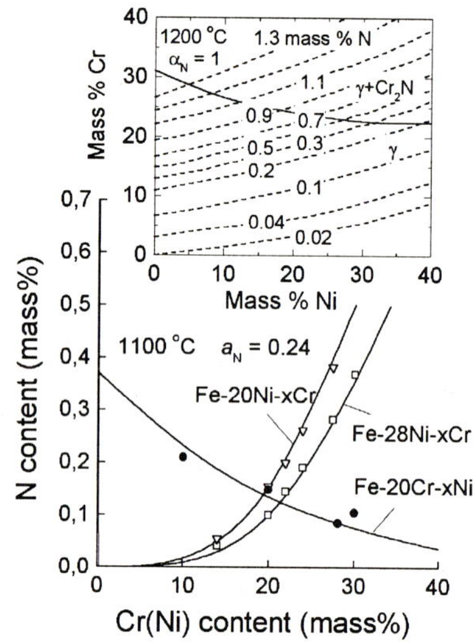


**Fig. 2.13.** Effects of chromium and nickel on the gaseous nitrogen solubility at 1100 °C and activity $a_N$ = 0.24 in 20(28)Ni and 20Cr austenitic steels, respectively (after Kikuchi et al. 1989). The insert shows effects of chromium and nickel on the nitrogen solubility at 1 bar pressure of nitrogen and temperature 1200 °C. Iso-concentration curves (dashed lines) and γ-(γ + $Cr_2N$) phase boundary (solid line) were calculated by Kikuchi et al. (1989) based on the experimental data

Some fundamental studies in this relation were performed by Kikuchi et al. (1989). They used a seal capsule technique where the samples together with some nitrogen-containing material were held at different temperatures for a long time to provide equilibrium. The nitrogen activity was easily varied in these experiments by choosing the type and amount of the nitrogen supplier. The effect of chromium and nickel on the solubility of gaseous nitrogen at 1100 °C at a nitrogen activity of $a_N = 0.24$ is shown in Fig. 2.13.

Chromium increases and nickel decreases the nitrogen solubility in Fe-Cr-Ni austenite. The solid iso-activity lines in Fig. 2.13 represent the results of thermodynamic calculations which are consistent with the experimental data. Kikuchi et al. (1989) have also calculated the iso-concentration curves of nitrogen solubility in the Cr-Ni steel at 1200 °C, the chromium and nickel contents being varied (see the insert in Fig. 2.13). It is shown that, at some chromium and nickel contents, more than 1 mass % of nitrogen can be introduced into steel in equilibrium with gaseous nitrogen at 1 bar pressure, however, there is some limit for the dissolution of nitrogen in austenite and, at high nitrogen contents, the chromium ε-nitride precipitates. Therefore, it is the solubility of the nitride, rather than that of gaseous nitrogen, that determines the nitrogen content in austenitic matrix.

The method of lattice parameter measurements is shown to be also an effective tool for the study of nitrogen solubility in austenite in equilibrium with nitrides (see e.g. Kikuchi et al. 1981, 1989). After some solution treatment providing the dissolution of nitrogen in austenite, samples were exposed to lower temperatures for a time satisfying the equilibrium conditions, and a decrease in the lattice parameter versus the ageing time represents the precipitation of nitride from the solid solution.

The experimental data obtained in such a way for two austenitic Cr-Ni steels, Cr25Ni20 and Cr25Ni28, are shown in Fig. 2.14.

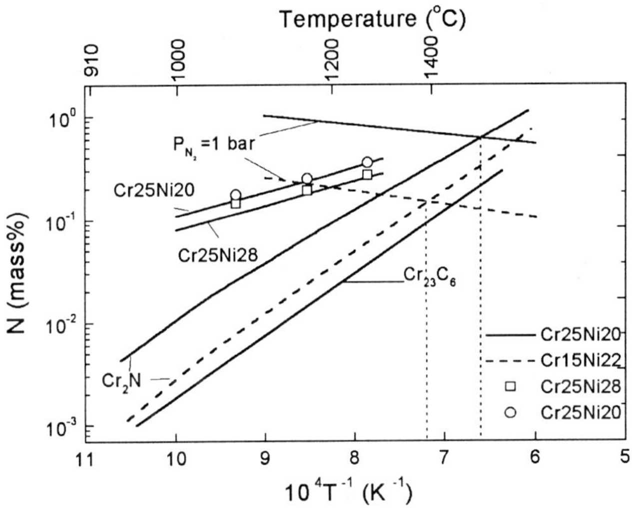


**Fig. 2.14.** The temperature dependence of nitrogen solubility in equilibrium with $Cr_2N$ nitride in the steels Cr25Ni20 and Cr25Ni28. Experimental data after Kikuchi et al. (1989), calculated curves after Frisk (1991). The calculated data for the temperature dependence of the nitrogen solubility in the steels Cr25Ni20 (solid lines) and Cr15Ni22 (dashed lines) in equilibrium with $Cr_2N$ nitride and with the nitrogen gas at 1 bar pressure are also presented after Kikuchi et al. (1989). The solubility of $Cr_{23}C_6$ carbide in Cr25Ni20 (solid line) is shown for the comparison

One can conclude that, with an increase of the nickel content, the solubility of nitrogen in austenite in equilibrium with ε-nitride changes in the same manner as in the liquid, i.e. it decreases. At the same time, a marked difference from the binary Fe-N diagram (Fig. 2.4) can be noted: the solubility of ε-$Cr_2N$ nitride in the Fe-Cr-Ni austenite increases with temperature in contrast to the solubility of ε-$Fe_3N$ in the iron.

The data calculated by Frisk (1991) for the fcc/fcc+ε phase boundary in the steels Cr25Ni28 and Cr25Ni20 are shown in the same Figure as solid lines. The agreement with the experiment is quite good for high temperatures, but, as the temperature decreases, the discrepancy between the calculations and the measurements becomes significant.

Figure 2.14 contains also the comparison between the solubilities of nitrogen in austenite in equilibrium with gaseous nitrogen and ε-nitride at different temperatures for the steels Cr25Ni20 (solid lines) and Cr15Ni22 (dashed lines). The conclusion can be made that, at high temperatures, the equilibrium with gaseous nitrogen determines the nitrogen content in austenite, whereas the solubility of ε-nitride is decisive at lower temperatures, and the transition temperature (shown by vertical dotted lines) falls with a decrease in the chromium content.

Finally, it is interesting to compare the solubilities of chromium nitrides and carbides in austenite, because it is this point that explains a reason for the substitution of carbon by nitrogen in steel. The results of calculations performed by Kikuchi et al. (1989) for the austenitic steel Cr25Ni20 are presented in Fig. 2.14. In the entire practically important temperature range, the solubility of ε-$Cr_2N$ nitride exceeds that of $Cr_{23}C_6$ carbide. For this reason, austenitic nitrogen steels reveal a higher resistance to sensitisation treatments at critical temperatures.

The π-nitride in high-chromium high-nickel austenitic nitrogen steels was shown to be formed from the ε nitride during a long-time ageing at temperatures below 1000 °C (Kikuchi et al. 1989, Kajihara et al. 1993). It was never observed in Cr18Ni8 nitrogen-containing steel. An influence of alloying by Mo, W and Si was studied and an increased stability of π-nitride due to these elements has been shown.

## 2.3.3 The Fe-Cr-Mn-N phase diagram

The substitution of Ni by Mn in stainless austenitic steels eases alloying by nitrogen because both Cr and Mn increase the solubility of nitrogen in liquid iron and in austenite. That is why Cr-Mn-N compositions represent an important group of high nitrogen steels. We describe the Fe-Cr-Mn-N system briefly following the thermodynamic analysis given by Frisk and Qiu (1994) based on the previous descriptions of lower order systems. The Mn-N system was studied by Qiu and Fernández Guillermet (1993). Thermodynamical assessments of the Cr-Mn-N and Fe-Mn-N systems were performed by Frisk (1993) and Qiu (1993), respectively.

***The liquid phase*** was supposed to be a substitutional solution. Experimental data by Satir-Kolorz et al. (1991), Pomarin et al. (1983), Nemchenko et al. (1968) were used to adjust the interaction parameter $L^{liq}_{Cr,Fe,Mn,N}$. The calculated solubility of nitrogen at two temperatures is shown, after Frisk and Qiu (1994), as a function of the pressure in Fig. 2.15 in comparison with experimental data. The solubility of nitrogen decreases with an increase in temperature and increases with the nitrogen gas pressure. As expected, chromium assists a higher equilibrium content of nitrogen in the liquid steel.

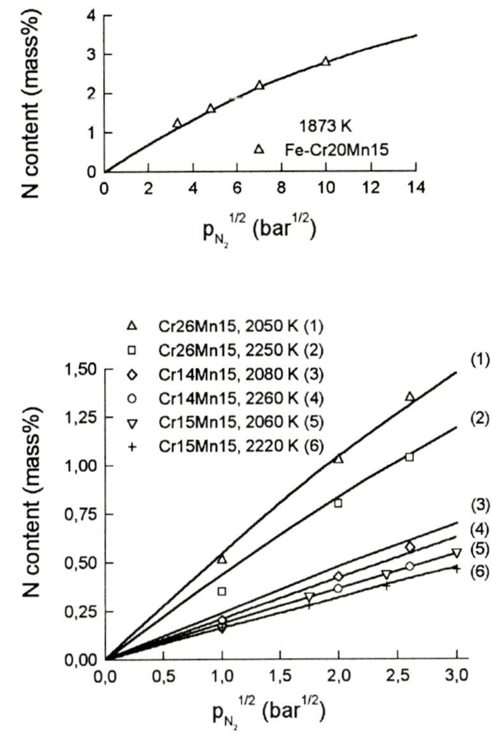


**Fig. 2.15.** Nitrogen solubility in Fe-Cr-Mn liquid alloys as calculated by Frisk and Qiu (1994) and measured by Satir-Kolorz et al. (1991) for the steel Cr20Mn15 at 1873 K and by Pomarin et al. (1983) for steels Cr(x)Mn15 at temperatures above 2000 K

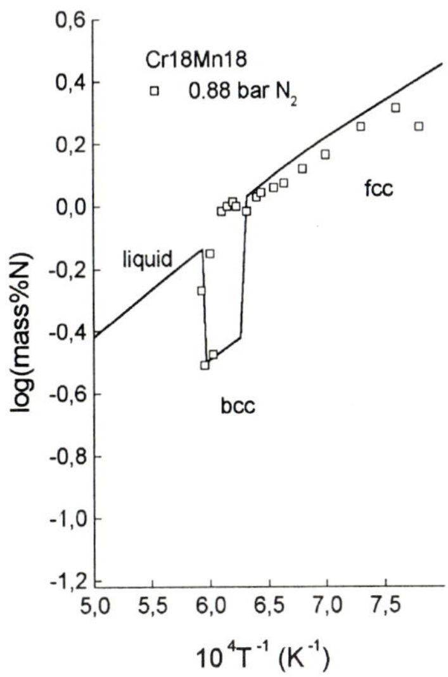


**Fig. 2.16.** Temperature dependence of nitrogen solubility in different phases for the steel Cr18Mn18 as calculated by Frisk and Qiu (1994) and measured by Zheng (1991) in equilibrium with gaseous nitrogen at $p_{N2} = 0.88$ bar

***The solid state.*** The temperature dependence of nitrogen solubility in different phases is shown in Fig. 2.16 according to Frisk and Qiu (1994) for the steel

Cr18Mn18. The authors have used the experimental data by Zheng (1991) for the comparison with calculations.

It is important to note that, according to the experimental data, the δ-ferrite gap in nitrogen solubility seems to disappear with increasing chromium and nitrogen contents, as is the case for Fe-Cr-Ni-N compositions (see Fig. 2.12), whereas the calculated curves in Fig. 2.16 show a significant decrease in the nitrogen content of the bcc phase as compared to the liquid and fcc phases. This disagreement between the calculations and the experiment was explained by Frisk and Qiu (1994) as a result of the underestimated nitrogen solubility in the bcc phase of the ternary Fe-Mn-N and Cr-Mn-N systems in the calculations.

## 2.4 Precipitates in austenitic steels

Austenitic steels are usually constructed as single-phase γ solid solutions where precipitates are undesirable, exept for some special cases. Unfortunately, service of stainless austenitic steels at elevated temperatures is accompanied by precipitating thc phascs many of which dctcrioratc mcchanical and corrosion properties. In this section, we discuss the precipitation reactions in austenitic nitrogen steels. Two kinds of precipitates are met in austenite: intermetallic phases and nitrides/carbides.

### 2.4.1 Intermetallic phases

A detailed review of the intermetallic phases formed between transition elements belonging to so-called B-group (manganese, iron, cobalt and nickel) and A-group (titanium, vanadium, chromium, molybdenum etc.) was performed by Nevitt (1963). An array of these phases for the iron-based alloys includes ordered structures $B_3A$ (γ′), Laves phases $B_2A$ (η), CsCl type phases BA (β), mu-phases $B_7A_6$ (μ), chi (χ)- and sigma (σ)-phases of complicated compositions. In stainless austenitic steels, one meets σ, χ and η phases mainly. The temperature range of their precipitation is expanded from 600 to 1150 °C. The structure and properties of these phases were reviewed by Pridantsev et al. (1969), Novak (1977), Krauss (1990) etc. Conditions of their formation in austenitic steels can also be found in some original papers (see e.g. Wiegand and Doruk 1962, Blenkinsop and Nutting 1967, Weiss and Stickler 1972, Barcik 1986, Machado and Padilha 1996 etc.).

***The σ phase*** is studied extensively because of its wide occurrence and notorious effect on the toughness of stainless steels. Its other harmful effect results in intercrystalline corrosion as σ is not chemically resistant.

The σ phase is described by formula $(Fe,Ni)_x(Cr,Mo)_y$ or, based on a three-sublattice model (see Sect. 2.3.2), one proposes its more complicated structure in austenitic Cr-Ni steels, $(Fe,Ni)_8Cr_4(Cr,Fe,Ni)_{18}$. It has a body-centered tetragonal crystal lattice and the values of the lattice parameters are found to vary from 0.828

to 0.921 nm (*a*) and between 0.4597 and 0.478 nm (*c*) (Blenkinsop and Nutting 1967, Weiss and Stickler 1972). The chemical compositions of σ precipitates are given in Table 2.3 along with the data for other intermetallic phases.

**Table 2.3.** Compositions (mass %) of intermetallic phases precipitated in austenitic steels

| Phase | Steel | Heat treatment | Cr | Ni | Mo | Si | Mn | Fe |
|---|---|---|---|---|---|---|---|---|
| σ | Cr18Ni11Mo1[(1)] | 700 °C 10000 h | 33.44 | 4.48 | 5.42 | 0.74 | - | - |
| σ | Cr17Ni13Mo2[(2)] | 815 °C 3000 h | 29 | 5 | 11 | - | - | 55 |
| σ | Cr12Ni12Mo2[(3)] | 675 °C 100 h | 33.7 | 4.7 | 5.8 | 0.6 | 4.5 | 47.9 |
| σ | Cr20Ni(18-25)MoMn[(4)] | 700-1050 °C 1000-4500 h | 27-32 | 8-15 | 10-16 | 0-1 | 1-9 | 35-43 |
| χ | Cr16Ni15Mo2[(5)] | | $(Fe,Ni)_{36}Cr_{18}Mo_4$ | | | | | |
| χ | Fe-Cr-Mo alloys[(6)] | 810-870°C | 16.8 | - | 26.9 | - | - | 56.2 |
| χ | Cr17Ni13Mo2[(2)] | 815 °C 3000 h | 21 | 5 | 22 | - | - | 52 |
| χ | Cr20Ni(18-25)MoMn[(4)] | 700-1050 °C 1000-4500 h | 25-32 | 7-13 | 18-25 | 0-1 | 1-8 | 29-45 |
| η | Cr17Ni13Mo2[(2)] | 815 °C 3000 h | 11 | 6 | 45 | - | - | 38 |
| η | Cr20Ni(18-25)MoMn[(4)] | 700-1050 °C 1000-4500 h | 14-20 | 7-12 | 39-44 | 1-2 | 0-5 | 27-35 |
| η | Cr8Ni15Mo10[(7)] | - | 3-6 | 4 | 44 | - | - | 46 |

(1) Wiegand and Doruk (1962)
(2) Weiss and Stickler (1972)
(3) Ritter and Henry (1985)
(4) Jargelius-Pettersson (1998)
(5) Kautz and Gerlach (1968)
(6) Kasper (1954)
(7) Bechtoldt and Vacher (1953)

In CrNi steels, the σ phase occurs usually at contents of more than 18 mass % of chromium and of nickel in the limits between 8 and 30 %. The concentration range of its appearance is often analysed in terms of the electron/atom ratio which means the number of $(s+d)$ electrons in the external shell of a free atom (7 e/a for Mn, 8 e/a for Fe etc.). The σ phase is supposed to exist within 5.6 to 7.6 e/a ratio limits. The other constraint is concerned with the atom size, although it is less important and even silicon was found to have a slight solubility in σ (see Wiegand and Doruk 1962, Ritter and Henry 1985) in spite of the large difference in atomic radii as compared to other constituents.

Alloying elements can assist the σ phase formation and(or) take part in its constitution. Silicon notoriously accelerates the precipitation of σ, which is explained by its action as an electron acceptor (Nevitt 1963, Hall and Algie 1966). Other unfavourably affecting elements are molybdenum and titanium, the latter participating more in χ and η phases. At (Cr+Ni) content about 30%, molybdenum is found to assist σ and χ formation, the former being substituted by η at an increase in the molybdenum content (see Bechtoldt and Vacher 1953).

Carbon suppresses the formation of σ, and this effect is often attributed to the formation of carbides, which is accompanied with a decrease in the content of σ-promoting elements (chromium and molybdenum) in the matrix and an enhancement of the σ-suppressing effect of nickel (Gow and Harder 1942, Morley and Kirkby 1952, Lismer et al. 1952). Nevertheless, carbon assists the formation of χ phase, of which the composition does not decisively differs from that of σ (Wiegand and Doruk 1962, Andrews and Brookes 1951). The reason for such a different behaviour can lie in no solubility of carbon in σ and a remarkable content of carbon in χ (Wiegand and Doruk 1962). Therefore, the formation of σ requires a preliminary removal of carbon atoms from the sites of its nucleation, which is achieved by the precipitation of carbides. This explanation is consistent with the experimental observations that the formation of the $M_{23}C_6$ carbides always precedes the σ precipitation in carbon-containing austenitic steels and, at the same time, carbides do not nucleate the σ phase (see e.g Duhaj et al. 1968).

The σ-forming tendency of steels was estimated by a simple empirical relation (Gow and Harder 1942):

$$(\%Cr - 16 \cdot \%C)/\%Ni.$$

The steels were considered liable to σ formation if this ratio was higher than 1.7. An effect of other elements was proposed to be accounted for using their σ-forming capability in relation to that of chromium (Hull 1973):

$$\text{Cr equivalent} = Cr + 0.31Mn + 1.76Mo + 0.97W + 2.02V + 1.58Si + 2.44Ti + 1.70Cb + 1.22Ta - 0.266Ni - 0.177Co.$$

Sometimes, in analogy with nickel-base superalloys, one uses an electron vacancy number $N_v$ where the contribution of each element is given by the product of its electron vacancy (a number of electrons which has to be added in order to fill up the $3d$-shell of the corresponding atom in the metallic state) and its atomic percentage (Woodyatt et al. 1966, Mihalisin et al. 1968):

$$N_v = 0.66Ni + 1.71Co + 2.66Fe + 4.66Cr + 9.66(Mo + W) + 5.66V + 6.66Zr + 10.66Cb.$$

The σ phase can be formed in a steel if $N_v$ is above 2.52. The value of $N_v$ = 2.88 was calculated for AISI 310 type stainless steel Cr25Ni20 (Woodyatt et al. 1966), which is consistent with an occurrence of the σ phase at this composition. Nevertheless, the practical use of "the electron vacancy number" is quite complicated because of uncertain concentrations of elements in solid solution arising from pre-precipitation phenomena.

The precipitation conditions of σ were studied in relation to the chemical compositions of steels and the occurrence of crystal lattice imperfections. The crystal structure of the σ phase is incompatible with that of the matrix and its nucleation requires a high-energy interface. That is why the nuclei of σ have no coherency with the matrix (Duhaj et al. 1968).

The σ phase is precipitated during the ageing of austenitic steels. As the annealing time is increased the following sequence of its appearance is observed: at grain boundary triple points → at grain boundaries → at incoherent twin boundaries → at the surface of oxide inclusions.

The σ phase was found to be the predominant grain boundary precipitate after long-term treatments (1000-4500 h) at 700-1050 °C in steels containing (mass %) 20Cr, 18-25Ni, 4,5-6Mo, 0-10Mn, 0-0.49N (Jargelius-Pettersson 1998). Moreover, in nitrogen-free steels the σ phase was the only precipitate in the range of 700-1200 °C and no single-phase γ field was observed (Jargelius-Pettersson 1994).

In austenitic high (Cr + Ni) steels containing more than 25% of Cr and 25 % of Ni, the precipitation of ferrite at grain boundaries is found to precede the σ formation (Duhaj et al. 1968, Singhal and Martin 1968) and, in this case, the σ phase is claimed to be transformed *in situ* from ferrite, whereas, at lower (Cr + Ni) content (16-21 % of Cr and 13-24 % of Ni), the σ phase forms directly from austenite (Singhal and Martin 1968, 1969).

In AISI 304 type of steel containing about 5 % of δ ferrite in hot rolled or solution treated states, the δ ferrite is decomposed during ageing by two eutectoid reactions (Tseng et al. 1994). Initially, the δ ferrite decomposes into $M_{23}C_6 + \gamma$. This reaction occurs along the δ ferrite/matrix interface forming a thin layer of the eutectoide and it is limited by the low content of carbon in the AISI 304 steel. The balance of the δ ferrite transforms into σ phase and austenite. An enhanced chromium concentration for σ is provided by the simultaneous formation of austenite due to the diffusion along the transformation interface.

Prior cold work accelerates the precipitation of the σ phase (Blenkinsop and Nutting 1967, Weiss and Stickler 1972, Duhaj et al. 1968, Morley and Kirkby 1952) due to an increase in the diffusion rate, although the amount of σ does not markedly changes and it increases after recristallisation only, when a fine grain structure is formed. Such a behaviour is consistent with the observation that neither dislocations nor coherent twins can serve as points for the nucleation of the σ phase.

***The χ phase*** is a carbon-dissolving compound that can be interpreted either as an intermetallic phase or as a carbide $M_{18}C$ (Goldschmidt 1967). It has a bcc α-Mn type crystal structure with a lattice parameter $a_0$ = 0.892 nm and a composition of $Fe_{36}Cr_{12}Mo_{10}$ in the Fe-Cr-Mo system (Kasper 1954) or $(Fe,Ni)_{36}Cr_{18}Mo_4$ in austenitic CrNiMnMoN steel (Kautz and Gerlach 1968, see also Table 2.3). It is also formed in the Fe-Cr-W system.

The crystal structure of χ phase has sufficiently large holes to accomodate the dissolved carbon and, in such a way, to be transformed into a carbide. Carbon stabilises this structure and assists an increasing content of stronger carbide formers (Mo, W) at the expense of Fe and Cr. A close link between the χ phase and the η-carbide $M_6C$ has to be noted. They can exist in a mutual equilibrium or substitute each other depending on the heat treatment.

Like the transition from α-manganese to β-manganese by an increase of temperature, χ-carbide can exist along with π-carbide which has a structure of β-manganese (Sect. 2.3.2). The latter contains less carbon ($M_{20}C$ or $M_{21}C$) and its decomposition results in the precipitation of χ- or η-carbides (Goldschmidt 1967).

The precipitation of χ starts at grain boundaries independent of carbides, however, after long ageing times, an increase in the amount of χ phase is accompanied with a decreasing number of carbide particles due to the redissolution of carbon in χ (Weiss and Stickler 1972). A remarkable depletion of the chromium and molybdenum contents in the matrix accompanies the grain boundary precipitation of χ. Incoherent twin boundaries as nucleation sites and intragranular precipitates were also observed after prolonged ageing.

Cold work accelerates the χ precipitation (Duhaj et al. 1968). Intragranular rod shape χ particles were observed to precipitate in the previously deformed AISI 316L steel after annealing at 800 °C for 300 h. The orientation relationship between the χ phase and austenite was determined as $(111)_\gamma \| (110)_\chi$, $[\bar{2}11]_\gamma \| [001]_\chi$.

***The η (Laves) phase*** was identified in molybdenum-containing stainless austenitic steels (Wiegand and Doruk 1962, Gerlach and Schmidtmann 1968, Thier et al. 1969, Weiss and Stickler 1972 etc.). It has a $Fe_2Mo$ composition (see also Table 2.3) and a hexagonal lattice with parameters $a$ = 0.4744 nm, $c$ = 0.7725 nm (Bechtoldt and Vacher 1953) or $a$ = 0.4755 nm, $c$ = 0.7740 nm (Thier et al. 1969).

There is some contradiction in relation to the carbon solubility in η. Wiegand and Doruk (1962) have found a significant carbon content in the chemically extracted χ+η mixture and concluded that both phases dissolve carbon from which it follows that carbon does not prevent the formation of η phase. According to Thier et al. (1969), the η phase does not contain carbon and, therefore, it can precipitate only if carbon is removed from the solid solution by carbide formation.

A variation in the ageing treatments affects insignificantly the amount and size of the η phase particles. If the solution treatment results in the formation of δ ferite, σ and χ phases will be favoured instead of the η phase.

Cold work is believed to accelerate the precipitation of η phase due to the increased diffusivity of the η forming elements (Weiss and Stickler 1972).

***A sequence of precipitated phases*** is shown in the time-temperature precipitation diagram for AISI 316 type austenitic steel Cr17Ni13Mo2 (Fig. 2.17) containing 0.023 (a) and 0.066 (b) mass % of carbon, according to Weiss and Stickler (1972). The C-curves characterise the start of the precipitation reactions.

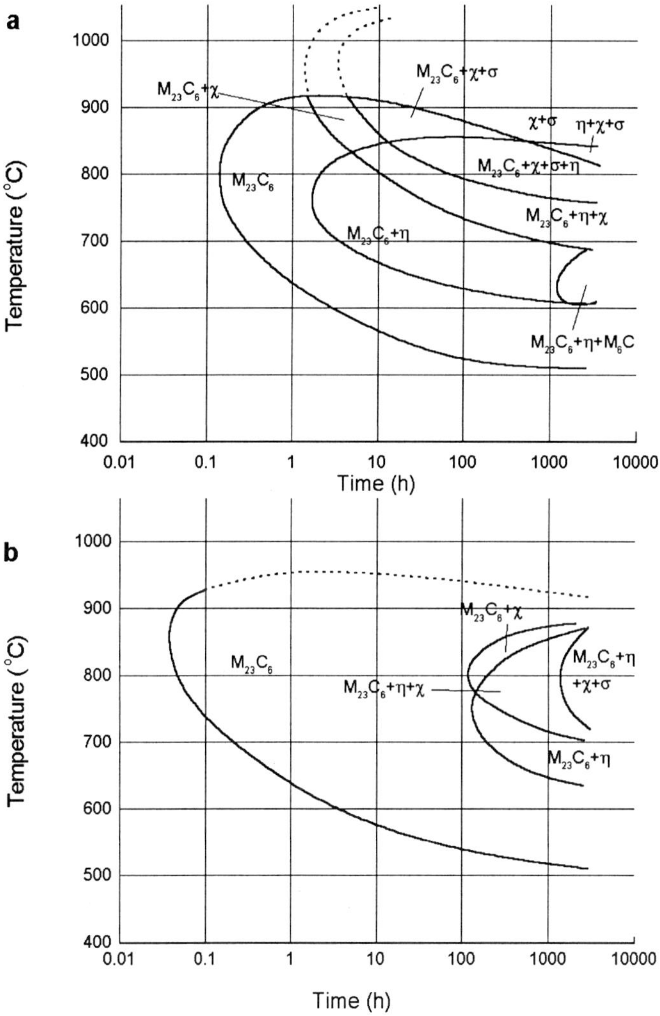


**Fig. 2.17.** Time-temperature precipitation diagrams of the steel Cr17Ni13Mo2 containing 0.023 **a** and 0.066 **b** mass % of carbon, solution treated at 1260 °C (after Weiss and Stickler 1972)

The precipitation of intermetallic phases follows that of $M_{23}C_6$ carbide in the whole temperature range, and, at low ageing temperatures, below 750 °C, the sequence $M_{23}C_6 \rightarrow \eta \rightarrow \chi$ is observed. At temperatures of about 800 °C, $\chi$ and $\eta$ are precipitated simultaneously (after ~10 h) and $\sigma$ appears after ~100 h of exposure. Above 900 °C, $\chi$ is the first intermetallic phase followed by $\sigma$. As was stated above, a variation in the solution treatment does not significantly affect the precipitation of intermetallic phases, resulting in a shift of carbide precipitation only. An increase in the carbon content retards the formation of all intermetallic phases (Fig. 2.17b).

## 2.4.2 Carbides

***$M_{23}C_6$*** is the main carbide precipitated in austenitic CrNi steels, chromium being its main metallic constituent. Other strong carbide-formers (molybdenum) can partially substitute chromium. First of all, $M_{23}C_6$ deteriorates the resistance to intercrystalline corrosion by depleting the chromium content at grain boundaries where it mainly precipitates.

A time needed for $M_{23}C_6$ precipitation at different interfaces increases in the following sequence: at grain boundaries → at non-coherent twin boundaries → at coherent twin boundaries → within grains (Čihal 1968). The morphology of $M_{23}C_6$ at grain-boundaries changes from a sheet-like one at low temperatures (below 600 °C) via feathery dendrite particles at temperatures of about 700 °C to discrete coarse particles at higher temperatures and increasing time. Non-coherent twin boundary carbides are lamellar or rod-like (Lewis and Hattersley 1965). Coherent twin precipitates are plate-like and intragranular precipitation occurs along dislocation lines.

Low-temperature morphologies are most harmful in relation to intercrystalline corrosion.

***MC carbides*** having a face-centered cubic lattice are formed in austenitic CrNi steels containing Ti or Nb. They reduce the carbon content in the matrix and prevent $M_{23}C_6$ precipitation. Along with a stabilising effect, their precipitation can be used for strengthening. After preliminary high temperature solution treatment, intragranular MC precipitates can be formed on dislocations or stacking faults, increasing creep resistance (Van Aswegen et al. 1964, Irvine et al. 1960).

TiC and TiN are isomorphous and Ti(CN) carbonitride exists in a wide range of interstitial contents.

***$M_6C$ carbide*** is formed in molybdenum-containing stainless steels and its suggested formula is $Fe_3Mo_3C$. It is supposed to be formed from $M_{23}C_6$ (Goldschmidt 1967) and has a cubic lattice with a lattice parameter $a_0 = 0.1111$ nm (Koch 1965).

In type AISI 316 steels, $M_6C$ was found only after more than 1000 h of ageing at 650 °C (Weiss and Stickler 1972). Its precipitation is accelerated by an increase in the molybdenum content, nevertheless, in nitrogen-free steel Cr17Ni13Mo5 containing 0.039 mass % N, it was found only after decomposition of the $\sigma$ phase (Thier et al. 1969).

## 2.4.3
## Effect of nitrogen on precipitation

Before discussing the precipitation behaviour of austenitic steels with nitrogen, let us make some remarks concerning an effect of alloying elements on the solubility of nitrogen and carbon in austenite. It is well known that Cr and Mn decrease while Ni increase the thermodynamical activity of nitrogen and carbon in ferrous austenites, which, in fact, means that Cr and Mn atoms attract C and N ones and Ni pushes them out of the solution. A comparative analysis of the interaction of carbon and nitrogen with the same alloying element in solid solution is difficult as standard states for both interstitial elements are completely different (see Sects. 1.1.2.1 and 1.1.2.2). This obstacle was avoided in Sect. 1.1.3.1 by analysing the thermodynamical equilibrium between binary Fe-i and ternary Fe-M-i alloys under suggestion that the equilibrium is set between interstitial atoms only.

Starting from a binary Fe-i alloy and introducing an alloying element M at a fixed concentration *i*, one can expect that, with an increasing content M, the content *i* will be decreased for repulsive *M-i* interactions and increased for attracting ones. By such an approach, an effect of M on the *i* content in Fe-M-i alloys does not depend on the choice of standard states, as the chemical potential and the activity of interstitials are fixed and are the same in binary and ternary alloys.

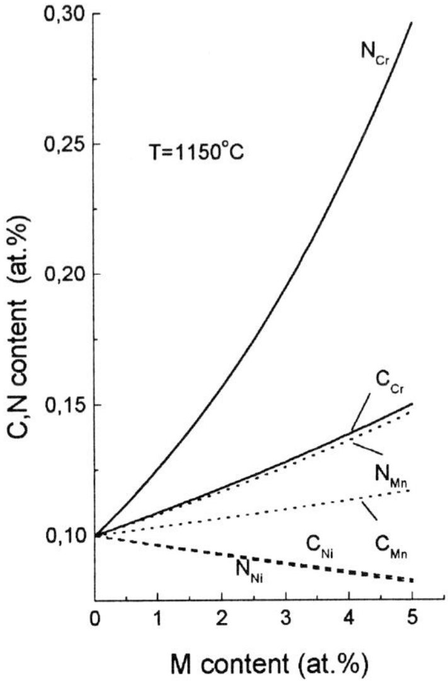


**Fig. 2.18.** Effects of chromium (solid lines), manganese (dotted lines) and nickel (dashed lines) on the solubility of nitrogen and carbon in iron at 1150 °C

Therefore, one can compare an effect of an alloying element M on the solubility of nitrogen and carbon in austenite, choosing equal concentrations in binary Fe-i alloys and analysing the *i* concentrations as functions of the M content in a ternary alloy. In order to neglect *i-i* interactions, it is better to perform this analysis for small *i* contents.

Such concentration dependences of C and N obtained using the ThermoCalc programme are presented in Fig. 2.18 for 0.1 at.% of C(N) in binary alloys at temperature of 1150 °C. It is seen that Ni decreases the nitrogen and carbon contents to an equal extent. The solubility of C(N) in austenite is increased by Cr and Mn, the Cr effect being stronger. Most important for the temperature stability of austenitic steels is that Cr and Mn provide a higher solubility of nitrogen in austenite as compared to that of carbon.

The precipitation behaviour of austenitic steels becomes completely different due to nitrogen. Along with its effect on the precipitation of carbides and intermetallic phases, chromium nitrides become an important constituent of aged austenitic steels.

***Carbides.*** Nitrogen shifts the precipitation of $M_{23}C_6$ carbide to longer times (Fig. 2.19). One can see that the stabilisation of austenite by nitrogen becomes notable at temperatures above 600 °C whereas, below 550 °C, the nitrogen effect can be even opposite.

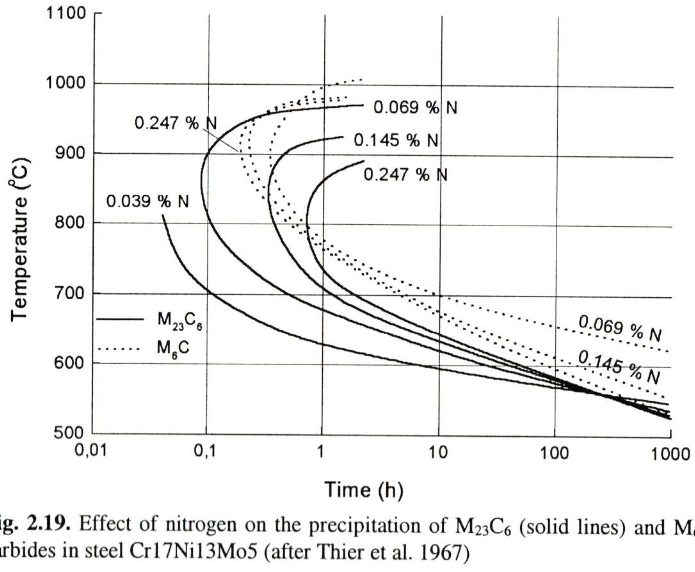


**Fig. 2.19.** Effect of nitrogen on the precipitation of $M_{23}C_6$ (solid lines) and $M_6C$ (dotted lines) carbides in steel Cr17Ni13Mo5 (after Thier et al. 1967)

The measurements of precipitation kinetics have led to the conclusion that nitrogen accelerates the chromium diffusion (Thier et al. 1969). In view of this result, a reason for the suppression of $M_{23}C_6$ precipitation by nitrogen can lie in the retardation of $M_{23}C_6$ nucleation because this chromium carbide does not contain nitrogen (Thier et al. 1969). Experimental data of zero solubility of nitrogen in $M_{23}C_6$ carbide are supported by *ab initio* calculations of the electronic structure (Antonov et al. 1995) showing that the crystal structure of chromium carbides is destabilised if some of carbon atoms are substituted by nitrogen atoms (we will also discuss this topic in Sect. 2.5). In this relation one has to remark that observations by Hsiao and Dulis (1957) of nitrogen in the carbide residuals of austenitic CrMnCN steels are perhaps due to a mixture of $M_{23}C_6$ and $M_6C$ as the latter dissolves nitrogen and progressively substitutes $M_{23}C_6$ as the nitrogen content increases.

At temperatures below 550 °C (Fig. 2.19), the acceleration of chromium diffusion due to nitrogen prevails over the retardation of $M_{23}C_6$ nucleation, as pipe diffusion contributes much to the chromium transfer, and, as a result, nitrogen assists $M_{23}C_6$ precipitation.

A conclusion of nitrogen-enhanced diffusion of chromium is consistent with the phenomenon of the formation of excessive vacancies due to interstitial nitrogen atoms discussed above (see Sect. 1.1.4.1), however, at a first sight, it contradicts with the nitrogen-caused decrease in the activity of chromium (see e.g. Jargelius-Pettersson 1998). One should note that diffusivity and activity do not always correlate to each other. The former is determined by the value of the energy barrier for a jump of chromium atoms, and an increase in the concentration of vacancies decreases this barrier. The latter depends on the relation between the free energies of phases, the surface energy of the nucleated phase etc.

An effect of nitrogen on the precipitation of $M_6C$ carbide is also controversial issue. As follows from Fig. 2.19, $M_6C$ precipitation in steel Cr17Ni13Mo5 is favoured by nitrogen, according to Thier et al. (1967). However, an opposite effect was obtained by Tuma et al. (1981) who observed a delay in the formation of $M_6C$ and $M_{23}C_6$ in steel Cr16Ni14Mo2.5 alloyed by 0.2 mass % N. According to calculations performed by Jargelius-Pettersson (1998) for the austenitic steel Cr20Ni25Mo4.5Mn1.5, nitrogen decreases the driving force for the precipitation of $M_6C$ carbide, which is confirmed by the experimental observation that $M_6C$, as well as $M_{23}C_6$, were precipitated in very small amounts in this steel.

***Intermetallic phases.*** A profound study of the $\sigma$, $\chi$, and $\eta$ precipitation in nitrogen-containing austenitic Cr17Ni13Mo5 steel was performed by Thier et al. (1967, 1969) in the sixties. Their results for a conventional nitrogen content are shown in Fig. 2.20. Unlike a similar steel containing less molybdenum (compare with Fig. 2.17), the precipitates of the $\sigma$ phase exist even in the as-quenched state because of an occurrence of $\delta$ ferrite. The start of its decomposition into austenite and $M_6C$ carbide is shown by a dashed line. The decomposition of the $\sigma$ phase results in an early formation of $\eta$ phase. Alloying by nitrogen suppresses the formation of the $\sigma$ phase.

Along with this effect, the precipitation of the $\chi$ phase is shifted by nitrogen to longer times and the temperature area of its occurrence becomes narrow (see Fig. 2.21). The $M_6C$ carbide was observed to appear instead of the $\chi$ phase with increasing nitrogen content. The $\eta$ phase precipitation in the steel Cr17Ni13Mo5 was found to be shifted to higher temperatures and accelerated by nitrogen (Fig. 2.21). This effect was attributed to the precipitation of chromium nitrides and the substitution of $\chi$ by $M_6C$ at higher temperatures (~900 °C) resulting in the removal of nitrogen from the solid solution. As a general trend, nitrogen suppresses the formation of σ, χ and η phases and such a favourable effect was ascribed by Thier et al. (1967, 1969) to the non-solubility of nitrogen in these phases.

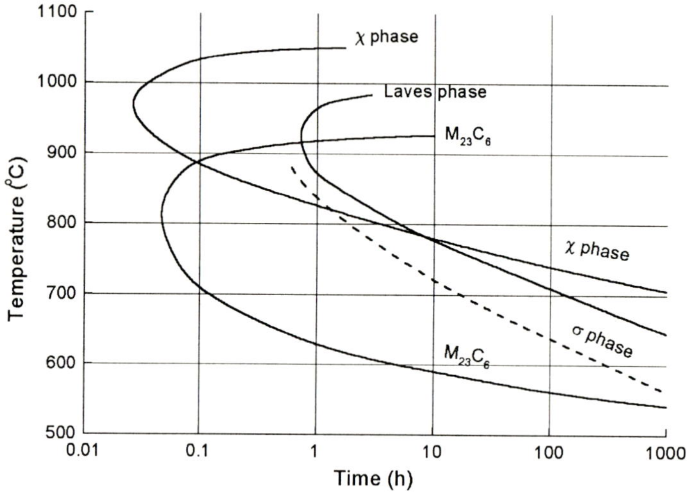


**Fig. 2.20.** Precipitation diagram of steel Cr17Ni13Mo5C0.05 containing conventionally 0.039 (mass %) of nitrogen. A dashed line shows the start of σ decomposition (after Thier et al. 1969)

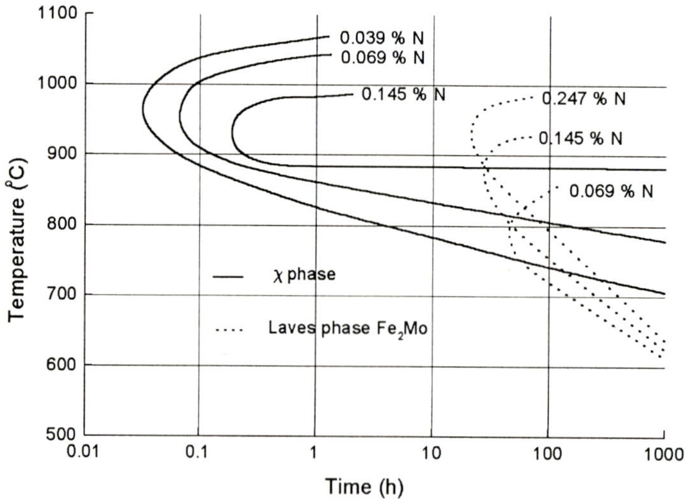


**Fig. 2.21.** Nitrogen effect on the precipitation of χ and η phases in Cr17Ni13Mo5 steel (after Thier et al. 1969)

This simple explanation was not supported in further studies. Moreover, some variety in the experimental data occurs. According to Grobner (1981), nitrogen prevents the formation of the $\sigma$ phase but assists the $\chi$ phase in steel Cr20Ni(20-30)Mo5. In contrast, Heubner et al. (1989) observed the retardation of the $\chi$ phase in steel Cr21Ni25Mo6 if the nitrogen content was increased from 0.14 to 0.19 mass %.

Brandis et al. (1976) ascribed the suppression of the $\sigma$ precipitation to a nitrogen-increased free energy of its formation, whereas the free energy of formation of $\chi$ phase was found to be independent on the nitrogen content.

The solubility of the $\sigma$ phase in austenite was calculated by Hertzman (1995) to be shifted to higher chromium contents due to nitrogen. Based on ThermoCalc calculations, Hertzman showed that nitrogen decreases the driving force for the precipitation of both phases $\sigma$ and $\chi$ in the austenitic steel Cr20Ni18Mo6. However, the $\chi$ phase was found to be less affected by nitrogen (by factor of 3, according to the calculated decrease in precipitation temperature).

Jargelius-Petterson (1998) observed, as a general trend in Cr20Ni(18-25)Mo(4.5-6)Mn(0.2-11) steels, a nitrogen-caused delay in the onset of sensitisation and a decrease in the maximum temperature of precipitation. According to her ThermoCalc calculations, nitrogen decreases the activity of chromium in austenitic steels and decreases the driving force for the precipitation of the $\sigma$, $\chi$ and $\eta$ phases.

This interpretation of the positive nitrogen effect is acceptable but cannot be exhaustive because carbon also decreases the activity of chromium, nevertheless it stabilises the $\chi$ phase. The nitrogen-increased stability of austenite due to ordering in the distribution of susbstitutional atoms can be additional reason for retarding the precipitation of intermetallic phases.

***$Cr_2N$*** substitutes $M_{23}C_6$ in nitrogen-containing austenitic steels at temperatures above 600 °C (see e.g. Kulmburg et al. 1972, Grützner 1973, Mielitynen 1979, Rashev et al. 1981, Presser and Silcock 1983 etc.). Time-temperature diagrams of $Cr_2N$ precipitation are presented in Fig. 2.22 for Cr-Ni-N and Cr-Mn-N steels as was summarised by Uggowitzer et al. (1985).

In consistency with the data of the influence of Cr, Ni and Mn on the solubility of nitrogen in ferrous austenites (see Fig. 2.18), the starting time for the precipitation of $Cr_2N$ nitride in Cr-Ni-N steels is shorter than in Cr-Mn-N ones, at equal nitrogen contents.

The precipitation kinetics and mechanisms were investigated in a number of studies. Simmons (1996) has shown that the nitride $Cr_2N$ is an initially precipitated phase in the steel Cr19Mn5Ni5Mo3N0.7 and proposed corresponding TTP diagrams. Rawers et al. (1992) alloyed 304 and 316 steels by nitrogen using the HIP process and constructed a phase diagram containing boundaries for the precipitation of $Cr_2N$ and CrN nitrides. Vanderschaeve et al. (1995) investigated mechanisms of $Cr_2N$ precipitation in high nitrogen Cr18Mn19 steel by means of transmission electron microscopy. They found that the reaction of discontinuous $Cr_2N$ precipitation is non-steady state and governed by two processes: at first by the intergranular diffusion of chromium and than by its bulk diffusion.

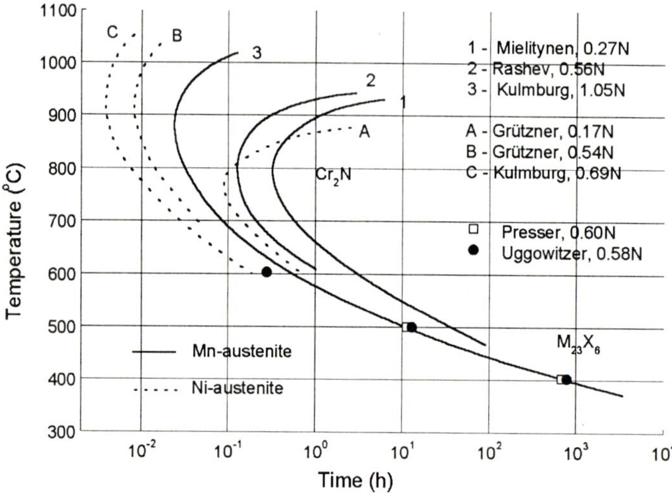


**Fig. 2.22.** Precipitation behaviour in austenitic Cr-Ni-N and Cr-Mn-N steels (after Uggowitzer et al. 1985)

A peculiarity of $Cr_2N$ precipitation in high-nitrogen austenitic steels is its thin-plate morphology due to a mechanism of discontinuous decomposition of nitrogen austenite (Kulmburg et al. 1972, Harzenmozer and Uggowitzer 1987, Rayaprolu and Hendry 1989 etc.). As a result, a cellular structure is formed starting at grain boundaries. Two different mechanisms control its formation in CrMn and CrNi nitrogen austenites.

According to Presser and Silcock (1983), the bulk diffusion of chromium governs the volume fraction of cellular precipitates in the chromium-manganese steel Cr18Mn18N0.6. This conclusion is also confirmed with some correction in the above cited study of Vanderschaeve et al. (1995). The nucleation, with a rate depending on the volume diffusion of substitutional solutes, rather than the growth of cellular precipitates was found to be a limiting step of this reaction. The orientation relationship between austenite and $Cr_2N$ precipitates was determined as

$$[110]_\gamma \parallel [0001]_{Cr_2N}, \quad [1\bar{1}2]_\gamma \parallel [2\bar{1}\bar{1}0]_{Cr_2N}.$$

In the CrNi(Mn) steels Cr25Ni20N0.59 (Kajihara et al. 1986) and Cr21Ni12Mn5Mo2N0.24 (Rayaprolu and Hendry 1989), the long range diffusion of nitrogen, not that of chromium, was confirmed to control the reaction of cellular decomposition, in accordance with the Kikuchi model. The growth of cells in these steels becomes possible due to the redistribution of nitrogen, so that the nitrogen content averaged over a cell is higher than that of the untransformed matrix. The orientations of precipitates and $\gamma$ matrix were determined by Rayaprolu and Hendry (1989) as

$$\{111\}_\gamma \parallel \{0001\}_{Cr_2N}, \quad [110]_\gamma \parallel [\bar{1}100]_{Cr_2N}.$$

This relationship between close packed planes of γ and ε structures is consistent with the splitting of dislocations in austenite needed to create the ε lattice as a sequence of stacking faults, which does not require long range diffusion of substitutional solutes for the nucleation of precipitates.

***Mechanical properties.*** Embrittlement of stainless steels caused by the precipitating intermetallic phases was studied in detail (Hull 1973) and is well known in practice. Impact toughness was found to catastrophically decrease when intermetallic phases are formed while hardness remained unchanged (Nilsson et al. 1998).

$M_{23}C_6$ precipitates at grain boundaries were found by Uggowitzer et al. (1985) to be responsible for a significant decrease in the fracture toughness of steel Cr18Mn18N0.58. An abrupt drop in fracture toughness was observed after long-time annealing in the temperature range of $M_{23}C_6$ precipitation.

The cellular pearlite-like structure of $Cr_2N$ precipitates is also unfavourable for the mechanical properties. A correlation between the extent of transformation and the deterioration of toughness, as shown in Fig. 2.23, was found by Bannykh and Blinov (1991). They proposed alloying by vanadium and a thermomechanical treatment to avoid the thin-plate morphology of nitride precipitates.

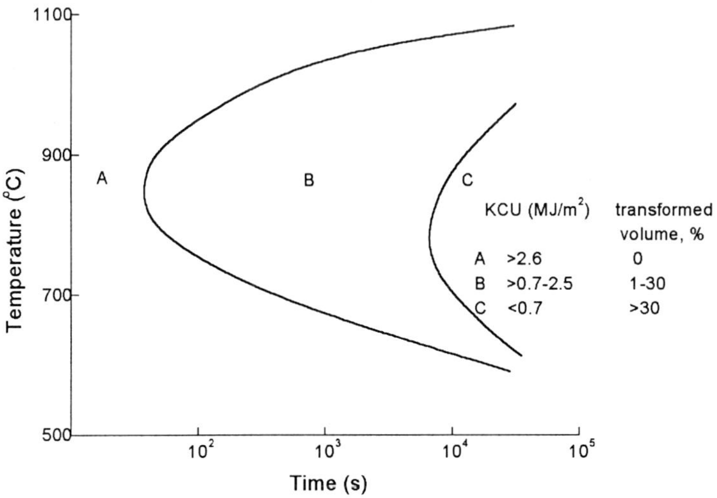


**Fig. 2.23.** Correlation between toughness and fraction of cellular structure in steel Cr22Mn14Ni8Mo1N0.7 quenched from 1200 °C and annealed at different temperatures. Areas A, B, C correspond to different fractions of the pearlite-like structure that determine the values of impact toughness (after Bannykh and Blinov 1991)

Grain refinement of high-nitrogen austenitic steels utilizing a grain-boundary precipitation reaction was studied by Tanaka (1994). Thermal cycling that was composed of the ageing at 700 and 900 °C and short-time resolution heating at 1200 °C led to a higher number of pearlite-like nodules along grain boundaries and, as a result, to an about three times smaller grain size.

## 2.4.4 Electron concept for stability of austenitic steels

Traditionally, electronegativity and atomic size of the alloying components are used to determine their solid solubility in alloys. The first one is obtained from ionisation energy and electron affinity or from thermochemical data. It characterises the electron transfer between solute atoms in solid solution. It is generally accepted that a difference of more than ±0.4 units of electronegativity or ±15% in atomic size restricts solid solubility. These empirical criteria are not always valid, e.g. in relation to transition metal alloys, which is just the case of austenitic steels.

The concept of an electron vacancy number, i.e. of the number of vacancies (holes) for electrons above the Fermi level in the *d*-band, is used to predict the tendency for precipitation of intermetallic phases in transition metal based alloys (see Sect. 2.4.1). In its physical meaning, it is close to the approach based on the difference of electronegativity, and uncertainties in the practical applications arise from the missing size factor.

A new method for the evaluating of phase stability in transition metal alloys was proposed by Morinaga et al. (1984, 1985) (see also Ezaki et al. 1986) based on the *d*-orbital levels of pure alloying metals. Their concept has its origin in the fact that, in most transition metals, some peak of state density exists just above the Fermi level (see, as an example, Fig. 2.24).

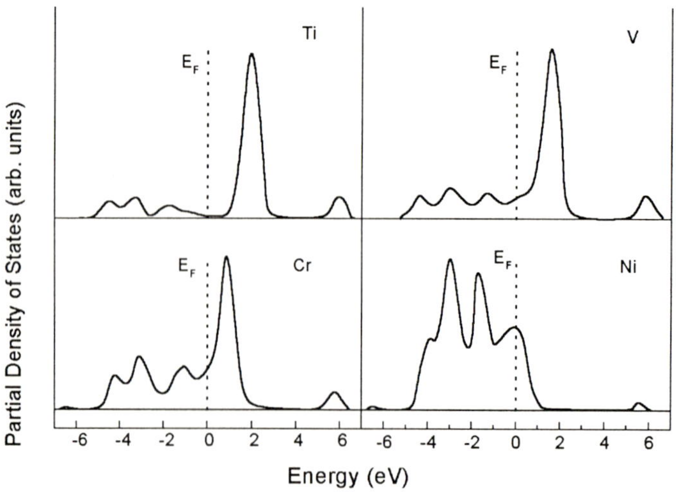


**Fig. 2.24.** Density of states for *d* electrons in 3*d* transition elements Ti, V, Cr, Ni as calculated by Morinaga et al. (1984). Positions of the Fermi level are shown by dotted lines

Energy levels within this peak can be virtually occupied by electrons forming so-called resonance states: the same electrons can partially fill the states above and below the Fermi level. Such states are a source of crystal lattice instability. If an alloying element like Ti, V, Cr, Mo etc. is added to Ni or Fe, such an "impurity" level above the Fermi one is being created in the alloy ($M_d$ level, according to the denotion by Morinaga et al.). If the position of this level is critically high, it increases the energy of the crystal lattice. Its energy can be decreased due to a change in the crystal structure removing extra-high energy levels, i.e. due to a phase transition or precipitation.

**Table 2.4.** Values of $M_d$ level for different elements as calculated by Morinaga et al. (1985)

| Element | Al | Si | Ti | V | Cr | Mn | Fe | Co | Ni | Cu |
|---|---|---|---|---|---|---|---|---|---|---|
| $M_d$ (eV) | 1.900 | 1.900 | 2.271 | 1.543 | 1.142 | 0.957 | 0.858 | 0.777 | 0.717 | 0.615 |
| Element | Zr | Nb | Mo | Hf | Ta | W | Re | | | |
| $M_d$ (eV) | 2.944 | 2.117 | 1.550 | 3.020 | 2.224 | 1.655 | 1.267 | | | |

Morinaga et al. have calculated the positions of electron levels above the Fermi level for pure transition metals (Table 2.4) and have shown that the energy of $M_d$ level correlates with the values of electronegativity and metallic radius. In other words, the proposed approach takes into account both electronegativity and atomic size.

Next, the authors have introduced some empirical value of $M_d$ for alloys as a compositional average of its values for the alloy components

$$\overline{M}_d = \sum_i^n x_i (M_d)_i \quad ,$$

where $x_i$ is the atomic fraction and $(M_d)_i$ is the $M_d$ value of the $i$-th component. If $M_d$ increases above some critical value, i.e. if the energy of electrons at the "impurity" level above the Fermi one becomes critically high, a phase instability occurs and precipitation takes place.

As an example, the application of the $M_d$ concept for Fe-Cr-Ni alloys is shown in Fig. 2.25. A dash-dotted line in the ternary diagram represents an iso $M_d$ curve corresponding to the changed alloy composition. It is seen that the value of $M_d$ = 0.905 describes the $\gamma/\gamma+\sigma$ phase boundary satisfactorily, whereas the electron vacancy concept (the dashed iso $N_v$ line for a critical value of $N_v$ = 2.49) is hardly consistent with the experimental data.

The $M_d$ concept was used by Uggowitzer et al. (1993) to predict $Cr_2N$ and $\sigma$ phase precipitations in austenitic nitrogen steels (see Fig. 2.26). An area of precipitation-free compositions is restricted by the lines showing $Cr_2N$ and $\sigma$ precipitations. To avoid the precipitation of $Cr_2N$ nitrides, the contents of Cr, Mn and Mo have to be increased and the Ni content limited, in accordance with the $M_d$

values of these elements. On the other hand, the critical values for σ precipitation restrict the Cr, Mn and Mo contents whereas the Ni content can be increased. As follows from Fig. 2.26, the allowed area of compositions is shifted towards higher $M_d$ values, i.e. to higher Cr and Mo contents, if the nitrogen concentration is increased.

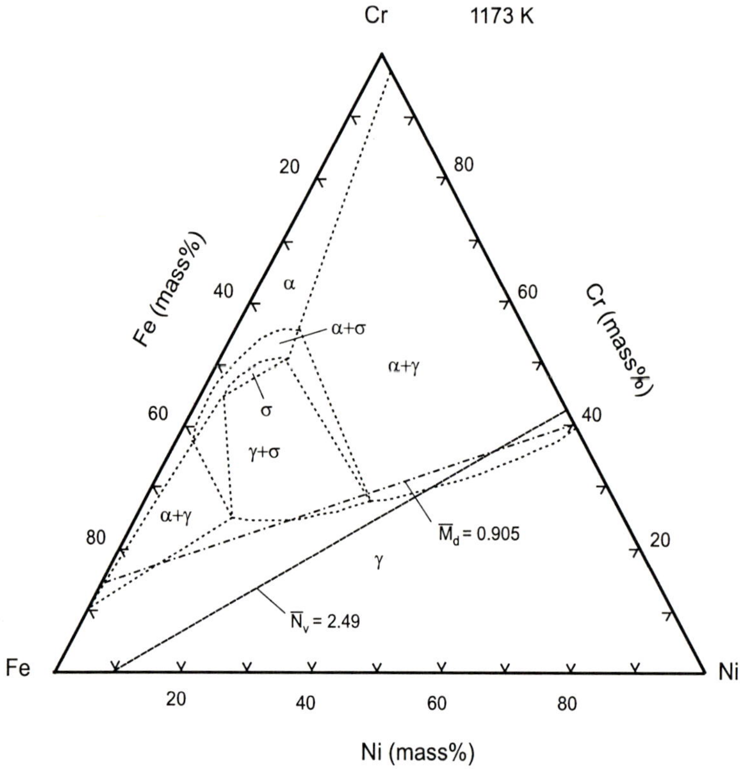


**Fig. 2.25.** Ternary Fe-Cr-Ni phase diagram at 1173 K. The dash-dotted curve represents chemical compositions corresponding to the value of $M_d$ = 0.905 which is critical for the stability of the γ phase. The iso $N_v$ line for a critical value of the electron vacancy number $N_v$ = 2.49 is also shown (dashed line). After Morinaga et al. (1985)

Thus, practical applications of the $M_d$ concept are shown to be successful. At the same time, some uncertainties arising from this model are worth noting.

First, the DV-Xα cluster method used by Morinaga et al. in their calculations is not an *ab initio* one, and some problems with the results of calculation may be caused by discontinuities of the wave functions at boundaries of the chosen atomic cluster. Second, the $M_d$ values for Si and Al are just derived from the correlation between $M_d$ levels and metallic radii and have no physical meaning as these elements have no *d*-electrons. For the same reason, the applicability of this concept to nitrogen is doubtful. Some shift of *d*-levels in iron-based alloys can be only caused by a dilation of the crystal lattice if nitrogen is introduced.

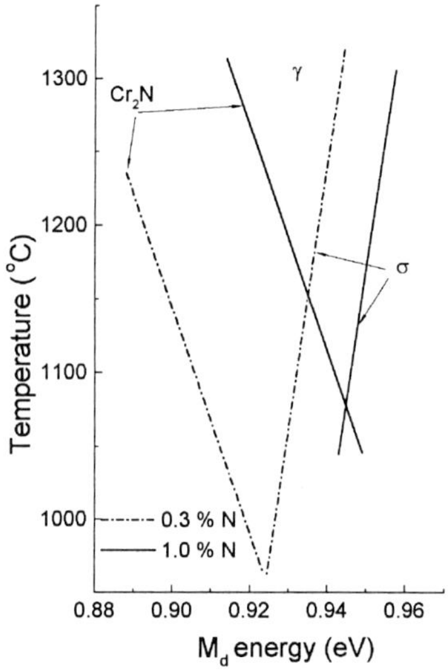


**Fig. 2.26.** Use of the $M_d$ concept for the prediction of instability of austenitic nitrogen CrNiMnMo steels at solution temperature 1150 °C. With increasing nitrogen content, the "allowed" area of chemical compositions providing single γ phase is shifted towards higher $M_d$ values, which means higher Cr and Mo contents. After Uggowitzer et al. (1993)

Nevertheless, semi-empirical rules successfully tested on practically important alloy systems are likely to be very useful for the design of alloys having complicated chemical compositions.

## 2.5 Precipitates in martensitic steels

Nitrogen is found to provide excellent properties for corrosion, temper and creep resistant martenistic steels (see Sect. 5.1). A partial substitution of carbon by nitrogen is used in practice to stay within the pressure limit of existing melting units during ferritic solidification, which is connected with a reduction of nitrogen solubility.

A reason for improved mechanical and corrosion properties of martensitic nitrogen steels lies in the fine and homogeneous distribution of precipitates having a smaller size as compared to carbides. It is obviously concerned with diferent atomic distributions in as-quenched and tempered martensites.

The following microstructural peculiarities are caused by nitrogen in martensitic steels alloyed by 9 to 12 mass % of chromium.

A mixed twinned and lath martensitic structure is obtained after quenching. If solution temperatures are below 1200 °C, undissolved Nb- and V-rich cubic MX type precipitates exist in the as-quenched state. At temperatures above 500 °C, Cr-rich hexagonal $M_2X$ phase is precipitated, while fine (Nb,V)X nitrides appear after tempering at 600–700 °C, according to Anthamatten et al. (1989, 1990). The authors do not mention the precipitation of cubic CrN nitrides.

In some contrast to these data, along with (NbV)(CN) precipitates, very fine (few nanometers) cubic (FeCrV)(CN) particles were identified by Vanderschaeve et al. (1993) in steel tempered at 500 °C. They observed also $Cr_2N$ and $(FeCr)_{23}C_6$ precipitates after tempering above 600 °C.

Cubic MN nitrides were found by Goecmen et al. (1996) in carbon-free 9–12%Cr steels tempered at ~700 °C.

Alloying with nitrogen is shown to retard the precipitation of $M_{23}C_6$ carbides and, at an N/C ratio of more than 3.0 in 9%Cr steel, $M_{23}C_6$ precipitates do not appear (Ernst et al. 1990). Steels studied by these authors contained also about 2% of W, which caused an undesired precipitation of a $Fe_2W$ type Laves phase during creep tests, and nitrogen was found to accelerate its formation. $M_6C$ carbide was precipitated in nitrogen-free steel after several hundred hours of creep exposure at 700 °C, whereas nitrogen completely suppressed its precipitation. Thus, it seems that the effect of nitrogen on the precipitation of Laves phase and $M_6C$ carbide in martensitic steels does not always correlate with its influence in austenitic steels (see Sect. 2.4.3).

Some discrepancies between above mentioned results may be caused by the difference in the chemical compositions.

A new Cr15(C+N) martensitic steel was proposed for corrosion-resistant rolling contact bearings (Berns et al. 1990, see in more detail Sect. 5.1.1). Based on the data of nitrogen and carbon distribution in austenite, as described in Chap. 1, one can expect that the tempering behaviour of nitrogen and carbon atoms in martensite has to be different. As was mentioned in Sect. 1.2.2, the redistribution of nitrogen atoms in binary Fe-N martensite during the early stages of ageing is characterised by short range ordering, whereas clustering of carbon atoms precedes the precipitation in Fe-C martensite. Moreover, the ageing of Fe-(C,N) martensite results in a separate precipitation of low-temperature carbide and nitride (Van Genderen et al. 1992). One can assume that the precipitation sequence depends on the N/C ratio. Along with this, the presence of strong carbide and nitride formers in alloyed Fe-(C,N) martensites (like chromium, vanadium etc.) has to affect the partitioning of nitrogen and carbon between the precipitates and their final type. Thus, one can suggest a very complicated competition between carbon and nitrogen atoms in their redistribution during the low-temperature ageing and in the formation of precipitates at higher tempering stages.

Based on the concept of nitrogen-induced short range atomic ordering (see Chap. 1), we will consider a correlation between short range atomic order and the precipitation behaviour in martensitic nitrogen and carbon chromium steels in an attempt to clarify a physical reason for the favourable influence of nitrogen on the properties of tempered martensite (see also Berns et al. 1994, 1995, 1997).

***As-tempered hardness.*** The evolution of hardness due to tempering reveals a notable difference between carbon, nitrogen and carbon+nitrogen martensites (see

Fig. 2.27). In the as-quenched state, the hardness of carbon martensite is the highest. However, it decreases monotonously after tempering and the effect of secondary hardening is not significant. In contrast, the hardness of as-quenched nitrogen martensite is lowest, whereas it increases due to tempering and reaches a maximum value after exposure at 450 °C. A two-stage tempering behaviour occurs in the carbon+nitrogen martensite. Between 100 and 300 °C the hardness drops as in the carbon grade and above 300 °C it increases as in the nitrogen grade.

This effect of nitrogen on secondary hardening of martensitic stainless steels can be significantly increased by additional microalloying with vanadium and niobium which cause a complicated picture of precipitation during tempering (see e.g. Berns and Xie 1992).

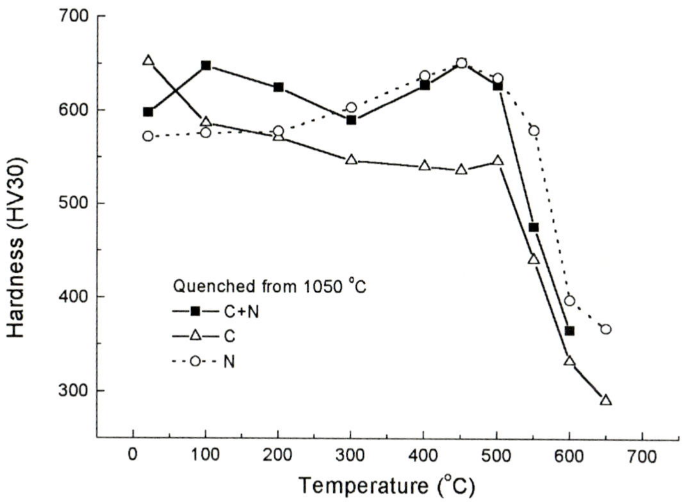


**Fig. 2.27.** Effect of tempering (consecutively 2 h at each temperature) on the hardness of Cr15Mo1 steels alloyed with 0.6C, 0.62N or 0.29C+0.35N. Quenched after solution treatment at 1050 °C

***Mobility of dislocations.*** Some knowledge about the properties of dislocations in martensites can be obtained by means of internal friction measurements. The interaction of dislocations with interstitial atoms in metals and alloys results in the so-called Snoek-Köster (S-K) relaxation. The S-K relaxation peak displayed during tempering of the studied martensites is presented in Fig. 2.28. It is significantly higher in Cr15Mo1N0.6 than in Cr15Mo1C0.6 martensite. The S-K relaxation takes place in the martensite and does not occur in ferrous austenite. Thus, if we take into account a higher fraction of retained austenite in the nitrogen steel (see Figs. 2.36 to 2.38 and Table 2.6) and normalise the curves in Fig. 2.28 to 100 % of martensite, the difference between the S-K peak intensities are even more distinct.

A detailed mechanism of S-K relaxation is still a controversial issue. Schoeck (1963) has described it in the frame of spring model and derived, as a source of relaxation, the movement of dislocations dragging interstitial clouds. The relaxation occurs if dislocations are sufficiently mobile while interstitial atoms can follow dislocations causing their viscous movement.

According to Seeger (1979), the formation of paired kinks on screw dislocations followed by the migration of interstitial atoms is responsible for the absorption of energy. In this model, the S-K relaxation strength is just equal to that of $\gamma$ relaxation, i.e. to the area swept by screw dislocations in the course of formation of the kink pairs.

Gavriljuk and Yagodsinski (1986) proposed a model of the S-K relaxation containing, as controlling mechanism, a movement of preexisting kinks along screw dislocations followed by the pipe diffusion of interstitial atoms, which results in the viscous transverse slip of dislocation segments.

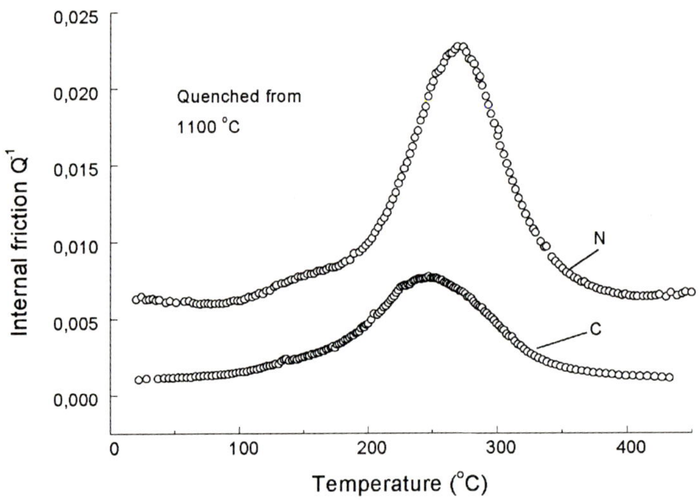


**Fig. 2.28.** Internal friction in steels Cr15Mo1N0.62 and Cr15Mo1C0.6 quenched after solution treatment at 1100 °C. The intensity of S-K relaxation peak characterises the area swept by dislocations under an applied stress, i.e. the dislocation mobility at the temperature of the peak

A common base in these theories is that the S-K relaxation arises from the movement of dislocation segments, while interstitial atoms dragged by dislocations restrict their movement and make it viscous. Therefore, the strength of S-K relaxation (the intensity of S-K peak) is determined by the area swept by dislocation segments for one vibration cycle, i.e. it characterises a mobility of dislocations.

It follows from the S-K peak intensities in Fig. 2.28, as well as from the values of internal friction background, which is mainly due to the dislocation vibrations, that dislocations in tempered nitrogen martensite possess a higher mobility as compared to that in the carbon one. This dislocation property may be important in the providing a higher impact toughness of tempered martensitic nitrogen steels.

***Precipitation sequence.*** A conclusion of the different precipitation behaviour during tempering of martensitic nitrogen and carbon steels follows unambiguously from the dilatometric curves that allow to study the temperature ranges of phase transformations in tempered martensite (see Fig. 2.29).

The temperature ranges of the 1st and 2nd phase transformations during tempering of carbon martensite with 15 mass % Cr are nearly the same as in plain carbon martensite, while the retained austenite is decomposed at higher temperatures. Some contraction attributed to the precipitation of the chromium carbide occurs after decomposition of retained austenite.

In nitrogen martensite, the third stage of tempering, i.e. the formation of stable interstitial phases, is extended over a broad temperature range. The retained austenite is also more stable and its decomposition (or secondary martensitic transformation) is seen in the dilatometric curve during cooling.

In the mixed C+N martensite, there is no sign of the 1st and 2nd transformation, while the 3rd stage is notably shifted to higher temperatures and smoothed within the broad temperature area.

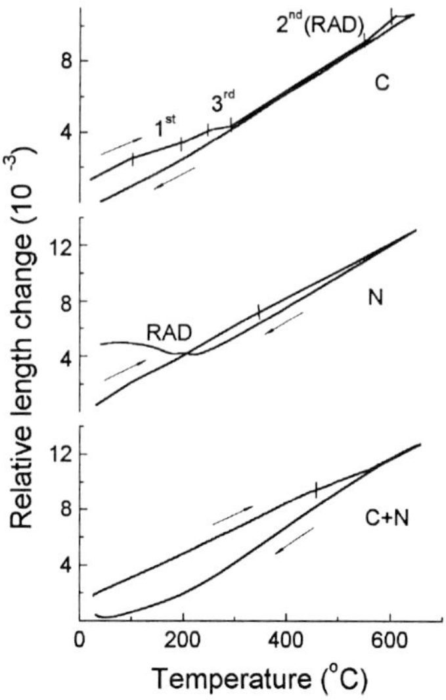


**Fig. 2.29.** Length change during tempering of martensitic steels Cr15Mo1N0.62, Cr15Mo1C0.6 and Cr15Mo1N0.35C0.29 quenched in water from 1100 °C. The temperature ranges of the 1st (precipitation of low-temperature carbides or nitrides), 2nd (decomposition of retained austenite) and 3rd (precipitation of stable carbide/nitride phases) transformations during tempering are marked

TEM studies allow one to clarify the nature of precipitated phases. As-quenched N, C, (C+N) martensites have a mixed substructure of dislocations and twins, the latter being of the [112]<111> system (see Fig. 2.30 ). Coarse undissolved $Cr_{23}C_6$ particles are present in the as-quenched chromium-carbon martensite. The substitution of carbon by nitrogen allows to avoid any precipitates in the as-quenched state. This is the main advantage of martensitic nitrogen steels, because the higher content of chromium is kept in the solid solution the better is corrosion resistance. No chromium carbides are observed after quenching the carbon+nitrogen martensite, but, unexpectedly, coarse cementite precipitates can be met in its structure (Fig. 2.31). They were obviously precipitated in the course of previous treatments including hot forging and intermediate thermal cycles.

After tempering at 100 °C the martensite structure is not changed, however, diffraction patterns of nitrogen martensite reveal some peculiarities, namely, diffusion streaks in <100> direction, which is an evidence of plane ordered nitrogen-rich domains located parallel to (100) atomic planes.

A patched contrast arising from fine precipitates is observed in the structure after tempering at 200 °C (Fig. 2.32). Single crystal microdiffraction patterns reveal the presence of hexagonal ε-nitride $(FeCr)_2N$ characterised by the crystal lattice parameters $a$ = 0.276 nm and $c$ = 0.442 nm in the nitrogen martensite and of hexagonal ε-carbide $(FeCr)_2C$ with the lattice parameters $a$ = 0.275 nm and $c$ = 0.435 nm in the carbon martensite.

Precipitates in the C+N martensite have also a hexagonal close packed lattice, however, the iron carbide, nitride or carbonitride cannot be precisely identified in the diffraction patterns because of the very close values of lattice parameters for the ε-nitride and ε-carbide.

Tempering at 300 °C causes some increase in the contrast of fine precipitates. According to diffraction patterns, the hexagonal ε-nitride $(FeCr)_2N$ in the nitrogen martensite is transformed into the orthorhombic ζ-nitride $(FeCr)_2N$ with the lattice parameters $a$ = 0.551 nm, $b$ = 0.829 nm, $c$ = 0.443 nm. Its precipitates are coarsened with further increase in temperature (see Fig. 2.33 for tempering at 550 °C) and, in the structure of nitrogen martensite tempered at 600–650 °C, they are formed as thin plates (Fig. 2.34). At the same time, the stable nitride $(CrFe)_2N$ having the hexagonal close packed lattice is precipitated and its coarse globular particles are seen along with the plates of the ζ-nitride.

The local X-Ray spectral analysis shows (see Table 2.5) that hexagonal nitride is enriched in chromium, whereas the chromium content in the plates of orthorhombic nitride does not differ from that in the matrix within the limits of the experimental error.

**Table 2.5.** Chromium and molybdenum contents (mass %) in different phases of nitrogen and carbon martensites after tempering at 650 °C (the data are obtained from areas of 20 nm in diameter and, for particles of smaller size, include some contribution of matrix)

| | Cr15Mo1N0.62 | | | Cr15Mo1C0.6 | | | |
|---|---|---|---|---|---|---|---|
| Element | matrix | $(FeCr)_2N$ | $(CrFe)_2N$ | matrix | $(CrFe)_{23}C_6$ | $Fe_3C$ | $(FeCr)_7C_3$ |
| Cr | 14.2 | 16.1 | 27.6 | 11.8 | 53.3 | 12.6 | 19.3 |
| Mo | 1.18 | 0.97 | 0.86 | 0.12 | 1.57 | 0.48 | 0.78 |

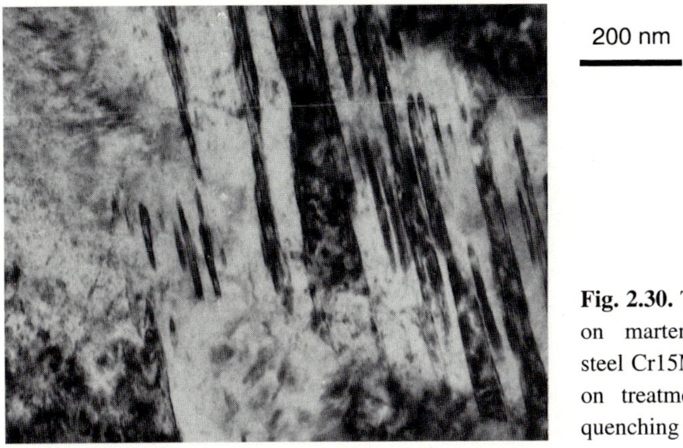

**Fig. 2.30.** Twinned and dislocation martensite in as-quenched steel Cr15Mo1N0.62 after solution treatment at 1100 °C and quenching in water

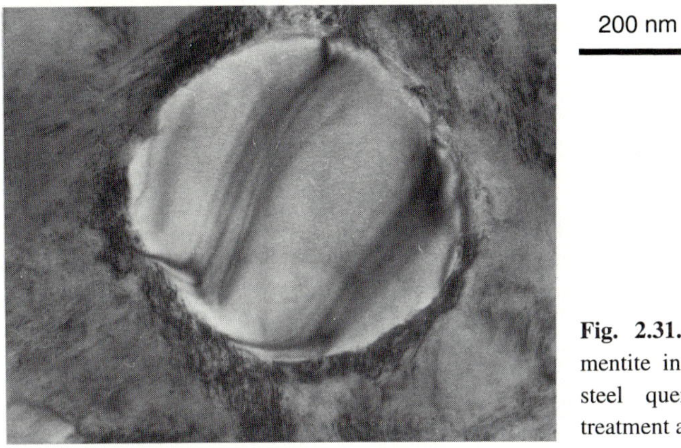

**Fig. 2.31.** An undissolved cementite in Cr15Mo1N0.35C0.29 steel quenched after solution treatment at 1100 °C

**Fig. 2.32.** Structure of martensite in Cr15Mo1N0.35C0.29 steel after tempering at 200 °C for 2 hours

50 nm

**Fig. 2.33.** $(Fe,Cr)_2N$ particles in steel Cr15Mo1N0.62 after tempering at 550 °C for 2 hours

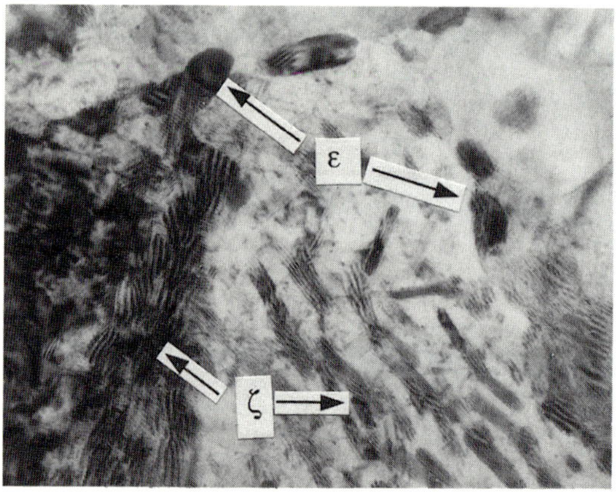

50 nm

**Fig. 2.34.** Precipitates of orthorhombic $\zeta$-$(FeCr)_2N$ (thin plates) and hexagonal $\varepsilon$-$(CrFe)_2N$ (round particles) nitrides in steel Cr15Mo1N0.62 after tempering at 650 °C for 2 hours

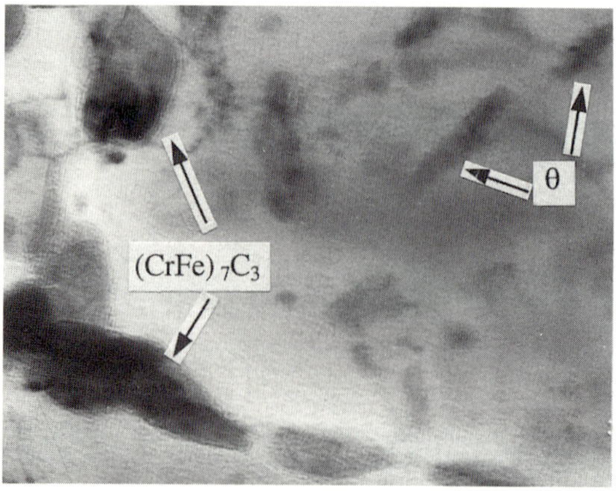

50 nm

**Fig. 2.35.** Globules of $(CrFe)_7C_3$ carbide and $\theta$-$(FeCr)_3C$ cementite plates in the structure of steel Cr15Mo1C0.6 after tempering at 600 °C for 2 hours

In the carbon martensite, the ε-carbide transforms into cementite in the course of tempering at 300 °C. A further increase in temperature leads to the coarsening of plate-like cementite particles. Tempering at 600–650 °C causes the precipitation of hexagonal carbide $(FeCr)_7C_3$ at the boundaries of former martensite crystals mainly (see Fig. 2.35, its precipitation corresponds to the effect of contraction in the dilatometric curve in Fig. 2.29C after dilatation caused by the decomposition of the retained austenite). Thus, three carbide phases coexist in the high-tempered carbon martensite: undissolved $(CrFe)_{23}C_6$ globules, cementite plates and particles of hexagonal chromium carbide. Higher temperatures or longer exposition at 600-650 °C would lead to the full substitution of cementite by the more stable $(FeCr)_7C_3$ carbide and its subsequent transformation into $(CrFe)_{23}C_6$. According to the data of X-ray spectral analysis (see Table 2.5), the molybdenum content in the matrix is decreased in the vicinity of carbide particles. Some enrichement of carbides in molybdenum occurs, which is not the case for nitrides. A depletion of the matrix in molybdenum and chromium has to result in a decrease of the pitting potential of tempered carbon martensite (see Sect. 3.2.2.2).

Tempering of the nitrogen+carbon martensite in the range between 200 °C and 400 °C does not lead to the precipitation of new phases and the growth of previously formed low-temperature precipitates occurs only, which is consistent with the character of the dilatometric curve (Fig. 2.29, C+N). Tempering at 500 °C causes the transformation of low-temperature nitride/carbide into two more stable phases: orthorhombic cementite $(FeCr)_3C$ and cubic chromium nitride (CrFe)N (see also Berns et al. 1997). This result allows to conclude that chromium carbonitrides are not formed at all in martensitic steels if they do not contain strong carbide/nitride formers like vanadium, niobium or titanium. This conclusion is supported by theoretical *ab initio* calculations of the electronic structure for chromium nitrides and carbides (see Antonov et al. 1996) where it is shown that a partial substitution of carbon by nitrogen in the crystal lattice of chromium carbides and, *vice versa*, an addition of carbon instead of nitrogen atoms to the crystal lattice of nitrides shifts the Fermi level towards higher state densities, i.e. destabilises the crystal structure.

A separate formation of the cementite and chromium nitride during tempering of nitrogen+carbon martensite provides one also with a strong emphasis that, at 200 °C, the ε-nitride and the ε-carbide were precipitated separately and, at higher temperatures, the former is transformed into cubic chromium nitride while cementite is formed from the latter. Let us remark that such a precipitation behaviour is consistent with the results obtained by Mittemeijer and co-workers as discussed in Sect. 1.2.2: ε(η)-carbide and $\alpha''$-nitride are precipitated separately during low-temperature ageing of ternary Fe(CN) martensite. One can suppose that a delay in the precipitation during tempering of nitrogen+carbon martensite is due to some redistribution of nitrogen and carbon atoms which has to occur before the separate nucleation of nitrides and carbides becomes possible.

Thus, the following precipitation sequence is observed during tempering of carbon, nitrogen and carbon+nitrogen martensites:

Cr15Mo1C0.6

| | | |
|---|---|---|
| 200 °C | 300 °C | 600 °C |
| ε-carbide | θ $(FeCr)_3C$ | θ $(FeCr)_3C$ + hcp $(CrFe)_7C_3$ |

Cr15Mo1N0.62

| | | |
|---|---|---|
| 200 °C | 300 °C | 600-650 °C |
| ε-nitride | ζ $(FeCr)_2N$ | ζ $(FeCr)_2N$ + hcp $(CrFe)_2N$ |

Cr15Mo1N0.35C0.29

| | | |
|---|---|---|
| 200 °C | 500 °C | 650 °C |
| ε-carbide/nitride | θ $(FeCr)_3C$ + fcc (CrFe)N | θ $(FeCr)_3C$ + hcp $(CrFe)_2N$ |

Coarse globules of $(CrFe)_{23}C_6$ in the carbon martensite and of $(FeCr)_3C$ in the carbon-nitrogen one, undissolved at previous solution treatments, are present at each tempering stage.

It is remarkable that the cubic chromium nitride (CrFe)N is not present in the tempered nitrogen martensite, whereas it precipitates as an intermediate phase between ε-nitride and hexagonal $(CrFe)_2N$ if the steel contains both nitrogen and carbon.

A higher stability of retained austenite in nitrogen and nitrogen+carbon martensites is also a feature of their tempering behaviour. As follows from Table 2.6, the retained austenite decomposes at 400 °C in the carbon steel, whereas higher temperatures are needed for its decomposition in nitrogen and nitrogen+carbon steel.

**Table 2.6.** Fractions of retained austenite (RA), ±2%, after quenching in water from 1100 °C and tempering for 2 hours

| Steel | as-quenched | 400 °C | 550 °C | 600 °C | 650 °C |
|---|---|---|---|---|---|
| Cr15Mo1C0.6 | 27.4 | 7.7 | 0 | | |
| Cr15Mo1N0.62 | 51.1 | 35.7 | 10 | 0 | |
| Cr15Mo1C0.29N0.35 | 70.9 | 70.9 | 58.1 | 50.5 | 1.8 |
| Cr15Mo1C0.29N0.35* | 31.2 | 31.2 | 28.8 | 0.82 | 0.66 |

*cooled in liquid nitrogen after water quenching

Moreover, the initial carbon, nitrogen and nitrogen+carbon austenites are characterised by a strikingly different stability to martensitic transformation (Figs. 2.36 to 2.38, see in more detail Berns et al. 1995).

Mössbauer spectra allow to detect simultaneously ferromagnetic and paramagnetic phases in alloys and determine their fractions with a high accuracy. Less than 30 % of the retained austenite is left in the steel Cr15Mo1C0.6 after quenching in water according to Mössbauer data (Fig. 2.36 and Table 2.6). The substitution of carbon by nitrogen strongly increases the fraction of the retained austenite, which is obviously concerned with the decreasing start temperature of martensitic transformation. The highest stability of austenite is observed in C+N steel (see Fig. 2.38, Table 2.6)

The reason for such a difference in the resistance to martensitic transformation and tempering can be understood on account of a different atomic distribution.

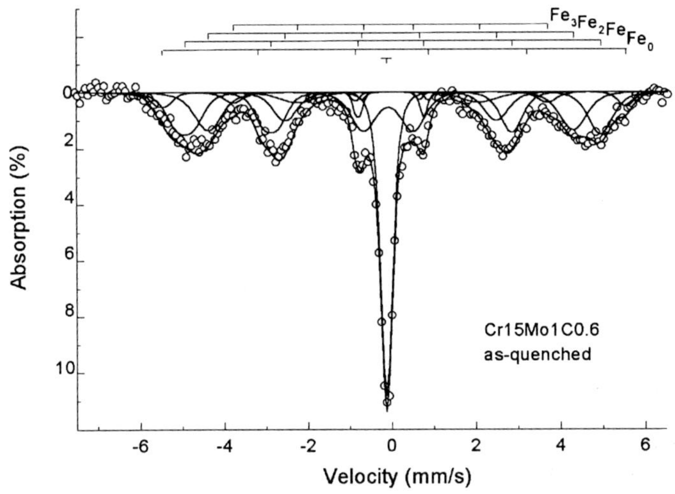


**Fig. 2.36.** Mössbauer spectrum of as-quenched steel Cr15Mo1C0.6. The single line belongs to the iron atoms in paramagnetic retained austenite. A sextet consisting of four components $Fe_0$ to $Fe_3$ arises from the iron atoms in ferromagnetic martensite having a different chromium neighbourhood

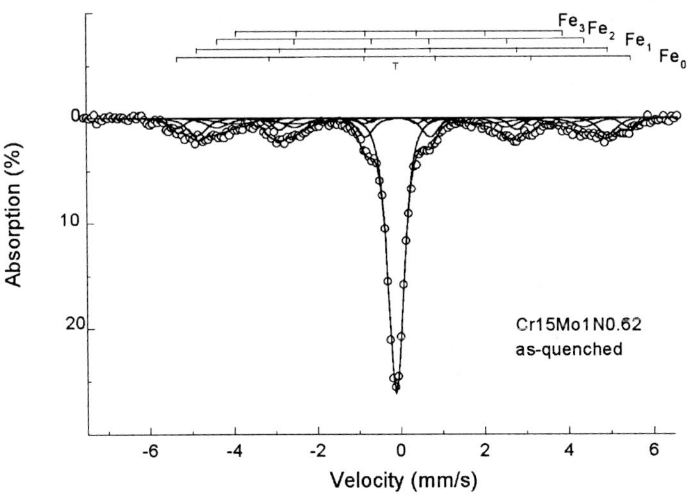


**Fig. 2.37.** Mössbauer spectrum of as-quenched steel Cr15Mo1N0.62. A higher fraction of retained austenite is left after quenching as compared to the carbon steel (see Fig. 2.36)

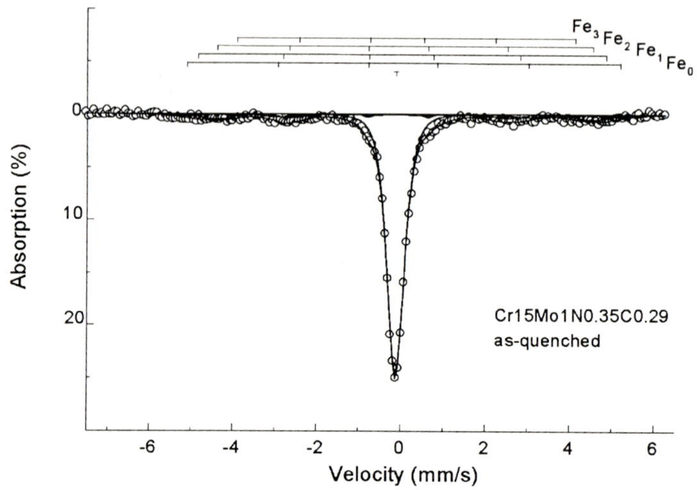


**Fig. 2.38.** Mössbauer spectrum of as-quenched steel Cr15Mo1N0.35C0.29 which possesses the highest stability of austenite to martensitic transformation (compare with Figs. 2.36, 2.37)

***Short range atomic order.*** The distribution of solute atoms in iron-based solid solutions is being successfully studied using Mössbauer spectroscopy (see e.g. Amigood and Litvinov 1973, Solomon and Levinson 1978). Mössbauer spectra of the quenched Cr15Mo1C(N) steels consist of a single paramagnetic line arising from iron atoms in the retained austenite and of a ferromagnetic sextet belonging to iron atoms in martensite (Figs 2.36 to 2.38). Solute atoms (Cr, Mo etc.) affect the magnetic properties of neighbouring iron atoms, that is why the sextet contains different components ($Fe_0$ to $Fe_3$) depending on the number of solute atoms in the nearest neighbourhood of an iron atom. Within the frame of experimental errors, one can neglect an effect of 1% molybdenum on the spectra as compared to that of 15% chromium. It is also important for the analysis that the spectra of high-tempered states, like that in Fig. 2.39, do not contain an effect of interstitial atoms which are mainly in the precipitates, and the intensity of spectra components is affected by the distribution of substitutional elements only.

It follows from Mössbauer studies of the iron alloys with the bcc crystal lattice that solute atoms in the first and second coordination spheres (nearest and next to nearest neighbours) around an iron atom decrease its magnetic moment by the same value. The hyperfine field $H_e$ at the iron nuclei, which is proportional to the splitting of sextets, i.e. to the distance between their outer lines, is decreased by about 2.69 $T$ due to each Cr atom as nearest neighbour of an Fe atom in the first or second coordination sphere (Vincze and Campbell 1973). This result reflects peculiarities of atomic interactions in the bcc crystal lattice, so that the coordination number 14 not 8 has to be accounted for while treating the experimental data.

An example of an $Fe_2$ atomic configuration with one Cr neighbour in the first and second one in the second coordination sphere is given in Fig. 2.40.

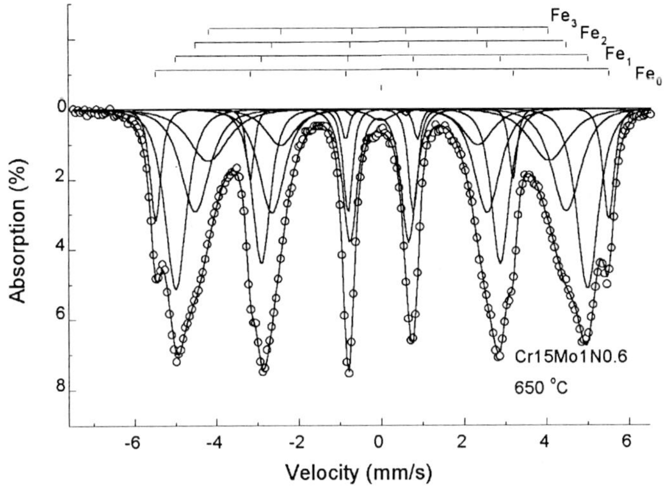


**Fig. 2.39.** Mössbauer spectrum of steel Cr15Mo1N0.62 hardened and tempered at 650°C for 2 hours. Components $Fe_0$ to $Fe_3$ belong to iron atoms having different numbers of chromium atoms as nearest neighbours

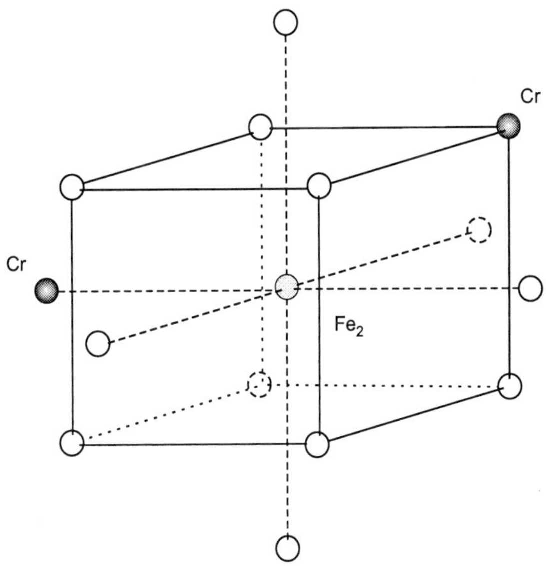


**Fig. 2.40.** An iron atom $Fe_2$ in the bcc iron lattice having two solute chromium atoms (one in the first and one in the second coordination spheres) as nearest neighbours

The integral intensity of components, i.e. the area under the lines of Mössbauer sextets, is proportional to the abundance of corresponding iron sites. Therefore, it is possible to obtain knowledge of the chromium distribution in iron martensites, by comparing the intensities of spectra components with the probabilities $P_i$ of an iron atom to have $i$ chromium atoms as nearest neighbours at random atomic distribution

$$P_i = C_{14}(i) \cdot c_{Cr}^{\ i}(1 - c_{Cr})_{14-i} \tag{2.7}$$

where $C_{14}(i)$ is the number of combinations of $i$-th class from 14 atoms, $c_{Cr}$ is the atomic chromium concentration.

As previously in Sect. 1.1.2.6, we will discuss the atomic distribution in terms of the Cowley parameter of short range atomic order to distinguish between clustering and short range atomic ordering. For the binary FeCr alloy, the Cowley parameter for the atomic distribution within the limits of two nearest coordination spheres in relation to an iron atom in the origin of coordinates is

$$\alpha_1 = 1 - N_1^{FeCr} / Nc_{Cr}(1 - c_{Cr})Z_1$$

where $N_1^{FeCr}$ is the number of iron-chromium pairs in the first and second coordination spheres around an iron atom, $c_{Cr}$ is the chromium atomic concentration, $Z_1 = 14$ is the number of sites in the first and second coordination spheres, $N$ is the total number of atoms, $Nc_{Cr}(1\text{-}c_{Cr})Z_1$ is the number of iron-chromium pairs in the first and second coordination spheres for random atomic distribution.

If short range atomic order exists, Eq. (2.7) can be rewritten as

$$P_i(\alpha_1) = C_{14}(i) \cdot [c_{Cr}(1 - \alpha_1)]^i [1 - c_{Cr}(1 - \alpha_1)]^{14-i}.$$

For practical use, this equation can be simplified due to expansion of $P_i(\alpha_1)$ with respect to small $\alpha$:

$$P_i(\alpha_1) = P_i^{random} \cdot [1 - \alpha_1 \frac{i - 14c_{Cr}}{1 - c_{Cr}}]. \tag{2.8}$$

The values of $P_i(\alpha_1)$ are equal to the experimental data of $Fe_i$ abundances. Using Eq. (2.8) and intensities of $Fe_0$ components which are more precisely determined (see Fig. 2.40), the values of $\alpha_1$ were obtained to be equal to 0.17, –0.02 and –0.2 for the carbon, nitrogen and nitrogen+carbon martensites, respectively. This evaluation of short range atomic order is not completely precise because, during the fitting of Mössbauer spectra, one has to account also for $Fe_4$ and $Fe_5$ components the contributions of which to the spectra are not negligible. Nevertheless, it shows that the properly established tendency to short range decomposition ($\alpha > 0$) in FeCr solid solutions (see Solomon and Levinson 1978 and data in Sect. 1.1.2.6) is enhanced by carbon and diminished by nitrogen. Alloying with nitrogen+carbon distinctly favours short range atomic ordering of chromium atoms ($\alpha < 0$). Thus, nitrogen in the solid solution of martensitic steels prevents clustering of chromium atoms.

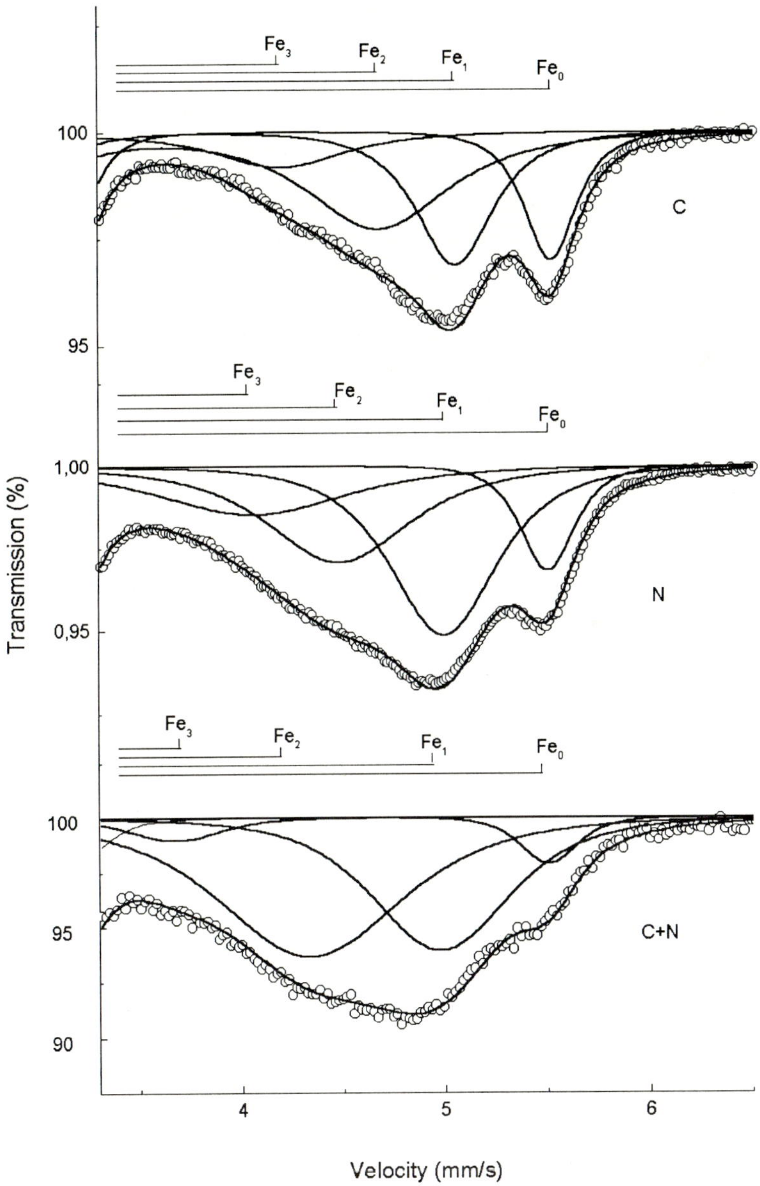


**Fig. 2.41.** Outer lines of Mössbauer spectra of steels with (mass %) 15 Cr, 1 Mo and 0.6 C or 0.62 N or 0.35 N + 0.29 C after quenching from 1100 °C and tempering at 650 °C for 2 hours

The nature of such a favourable nitrogen effect can be explained based on the atomic distribution in austenitic nitrogen steels discussed in Chap. 1. Due to the effect of nitrogen on the electronic structure, namely due to an increased concentration of free electrons, and due to a correlation between the interatomic bonding and atomic distribution, short range atomic ordering of substitutional solutes in austenite is promoted in the nitrogen-containing steels. It is obvious that the atomic distribution in as-quenched martensite is inherited from austenite. The nitrogen-induced tendency to short range atomic ordering in the distribution of chromium atoms is a reason for the observed delay in the precipitation and growth of chromium-rich nitrides. Even after high-temperature ageing, when nitrogen atoms are mainly bound in precipitates, the short range atomic ordering of chromium atoms is still retained because the decomposition of a solid solution needs a longer time compared to that applied for tempering.

A reason for enhanced short range atomic ordering in the nitrogen+carbon steel is not completely clear. Along with the stability of austenite to martensitic transformation, it provides a higher resistance of nitrogen+carbon martensites to tempering, which results in their excellent mechanical and corrosive properties.

In addition, it is worth noting that studies of Fe-(C+N) alloys which do not contain substitutional alloying elements (see Rawers et al. 1996, Rawers and Uggowitzer 1998) have also shown that combined alloying with nitrogen+carbon leads to a decrease in the start temperature of martensitic transformation, i.e. to an increasing stability of austenite to the phase transition, as compared to alloying by carbon and to a high stability of both austenite and ferrite to the precipitation of nitrides and carbides. These results mean that the Fe-(C+N) solid solution is characterised by a higher degree of short range ordering as compared to the Fe-N austenite (see also Sect. 1.1.2.5) and as a result by a higher thermodynamical stability.

# 3 Key properties

Nitrogen as an alloying element in steel causes non-trivial and even extraordinary changes in properties. In particular this statement relates to an unusual combination of the yield strength and fracture toughness. For the first time perhaps, mechanical properties of nitrogen steels were tested by Andrew (1912) who constructed a furnace for melting under high pressure, obtained Fe-N samples and revealed the nitrogen-caused increase in yield strength as well as an austenitizing nitrogen effect. Frehser and Kubisch (1963) were the first to discover that, with increasing nitrogen content, an increase in the yield strength of austenitic steels is not accompanied by the expected deterioration of toughness. In fact, this result meant that nitrogen steels represent a new promising class of engineering materials. It was also shown later in a number of studies that nitrogen in steel can improve fatigue life, strength at low and elevated temperatures, cold work hardening, wear resistance etc.

Another area of successful application of nitrogen steels is provided by an improvement of the chemical properties, i.e. by an increasing time of service in aggressive environments. Being dissolved in the matrix, nitrogen improves the resistance to local kinds of corrosion and, at least, does not increase the rate of general corrosion. In principle, such a combination of properties allows to construct high-strength stainless steels – quite a complicated task on account of the low strength of the austenitic solid solutions as a base for design of stainless steels.

We will try to describe the main properties of nitrogen steels and discuss the nature of the nitrogen effect using the data on the structure and constitution presented in the two previous chapters.

## 3.1 Mechanical properties

We start from discussing the yield strength of nitrogen austenitic steels will be discussed paying due attention to the nitrogen-caused solid solution strengthening. Both the athermal and the thermally activated contributions to the flow stress will be considered, and an explanation for the nitrogen-induced anomalous strengthening of austenitic steels at low temperatures will be proposed using data of the effect of nitrogen on the electronic structure and stacking fault energy. Next, we will discuss the role of crystal lattice defects in strengthening and present available mechanisms of the strengthening by grain boundaries on account of the nitrogen-caused increase in the barrier effect of grain boundaries concerning the transfer of dislocation slip from one grain to the adjacent one. The role of dislocations and twins will be considered in relation to cold work hardening with a reference to data of the effect of nitrogen on the dislocation structure and twinning.

Finally, a review of the experimental data concerning the influence of nitrogen on fracture toughness, fatigue, creep and wear resistance will be presented.

### 3.1.1 Solid solution strengthening

Irvine et al. (1969) described an effect of different solutes on the mechanical behaviour of austenitic steels. Based on the results of mechanical tests of a number of steels, they derived regression equations for the dependence of the yield strength $R_{p0.2}$ and ultimate strength $R_m$ on the microstructure and composition in mass % (we present their data in MPa):

$$R_{p0.2} = 63.5 + 496N + 356.5C + 20.1Si + 3.7Cr + 14.6Mo + 18.6V + 4.5(W,\%) + 40.3Nb + 26.3Ti + 12.7Al + 2.5\delta + 7.1d^{-1/2}$$

and

$$R_m = 449.5 + 852.5N + 542.5C + 37.2Si - 1.5Ni + 18.6Mo + 77.5Nb + 46.5Ti + 18.6Al + 2.2\delta + 7.1t^{-1/2},$$

where $\delta$ is the content of $\delta$-ferrite in vol. %, $d$ is the grain size in mm and $t$ is the number of twins per mm. According to the authors, a free dislocation path determined by the number of twins within the grain has no significant effect on the yield stress, but affects the ultimate tensile strength.

It followed from the above equations that, at ambient temperature, nitrogen is about 1.5 times more effective in strengthening the austenite as compared to carbon. In the preliminary discussion of the nature of solid solution strengthening of austenitic steels it is useful to compare, after Irvine et al., the strengthening effect of different elements and the crystal lattice distortions (Table 3.1).

Despite the scattering of the experimental data obtained by different researchers (see also the data for lattice distortions in Table 1.2) nitrogen clearly causes higher distortions in austenite than carbon, which results in the higher strengthening effect. The correlation between distortion and strengthening holds also true for ferrite-stabilising elements, whereas austenite-stabilising elements cause no detectable lattice distortions and strengthening in austenitic steels.

The difficulties in explaning the strengthening mechanism of nitrogen as well as of carbon in austenitic steels originate from the fact that interstitials in the octahedral sites of the fcc lattice do not cause non-cubic elastic distortions and, as a result, do not interact with screw dislocations and show a weak interaction with edge dislocations. However, the interaction between interstitials and substitutional solutes results in the formation of *i-s* complexes which have a non-cubic crystalline symmetry (see Sect. 1.1.3), and these complexes contribute to the interstitial-dislocation interaction in austenitic steels. As shown in Sect. 1.1.4.2, the energy of binding between nitrogen atoms and dislocations is higher than that between carbon atoms and dislocations. Therefore, we can expect a higher effect of nitrogen on the solid solution strengthening of austenitic steels.

**Table 3.1.** Lattice distortions and strengthening (MPa) due to solute elements (after Irvine et al. 1969)

| Solute element | Type | $\Delta a/a_0$ per 1 at.% $\times 10^3$ | $\Delta R_{p0.2}$ per 1 at.% | $\Delta R_m$ per 1 at. % |
|---|---|---|---|---|
| N | Interstitial | +2.34 (2.40*) | 124 | 213.1 |
| C | Interstitial | +1.67 (2.18*) | 76.6 | 116.2 |
| Cr | Substitutional | +0.75* | 3.4 | 0 |
| Si | (δ-stabiliser) | -1.39 | 10.1 | 18.6 |
| Nb | " | - | 66.9 | 131.7 |
| Ti | " | - | 22.6 | 40.3 |
| V | " | +0.42 | 16.9 | 0 |
| W | " | +0.83 | 14.9 | 0 |
| Mo | " | +0.92(+1.5**) | 24.9 | 0 |
| Al | " | - | 6.2 | 9.3 |
| Ni | (γ-stabiliser) | -0.05(-0.03**) | 0 | -1.55 |
| Mn | " | +0.05(+0.22**) | 0 | 0 |
| Cu | " | +0.64 | 0 | 0 |
| Co | " | -0.11 | 0 | - |

* Ledbetter and Austin (1987)
** Byrnes et al. (1987)

The yield strength of solid solutions is controlled by two main contributions, the first one $R_a$ being athermal and the second one $R_T$ exhibiting a temperature dependence originating from a thermally activated overcoming of obstacles by moving dislocations:

$$R_{p0.2} = R_a + R_T.$$

In iron-based fcc solid solutions (stainless austenitic steels), the flow stress is temperature-dependent below 500 K. We will discuss these two components of the solid solution strengthening separately.

### *3.1.1.1 Athermal strengthening*

It is usually treated in terms of the theories proposed by Nabarro (1946) and Mott (1950), Fleischer and Hibbard (1963), Friedel (1963) and Labusch (1972). All these theories are based on the statistical distribution of the interstitials which act as obstacles for the dislocation movement. We shall briefly discuss them in order to test whether the contribution of nitrogen to athermal solid solution strengthening can be significant.

The interaction force $f_0$ acting upon the dislocation line from a single obstacle is so weak that the angle $\Theta$, at which the direction of the dislocation is changed because of this interaction, is much less than the unity: $\Theta = f_0 / T << 1$, where $T$ is the line tension of the dislocation. Mott and Nabarro suppose that the statistical internal stress field $\sigma$ created by the interstitial atoms has some characteristic wavelength $\lambda$ (see Fig. 3.1a).

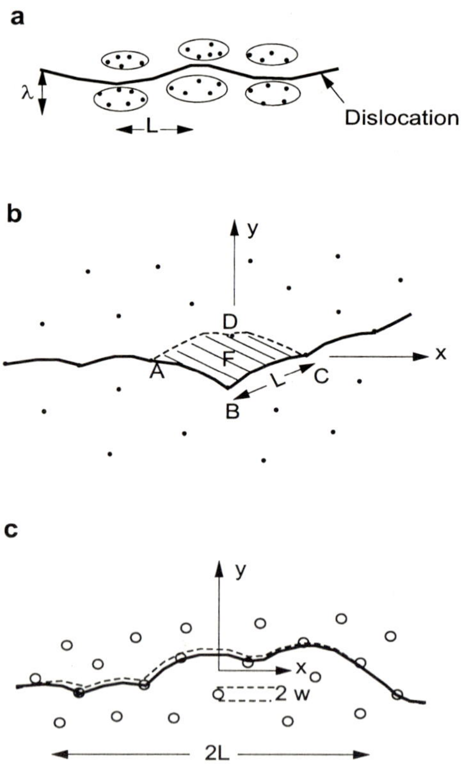


**Fig. 3.1.** Representation of the interaction between dislocation and solutes: **a** – effective obstacles according to Mott-Nabarro; **b** – critical state for the dislocation according to Fleischer-Friedel, **c** – forest of attractive potential wells of width $2w$ and movement of the dislocation line in the Labusch limit as presented by Nabarro (1977)

As $\Theta << 1$ the dislocation line cannot accommodate itself to this stress field and, for this reason, some average value, $\sigma_0$, acts on the length $L >> \lambda$. The statistical deviations $\sigma_L$ of this average stress are considered as "effective obstacles" and the value of $L$ is chosen so that a complete accommodation of the dislocation line to the effective obstacles takes place. The latter results in the following dependence of the shear stress $\tau_c$ applied to overcome the array of the interstitials on the atomic concentration $c_{at}$:

$$\tau_c \sim \sigma_0^{4/3} c_{at}^{11/9} T^{-1/3} \log c_{at}. \tag{3.1}$$

Such a concentration dependence of the yield strength was never observed for solid solutions. A weakness of the Mott-Nabarro theory, as shown by Labusch (1972), is concerned with an unreasonable choice of the wavelength $\lambda$ and with the way of the calculation for the mean deviation amplitude $\sigma_L$ of the internal stress field.

An essence of the Fleischer-Friedel theory consists of the analysis of the dislocation slip through the array of interstitials (Fig. 3.1b). The interstitials are represented as point-like obstacles. The conditions at the obstacles touched by the dislocation and at those that are not touched are considered to be quite different as the radius of the interstitial-dislocation action $w$ is supposed to be arbitrarily small. If an external shear stress is applied, the dislocation remains at the locations where the internal stress $\sigma$ is strongly positive, i.e. sufficient to counteract the applied stress, whereas in between the dislocation line is bent out and leaves the areas of negative $\sigma$ until it finds some new locations with a strongly positive $\sigma$. The interstitial atoms play the role of these locations with a sufficiently strong positive $\sigma$, i.e., they are the pinning points. A peculiarity of the dislocation movement under an applied stress is that the full adaptation of the dislocation line to the new positive maxima of $\sigma$ is not necessary: already in the course of the adaptation the dislocation breaks away from the previously touched maxima if a critical state is achieved there.

The average distance between the touched obstacles is $L$ (Fig. 3.1b). In the critical state the equality of the applied and internal actions on the dislocation is achieved: $\tau b \approx f_0/L$ where $b$ is the Burgers vector. The expression for the concentration dependence of the shear stress $\tau_c$ is derived from the calculation of the value of $L$. If the maximum interaction force between the dislocation and the obstacle is exceeded, the former breaks away, sweeps the square F and takes up the dashed state (Fig. 3.1b). This process is accompanied by an increasing force at the neighbouring obstacles, they are overloaded and broken, so that this action propagates until the moving dislocation is captured by other obstacles. The condition of the critical state, when the breaking off starts, is $F \approx 1/c$. From the formula for the line tension $T$, one obtains $F = L^3\tau b/2T$, and, finally, the condition $\tau b \approx f_0/L$ transforms into equation

$$\tau_c b = f_0^{3/2} c^{1/2} (2T)^{-1/2}. \tag{3.2}$$

Another way to obtain the concentration dependence of the critical shear stress at the point-like obstacles consists in starting from $\tau = 0$ and studying how the average distance between the touched obstacles changes with increasing $\tau$. It occurs in such a manner that the dislocation line bends out between the earlier touched obstacles and at the same time touches new obstacles. A simple derivation proposed by Friedel (1964) leads to the formula $L = (T/\tau bc)^{1/3}$ for such a behaviour. Entering this in $\tau b \approx f_0/L$, one obtains essentially the same result as given by Eq. (3.2).

One shortcoming of this theory is that the distribution functions of the forces $f_0$ and the distances between the obstacles $L$ are not taken into account. However, the most serious doubt is caused by the postulate of point-like obstacles. To satisfy this supposition, the action radius of the obstacle-dislocation interaction has to be smaller than the mean bend-out between two obstacles. Otherwise, the difference between the touched and untouched obstacles (the first ones withstand the applied stress, the second ones assist it) looses sense. Nevertheless, the concentration dependence (3.2) for the yield strength of solid solutions is quite satisfactorily consistent with the experiment.

The theory proposed by Labusch (1972) is a modification of the Mott-Nabarro theory (Fig. 3.1c). Unlike the Friedel limit shown in Fig. 3.1b, when the dislocation breaks away from obstacle B, it is shifted in many points outside the segment AC. Labusch has performed a more precise calculation of the mean deviation amplitude $\sigma_L$ of the internal stress field taking into account an effect of the nearest atomic planes neighbouring the slip plane. Having obtained also a more reasonable expression for the characteristic wavelength $\lambda$ of the internal stress field ($\lambda = nBw$, where $B$ is a numerical factor of about unity, instead of $\lambda = c_v^{-1/3}$ supposed by Mott and Nabarro), Labusch has proposed the following equation for the critical shear stress:

$$\tau_c b = A' \cdot f_0^{4/3} c^{2/3} w^{1/3} T^{-1/3} \quad , \tag{3.3}$$

where A' is again a numerical factor of about unity. Such a concentration behaviour of the yield strength of solid solutions is also observed in experimental studies.

Later on, Nabarro (1977) has also proposed some new modification of the theory representing the obstacles as a forest of the potential wells of width $2w$. An essential point in this theory is the ratio $\beta = f_0/4Tcw^2$. If $\beta >> 1$ (the point-like obstacles), one obtains the Friedel limit (Fig. 3.1b) and $\tau_c \sim c^{1/2}$. For $\beta << 1$ (the obstacles have some finite size), the Labush limit (Fig. 3.1c) is valid and $\tau_c \sim c^{2/3}$.

In a more detailed analysis, the contributions of the modulus and size misfits to strengthening have to be also distinguished. The interaction force $f_0$ in Eqs. (3.2) and (3.3) is proportional to the interaction parameter $\varepsilon_L$, and $\varepsilon_L^2 = \eta^2 + D^2\delta^2$, where the modulus misfit $\eta$ and the size misfit $\delta$ are defined by the shear modulus $G$, the lattice parameter $a$ and the concentration $c$: $\eta = (1/G)(dG/dc)$ and $\delta = (1/a)(da/dc)$, respectively. D is the constant of about 20 (Haasen 1976). The elastic properties of austenitic stainless steels (the Young's and shear moduli) were studied by Ledbetter and Austin (1987) and shown to be not noticeably different for the austenitic steels wth carbon or nitrogen, whereas the size misfits caused by these two elements in austenite differ clearly (see Tables 3.1 and 1.2). Another argument against a possible role of the elastic moduli in the solid solution strengthening is given by Reed and Simon (1989) based on the fact that the misfits of the shear and Young's modulus in austenitic steels with carbon and nitrogen change their sign on cooling from 295 to 4 K, which is not accompanied by any discontinuity in the strengthening effect.

Let us now discuss the consistency of the experimental data available for austenitic nitrogen steels with the above theoretical expressions. The mechanical properties of austenitic nitrogen steels have been measured by Norström (1977), Tobler and Reed (1980), Takahashi et al. (1982), Sakamoto et al. (1984), Nilsson and Thorvaldsson (1985), Reed et al. (1986), Reed and Simon (1989), Byrnes et al. (1987) etc. As an example, some data are shown in Fig. 3.2.

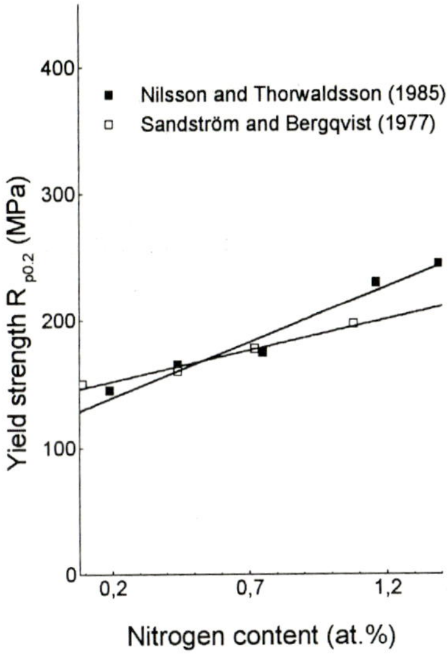


**Fig. 3.2.** Yield strength *versus* the nitrogen content at 673K in steels Cr(19-22)Ni10N (after Sandström and Bergqvist 1977) and Cr26Ni32Mo3N (after Nilsson and Thorvaldsson 1985)

A linear concentration dependence can be stated, exactly as obtained by Irvine et al. (1969) from measurements at room temperature.

Reed and Simon (1989) performed a detailed analysis of the mechanical properties of ten AISI 304 type steels alloyed by nitrogen and having some small nearly constant content of carbon. They fitted experimental data obtained at 295 K using three different methods for their treatment: a linear least-square procedure, an iterative non-linear least-square procedure and a treatment accounting for an interactive effect of the grain size and of the nitrogen content. The obtained values of the yield strength were proportional to (C,N), $(C,N)^{0.785}$ and $(C,N)^{0.5}$, respectively.

In contrast to the data of Irvine et al. (1969), the strengthening effects of nitrogen and carbon at 295 K, as obtained by Reed and Simon (1989), coincided within the limits of error, though some preference of nitrogen occurred.

The linear concentration dependence of the athermal component of the yield stress in austenitic nitrogen steels was also observed by Byrnes et al. (1987). Thus, the experiment does not give an unequivocal confirmation that the existing theories of solid solution strengthening can be correctly applied to the nitrogen austenite. A linear concentration dependence of the yield strength observed in most experiments suggests that some other factors, beside the size misfit, affect the athermal component of the yield strength.

The effect of nitrogen on the stacking fault energy cannot be responsible for the disagreement between theory and experiment because, as was shown in Sect. 1.1.5.1, nitrogen can decrease as well as increase the value of SFE at room temperature depending on the basic chemical composition of steels.

The short range order in the distribution of nitrogen atoms is thought to give a complementary contribution to the athermal strengthening. According to Fisher (1954), any type of short range departure from randomness in the distribution of solute atoms can provide the contribution to the shear stress as large as $\gamma/b$, where $\gamma$ is the energy of the disordered interface produced by the passage of a dislocation through a short range ordered material, and $b$ is the magnitude of the Burgers vector. As shown in Sect. 1.1.5.2, short range atomic ordering is responsible for the planar slip in austenitic nitrogen steels.

Fischer's idea was used by Gruicic et al. (1989) in their theoretical assessment of the solid solution strengthening of austenitic nitrogen steels. They supposed a random distribution of Fe, Ni and Cr in the fcc lattice and analysed the occupation of the octahedral sites in the interstitial sublattice by nitrogen depending on the type of the metallic atoms constructing the octahedral cell. Thermodynamic studies of the nitrogen solubility in Fe-Cr-Ni alloys performed by McLellan (1982) and Ko and McLellan (1983) were used as a basis for such an analysis. It is natural that nitrogen occupies the sites in cells with an excess of Cr atoms. The slip of dislocations on (111) planes in the fcc lattice removes some metallic atoms from the nearest surrounding of the N atoms and substitutes them by others. In the average, this process will result in a decrease of the excess of Cr atoms in the cells occupied by N atoms and, therefore, N atoms turn up in higher energy sites. The second dislocation will preferentially traverse the same slip plane because the energy needed to move it through the partially disordered crystal is smaller (see also discussion of planar slip in Sect. 1.1.5.2). Thus, the N occupation of high-energy interstitial sites becomes significant after passage of many dislocations. The work needed to change the equilibrium atomic distribution and increase the free energy of the crystal originates from the applied force acting on dislocations and moving them, i.e. from an increased shear stress.

This theory predicts a linear concentration dependence of the yield strength. In fact, it is close to that by Fleischer-Friedel, but the obstacles are supposed to have a different strength. However, chromium and nickel atoms are not randomly distributed in iron-based solid solutions as supposed by Gruicic et al. (1989). Chromium atoms are prone to clustering while the distribution of nickel atoms is characterised by short range ordering (see Sect. 1.1.2.6)

Later on Grujicic and Owen (1995) proposed a model of short range order in CrNi austenitic nitrogen steels according to which alloying with nitrogen changes the distribution of chromium atoms from a random to a clustered one resulting in a larger fraction of the nitrogen atoms occupying interstitial sites within high chromium clusters. This approach is at variance with the experimental data of small angle neutron scattering and of magnetic studies of atomic clusters (Sect. 1.1.2.6) showing that, on the contrary, nitrogen in austenitic steels prevents clustering of the substitutional solute atoms and promotes short range atomic ordering.

One has also to note that, if the contribution of nitrogen to strengthening of austenitic steels arises from the clustering of chromium atoms and strong interaction between nitrogen and chromium atoms, alloying with carbon would result in the same or even stronger effect than that with nitrogen. This is to be expected because carbon atoms in austenite also prefer low-energy interstitial sites with an excess of chromium atoms. Moreover, they clearly assist clustering of chromium atoms, which results in an easy precipitation of chromium carbides (see also Figs. 1.24 and 1.25 concerning carbon-enhanced clustering of substitutional atoms). Thus, carbon atoms have also to create localised obstacles to be overcome by moving dislocations as was proposed by Gruicic et al. for nitrogen in Fe-Cr-Ni austenite.

It is also remarkable that, in contrast to nitrogen, carbon in austenitic steels does not promote planar slip which seems to be caused by short range ordering, not by clustering.

Thus, the effect of nitrogen on the athermal solid solution strengthening of austenite in comparison to carbon can arise from the ordered distribution of nitrogen atoms and from the nitrogen-promoted tendency to short range ordering in the distribution of substitutional solute atoms. A larger misfit factor caused by nitrogen atoms is also important.

### 3.1.1.2 Temperature dependence of the yield strength

A large increment of the flow stress of austenitic nitrogen steels is observed in the temperature area below 500K and, in particular, below room temperature (see e.g. Norström 1977, Sandstrom and Bergqvist 1977, Nilsson and Thorvaldsson 1985, Byrnes et al. 1987, Reed and Simon 1989).

It is important that the anomalous strengthening effect caused by nitrogen at low temperatures is accompanied by a sufficiently high ductility and toughness if the content of nitrogen is not critically high. For this reason, austenitic nitrogen steels represent an excellent engineering material for the cryogenic industry. For example, the structural constraint in large magnetic fields requires a material of 1200 MPa yield strength and 200 $MPa \cdot m^{1/2}$ fracture toughness at 4 K (Shimamoto et al. 1986). Austenitic nitrogen steels can withstand such a large applied stresses and, at the same time, retain the necessary ductility and toughness (see Sect. 5.2.6 about details).

The experimentally observed difference between nitrogen and carbon effects on the temperature dependence of the yield strength is significant. According to Reed and Simon (1989), the effects of nitrogen and carbons on strengthening at 4 K are represented by the coefficients in the following equations:

$$R_{p0.2} = (325 \pm 39) + (592 \pm 32)[N] + (426 \pm 80)[C] + (612 \pm 245)d^{-1/2}$$

for a linear least-square procedure,

$$R_{p0.2} = (-206 \pm 58) + (1045 \pm 54)[N]^{0.5} + (392 \pm 69)[C]^{0.5} + (633 \pm 240)d^{-1/2}$$

for an iterative, non-linear least-square procedure.

In contrast to the data of the same authors for 295 K (see the previous Section), the difference between carbon and nitrogen effects at 4 K exceeds many times the limits of error. At 4 K, nitrogen is by a factor of 1.4 a more effective strengthener of austenitic steels as compared to carbon.

This striking effect leads sometimes to the conclusion that nitrogen assists a bcc-like behaviour of the flow stress of austenitic steels with temperature, which is strongly different from the behaviour of fcc substitutional solid solutions (Fig. 3.3). In a support of this idea, theoretical studies were performed, according to which the alloying of austenitic steels by nitrogen causes a splitting of the core of screw dislocations on two or more non-parallel planes, exactly as it takes place in the crystals with a bcc lattice (see Grujicic 1994, 1995).

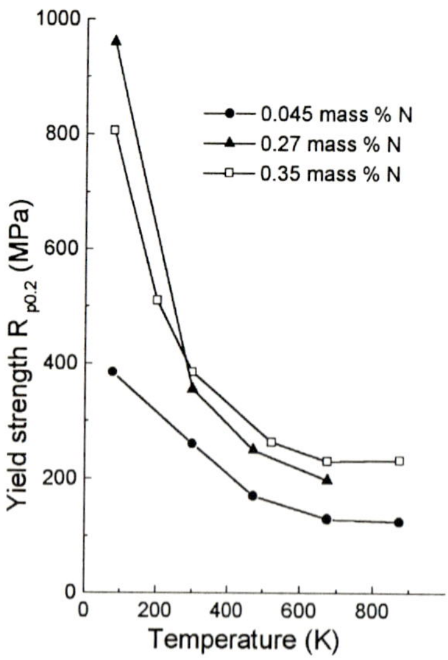


**Fig. 3.3.** Effect of nitrogen on the temperature dependence of the yield strength of the austenitic steels Cr(19-22)Ni10N0.27 (after Sandström and Bergqvist 1977) and Cr26Ni32Mo3N0.045 and N0.35 (after Nilsson and Thorvaldsson 1985)

Sandstrom and Bergqvist (1977) were perhaps the first to ascribe the nitrogen effect at low temperatures to the control of plastic flow by a thermally activated process. The latter means that nitrogen creates such obstacles for dislocation slip or assists a higher effectiveness of the already existing obstacles which need the thermal energy to be overcome.

The theories of solid solution strengthening presented in the previous section cannot be applied to the temperature dependent part of the yield strength because they do not include in any way the thermally activated overcoming of the obstacles by dislocations and, therefore, their application is strongly limited to 0 K.

A real breakthrough in the understanding of the low temperature mechanical properties of austenitic nitrogen steels was achieved due to precise mechanical tests performed by Obst and Nyilas (1991, 1993), see also Obst (1998), who developed a highly-sensitive clip-on extensometer and showed (see Fig. 3.4) that the temperature dependence of the yield strength of austenitic nitrogen steels is typical for fcc crystals with a low stacking fault energy and has a step-like behaviour predicted by Seeger's theory of the flow stress for fcc metals (Seeger 1954, 1955).

Therefore, there is no need to suggest a nitrogen-caused transformation to the bcc-like mechanical behaviour at low temperatures because such a temperature behaviour does not really exist.

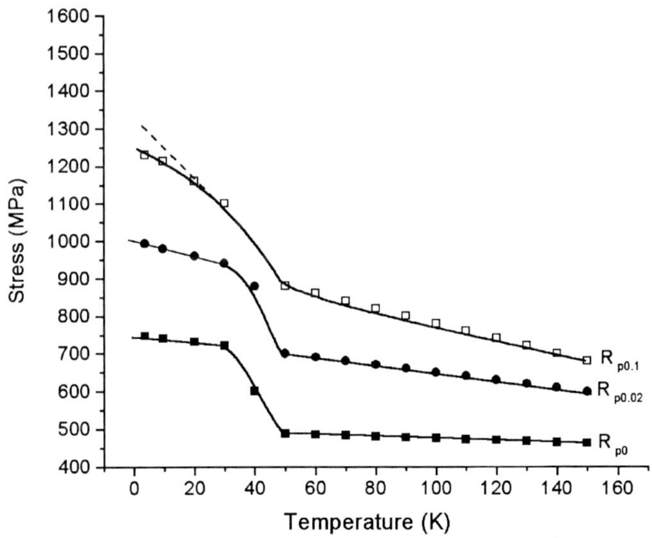


**Fig. 3.4.** Effect of temperature on the tensile properties of the austenitic steel Cr25Ni15N0.35 (after Nyilas et al. 1993)

According to Seeger (1955), the temperature dependence of the yield (flow) stress in fcc crystals is determined by the intersection of dislocations and the generation of point defects by jogs formed by intersecting dislocation lines (Figs. 3.5 and 3.6). Seeger analysed the following three processes which can contribute to the temperature-dependent flow stress in fcc crystals below the temperature of selfdiffusion: (1) interaction between gliding edge dislocations and a forest of dislocations threading the slip plane, (2) interaction between gliding screw dislocations and this forest, (3) thermally activated generation of vacancies by jogs in screw dislocations.

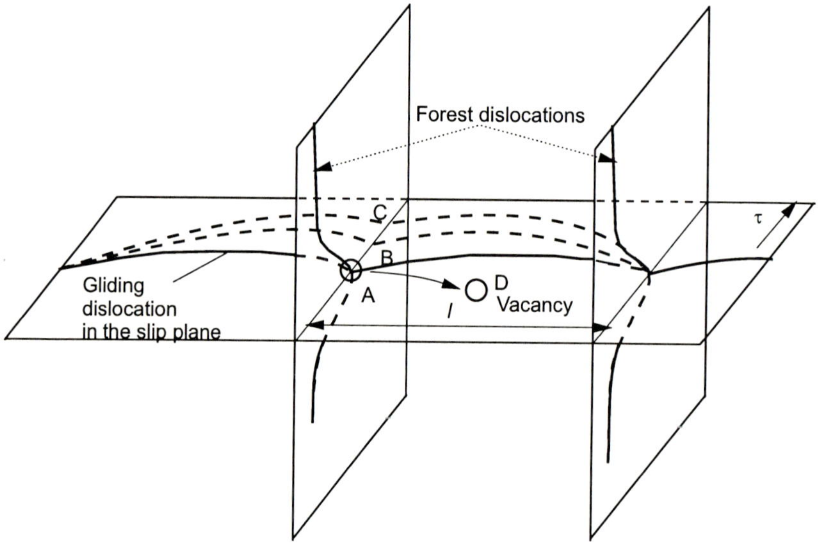


**Fig. 3.5.** Interaction of gliding and forest dislocations producing a thermally activated generation of the vacancies

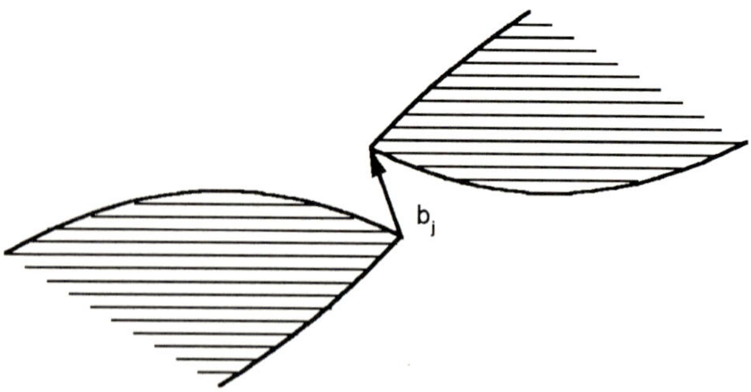


**Fig. 3.6.** Constriction and formation of a jog during intersection of gliding extended dislocation with a forest dislocation

If the temperature is sufficiently low, so that selfdiffusion and therefore climb of dislocations for the time of deformation are impossible, the jogs that are being formed during the intersection of gliding dislocations leave vacancies or interstitials behind them. As the energy necessary to add a vacancy to the end of a row of lattice defects, generated by a gliding jog, is about one order smaller compared to the case of interstitials, the generation of vacancies can be a preferential controlling mechanism for the flow stress in fcc crystals at low temperatures. Figure 3.5 shows a gliding dislocation pinned by the intersected dislocations of the forest, $l$ being an average distance between the neighbouring jogs. Stages in the movement of a screw dislocation at the jog pinning point include an intermediate activated state $B$, a vacancy formation and jog advance $C$, and a jump of the vacancy into bulk $D$. In position $C$, the situation is not stable, as the vacancy and the jog can recombine, restoring the initial state $A$. After the jump of the vacancy into bulk (D), the new position $C$ for the gliding dislocation becomes stable.

The situation is a bit different if the gliding dislocation is extended. At the point of intersection (the position of a jog) the stacking fault forms a constriction (Fig. 3.6). After the generation of a vacancy, the jog and the constriction move away, while the vacancy remains attached to one of the partial dislocations, which decreases the activation energy by an interaction energy between dislocation and vacancy.

The temperature dependence of the yield strength is derived from the following equation for the strain rate in thermally activated processes (Seeger 1954):

$$\dot{\varepsilon} = NFb\nu_0 \exp\left\{-\frac{U_0 - V(\tau - \tau_G)}{kT}\right\} , \tag{3.4}$$

where $N$ is the density of the sites at which the activation takes place, $F$ is the area swept by the dislocation during overcoming an obstacle, $\nu_0$ is a frequency in the order of magnitude of Debye's frequency, $U_0$ is the free activation energy at zero stress, $V$ is the activation volume, $\tau$ is the shear stress and $\tau_G$ is the athermal component of the shear stress the temperature dependence of which is that of the shear modulus $G$.

The activation volume is $V = bld$, where $b$ is the modulus of the Burgers vector, $l$ is the distance between the sites at which the dislocation line is being held up and $d$ is an activation distance (about the size of the obstacle to be overcome). If the generation of vacancies is the controlling thermally activated process, $d$ is of about the interatomic distance $b$. For extended dislocations $d$ is about equal to the separation of the half-dislocations (~ 10$b$).

If Eq. (3.4) is solved for the shear stress one obtains (see Seeger 1955):

$$\tau = \begin{cases} \tau_G + \left[U_0 - kT\ln\left(NFb\nu_0/\dot{\varepsilon}\right)\right]/V & T < T_0 \\ \tau_G & T > T_0 \end{cases} \tag{3.5}$$

with

$$T_0 = U_0 / k \ln(NFb\nu_0/\dot{\varepsilon}).$$

Equation (3.5) is represented in Fig. 3.7a. The temperature dependence of the flow stress is characterised by the athermal component $\tau_G$, a transition temperature $T_0$ and the thermally activated stress $\tau_T$ which has to be added to thermal fluctuations in order to push the dislocation line through the obstacle. At $T = 0$ K, where thermal fluctuations are frozen, $\tau_T = U_0/V$ is obtained.

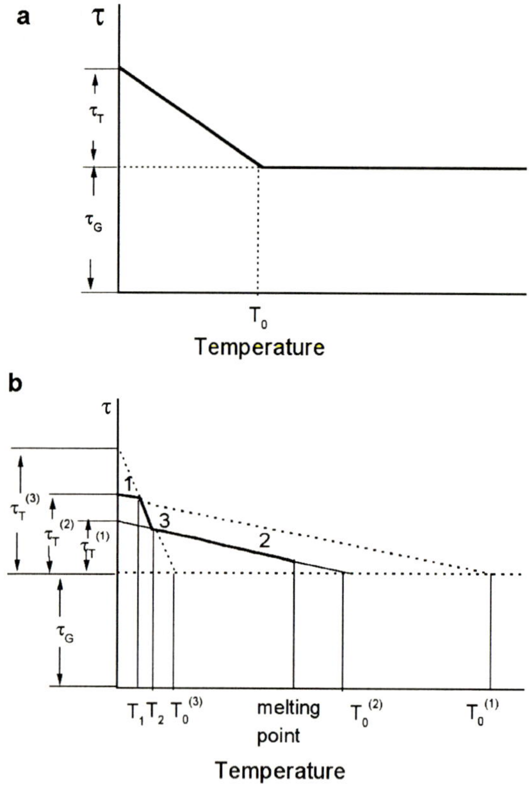


**Fig. 3.7.** Flow stress *versus* temperature for fcc crystals with a low value of stacking fault energy, according to Seeger's theory (Seeger 1955). **a** a representation of Eq. (3.5) if only one thermally activated process is operating, **b** the temperature dependence of the flow stress for the three controlling mechanisms of plastic deformation: cutting of edge (1) and screw (2) dislocations through the dislocation forest and generation of vacancies by jogs in screw dislocations (3), as predicted to occur in crystals with low stacking fault energy

Seeger has shown that for fcc crystals of a low stacking fault energy, the temperatures $T_0^{(1)}$ and $T_0^{(2)}$ for processes (1) and (2) are extremely high (for Cu they are higher than the melting point). At a given temperature the plastic flow will be controlled by those processes which require the lowest stress for the movement of dislocations. As the process (2) is a prerequisite for the process (3), the one that requires the higher shear stress will control the slip of screw dislocations. Superimposing the curves representing Eq. (3.5) for the above three processes results in the full parts of straight lines observed in the experiment (see Fig. 3.7b). Just such a mechanical behaviour of austenitic nitrogen steels was observed by Nyilas et al. (Fig. 3.4). Therefore, in accordance with Seeger's theory, part 2 of the flow stress in Fig. 3.7b represents, as a controlling mechanism, the cutting of screw dislocations through the forest of dislocations threading the slip plane, part 3 is controlled by the generation of vacancies by jogs in screw dislocations and part 1 corresponds to the cutting of edge dislocations through the dislocation forest.

We can now discuss the question why nitrogen abnormally increases the flow stress of austenitic steels at low temperatures (see about details Gavriljuk et al. 1998 I). Figure 3.8 shows the yield strength versus temperature of three austenitic Cr18Ni16Mn10 steels containing no interstitials (within the limits of 0.05 mass %) or alloyed by carbon or nitrogen. The temperature behaviour of the yield stress is obviously consistent with Seeger's theory.

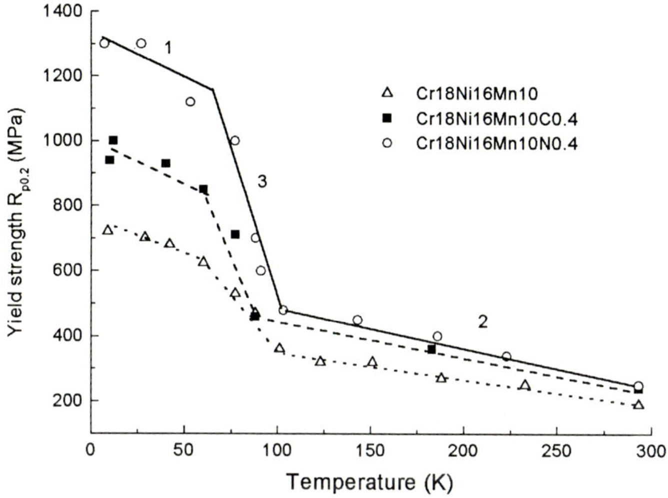


**Fig. 3.8.** Effect of nitrogen and carbon on the temperature dependence of the yield strength of the austenitic steel Cr18Ni16Mn10. Stages 1 to 3 controlling different mechanisms of plastic flow are shown, according to Seeger's theory

It follows from the comparison with Fig. 3.7b that the third and first stages of Seeger's scheme, the generation of vacancies and the cutting of edge dislocations through the forest, control the plastic flow at temperatures below 100 K. An increasing slope of the lines representing the third stage is mainly responsible for the increase in the low temperature strength if nitrogen is added to the austenitic CrNiMn steel. Thus, it is the movement of screw dislocations and, namely, the generation of vacancies by the gliding screw dislocations, that discerns the low temperature mechanical behaviour of the three steels under investigation. As shown above, the generation of vacancies by jogs in case of a low stacking fault energy contains, as a preceding stage, the formation of a constriction in the extended dislocations. The more split dislocations there are, the higher the applied stress has to be to form the constriction. Therefore, one can suggest that the nature of the different mechanical behaviour of the studied steels in stage 3 can be connected with a different splitting of dislocations at low temperatures. In other words, carbon and nitrogen can change the temperature dependence of the stacking fault energy to a different degree.

This suggestion is confirmed by the results of low temperature TEM studies (Fig. 3.9). At room temperature, alloying of the steel Cr18Ni16Mn10 by nitrogen causes a significantly higher value of SFE as compared to carbon alloying (see also Fig. 1.36 for the dependence of SFE on the content of nitrogen in this steel). Nevertheless, the SFE in nitrogen austenite decreases sharply with lowering the temperature, whereas the carbon austenite reveals a shallow temperature dependence of SFE.

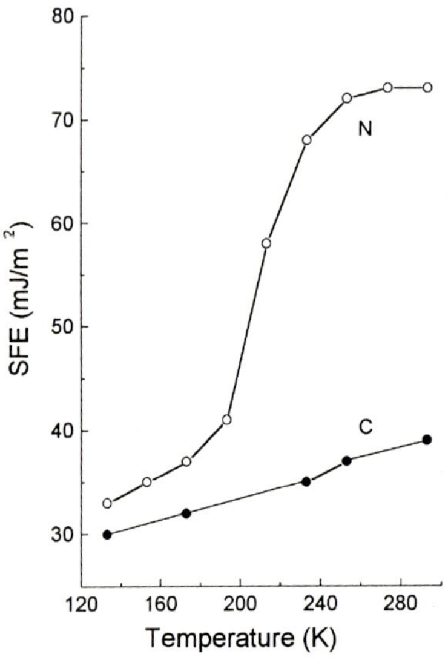


**Fig. 3.9.** Effect of nitrogen (0.4 mass %) and carbon (0.4 mass %) on the temperature dependence of the stacking fault energy (SFE) of the austenitic steel Cr18Ni16Mn10

There is also the opportunity to assess the splitting of dislocations at temperatures as low as that of liquid nitrogen using for this purpose the effects of strain-enhanced migration of interstitials and of their segregation at dislocations. It is generally known that plastic deformation accelerates the migration of solute atoms and causes the so-called "abnormal mass transfer" (see Larikov et al. 1975). For example, an enhanced migration of carbon in α-iron during deformation in liquid nitrogen was studied by Gavriljuk et al. (1981) using the radioactive isotope $C^{14}$.

The segregation of nitrogen and carbon atoms at dislocations during deformation at 77 K was used to preserve the low temperature splitting of dislocations and observe it at room temperature. Samples of austenitic steels Cr18Ni16Mn10 alloyed by carbon or nitrogen were deformed in tension by 2 % in liquid nitrogen. The dislocation structure is presented in Fig. 3.10.

Tangled dislocations are observed in the steel alloyed by carbon, whereas a clear splitting occurs in the nitrogen steel. This experiment allows one to conclude that the tendency of a stronger temperature dependence of the SFE in the nitrogen austenite revealed by the measurements down to 130 K (Fig. 3.9) continues also at lower temperatures. The value of SFE in the nitrogen austenite at 77 K is much lower as compared to that in the carbon austenite.

It is worth of noting that a similar effect was observed in the studies of the dislocation structure of austenitic nitrogen steel 316LN after cyclic deformation at 77 K (Vogt et al. 1993). In their TEM studies at room temperature the authors observed stacking faults after cyclic deformation of austenitic nitrogen steel at 77 K, whereas the structure of irregular walls and starting cell formation (i.e. the structure corresponding to narrow dislocations) was formed after fatigue tests at 300 K. No stacking faults or split dislocations were found after cyclic deformation at 77 K of the nitrogen-free steel 316L.

a

b

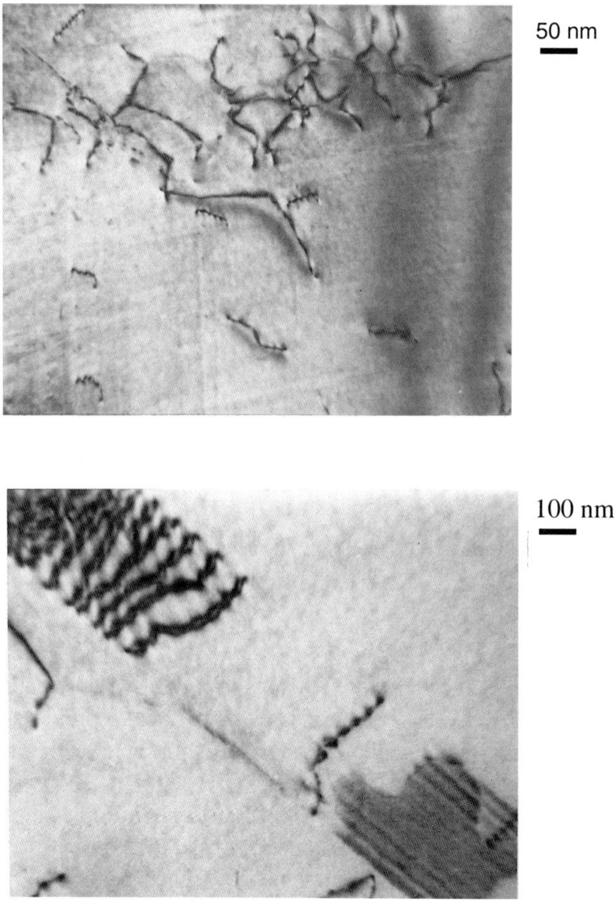


**Fig. 3.10.** Dislocations in steel Cr18Ni16Mn10 containing **a** 0.4 mass % of C and **b** 0.4 mass % of N after 2 % elongation at 77 K

Thus, the beneficial influence of nitrogen on the low temperature strength of austenitic steels can be attributed to its effect on the stacking fault energy at low temperatures. The constriction of split screw dislocations preceding the following generation of vacancies by jogs (see Figs. 3.5 and 3.6) needs an increasing applied stress as compared to narrow dislocations.

An explanation of this nitrogen effect can be proposed based on the results of studies of the electronic structure of austenitic nitrogen steels (Sect. 1.1.1.2). The alloying of austenitic steels by nitrogen increases the concentration of free electrons, i.e. leads to a higher density of states at the Fermi surface, whereas in carbon-containing steels the concentration of free electrons is decreased and a covalent interatomic bonding is promoted. A correlation occurs between the values of SFE and the state density at the Fermi surface (Sect. 1.1.5.1), although the nature and character of this correlation is not yet properly understood. The Fermi statistics predict a temperature dependence of the density of states at the Fermi surface (see scheme in Fig. 3.11).

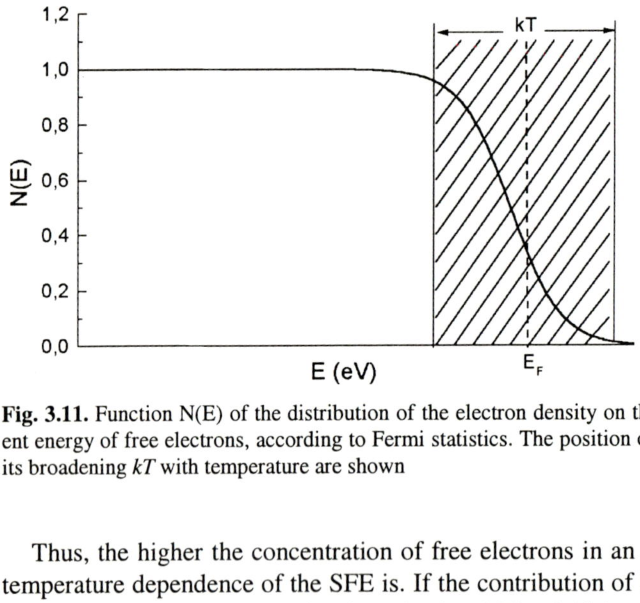


**Fig. 3.11.** Function N(E) of the distribution of the electron density on the states having a different energy of free electrons, according to Fermi statistics. The position of the Fermi level $E_F$ and its broadening $kT$ with temperature are shown

Thus, the higher the concentration of free electrons in an alloy, the clearer the temperature dependence of the SFE is. If the contribution of localised electrons to the atomic interaction prevails, the scheme of Fig. 3.11 is not applicable and no significant temperature dependence of the SFE can be observed.

An increase of the density of states by nitrogen in austenite and the nitrogen-enhanced temperature dependence of the SFE confirm that the values of the SFE in austenitic steels are essentially determined by their electronic structure.

## 3.1.2
## Strengthening by grain boundaries

Grain boundaries in austenitic steels are not effective strengtheners. They contribute to strengthening about one half of that in ferrite (Irvine et al. 1969). In this relation alloying of austenitic steels with nitrogen is of particular interest.

Nitrogen-induced grain boundary strengthening of austenitic steels is the second (after low temperature strengthening) enigmatic nitrogen effect. It is usually expressed as an increase of the coefficient $k_y$ in the Hall-Petch equation describing the effect of the grain size on the stress corresponding to the yield strength:

$$R_{p0.2} = \sigma_0 + k_y d^{-1/2},$$

where $\sigma_0$ is a friction stress, $k_y$ is a coefficient characterising the transfer of slip through the grain boundaries and d is the grain size. The yield strength of austenitic nitrogen and carbon steels in the Hall-Petch coordinates is presented in Fig. 3.12.

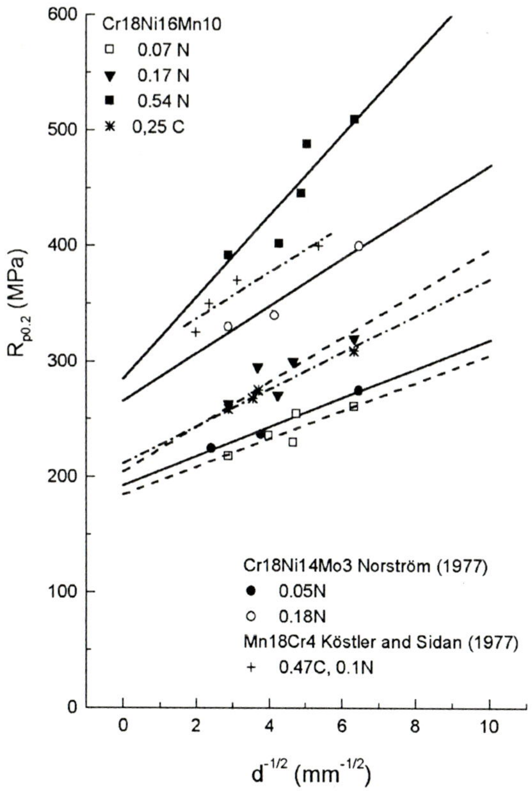


**Fig. 3.12.** Effect of nitrogen and carbon on grain boundary strengthening of steel Cr18Ni16Mn10. Data of Norström (1977) for the steel Cr18Ni14Mo3 alloyed by nitrogen and of Köstler and Sidan (1977) for the steel Mn18Cr4 alloyed by carbon are also shown for the comparison

With increasing nitrogen content, the effectiveness of grain boundaries in strengthening grows. Carbon is also found to promote grain boundary strengthening, however, its effect is much smaller.

Norström (1977) was the first to show that nitrogen in AISI 316L type austenitic steels causes, beside the solid solution strengthening, a significant grain size dependent increase of the yield strength (see his data in Fig. 3.12). He has also shown that the nitrogen-caused component of grain boundary strengthening does not depend on temperature up to 600°C.

Before discussing the mechanism of the contribution of nitrogen to the grain boundary strengthening, it is expedient to describe briefly the available interpretations of the Hall-Petch equation. A number of proposed theories can be divided into three main groups based on the critical role of planar dislocation arrays (Hall 1951, Petch 1953, Koehler 1952, Cottrell 1958, Armstrong et al. 1962), of work hardening (see e.g. Johnson 1962, Conrad 1963, Conrad et al. 1967) and of grain boundary sources (e.g. Li 1963, Li and Chou 1970).

The first one was proposed by Hall (1951) and Petch (1953) who postulated the formation of dislocation pile-ups at the grain boundaries, causing a stress concentration $n\sigma$ at the grain boundary or within it. A modification of this pile-up model was proposed by Cottrell (1958) who determined as critical a stress needed to unlock the dislocation sources in the adjacent grain. The main problem concerned with the pile-up model is that the planar slip is not a general case. Pile-ups are not observed in metals and alloys with a bcc crystal lattice, although the Hall-Petch equation is always correct.

In the cold work hardening model one postulates a stress-independent $\sigma_0$, a parabolic stress-strain relation $\sigma \sim \sqrt{\varepsilon}$ and a linear relation between the Hall-Petch slope $k_y$ and the square root of strain, $k_y \sim \sqrt{\varepsilon}$. A linear relation between the yield stress and the square root of the dislocation density is accepted as an experimental fact.

According to the work hardening model, the dislocation density is increased with decreasing grain size expressed as $d^{-1}$, which automatically leads to the Hall-Petch relation and, therefore, the term $kd^{-1/2}$ is attributed to work hardening. The relation $k_y \sim \sqrt{\varepsilon}$ was found in niobium (Conrad et al. 1967) but was not confirmed in Fe-Co alloys (Marcinkovski and Fisher 1965). Moreover, the Hall-Petch slope was observed to decrease with strain in copper (Carreker and Hibbard 1953), silver (Carreker 1957), aluminium (Carreker and Hibbard 1957) and iron (Li and Chou 1970).

The work hardening model has been strongly supported by the theory proposed by Ashby (1970) who was the first to introduce a definition of geometrically necessary dislocations stored during deformation of plastically non-homogeneous materials and showed that an increase in their density near the grain boundary with strain causes a grain-size-dependent contribution to the work hardening of polycrystals. Kashyap and Tangri (1995) who studied the evolution of the substructure of type 316L stainless steel in a wide range of strains found a higher dislocation density at the grain boundaries as compared to the grain core and explained strengthening by a barrier effect of these grain boundary layers of increased dislocation density.

A different role of deformation was supposed by Fischmeister and Karlsson (1977) who observed that an increase in the value of the stacking fault energy decreases the coefficient $k_y$ and attributed this effect to the easy formation of a cell structure where the cell boundaries, instead of the grain boundaries, limit the length of free slip of dislocations. Let us remark, however, that, according to the data in Fig. 3.12 for the steel Cr18Ni16Mn10, the coefficient $k_y$ is increased with increasing nitrogen content in spite of a nitrogen-caused increase in the stacking fault energy of this steel (see Fig. 1.36).

In the grain boundary source theory the Hall-Petch relation is determined by the capacity of grain boundaries to emit dislocations under loading. This approach does not require a stress concentration created by pile-ups, and, in contrast to the work hardening model, the Hall-Petch slope $k_y$ does not depend on the strain. Li and Chou (1970) consider, as experimental confirmation of their theory, observations of dislocation loops emitted by grain boundaries, of slip lines at grain boundaries and of grain boundary ledges which are supposed to be the dislocation donors. An important prediction which follows from the grain boundary source theory is that the segregation of atoms at grain boundaries has to affect the Hall-Petch slope $k_y$ through the stabilisation of the ledge structure and an increase in their density. It was shown by Dingley and McLean (1967) that ageing of iron increases $k_y$, in consistence with the grain boundary source theory.

The development of materials with ultrafine-grain size has caused a new wave of studies of the Hall-Petch relation (see Proceedings of the symposium "Grain size and mechanical properties – fundamentals and applications", Pittsburgh, Pensilvania, 1995). The pile-up approach was used for the interpretation of experimental data obtained for ultra-fine grain sizes and a remarkable decrease in the Hall-Petch slope was established (Armstrong 1995, Smith et al. 1995). The Hall-Petch strengthening was also found to be insignificant for nanophase materials (Siegel and Fougere 1995).

Let us now consider the experimental data of grain boundary strengthening in austenitic nitrogen steels. As stated above, Norström (1977) has described the effect of nitrogen on grain boundary strengthening and attributed it to a nitrogen-induced increase in the twin density that appears as an increased Hall-Petch coefficient $k_y$ for a normal grain size.

The experimental data of Norström were confirmed by Degallaix et al. (1986) studying austenitic steels of higher nitrogen content. The twin hypothesis was also used in their study as an explanation for the effect of nitrogen.

This hypothesis was tested by Varin and Kurzydlowski (1988) who varied grain size and density of coherent twin boundaries by a thermomechanical treatment of type 316 austenitic steels. They have shown that an increasing twin density has no measurable effect on the 0.2% yield strength. Based on the data of Briant (1987) on nitrogen segregation at the surface of intergranularily fractured austenitic steels, the authors interpreted their results in terms of the grain boundary segregation, i.e. in accordance with the grain boundary source theory.

Werner (1988) studied austenitic CrNi and CrMn steels alloyed by nitrogen up to 0.6 mass % and illustrated a large increase in the Hall-Petch slope $k_y$ due to nitrogen. He observed a decrease in the value of $k_y$ with increasing nickel content in steels containing the same amount of nitrogen and explained this effect by a nickel-caused increase in the value of the stacking fault energy. In view of the following discussion concerning the role of the locking of dislocation sources by nitrogen atoms in grain boundary strengthening, it is important to note that nickel decreases the interaction between dislocations and interstitials in iron-based alloys (Gavriljuk 1987).

Combining the nitrogen-induced grain-boundary strengthening with work hardening, Uggowitzer and Speidel (1990) succeeded to obtain the yield strength of about 3000 MPa in the austenitic steel Cr18Mn18N0.6. The effect of nitrogen on the Hall-Petch slope was explained in terms of the pile-ups theories (see Uggowitzer and Harzenmozer 1989 and also Sect. 5.2.3).

An increasing planar slip was also considered as a reason for a more effective grain boundary strengthening in austenitic nitrogen steels by Nyström et al. (1997).

Before discussing the increase of grain boundary strengthening by nitrogen in terms of existing theories, let us note that, in view of experimental data of the dislocation structure and mechanical properties of austenitic nitrogen steels, it seems that they are a unique object for testing different theoretical models because nitrogen assists planar slip in austenitic steels, increases work hardening of austenite and, in contrast to carbon, has no detectable or at least smaller affinity to grain boundaries.

As follows from TEM studies (see Sect. 1.1.5.2), planar slip is a peculiarity of deformed nitrogen CrNiMn steels. Planar dislocation arrays were also observed in nitrogen CrNi steels (see e.g. Sassen et al. 1989) and CrNiMo steels (Müllner et al. 1993).

Nitrogen assists also the deformation twinning in austenitic steels. It was observed in CrNiMn steels of low nickel content (see Gavriljuk et al. 1990) and in the molybdenum-containing steel Cr18Ni14Mo3 (Müllner et al. 1993). Twinning also prevails with increasing nitrogen content in the deformed nickel-free steel Cr15Mn17 (see Fig. 1.39). At the same time, twinning as a deformation mode becomes less pronounced with increasing nitrogen content in high-nickel steel Cr18Ni16Mn10 (see Fig. 1.38), which arises from an increasing stacking fault energy (Fig. 1.36).

Thus, different substructures can be developed during plastic deformation of austenitic nitrogen steels, but nitrogen always promotes planar slip because of nitrogen-induced short range atomic ordering. An increase in the value of $k_y$ by nitrogen also occurs independent of the basic chemical composition of austenitic steels. So, planar slip and nitrogen-induced grain-boundary strengthening accompany each other.

However, one can conclude, based on the pile-up interpretation of the Hall-Petch equation (see e.g. Koehler 1952), that any factor assisting planar slip has to diminish grain-boundary strengthening, i.e. to decrease the koefficient $k_y$ because the easier a pile-up is formed, the earlier a critical stress at an obstacle (grain boundary) is achieved at the same applied stress. Therefore, in fact, the experimental observations of the nitrogen-caused planar slip in austenitic steels do not support the pile-up theories of grain-boundary strengthening.

As a rule, nitrogen increases cold work hardening of austenitic steels (see Sect. 3.1.3). Nevertheless, in high-nickel austenitic steels, where dislocations are narrow and nitrogen additionally increases SFE, a contribution of nitrogen to cold work hardening was not found (Fig. 3.13, see for details Glavatska 1998).

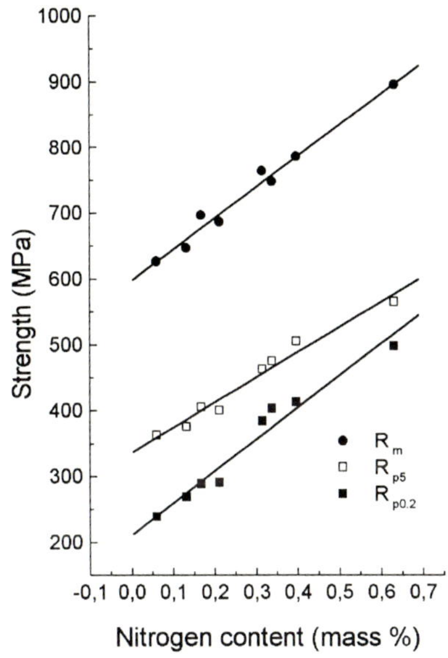


**Fig. 3.13.** Effect of nitrogen on the yield strength $R_{p0,2}$, flow strength $R_{p5}$ and ultimate tensile strength $R_m$ of steel Cr18Ni16Mn10 solution treated at 1150°C for 20 min followed by water quenching

A nearly parallel slope of the yield and ultimate strength with increasing nitrogen content and, moreover, a smaller at an intermediate strain ($R_{p5}$) mean that nitrogen in this steel contributes mainly to the solid solution and not to the work strengthening. Comparing this with the nitrogen effect on the grain boundary strengthening in the same steel it is clear that such a behaviour is not consistent with the work hardening theory.

It seems that work hardening, as it is represented in Ashby's model, can essentially contribute to the grain boundary strengthening in heavy deformed materials. However, not only the grain boundary strengthening factor but also its role in controlling the applied stress is diminished with decreasing strain.

A comment on the validity of the grain boundary source theory can be made based on the data of nitrogen and carbon effects on the Hall-Petch slope (Fig.

3.12). Carbon contributes to the grain boundary strengthening, however, its effect is significantly smaller than that of nitrogen. At the same time, carbon is known to have a strong affinity to grain boundaries in austenitic steels (see e.g. Petrov 1993), whereas no detectable segregation of nitrogen at grain boundaries was observed in austenitic (Petrov et al. 1999) as well as in ferritic steels (Rudy and Huggins 1966, see also Sect. 1.1.4.3).

Therefore, no marked nitrogen contribution to grain-boundary strengthening or at least smaller than that by carbon is expected according to the grain boundary source theory in contrast to the experimental evidence.

A comparative analysis of the effects caused by nitrogen and carbon in austenitic steels may serve as a key to the interpretation of the Hall-Petch equation. A relevant difference between nitrogen and carbon in austenitic steels is their different binding to dislocations. Taking into account that nitrogen atoms in austenite reveal a strong interaction with dislocations, let us test whether nitrogen and carbon effects on the grain boundary strengthening correlate with the locking of dislocation sources, which was supposed by Cottrell (1958) to be a controlling factor in the transfer of slip through the grain boundaries. Internal friction technique was used as described in Sect. 1.1.4.2 (see in detail Gavriljuk et al. 1999 I).

**Table 3.2.** Enthalpy of binding of nitrogen and carbon atoms to dislocations in the austenitic steel Cr18Ni16Mn10 as obtained from internal friction measurements

| Content (mass %) | 0.06N | 0.17N | 0.54N | 0.25C |
|---|---|---|---|---|
| $H_b$ ±0.05 eV | 0.56 | 0.66 | 0.76 | 0.62 |

As follows from Table 3.2, in accordance with the previous results (see Table 1.7), the binding of nitrogen atoms to dislocations is stronger than that of carbon atoms, and, what is particularly important, the pinning of dislocations grows with an increasing nitrogen content.

A clear correlation between the interstitial-dislocation interaction and the contribution of nitrogen and carbon to grain boundary strengthening (see Fig. 3.12) supports Cottrell's model.

Thus, based on the available experimental data that were used for the testing of different theories proposed for the explanation of the Hall-Petch relation, one can conclude that the contribution of nitrogen to grain boundary strengthening is provided by a strong interaction of nitrogen atoms with dislocations locking dislocation sources which have to be initiated in adjacent grains for the transfer of plastic deformation from one grain to others. The unlocking of dislocation sources requires an increased applied stress.

One remark to Cottrell's model is that planar dislocation arrays at grain boundaries are not needed to provide the necessary critical stress.

## 3.1.3 Cold work strengthening

The nitrogen-enhanced cold work strengthening of austenitic steels was described, perhaps first time, by Uggowitzer and Harzenmoser (1989). In combination with solid solution strengthening it allows to achieve high values of strength (Fig. 3.14). The slope of the yield and ultimate stresses of cold worked steels, as a function of strain, grows with increasing content of nitrogen, and, what is remarkable, nitrogen prevents the decrease in the coefficient of cold work hardening at large strains typical for nitrogen-free austenitic steels.

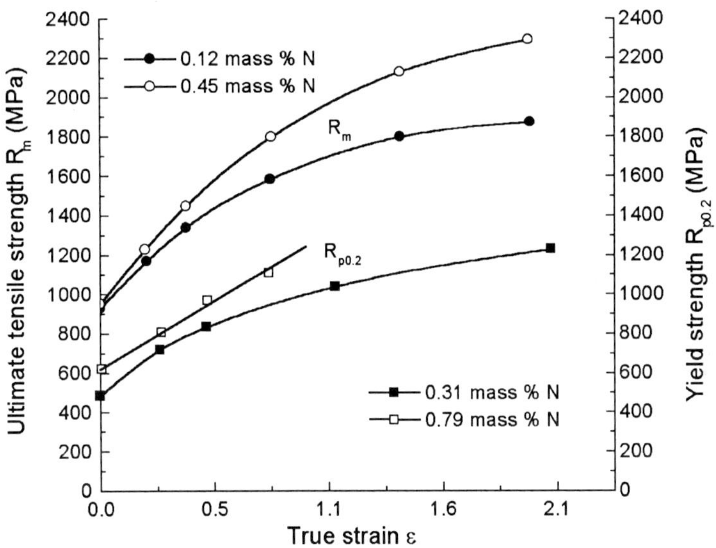


**Fig. 3.14.** Effect of nitrogen on the yield strength (after Uggowitzer and Harzenmoser 1989, steels Cr18Mn18N0.8 and Cr18Mn12N0.3) and ultimate strength (after Gavriljuk et al. 1990, steels Cr21Ni6Mn10N0.1 and ...N0.45) of cold drawn austenitic steels

Mechanical properties of cold worked austenitic nitrogen steels can be understood through a detailed analysis of their structure.

First of all, an important feature of the dislocation structure in deformed fcc crystals is its orientation dependence which arises from the effect of applied stress on the splitting of dislocations (see Copley and Kear 1968, Kestenbach 1977, Vergnol and Grilhe 1984, Chumlyakov et al. 1993). Under tension, as a result of interaction of the external stress field with the partial *a*/6<112> dislocations, an additional dislocation splitting occurs in the <111> and <110> orientations, while the splitting is diminished in the <100> orientation. The former orientations become "soft" while the latter is "hard". Thus, because of the difference in the effective splitting of dislocations in different crystallographic directions, an orientation dependence of the shear stress and work hardening occurs. Due to a change in the stacking fault energy and solid solution strengthening, nitrogen increases the crystallographic anisotropy of plastic deformation, which gives an important contribution to the mechanism of cold work strengthening during deformation by rolling or drawing when the crystallographic texture is formed.

As shown in Fig. 1.39, nitrogen enhances twinning in austenitic CrMn steels. A low SFE is needed for mechanical twinning as the latter comprises extended stacking faults. Thus, a decrease in the SFE due to nitrogen in austenite may be one of the reasons why nitrogen promotes twinning during cold work. Another reason lies in the nitrogen-caused solid solution strengthening that increases internal friction for dislocations passing over the slip plane and, therefore, prevents slip and increases the shear stress to the value needed for twinning.

An inhomogeneous twin structure in the cold drawn austenitic nitrogen steel was observed by Müllner et al. (1993): after heavy deformation the grains being twinned on all systems available coexisted with grains twinned on one system or containing no twins. The latter had usually <100> orientation along the wire axis. This feature was explained in accordance with the above mentioned effect of the applied stress on the dislocation splitting. A double fibre texture with <111> and <100> directions near to the wire axis is developed during cold drawing of fcc metals. Thus, extended stacking faults and enhanced twinning prevail in «soft» <111> crystals, while narrow dislocations and rare twins are typical for «hard» <100> crystals.

Müllner et al. (1994) studied also the phenomenon of second order twinning in deformed austenitic steels. They have shown that second order twinning is activated by the stress field of primary twins and provides a relaxation of energy of the primary twin-dislocation structure up to 30 % and more. Both primary and secondary twins are effective obstacles for the moving dislocations and further twinning and, therefore, contribute effectively to cold work hardening.

The influence of the crystallographic orientation on the structure of austenitic steels was studied by Akdut et al. (1997) using single crystals. They found a decrease in the scattering of the {111} orientations in cold rolled crystals due to nitrogen, which is a sign of a higher deformation homogeneity and results from the nitrogen-caused planarity of slip. Another important observation was the development of cracks after ~30% thickness reduction. This effect was enhanced by nitrogen. Further deformation led to healing of cracks which were closed again after 60% thickness reduction. The intermediate cracking of the austenitic phase caused by plastic deformation of high austenitic nitrogen steels or nitrogen duplex steels was also observed by the same authors (see Akdut and Foct 1996, Foct et al. 1993).

It is noteworthy that Akdut et al. (1997) did not find an effect of nitrogen on the microhardness of the cold rolled high-nickel steel Cr18Ni16Mo2.

Let us analyse the development of structure during cold work of austenitic nitrogen steels using, as representatives, the nickel-free steel Cr15Mn17 and the high-nickel steel Cr18Ni16Mn10. As shown in Sect. 1.1.5.2, the values of stacking fault energy and the effect of nitrogen on the SFE and substructure are quite different in these two types of steel. Different is also their work strengthening behaviour.

The steel Cr15Mn17N0.2 is characterised by an SFE of about 25 mJ/m$^2$. Plastic flow starts mainly due to dislocation slip (Fig. 1.39b) while twinning occurs at initial stages of deformation but does not play an essential role. After rolling by 30% thickness reduction, the twins on the (111) planes appear in the structure along with an increased dislocation density (Fig. 3.15a). The twins are deformed and broken by dislocation slip. Twinning on intersecting {111} planes can be observed after rolling by 45% thickness reduction (Fig. 3.15b). However, the twin boundaries are strongly blurred.

a

**Fig. 3.15.** Twin structure of the steel Cr15Mn17N0.2 after cold rolling: **a** twins crossed by dislocation slip after 30% reduction in thickness, **b**: twinning on intersecting {111} planes after 45% reduction in thickness

At the same time, even at higher deformations freshly formed twins are met (Fig. 3.16), which suggests that twinning appears periodically in the heavily deformed steel. Thus, dislocation slip and twinning can substitute each other during plastic deformation.

**Fig. 3.16.** Freshly formed twins in steel Cr15Mn17N0.2 after 60% reduction in thickness

Further deformation leads to the disappearance of twins. The most typical structure is shown in Fig. 3.17a where a curved band structure is seen. The same area is shown in Fig. 3.17b at a higher magnification. Domains resembling dynamic recovery have, in fact, a high density of dislocations and differ in the contrast conditions only.

One can conclude in view of the presented evolution of the structure that twinning and slip are concurrent during the plastic deformation of austenitic nitrogen steels. Their appearance depends on the crystallographic texture formed due to deformation: twinning as a prevailing mechanism of deformation is operating under conditions when slip is exhausted and vice versa. The areas of slip are limited by the twin boundaries.

A significant increase of internal stresses that occurs at large strains results in another mechanism of deformation, namely in rotation modes which arise from collective effects in the systems of interacting dislocations and cause the formation of the band structures consisting of disoriented areas with a high density of dislocations. These bands can be comparable in their size with the deformation twins, however, they have no definite crystallographic orientations. The contribution of the boundaries of deformation bands to cold work strengthening can be comparable with that from deformation twins. One can expect that the decrease in the stacking fault energy assists the earlier appearance of the deformation bands, which in turn has to result in a more effective cold work strengthening.

a

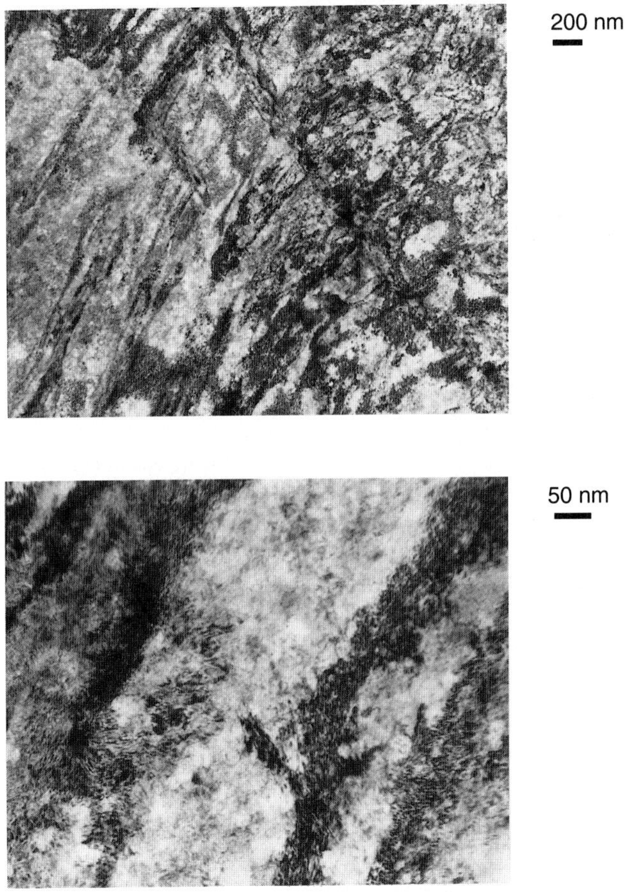


b

**Fig. 3.17.** Band structure of the steel Cr15Mn17N0.2 after cold rolling by 75% of reduction in thickness; **a** and **b** differ only in magnification

Thus, the plastic behaviour of austenitic nitrogen steels at large strains cannot be simply described by conventional dislocation slip or twinning. A reliable interpretation of the structural changes has to be based on the theory of disclinations which explains satisfactorily the fragmentation of structure due to rotation modes of deformation (see e.g. Rybin 1986, Vladimirov and Romanov 1986, Vladimirov 1988, Koneva and Kozlov 1990).

An increase in the nitrogen content to 0.45 mass % in the steel Cr15Mn17 results in a decrease of stacking fault energy (about 17 mJ/m$^2$) and, correspondingly, causes an early start of twinning during deformation (see Fig. 1.39). Further deformation leads to the breaking and disappearance of twins (see an example in Fig. 3.18 for 30% thickness reduction).

**Fig. 3.18.** Start of break-up of the twin structure in steel Cr15Mn17N0.45 cold rolled by 30% of reduction in thickness. Extinction contours give an evidence of significant internal stresses

**Fig. 3.19.** Band structure in steel Cr15Mn17N0.45 after cold rolling by 60% reduction in thickness

After large strains (60% thickness reduction) twins are practically not met in the structure. Bands of different orientations like those in Fig. 3.19 are the main structural elements. These bands contain a high density of dislocations and do not result from the dynamical recovery, which is clear from Fig. 3.20 where the same area of structure is shown with higher magnification. That is why the cold work strengthening remains sufficiently high although twinning does not much contribute to the plastic deformation.

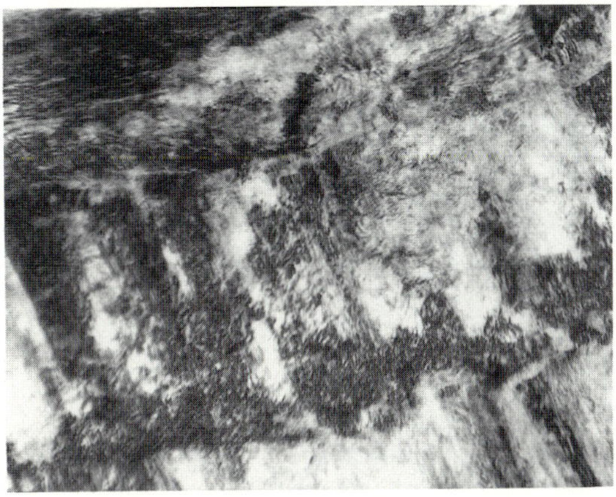

100 nm

**Fig. 3.20.** Crossing bands in the structure of steel Cr15Mn17N0.45 after cold rolling by 60% reduction in thickness. Fragment of structure presented in Fig. 3.19 is shown with higher magnification

The formation of the band structure prevails also during further deformation (Fig. 3.21) and, at the same time, some apparently non-textured areas with permanent and discrete disorientations appear (Fig. 3.22). They can be described as belonging to the dislocation-disclination substructures according to the classification proposed by Koneva and Kozlov (1990), see also Koneva et al. (1988). One has again to underline that no sign of dynamical recovery is observed although the cold work strengthening is decreased as compared to smaller deformations where the contribution of twins was essential.

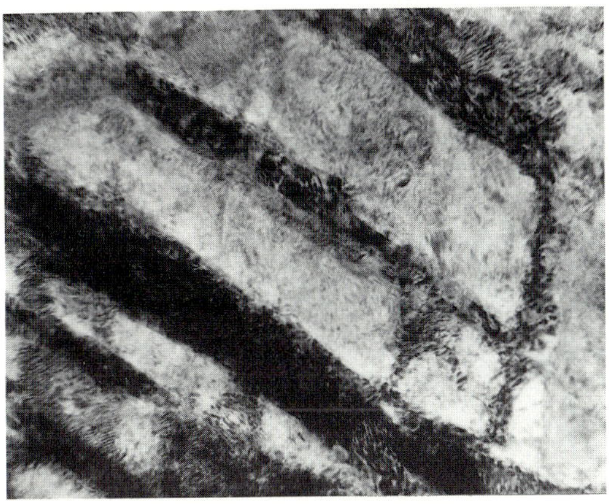

100 nm

**Fig. 3.21.** Structure of steel Cr15Mn17N0.45 after cold rolling by 75% of reduction in thickness

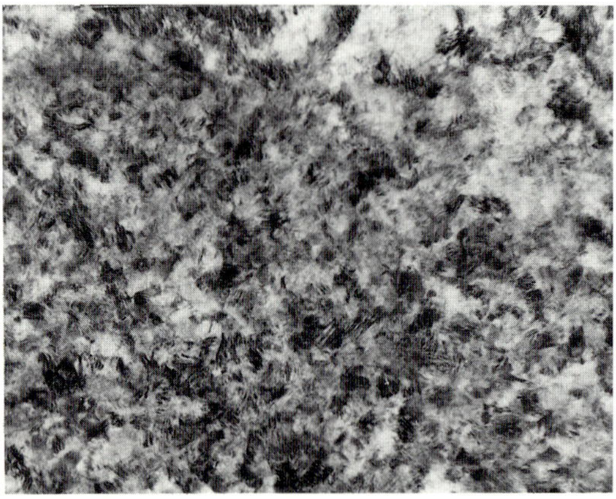


**Fig. 3.22.** Disoriented domains with high density of dislocations in the steel Cr15Mn17N0.45 after cold rolling by 75% of reduction in thickness

In contrast to Cr15Mn17, the high-nickel steel Cr18Ni16Mn10 does not reveal remarkable twinning during plastic deformation. Nitrogen causes a non-monotonous change in the values of stacking fault energy from 43 $mJ/m^2$ at 0.06 mass % N to its maximum value of 65 $mJ/m^2$ at 0.4 mass % N and down to 53 $mJ/m^2$ at 0.54 mass % N (see Fig. 1.36). In accordance with the concentration dependence of SFE, twins are present as a minor component of the substructure in the nitrogen-free deformed steel. They are practically not observed in the steel containing 0.4 % N and appear again if the nitrogen content is increased to 0.54 mass %.

**Fig. 3.23.** Island of the freshly formed twins in steel Cr18Ni16Mn10N0.54 after cold rolling by 45% reduction in thickness. High internal stresses have caused the cracking of the foil

In main features, the structure of cold rolled steel Cr18Ni16Mn10N0.54 resembles that of the heavily rolled steel Cr17Mn17N0.2 already after 45% thickness reduction. An example is shown in Fig. 3.23 where freshly formed twins coexist with the highly deformed non-twinned area. Again, one should state that this variety of structures cannot be understood in terms of the conventional slip or twinning, but only by taking into account the collective modes of deformation.

Measurements of the hardness show that nitrogen does not markedly contribute to cold work hardening of the high-nickel austenitic steel Cr18Ni16Mn10 (Fig. 3.24). The strain-induced increment of the microhardness is not increased by increasing the nitrogen content, whereas solid solution hardening was remarkable (see also Glavatska 1998). This behaviour of microhardness is consistent with the results of tensile tests (see Fig. 3.13) and with the measurements of microhardness performed by Akdut et al. (1997) on the steel Cr18Ni16Mo2N0.25.

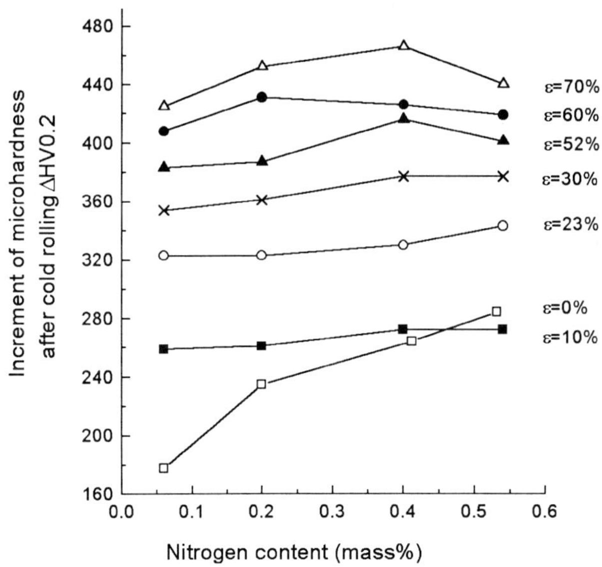


**Fig. 3.24.** Increment of microhardness of the steel Cr18Ni16Mn10N due to cold rolling by different thickness reduction (ε) as a function of the nitrogen content (after Glavatska 1998). Strain hardening does not clearly depend on the content of nitrogen

Thus, we can conclude that nitrogen provides a notable cold work strengthening in CrMn and low-nickel austenitic CrNiMn steels, but adds not much to that in high-nickel CrNiMn and CrNi steels. A reason for such a different behaviour is that the latter have high values of stacking fault energy and do not reveal a remarkable twinning during plastic deformation.

It is twinning that changes the course of cold work strengthening curves with increasing nitrogen content (see Fig. 3.14), so that a decrease in the increment of strength typical for highly deformed austenitic steels is shifted to higher strains.

Twins play in cold worked austenitic steels exactly the same role as cementite plates in pearlitic steels where cold drawing or rolling is used as an extremely effective way of obtaining high-strength wire or strip. However, as shown in the TEM study present above, twins unlike cementite plates do not survive at large strains. Thus, the role of nitrogen amounts to a prolongation of the twin life to higher degrees of deformation in CrMn and low-nickel CrNiMn steels where nitrogen decreases the stacking fault energy and assists mechanical twinning. If nitrogen does not affect the SFE or even increases it, the role of nitrogen is limited to solid solution strengthening.

In view of the presented results, one can conclude that the effect of nitrogen on cold work strengthening is quite moderate. Stable single phase austenitic steels, even being strengthened by nitrogen, are not very promising materials for high-strength long products such as wire, rope, strip. Based on the data of the structural evolution during cold work, one can suppose a higher success in the cold work strengthening of duplex (ferritic-austenitic) nitrogen steels or unstable austenitic steels with strain-induced martensite. The interphase boundaries have to be more effective barriers for plastic deformation as compared to twins (see Sect. 5.3.3).

Summing up, one should add that a strong interaction of nitrogen atoms with dislocations can be used for additional strengthening of cold worked austenitic steels due to strain ageing (Fig. 3.25).

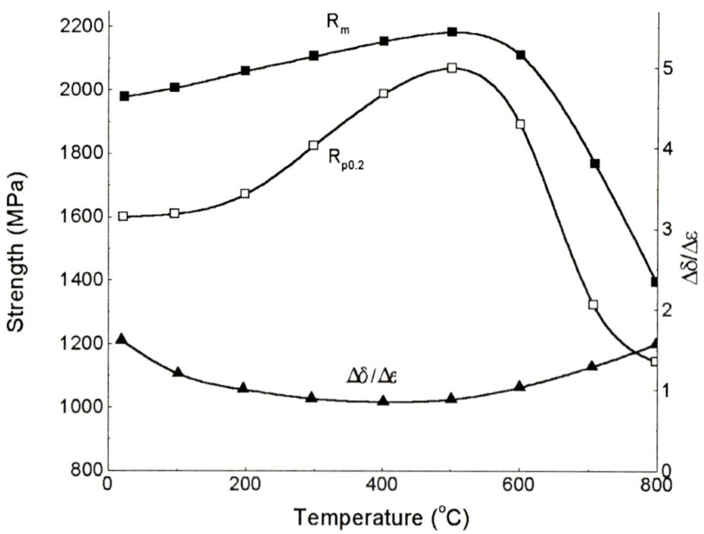


**Fig. 3.25.** Yield and ultimate tensile strength of cold drawn austenitic steel Cr21Ni5Mn10N0.45 after 70% reduction in the cross section as a function of the temperature of strain ageing (after Gavriljuk et al. 1990). The data of the pinning of dislocations by nitrogen atoms are also shown, characterised by the slope of strain-dependent internal friction $\Delta\delta/\Delta\varepsilon$ (area swept by dislocations during free decay of mechanical vibrations, see Sect. 1.1.4.2)

It follows from the presented data that the combination of cold work and strain ageing of austenitic nitrogen steels is in particular effective for an increase of the resistance to small strains. As an example, one can refer to studies performed by Holmberg et al. (1990) where austenitic nitrogen steels were proposed for the production of high-strength stainless springs.

## 3.1.4
## Toughness

A good combination of strength and toughness is a distinctive feature of nitrogen steels. Frehser and Kubisch (1963) were the first to find out that, unlike the usual mechanical behaviour of engineering materials, the toughness of austenitic nitrogen steels is not decreased while the yield strength is increased due to alloying by nitrogen. We illustrate this effect in Fig. 3.26 using the data of Harzenmoser (1990) who studied the effect of nitrogen on the fracture and impact toughness in a wide range of temperatures and basic chemical compositions.

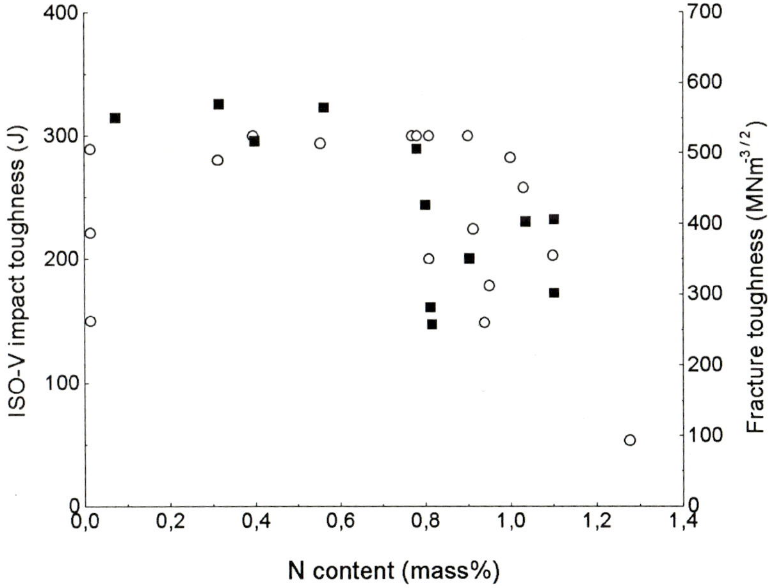


**Fig. 3.26.** Influence of nitrogen on the fracture toughness (solid squares) and ISO-V impact (open circles) of austenitic steels of Cr18Mn18 basic composition. Tests are performed at room temperature (after Harzenmoser 1990)

In accordance with other experimental data available, alloying by nitrogen up to 0.8-1.0 mass % does not deteriorate toughness, which allows to obtain a high product of yield strength and fracture toughness – a criterion that is often used by metallurgists for the evaluation of engineering materials.

The experimental data of the electronic structure of austenitic nitrogen steels (see Sect. 1.1.1) enable us to explain this nitrogen effect as a result of the increased concentration of free electrons, which enhances the metallic component of the interatomic bonding. One can state that the nitrogen-assisted metallic character of atomic interaction in austenitic steels withstands the embrittling influence of nitrogen expected for interstitial impurities. An opposite situation occurs for carbon in steels where both the carbon-induced elastic distortions and the enhanced covalent component of atomic interaction act as embrittling factors.

The effect of transition metals (Cr, Mn, Ni) on the toughness of austenitic steels is also consistent with their influence on the electronic structure. Manganese and chromium decreasing the metallic component of interatomic bonding in austenitic steels do not assist acceptable values of fracture toughness. At similar contents of nitrogen, a decrease of the chromium content in the CrMnN steel allowed to obtain at 4K a higher fracture toughness than in CrNiMnN steel (Harzenmoser et al. 1990). On the contrary, nickel, being a donor of free electrons in iron alloys, increases fracture toughness, in particular at low temperatures. The combination of nitrogen and nickel in austenitic steels is proven to provide excellent fracture toughness at low temperatures, as well as high strength. The steel Cr25Ni15N0.35 (CSUS-JN1) has at 4.2K a fracture toughness $K_{Ic}$ = 217MPa·m$^{1/2}$ and a yield strength $\sigma_{0.2} \approx$ 1400 MPa (Nyilas et al. 1993), which satisfies the requirements for structural materials intended for the design of large superconducting magnets in International Thermonuclear Experimental Reactor (ITER, see also Sect. 5.2.6).

At the same time, a high nitrogen content causes brittleness of austenitic steels even at room temperature. In nickel-free CrMn steels this is observed if the nitrogen content approaches about 1 mass % (Fig. 3.26). When the temperature is decreased, brittleness occurs at smaller nitrogen contents and even in nickel-containing CrNiMn steels (see Tobler and Meyn 1988). Cleavage is observed along the crystallographic planes {111}, i.e. along the slip planes, that is why it was characterised by Tobler and Meyn as a slip band cleavage-like cracking. Pure cleavage would result in {001} fracture planes.

This nitrogen-caused cleavage-like fracture is observed in impact testing as well as in tension tests. Impact tests reveal a clear ductile-to-brittle transition (DBT) that generally does not occur in the alloys with an fcc lattice. Typical results of ISO-V impact testing are presented in Fig. 3.27. A striking difference between nitrogen- and carbon-containing austenitic steels is that the latter are characterised by a gradual decrease of impact toughness with decreasing temperature.

The cleavage-like fracture of austenitic nitrogen steels is characterised by some features: (i) alloying elements have a strong effect (nickel suppresses whereas nitrogen, manganese and chromium assist cleavage, see Harzenmoser 1990, Ishizaki et al. 1992, Uggowitzer et al. 1992); (ii) the grain size has an insignificant effect (Tomota and Endo 1990), which is quite different from the behaviour of bcc materials where conventional cleavage is observed; (iii) preliminary deformation does not decrease the temperature of DBT as it occurs in bcc materials (Tomota and Endo 1990, Ishizaki et al. 1990); (iiii) an increase in the strain rate causes early necking, which results in a ductile fracture.

A number of studies were performed in attempts to clarify the mechanism of nitrogen-caused brittle fracture. Tobler and Meyn (1988) stressed an important role of planar slip and suggested the following sequence of the events for slip-band cracking: the formation of narrow slip bands, accumulation of shear strain in the narrow band because of difficult activation of the out-of-plane secondary slip, weakening of the slip band by accumulated dislocations to the extent that its final separation, i.e. cleavage-like fracture, becomes possible under the component of the applied stress.

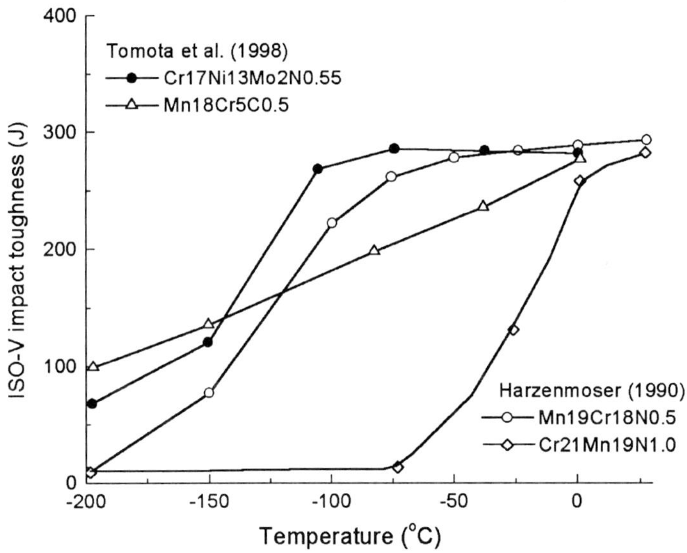


**Fig. 3.27.** Absorbed energy as a function of temperature during ISO-V impact testing of steels Cr17Ni13Mo2N0.55 and Mn18Cr5C0.5 (after Tomota et al. 1998) and steels Cr18Mn19N0.5 and Cr21Mn19N1.0 (after Harzenmoser 1990)

A decisive role of twinning was proposed by Müllner et al. (1994 I, 1997), namely the intersection of deformation twins as a reason for the nucleation of cracking. The main idea is that, if the dislocation density is sufficiently high, the stress relaxation at the crack tip is provided by the dislocation slip and the tip can blunt. Nitrogen, as well as manganese, shift the onset of twinning to lower strains where the density of dislocations is low and a critical stress for brittle fracture can be achieved before the crack tip blunts. This mechanism is at some variance with the observations by Tomota et al. (1998) of a high density of dislocations just under the fracture surface and the results of Tomota and Endo (1990), Tomota et al. (1998 I) that, with increasing strain rate, the fracture mode can be drastically changed from cleavage-like to ductile.

Brittleness of high nitrogen CrMn steels was studied by Liu et al. (1998) and Vogt et al. (1994) who observed both transgranular and intergranular fracture.

For steel Cr18Mn18N0.7 where stacking faults and planar slip bands ("planar deformation structures"), not twins, were introduced by tension up to rupture at 77K, Liu et al. (1998) proposed, as a mechanism of brittleness, the formation of cracks at the intersection of "planar deformation structures" and at grain boundaries on which "planar deformation structures" impinge. Finally, the coalescence of these cracks occurs causing trans- and intergranular cracking.

Vogt et al. (1994) analysed the role of nitrogen atoms on the cleavage stress $\sigma_c$ in terms of the Orowan relation

$$\sigma_c = E\gamma_{hkl}^{1/2}\, d_{hkl}^{-1/2},$$

where $E$ is the Young's modulus, $\gamma_{hkl}$ is the surface energy, $d_{hkl}$ is the interplanar spacing. Stiffness $E$ and surface energy $\gamma$ are slightly decreased by nitrogen in austenite, however, the main effect is thought to be provided by a local concentration of microstructural stresses because of the nitrogen-caused localisation of deformation. For the same reason, nitrogen in austenite increases the plastic flow stress so that the cleavage stress is achieved before the plastic flow stress. In other words, the cleavage is described using the well known engineering scheme of brittleness that was first proposed by Ioffe et al. (1924), see also the historical review presented by Finnie and Mayville (1990).

Cleavage-like fracture of austenite in high-nitrogen (0.6 mass % N) duplex steels was found by Foct and Akdut (1993) and described as a result of the nitrogen-increased short range effective stress.

A profound study of the mechanism of the cleavage-like fracture was performed by Tomota et al. (1998, 1998 I). They confirmed in main features the explanation given by Tobler and Meyn (1988), namely that the localisation of plastic flow in narrow slip bands leads to considerable slipping on the {111} planes. At the same time, some additional special features of the fracture were established. First of all, it was shown that the flat fracture surfaces (cleavage-like facets) are, in fact, non-active {111} slip planes which are composed of a combination of active slip planes. The fracture facets contain intrusions (extrusions in their counterparts, respectively) arising from slipping-off of the active {111} planes. Thus, this is, in fact, the macrobrittle fracture caused by the nitrogen-enhanced localised plasticity.

An examination of the structure right under the fracture surface revealed an extremely high density of dislocations, in contrast to the twinning model. Neither twinning nor ε-martensite have been detected to be induced by impact testing at 77K in austenitic Cr18Mn18N(0.5 to 0.8) and Cr17Ni13Mo2N0.5 steels.

As stated above, early necking with increasing strain rate leads to ductile fracture of high austenitic nitrogen steels at low temperatures. This effect was attributed by Tomota et al. (1998 I) to a local temperature rise generated by rapid motion of a considerable number of dislocations within narrow slip bands. However, the scale and significance of such a local heating during the low temperature plastic deformation of fcc metals and alloys with low stacking fault energy were in detail analysed by Obst and Nyilas (1991) who obtained conclusive evidence of the negligible effects of local heating.

Obst and Nyilas have shown that the low temperature softening caused by an increase in the strain rate of austenitic nitrogen steels arises from the breakdown of Lomer-Cottrell barriers by a critical number of edge dislocations in highly localised slip planes. The last catastrophic event in this case leads to macroscopic fracture. Moreover, if a "thermal spike" at high strain rates is sufficiently high so that the temperature rises above $T_1$ (see Fig. 3.7), screw dislocations glide easier than edge dislocations. Unlike edge dislocations, the planar arrangement of screw dislocations is not stable and, under a critical stress, they could leave the pile-up by cross slip and, in this way, bypass the Lomer-Cottrell barrier without giving rise to the catastrophic collapse of the obstacle, i.e. without softening.

## 3.1.5 Fatigue

Fatigue properties are evaluated using different criteria such as the fatigue strength (fatigue limit, endurance limit), fatigue life (the number of cycles to failure at a constant strain amplitude), fatigue crack growth rate (propagation of the initially induced crack during fatigue tests). The effect of nitrogen on fatigue will be discussed separately for austenitic, martensitic and duplex steels.

### *3.1.5.1 Austenitic steels*

As shown in Chap. 1, the strain behaviour of austenitic steels is characterised by planar slip. An effect of nitrogen on the fatigue is mainly attributed to the formation of planar dislocation arrays. The fatigue strength was found to be directly proportional to the difficulty of the cross slip of dislocations, that in turn was explained by the low values of stacking fault energy (Grosskreutz 1972).
Due to the cross slip, high stacking fault energy promotes the formation of coarse slip bands and large plastic zones at the crack tips, which assists the initiation and growth of the fatigue cracks.

Preventing the cross slip, the low stacking fault energy provides reversible gliding of dislocations on the same plane so that dislocations of opposite sign can annihilate when the compression cycle is replaced by the tension cycle (see Margolin et al. 1976).

Effect of stacking fault energy on the character of cyclic strain curves was analysed by Feltner and Laird (1967) in terms of wavy slip (high stacking fault energy) and planar slip (low stacking fault energy). The latter causes an asymmetry of the hysteresis loops with the flow stress in direction of the first applied stress being lower. In other words, when the test is started in tension, the flow stress in compression is diminished because back stresses are created by obstacles in the slip plane and the reverse gliding of the planar dislocation arrays in the opposite direction requires a lower applied stress.

In view of cited data and their interpretation, nitrogen is expected to affect fatigue whatever reason exists for planar slip in austenitic nitrogen steels: short range atomic ordering or decrease of the stacking fault energy.

***CrNi steels*** were studied in relation to an effect of nitrogen on fatigue strength, fatigue life and crack growth rate. Low cycle fatigue behaviour was tested by Nilsson (1983) in his studies of steels Sandvik 253MA (Cr20Ni10Si1N0.2) and AISI 316 (Cr17Ni12Mo3). It was shown that, at 600 °C, the nitrogen-containing steel 253MA exhibits nearly twice the fatigue life of AISI 316 in the whole range of strains. In consistency with the above mentioned, the planar slip prevailed in the former while wavy slip occurred in the latter. A similar result was obtained by Degallaix et al. (1986) at room temperature.

***Softening-hardening*** behaviour during fatigue tests was studied by Degallaix et al. (1986) who showed that, at low levels of strain amplitudes, nitrogen in austenitic CrNi steels causes the cyclic softening that is just expected if the reversibility of slip occurs, while cyclic hardening is promoted by the dislocation subcell structures. Softening results in a prolonged number of cycles to rupture, being increased with the nitrogen content.

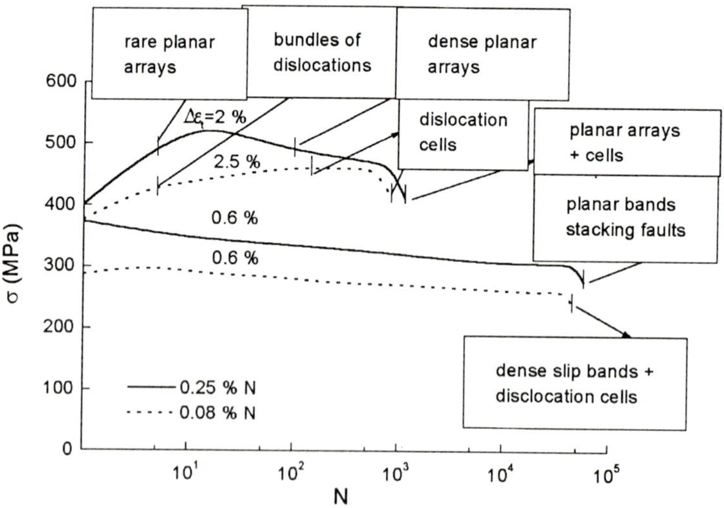


**Fig. 3.28.** Fatigue stress *versus* number of cycles for two austenitic steels AISI 316 with nitrogen contents 0.08 (dashed lines) and 0.25 mass % (solid lines) at the strain amplitudes 0.6 and 2(2,5)%, after Taillard and Foct (1989) and Degallaix et al. (1989). Dislocation substructures developed at different stages of fatigue tests are characterised in rectangles

The fatigue life of austenitic CrNi nitrogen steels, as evaluated on Coffin-Manson fatigue curves (logarithm of cyclic total strain amplitude versus logarithm of the number of cycle to rupture), was found to be increased up to 0.12 mass % N, while greater nitrogen contents up to 0.39% had no effect. Two kinds of fatigue behaviour in relation to hardening-softening, as observed in austenitic steels, are shown in Fig. 3.28 after Taillard and Foct (1989) and Degallaix et al. (1989) who examined dislocation substructures at different stages of fatigue tests.

At high nitrogen contents and high strain amplitudes, a short phase of hardening occurs (about 20 cycles), and widely spaced slip bands in {111} planes correspond to this stage. This short hardening stage is replaced by softening when the planar dislocation arrays become dense. In nitrogen-free and low nitrogen steels, the cell dislocation structure is prevailing during cyclic deformation, which results in a permanent hardening up to the rupture. Softening occurs in both steels at low strains in spite of different dislocation substructures: planar slip bands at high nitrogen contents and cell substructures at the low ones.

***Low temperature fatigue*** of austenitic nitrogen steels is of growing interest because the strengthening of austenitic steels by nitrogen finds practical applications at cryogenic temperatures.

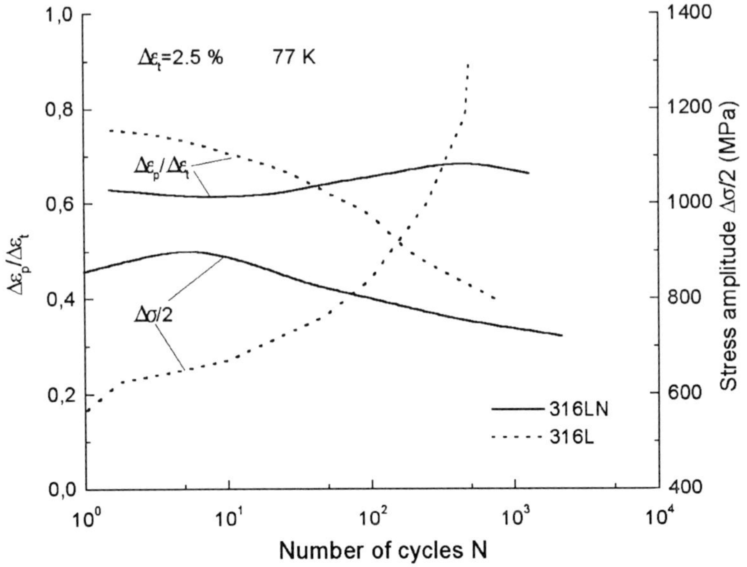


**Fig. 3.29.** Evolution of the stress amplitude $\Delta\sigma$ and of the plastic component in the hysteresis loop $\Delta\varepsilon_p$ during cycling of steels 316L and 316LN at 77 K (after Vogt et al. 1991)

An analysis of different contributions to the cyclic stress (see Vogt et al. 1993) led to a conclusion that both nitrogen and decrease in the temperature of tests promote planar slip, however effects of nitrogen and temperature were described to be of different nature: nitrogen is thought to induce planar slip mainly through the "pinning" effect (short range order), while the low test temperature causes a "lattice friction" effect. At the same time, the examination of the dislocation structure produced by the low temperature cyclic deformation revealed a nitrogen-caused decrease in the stacking fault energy at 77 K, which is obviously a reason for an increase in the slip planarity. As shown above (Sect. 3.1.1.2), nitrogen in austenitic steels causes the sharp temperature dependence of stacking fault energy due to an increase in the concentration of free electrons. A decrease of the stacking fault energy in austenitic CrNi steels with decreasing temperature was also observed by Thomas (1976). It means, that, in fact, both factors, enhanced short range ordering and decreased stacking fault energy at low temperatures, are caused by nitrogen and have the same physical nature.

An important observation was made by Vogt et al. (1991) in relation to the low cycle fatigue behaviour. They have found that alloying with nitrogen increases the plastic component of the hysteresis loop.

As follows from Fig. 3.29, fatigue tests of 316L steel at a constant total strain amplitude $\Delta\varepsilon_t = 2.5\%$ are accompanied by a strong increase in the cyclic stress amplitude and a corresponding decrease in the plastic strain $\Delta\varepsilon_p$. Alloying by nitrogen retains a high fraction of the plastic strain in the total strain amplitude.

***The fatigue crack growth rate*** at low temperatures was analysed by Tobler and Reed (1984) and Vogt et al. (1991) taking into account the stability of the austenitic phase. An effect of strain-induced martensitic transformation on the fatigue crack propagation behaviour was not sufficiently clarified. In unstable austenitic steels, like AISI 304 type Cr18Ni10, where the martensitic transformation occurs during fatigue tests at 77 and 4 K, even a detrimental nitrogen effect was observed (see Fig. 3.30) as compared to the same steel with low C+N contents and the fully nitrogen-free austenitic steel AISI 310 type Cr25Ni20 (Tobler and Reed 1984). A feature of this study was the observation of a change in the fracture mechanism from facet-free transgranular failure at low C+N contents to partially facetted behaviour at high C+N contents. The positive role of strain-induced martensite due to the absorption of energy during martensitic transformation or due to an increase in the strain hardening capability was discussed by Tobler and Reed as a reason for a decreased fatigue crack growth rate at low C+N contents.

However, in more stable steels, like 316L type (Cr17Ni14Mo2), where the fraction of martensite induced during low temperature strain is much smaller, the fatigue crack growth rate at 77 K was found to decrease with increasing nitrogen content (Dhers et al. 1989, Vogt et al. 1991). In Fig. 3.30 the comparison of data for steels Cr18Ni10 and Cr18Ni14Mo2 is presented.

In view of these results, the role of strain-induced martensite on the growth of the fatigue crack seems to be not decisive because, in both 304 and 316 steels nitrogen decreases the fraction of martensite at 77 K.

A noteworthy difference between the low temperature martensitic transformations in these steels is that $\alpha'$ martensite is formed in 304 steel during cyclic deformation at 77 K, whereas ε martensite is mainly induced in 316 steel. The absorption of energy during the $\gamma \rightarrow \varepsilon$ martensitic transformation and the strain hardening capability of the ε martensite are smaller than those of α′ martensite. So, these two factors cannot be the main reasons for the impeded fatigue crack propagation in the high-nitrogen 316 and low-nitrogen 304 steels at 77 K.

One can suggest that planar slip in the parent austenitic phase competes with martensitic transformation in controlling the rate of the fatigue crack growth at low temperatures. High contents of nickel and nitrogen (steel 316LN) provide an enhanced slip planarity at low temperatures due to short range atomic ordering and a strong temperature dependence of the stacking fault energy, which overcompensates a decrease in the extent of the strain-induced martensitic transformation. Such a compensation can be insufficient in steel 304 because of the lower nickel and nitrogen contents.

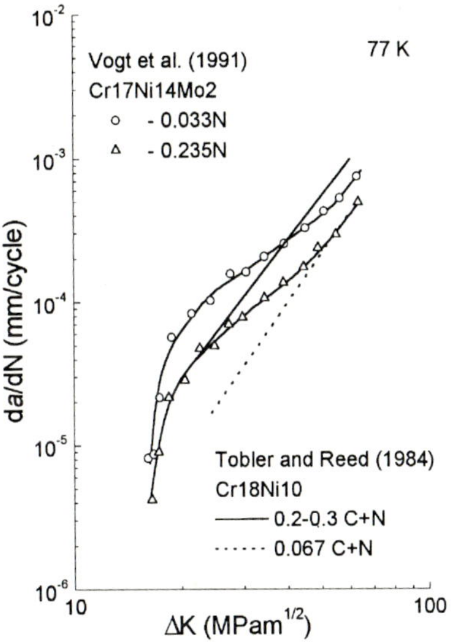


**Fig. 3.30.** Opposite effects of nitrogen on the fatigue crack growth rate at 77 K in steels Cr18Ni10 after Tobler and Reed (1984) and Cr17Ni14Mo2 after Vogt et al. (1991). In steel Cr18Ni10, the carbon content was varied from 0.03 to 0.09 and the nitrogen content from 0.04 to 0.24 mass %

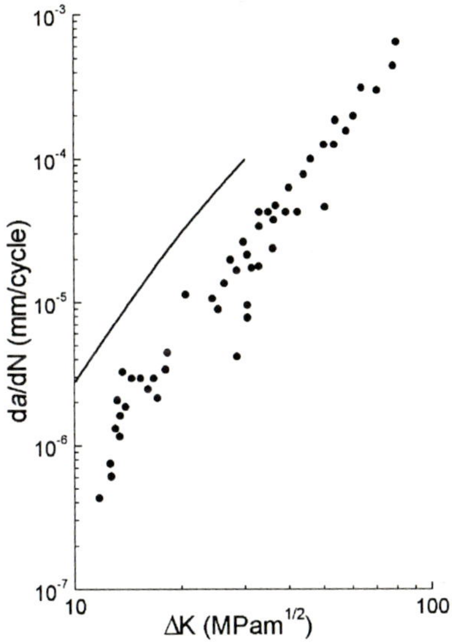


**Fig. 3.31.** Fatigue crack growth rate versus stress intensity factor range in steels Cr25Ni15Mn4N0.35 (filled circles) and Cr17Ni13Mo3N0.15 (the solid line) at 12 K (after Nyilas et al. 1993)

The validity of this explanation is demonstrated in Fig. 3.31 by a comparison of the fatigue crack growth rate in steel 316LN (0.15% N) and high nitrogen stable austenitic steel Cr25Ni15Mn4N0.35 (CSUS-JN1) at 12 K according to the data obtained by Nyilas et al. (1993). It is seen that a further increase in the nitrogen content as compared to that in 316LN provides a significantly lower fatigue crack growth rate even in the absence of the strain-induced martensitic transformation.

***CrMn steels*** reveal quite a different fatigue behaviour. No remarkable influence of nitrogen was observed by Sun et al. (1990) in CrMn steels in the range of nitrogen contents of 0.6 to 1.0 mass %, either in the stress amplitude or in the cycles to rupture. Moreover, the analysis of fatigue data in Coffin-Manson coordinates showed that, whatever nitrogen content or treatment (cold working, strain ageing), no improvement in the low cycle fatigue life can be achieved in austenitic CrMnN steels as compared to conventional AISI 316 CrNi steels. An interesting observation made by Sun et al. was the cyclic softening and the absence of the hardening in the whole range of fatigue life (compare with the data for CrNi steels in Figs. 3.28 and 3.29). Such a behaviour seems to be reasonable as, in high nitrogen CrMn steels, planar slip is developed more strongly than in CrNi steels containing less nitrogen.

The fatigue resistance of high nitrogen CrMn steels is decreased at low temperatures and brittle intergranular fracture occurs (Vogt et al. 1996). Low temperature ageing at 250 to 350 °C was found to cause a transition from intergranular to transgranular crack initiation, a more favourable distribution of plastic strain in numerous fine slip bands and, as consequence, an enhancement of the resistance to crack initiation.

***High temperature fatigue*** of nitrogen steels was studied aiming at the elaboration of advanced materials for service at elevated temperatures (see Escher 1999). An important result of these studies is that alloying with nitrogen+carbon provides a higher stability of the structure against precipitation at high temperatures as compared to fully nitrogen steels, which results in better fatigue properties in the range of large number of cycles (Fig. 3.32).

At both temperatures (700 and 800 °C) the fatigue strength of the nitrogen steel is higher for the low-cycle range whereas the advantage of the carbon+nitrogen steel is revealed in the range of high cycle numbers. This result can be explained by the higher degree of short range atomic ordering caused by nitrogen+carbon alloying as compared to alloying by nitrogen or carbon each, which was shown to increase the stability of austenite (see Sects. 2.5 and 5.2.7).

Finishing the review of the data on fatigue of austenitic nitrogen steels, we consider the effect of a nitrogen-enriched surface. As shown by Paul et al. (1996), the surface nitriding of the steel Nitronic 50 (Cr21Ni12Mn5Mo2N0.2) up to 0.6 mass % N leads to an increased tensile strength but not to more fatigue resistance. This result is consistent with the data of Degallaix et al. (1986) about the limit of the nitrogen content (0.12 mass %) up to which an increase of the fatigue strength can be obtained. Nevertheless, a good combination of tensile strength, fatigue resistance and corrosion properties can be achieved if nitrogen is present in solid solution, i.e. if nitrides in the surface layer are dissolved due to some high temperature treatment.

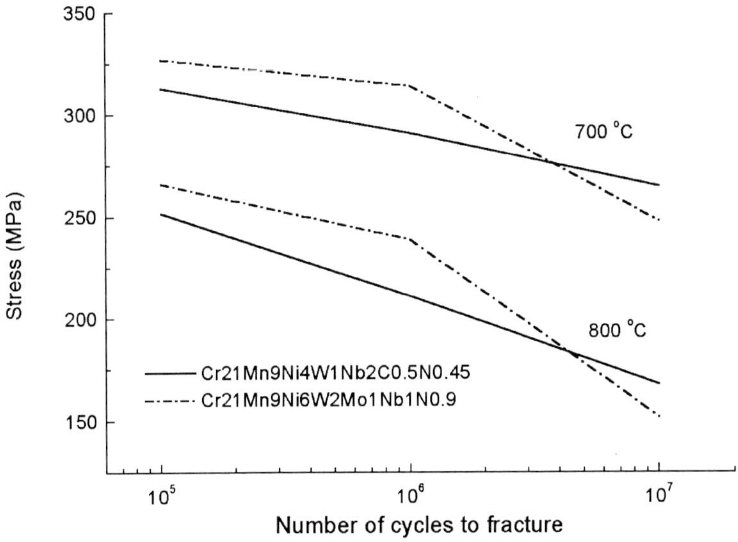


**Fig. 3.32.** Wöhler curves at 50 Hz at 700 and 800 °C for steels Cr21Mn9Ni4W1Nb2C0.5N0.45 and Cr21Mn9Ni6W2Mo1Nb1N0.9 (after Escher 1999)

### *3.1.5.2*
### *Martensitic and duplex steels*

Fatigue of martensitic steels was studied by Vogt et al. (1994 I) in relation to the combined effect of nitrogen and silicon because silicon nitride is used as nitrogen supplier in the pressurized electroslag remelting, PESR, (see Sect. 4.1.2). The substitution of carbon by nitrogen in martensitic 12%Cr steel was shown to provide a fine precipitation structure, which leads to an increase in the cyclic stress at constant cyclic strains. The fatigue life characterised by Coffin-Manson curves is nearly the same in C-Si and silicon-free N steels and higher than that in N-Si steel. If to evaluate the fatigue strength by the Wöhler plot of the cyclic stress amplitude versus number of cycles to failure, the N-Si steel reveals the best quality. Nitrogen in martensitic 12%Cr steels is thought to not improve the fatigue properties, which is attributed to the lack of soft areas for the accommodation of the plastic strain.

A clear advantage of martensitic nitrogen steels is observed if fatigue tests performed at elevated temperatures (see Fig. 3.33).

In the whole range of the number of cycles to the rupture, the fatigue strength of martensitic nitrogen steel developed by Berns et al. (1998), see also Escher (1999), exceeds that of standard martensitic carbon steels, which is a consequence of a highly-stable fine structure of the tempered nitrogen martensite (see also Sects. 2.5 and 5.1.6).

Duplex steels are not sufficiently studied in relation to their fatigue properties. In absence of nitrogen or at a small concentration of it, the fatigue behaviour is controlled by the plastic deformation of the austenitic phase in the range of small strain amplitudes, while no change is observed within the harder ferritic matrix (see e.g. Mateo et al. 1996). In an intermediate range of strain amplitudes ($10^{-4}$ to $6{\cdot}10^{-4}$), both phases are involved in cyclic deformation.

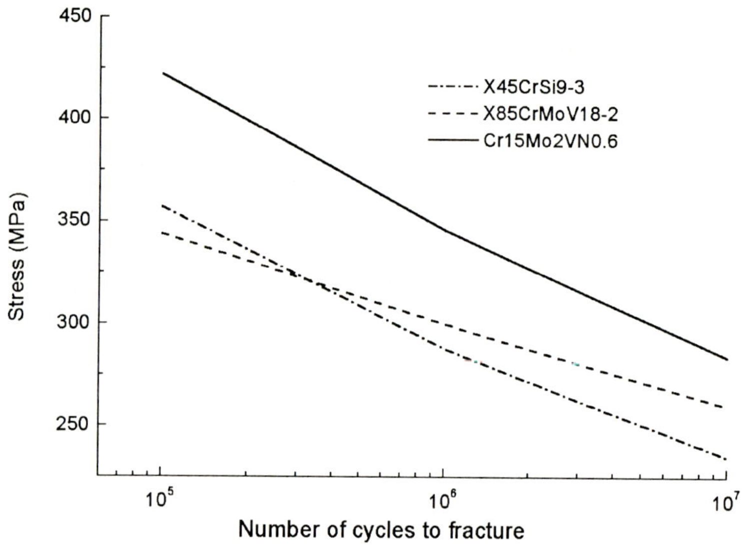


**Fig. 3.33.** Wöhler curves at 50 Hz for the martensitic nitrogen steel Cr15Mo2VN0.6 and carbon steels Cr9Si3C0.45 (X45CrSi9-3) and Cr18Mo2C0.85 (X85CrMoV18-2) at 600 °C (after Escher 1999)

The slip reversibility is increased in the austenitic phase due to high nitrogen content (see Degallaix et al. 1988, 1993). At the same time, because of hardening of austenite, nitrogen assists the plastic strain in ferrite. The fatigue strength in Coffin-Manson coordinates is increased by nitrogen (see also Fig. 5.29).

A new information about the fatigue behaviour of duplex nitrogen steels is provided by the recent study of Vogt et al. (1999). A change in the cyclic response and fatigue damage is found at the applied strain amplitude $\Delta\varepsilon_t = 0.8\%$. In the low strain regime the cyclic plasticity is controlled by the austenitic phase where nitrogen causes planar dislocation slip in contrast to the nitrogen-free duplex steels. Cracks are initiated in ferrite. In the high strain regime the cyclic plasticity is controlled by both the phases. Microtwinning appears in austenite along with planar dislocation slip. Cracks initiate in ferrite and later in austenite.

### 3.1.6
### Creep

***Austenitic steels.*** The effect of nitrogen on room temperature creep of austenitic steels was studied by Tendo et al. (1991) and Uggowitzer (1993). An increase in the creep strain of CrNi steels due to small addition of nitrogen (about 0.1 mass %) was observed by the former, whereas an improved room temperature creep behaviour (about 40% of the increase of the yield strength) of high nitrogen steels Cr18Mn18N(0.6-1.1) was shown by the latter. A higher resistance to room temperature creep was attributed to the nitrogen-increased elastic limit.

The effect of nitrogen on the creep of austenitic CrNi steels at high temperatures was studied by Matsuo et al. (1989, 1990) and Nakazawa et al. (1989).

Matsuo et al. obtained a strong evidence that solid solution strengthening of the steel Cr25Ni28 by nitrogen remains effective during creep at temperatures up to 1000 °C. The strengthening factor of nitrogen measured as a ratio of the minimum creep rate of a nitrogen-containing steel to that of a nitrogen-free steel was shown to be independent of the carbon content and to increase with increasing applied stress. An examination of the structure revealed the formation of well developed subgrains in the nitrogen-free steel while no subgrains were observed in nitrogen-containing steels in the whole range of stresses. A decrease of the minimum creep rate by more than one order of magnitude by the addition of 0.1 mass % of nitrogen was achieved at 700 °C and attributed to the nitrogen-caused suppression of subgrain formation.

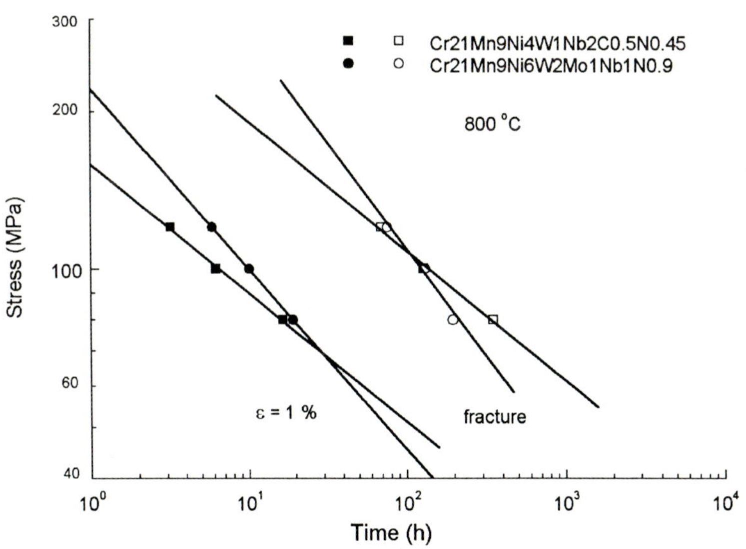


**Fig. 3.34.** Creep strength of austenitic steels alloyed by nitrogen or nitrogen+carbon at 1% elongation and at fracture. The temperature of tests was 800 °C (after Escher 1999)

Nitrogen-enhanced planar slip and climb of dislocations between planes of activated slip control the creep rate according to Owen (1990).

Nakazawa et al. (1989) observed a significant difference of the creep properties at 550 °C of steel Cr18Ni14Mo2 alloyed by nitrogen or carbon. In spite of higher strengthening nitrogen did not change the rupture ductility, whereas carbon reduced it. The authors found a quite different state of grain boundaries in the carbon and nitrogen steels after the creep tests. While the addition of carbon led to the precipitation of large carbide particles on grain boundaries, no grain-boundary nitrides were observed in the nitrogen steel. This result can be explained based on the different grain boundary segregation of carbon and nitrogen in austenitic steels (see Sect. 1.1.4.3). Thus, the deterioration of the creep properties by carbon can be attributed to grain-boundary embrittlement, and the beneficial effect of nitrogen is obviously due to its smaller tendency to form grain boundary segregations.

The effect of nitrogen and nitrogen+carbon on high-temperature creep was examined by Escher (1999) for austenitic steels of complicated compositions.

Nitrogen was found to provide better creep properties at high stresses and, correspondingly, short exposures (Fig. 3.34). In the range of small stresses nitrogen+carbon steels revealed a higher resistance to the creep deformation, which is obviously due to a higher structure stability of the nitrogen+carbon austenite as was shown in Sect. 2.5. The minimum creep rate at 700 and 800 °C is also smaller for the nitrogen steel in the range of high stresses, and a clear tendency to a lower creep rate for the nitrogen+carbon steel is observed at smaller stresses (Fig. 3.35).

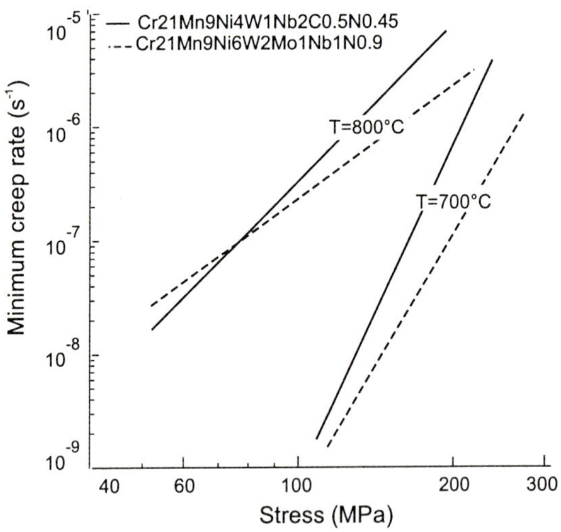


**Fig. 3.35.** Minimum creep rate of nitrogen and austenitic nitrogen+carbon steels at 700 and 800 °C represented by Norton's equation $\dot{\varepsilon} = A\sigma^n$ (after Escher 1999)

These results demonstrate a promising direction in the design of high nitrogen steels using the combined alloying by nitrogen+carbon.

***Martensitic steels.*** The substitution of carbon by nitrogen was proven to cause a beneficial effect in heat resistant martensitic steels. These steels contain 9-12 % of chromium and find a wide application in plants for power generation. A typical Cr12Mo1V steel containing ~0.2 % of carbon provides an upper limit temperature around 550 °C. The replacement of about two thirds of carbon by nitrogen results in an increase of fracture toughness and creep rupture strength (see e.g. Speidel and Uggowitzer 1985, Berns and Krafft 1989, Anthamatten et al. 1989, Speidel 1989). Such modified steels may provide a creep rupture strength of more than 100 MPa for $10^5$ hours at 600 °C and an impact toughness of about 27 J. For applications as gas turbine discs at 400-500 °C, martensitic nitrogen steels satisfy the requirements of more than 800 MPa of yield strength and an impact toughness of about 50 J.

As shown by Berns and Krafft (1989), substituting carbon by nitrogen in a 12%Cr steel results in an increasing time to the rupture in iso-stress creep tests and a higher flow stress under constant strain rate (Fig. 3.36, see also Sect. 5.1.9).

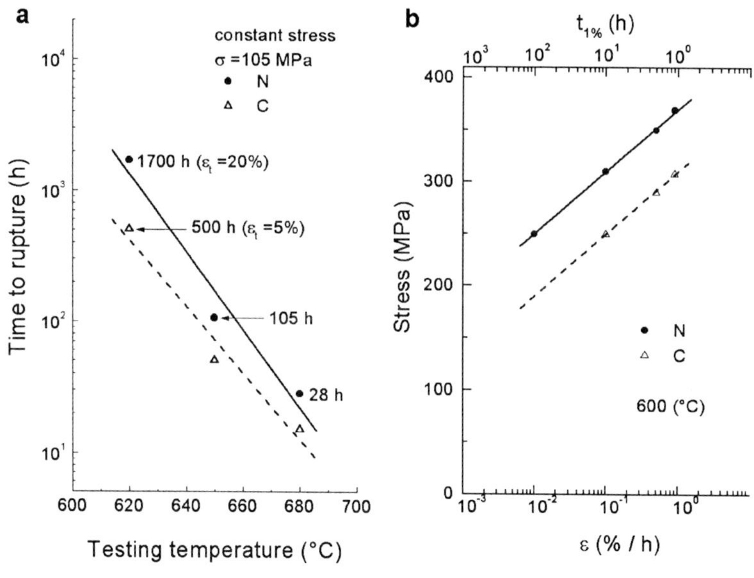


**Fig. 3.36.** **a** Time to rupture and **b** flow stress under constant strain rate of steel Cr12Mo1V containing (mass %) 0.21C or 0.25N (after Berns and Krafft 1989)

## 3.1.7 Wear

The effect of nitrogen on sliding wear, abrasive wear and impact wear was studied mainly. The term "sliding wear" is used instead of "adhesive wear" in order to take into account related phenomena on the counteracting surfaces (the formation of oxide films, wear debris etc.) which are not met in the case of pure adhesion.

***Sliding wear*** is known to be essentially determined by the initial hardness and work hardening of tested materials (see e.g. Archard 1953, Boas and Rosen 1977). That is why stable austenitic steels are less resistant in dry sliding contact than metastable ones which undergo the transformation to bcc martensite under load (Korshunov and Mints 1969). Alloying of austenitic steels by nitrogen causes both solid solution and work hardening, which has to result in increased resistance to sliding wear. Campillo Illanes and Sarkar (1986) studied the dry sliding wear of stable austenitic nitrogen CrNi and CrMn steels at room temperature using the pin-bush technique. Steels showed a high running-in wear but the steady state wear rate was significantly decreased due to nitrogen. This means that some stable work hardened layer has to be formed before austenitic nitrogen steels acquire resistance to the wear. Thus, the nitrogen-enhanced work hardening, not the contribution of nitrogen to the initial hardness, seems to be decisive in controlling the sliding wear resistance of stable austenitic steels.

Contributions of different factors to the sliding wear of cast and powder metallurgical austenitic nitrogen steels were analysed by Terwo (1997, 1998) and Romu et al. (1996) based on pin-on-disc tests.

It was shown that an increase in the nitrogen content increases the strength factor $K$ and decreases the strain hardening exponent $n$ in the Hollomon relation $\sigma = K\varepsilon^n$. Thus, nitrogen enhances work hardening at low strains. This suggests that a small amount of plastic deformation is needed in order to produce a certain hardness providing low wear.

The role of friction in wear was found to be related to lattice vibrations on the surfaces during sliding motion. Measurements of elastic constants led to the conclusion that nitrogen increases the shear modulus and decreases the bulk modulus and Poisson's ratio (Tervo et al. 1998) and may decrease or increase the friction of austenitic stainless steels depending on the basic chemical composition and the route of production (Tervo 1997). At the same time, the effect of nitrogen on the value of the friction coefficient is profoundly affected by the operating wear mechanism. In particular, friction is lowered if the counterface (chromium or high carbon martensitic steel) experiences some wear during the tests – a typical consequence of an increasing nitrogen content in austenitic steels.

Nitrogen ion implantation as a way to improve the wear resistance was examined by Cavalleri et al. (1986) in tests with a high speed steel (66 HRC) as wear partner. A content of nitrogen of about 20 at.% was found in the near-surface region (100 nm from the surface), which stabilised the austenitic matrix against strain-induced martensitic transformation and resulted in the formation of chromium and iron nitrides. The weight loss in wear tests was observed to be less than half for implanted steel.

***Abrasive wear*** was studied in unstable austenitic CrMnN steels by Lenel and Knott (1987). These steels combine strengthening effects caused by strain-induced $\gamma\rightarrow\alpha'$ martensitic transformation and nitrogen, and the wear resistance was shown to be slightly better than that of Hadfield steels and high carbon martensitic steels and significantly improved compared to commercial austenitic stainless steels. The difference between experimental and reference steels was more pronounced for the rounded than for the sharp abrasive. The role of the work hardening on the abrasive wear resistance was considered to be insignificant. The negligible effect of work hardening in the case of Hadfield steel when the wear was caused by large and hard abrasive particles (Avery 1974) is consistent with this conclusion.

No definite correlation between the initial hardness and abrasive wear was found by Romu et al. (1993) for cast and hipped austenitic nitrogen stainless steels. An increase in the nitrogen content had a slight effect on the volume loss during wear in steels where nitrogen was mainly in solid solution, whereas hipped materials containing large hard particles revealed the best resistance to abrasive wear.

Berns and Wang (1993), see also Wang (1992), studied the abrasive wear of powder metallurgical stainless martensitic composites of Cr15Mo1 basic composition containing particles of chromium nitride. A considerable improvement of the abrasion resistance was achieved if the size of nitride inclusions in matrix was higher than the width of the wear grooves. A comparison of the hardness of the abrasive materials and the nitrides was also found to be important. Wear resistance can be improved only if the abrasive material had a smaller hardness than the nitrides.

***Impact wear*** was shown to be improved by addition of nitrogen to high carbon martensitic die steels alloyed by Cr, W, V, Mo (Hu et al. 1990). Nitrogen was found to replace carbon mainly in carbides $V_4C_3$. Chromium carbides $M_7C_3$ and $M_{23}C_6$ remained unaffected. The solubility of carbide $M_6C$ and, correspondingly, the content of W and Mo in the matrix were increased. As a result, a good combination of secondary hardness, impact toughness and impact wear was obtained.

Impact abrasive wear of duplex CrMnN steels (about 17Cr, 8Mn, 0.2N in mass %) was studied by Whitefield and Bennekom (1996) in comparison with the reference duplex steel SAF 2205 (Cr21Ni6Mo3). Similar results were obtained for as-cast alloys while after heat treatment an improved wear resistance of experimental steels was found as a result of an increased fraction of austenite and enhanced work hardening.

## 3.2 Chemical properties

A great deal of studies were performed to clarify the effect of nitrogen on the corrosion properties of stainless steels. As an introduction to this topic one can say that the data about the nitrogen effect on general corrosion are contradictory while the resistance to local kinds of corrosion (intercrystalline corrosion, pitting, crevice corrosion and stress corrosion cracking) is unambiguously shown to be improved by nitrogen.

Four existing hypotheses interpreting the corrosion behaviour of nitrogen steels were summarised by Speidel (1991):

- nitrogen dissolves during corrosion reaction to form ammonia, which increases pH and renders the electrolytes at the pit sites less acidic;
- the surface is enriched by nitrogen because other important alloying elements also segregate there, so that passivity is improved;
- nitrogen takes part in the formation of the passive film to form a dense oxynitrided layer;
- nitrogen dissolves to form nitrites or nitrates that may produce an inhibiting effect locally at the pits.

In this section we discuss available experimental data for each kind of corrosion paying due attention to proposed mechanisms. Finally, the effect of nitrogen on the embrittlement of stainless steels in hydrogen as an extremely aggressive environment will be described.

### 3.2.1 General corrosion

Anodic dissolution of nitrogen steels in acids was reviewed in detail by Truman (1989) and Grabke (1996) covering a number of studies performed over several decades.

In austenitic steels containing (mass %) 23-25Cr, 7-14Ni and 0.5-1.5Mo, a noticeable reduction in the rate of general corrosion in 10% HCl at 25 °C and in boiling 5% $H_2SO_4$ was observed by Sakamoto et al. (1980) with increasing content of nitrogen from 0.1 to 0.38 mass %. Qualitatively the same result was obtained by Berns et al. (1990) for martensitic 15%Cr steels. The substitution of carbon by nitrogen led to a decrease in the critical anodic current density as well as passive current density by more than one order of magnitude.

A steel containing (mass %) 22Cr, 9Ni, 2.6Mo, 0.4N was shown to be more resistant in nitric, acetic and phosphoric acids compared to AISI 316, although this could result from a higher content of chromium than nitrogen (Truman 1989).

In steels Cr17Ni12Mo2.5 alloying by nitrogen (0.22 mass %) enhanced the anodic dissolution in 1-2.5 M hydrochloric acid, whereas with increasing concentration of acid (4 M) an opposite effect of nitrogen was observed and in 4.5-5 M HCl the rate of dissolution was high in both nitrogen-free and nitrogen-containing steels (Newman and Sharabi 1987).

An increase of the nitrogen content from 0.06 to 0.18 mass % decreased the passive current density for steels Cr20Ni25 and Cr20Ni25Mo4.5 tested in HCl and $H_2SO_4$, according to Jargelius (1986), which was suggested to result from the nitrogen accumulation under the passive film, and at the same time it was shown that the passi vation process was similar in both low- and high-nitrogen steels and no indication was obtained that any characteristic time constant was related to the presence of nitrogen in steel (see also Jargelius-Pettersson 1998 I).

Nickel-free austenitic CrMn steels alloyed by carbon or nitrogen were studied by Miyahara et al. (1991) aiming at the development of fusion materials of the reduced radioactivity. Both carbon and nitrogen in solid solution decreased the critical and passive current density. The precipitation of carbides at grain boundaries during ageing treatments increased the passive current density and did not affect the passivation potential and critical current density. In contrast, the precipitation of nitrides had a very small influence on the anodic dissolution.

In contrast to the above presented data, other authors described a detrimental effect of nitrogen on the general corrosion of stainless steels.

Nitrogen in austenitic AISI 316 steels, the content of which was varied from 0.03 to 0.25 mass %, was shown to cause a slightly adverse effect on the range of passivity and the rate of corrosion in sulphuric, hydrochloric, nitric, phosphoric and acetic acids (Truman 1989). After the same author, 0.2 mass % N make a copper bearing 20%Cr steel with a high nickel and molybdenum content less resistant in pure sulphuric acid, although the corosion behaviour was better in hydrochloric acid and in fluosilicic acid solutions.

The current density before passivation was increased by nitrogen in austenitic CrNiMo steels as well as in duplex austenitic-ferritic steels according to Werner et al. (1990). The detrimental effect was not pronounced but clearly proven.

An enhanced anodic dissolution of the steel Cr17Ni13N0.15 in 0.5 M $H_2SO_4$ as compared to Cr18Ni10 was observed by Lu et al. (1993). An addition of molybdenum was found to decrease the anodic dissolution current.

Grabke et al. (1993) have shown that the alloy Fe-0.7N (mass %) with a martensitic structure is characterised by an enhanced dissolution rate in acidic solutions compared to the pure iron, although this effect could be caused by large distortions in the crystal lattice of martensite. The same result was obtained by Lu et al. (1991) for nitrided iron samples containing $Fe_2N$ and $Fe_4N$ nitrides. A peculiarity of their observations was that the rate of anodic dissolution of nitrided iron was higher than in pure iron until a current limited by mass transport is reached, suggesting that, once a pit is formed, iron nitride dissolves and inhibits the pit development.

Experimental data concerning the formation of the passive film in acidic solutions were analysed by Grabke (1996). An enrichment of negatively charged N atoms is found at the passive film/metal interface. These cations must react with hydrogen forming $NH_x$ species, in order to be transferred through the passive layer, and ammonium ions $NH_4^+$ to be transferred into the electrolyte. The formation of $NH_4^+$ ions was observed in electrochemical tests performed by Chyou and Shih (1990). A segregation of nitrogen at the metal surface during anodic dissolution of 304 and 317 stainless steels was also detected by Clayton and Martin (1989). These authors observed an enrichment of the surface in Cr and Mo claiming the formation of surface nitrides. However, no chromium or molybdenum nitrides were found in the passive film based on a comparison of X-ray photoelectron spectra obtained for the passive films and for two-dimensional surface compounds CrN and MoN (see Grabke 1996).

Summarising the experimental data available, one can state that both the beneficial and the detrimental effects of nitrogen on the general corrosion are not large and hardly have any practical significance. One can also refer to an observation made by Kearns (1985) as typical for the practice. He did not find any effect of nitrogen on the general corrosion in his studies of corrosion resistance and electrochemical characteristics of stainless steels containing (mass %) 18 to 20Cr, 13 to 25Ni, 3 to 6Mo and up to 0.25N.

## 3.2.2 Localised corrosion

Studies of the effect of nitrogen on the resistance of steels to localised corrosion started along with the first attempts to develop high nitrogen steels. Electrochemical experiments were combined with a thorough metallographic examination. First of all these studies concerned austenitic steels as an important class of corrosion-resistant engineering materials. The use of nitrogen as an alloying element promised to overcome the main shortcoming of austenitic steels – their low strength, that could not be done using carbon because of its detrimental influence on the corrosion behaviour. For this reason it was important, beside the studies of mechanical properties, to test how nitrogen affects the resistance to intercrystalline corrosion, pitting, crevice corrosion and stress corrosion cracking.

### *3.2.2.1 Intercrystalline corrosion*

Sensitisation caused by the grain-boundary precipitation of chromium carbides is one of the most important factors limiting an increase of the service temperature of austenitic stainless steels. The mechanism was proposed by Bain et al. (1933). Depletion in chromium around the precipitates leads to the selective dissolution of grain-boundary areas in corrosive environments. It is clear from the data of the effect of nitrogen on the precipitation in austenitic steels (see Sect. 2.4) that nitrogen is expected to improve the resistance to intercrystalline corrosion because the precipitation of chromium carbides is prevented in nitrogen-alloyed steels and, if chromium nitrides $Cr_2N$ are precipitated, their formation has to be accompanied by a smaller chromium depletion of adjacent areas in comparison with $Cr_{23}C_6$ carbides.

Experimental studies of the sensitisation of austenitic nitrogen steels are consistent with this conclusion. Briant (1982) has shown that the sensitisation of steels containing (mass %) 17 to 18Cr, 9 to 13Ni, 0 to 2Mo is retarded by lowering the carbon content or by increasing the nitrogen content varying up to 0.124 mass % N in his study. He also demonstrated a retarding effect of molybdenum and manganese (up to 2 mass % each) on the sensitisation and explained it by an increase in the activity of chromium in the matrix in equilibrium with the carbide at the interface resulting in a decrease of the effective chromium content controlling sensitisation. It was supposed also that molybdenum eases the passivation, so that more chromium depletion is required to cause corrosion failure.

Measurements by analytical electron microscopy (Betrabet et al. 1987) have shown that alloying of austenitic CrNi steels by nitrogen increases the chromium concentration at the grain boundaries in the sensitised state while the volume diffusion coefficient of chromium, as calculated from the chromium profiles, is decreased by nitrogen. Both factors retard the chemical driving force for the growth of chromium carbides.

Mozhi et al. (1985, 1986) and Betrabet et al. (1987) constructed a sensitisation diagram for nitrogen-containing AISI 304 type steels using thermodynamic data for calculations and electrochemical potentiokinetic reactivation (EPR) technique for the experimental measurements. Some of their data are presented in Fig. 3.37 showing how nitrogen affects the TTS diagram.

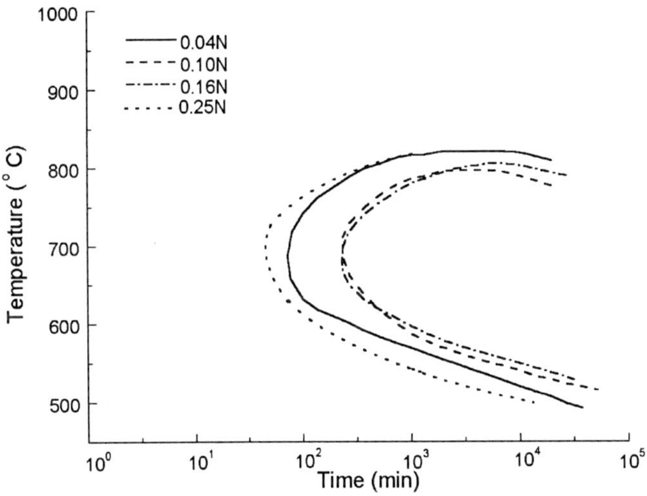


**Fig. 3.37.** Time-temperature-sensitisation diagram of AISI 304 steel alloyed by nitrogen as obtained by means of EPR technique. After Mozhi et al. (1985)

A saturation of the beneficial effect of nitrogen in this type of steels occurred at about 0.16 mass %. It was also obtained that, at 0.25 % N, the TTS diagram is shifted back to shorter times and this shift is accompanied by the precipitation of $Cr_2N$ nitrides as revealed in TEM studies. Thus, nitrogen retards the sensitisation of austenitic steels only if it is dissolved in the austenitic matrix.

High alloy austenitic CrNiMoN steels were examined by Kearns (1985) and Jargelius-Pettersson (1992). The addition of 0.18 mass % N to the steel Cr20Ni25Mo6.5 was shown to prevent intergranular precipitation after the exposure for 100 min at 800 °C (Kearns 1985). A critical nitrogen content of 0.13 mass %, above which no retardation and even more rapid sensitisation occurred, was observed by Jargelius-Pettersson (1992) for steels Cr20Ni25 while in the case of the steel Cr20Ni25Mo4.5N containing Mo there was no limit to the beneficial effect of nitrogen in the whole range of its content (up to 0.21%). The first result was attributed to the precipitation of $Cr_2N$ nitrides, the second one was ascribed to the retardation of intermetallic phase precipitation. A peculiarity of these steels is the formation of the $\sigma$ and $\chi$ phases along with carbides and nitrides during sensitisation treatments (see Sect. 2.4), and the effect of nitrogen depends strongly on the chemical composition.

The addition of 5 to 10 mass % of manganese to steel Cr20Ni18Mo4.5N resulted in a progressively enhanced intercrystalline corrosion (Jargelius-Pettersson 1996) in contrast to earlier data obtained by Briant (1982) for smaller manganese contents. Nitrogen was found to retard the onset of sensitisation and depress the maximum sensitisation temperature. The effect of nitrogen is in a good agreement with the complicated precipitation picture observed in these steels (see Jargelius-Pettersson 1994). Nitrogen retarded the formation of the $\sigma$ phase but stabilised the $\chi$ phase. At 0.5% N $Cr_2N$ nitride precipitated at an early stage while at smaller nitrogen contents it appeared only after the matrix was enriched in nitrogen as a result of the preliminary precipitation of $\sigma$ phase.

### *3.2.2.2*
### *Pitting*

The resistance to pitting corrosion is mainly described by the pitting potential or critical pitting temperature. A beneficial influence of nitrogen was observed in a number of studies using different experimental techniques. Nitrogen in austenitic steels is shown to shift the breakthrough potential of pitting to more positive values, i.e. to retard the breakdown of the passive film. An example is given in Fig. 3.38 for two high-alloy steels studied by Escher (1999).

An interrelation between pit attack and intercrystalline corrosion was shown by Kamachi Mudali et al. (1996) and Jargelius-Pettersson (1996). Sensitisation treatments enhanced both the intercrystalline and pitting corrosion. Pits were developed in chromium-depleted regions such as triple points, grain boundaries and precipitate-matrix interfaces. The beneficial effect of nitrogen was confirmed and, moreover, found to compensate a detrimental influence of manganese.

Potential-current-density measurements of steels Cr18Mn18 alloyed by nitrogen were performed by Hartline (1974), Speidel (1991), Srinivasan et al. (1995). The pitting potential was found to be linearly increased with the content of nitrogen. A comparison with the data available for the influence of other important elements in stainless steels was made by Speidel (1991) as presented in Fig. 3.39.

Chromium and molybdenum assist nitrogen in improving the resistance to the pitting attack, whereas manganese and carbon render a deteriorating effect. An interaction between the effects of nitrogen, chromium and molybdenum was shown resulting in a marked increase of the pitting resistance of the highly alloyed

(22% Cr and 3% Mo) austenitic steel in the range of the nitrogen contents between 0.15 and 0.3 mass % (Truman 1989).

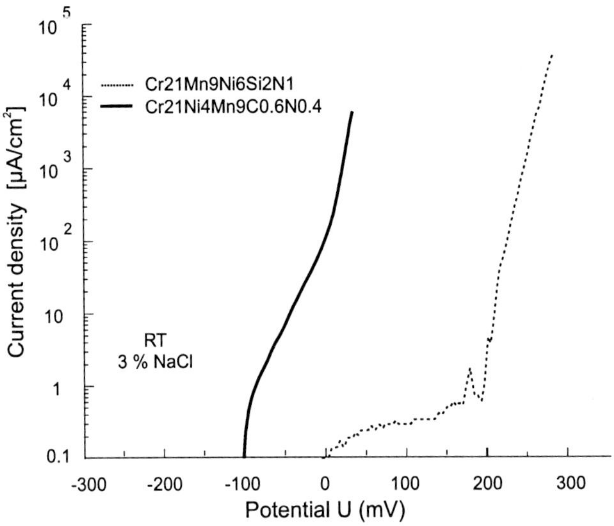


**Fig. 3.38.** Potentiodynamic curves of steels Cr21Mn9Ni4C0.6N04 and Cr21Mn9Ni6Si2N1 in 3% NaCl solution (after Escher 1999). Alloying with nitrogen shifts the value of the pitting potential to higher voltage

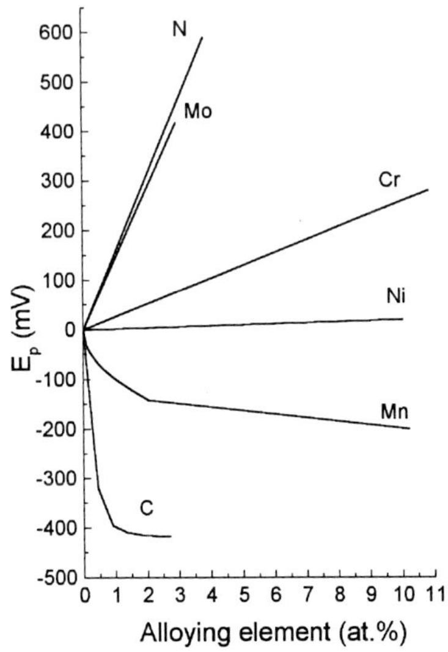


**Fig. 3.39.** Shift in pitting potential of austenitic steels containing up to 11 at. % Cr or basically 18 mass % Cr in dilute chloride solutions by different elements as summarised by Speidel (1991). The data are given in at. % for correct comparison

Based on this and other studies of the pitting corrosion in stainless austenitic steels (see e.g. Forchhammer and Engell 1969, Herbsleb 1982), the following formula was proposed to predict the pitting resistance equivalent number:

$$\text{PREN} \approx \%\text{Cr} + 3.3\%\text{Mo} + k\%\ \text{N}. \tag{3.6}$$

The values of the nitrogen factor $k$ equal to 13 (Gooh 1987), 16 (Lorenz and Medawar 1969), 20 (Uggowitzer et al. 1996), 25 (Menzel et al. 1996) and 30 (Speidel and Pedrazzoli 1992) were proposed.

This relation does not account for a beneficial synergetic effect of molybdenum and nitrogen on the pitting resistance. Such an attempt was made by Jargelius-Pettersson (1995) who proposed to change the formulation of the pitting resistance equivalent number as follows:

$$\text{PREN} \approx \%\text{Cr} + 3.3\%\text{Mo} + 37\%\ \text{N} + 4.5(\%\text{Mo})(\%\text{N}). \tag{3.6.a}$$

The effect of nitrogen in duplex steels is worth a special discussion because of two-phase composition and partitioning of alloying elements between austenite and ferrite. The austenitic phase is the weak phase in relation to the pitting resistance. However, nitrogen is mainly dissolved in austenite improving its corrosion properties. Beside this, due to a strong affinity to ferrite-forming chromium and molybdenum, nitrogen increases their concentration in austenite, i.e. changes the partitioning coefficients. As a result, the nitrogen factor k in the above mentioned formula for the PREN is close to its highest discussed value k ≈ 32 (see Bernhardsson 1991). Values of the PREN and the critical pitting temperature are given in Table 3.3 for some duplex nitrogen steels in comparison with high-alloy austenitic steels. The resistance of duplex nitrogen steels to pitting corrosion is not worse and even better than that of superaustenitic high-nickel high-molybdenum steels.

**Table 3.3.** Pitting resistance equivalent number calculated for duplex ($k$ = 32) and austenitic ($k$ = 16) steels and critical pitting temperature

| Alloy | N (mass%) | PREN | CPT (°C) | Ref. |
|---|---|---|---|---|
| Cr22Ni5Mo3 | 0.15 | 36.7 | 55 | Bernhardsson (1991) |
| Cr25Ni6Mo3 | 0.16 | 41.6 | 66 | ,, |
| Cr25Ni7Mo4 | 0.27 | 46.8 | 93 | ,, |
| Cr20Ni25Mo4.5 | - | 35 | 30 | Nilsson (1992) |
| Cr20Ni25Mo4.5 | 0.13 | 37 | 78 | Jargelius-Pettersson (1998 II) |
| Cr20Ni18Mo6 | 0.2 | 43 | 90 | Bernhardsson (1991) |

We will briefly describe the role of nitrogen in different stages of the pitting corrosion: breakdown of the passive film, growth of pits and repassivation.

***Pit nucleation.*** The passive oxide/hydroxide film is formed under a high electric field when a limit of the solubility of metal cations during anodic dissolution is reached (see e.g. Häfele et al. 1992). The breakdown of the passive film in an electrolyte is usually attributed to the presence of anions $Cl^-$ or other possible halides. Three main mechanisms are proposed to explain the nucleation of pits: penetration of aggressive anions through the passive layer, their local adsorption or breaking of the passive layer (see e.g. Strehblow 1976).

If the penetration mechanism is valid, the negatively charged nitrogen atoms (suggestively $N^{3-}$) segregated at the metal surface under the passive film may replace anions, which diminishes the transport and electric field gradient in the passive layer.

The passive film consists of small oxide crystals with a size of about 12 Å (Sewell et al. 1972). Occasional cracks or pores caused by stresses in the film can lead to the direct access of the electrolyte to the metal surface.

According to the adsorption mechanism, $Fe^{3+}$ ions interact with $Cl^-$ resulting in the local dissolution of the passive film.

There is no clear opinion concerning the breaking or adsorption mechanisms of the film. There is also no conclusive data about a possible role of nitrogen in counteracting these processes. Grabke (1996) suggested an important role of nitrogen in repassivation rather than in pit initiation. Negatively charged nitrogen atoms segregated beneath the passive film may repulsively interact with $Cl^-$ anions after critical thinning or breakdown of the film, which leads to a desorption of $Cl^-$. Thus, immediately after pit initiation its growth in nitrogen steels is suppressed by repassivation. Small current oscillations observed in Fig. 3.38 for the nitrogen-rich steel are consistent with this idea providing one with an indirect evidence of numerous cases of pit nucleation and repassivation before the breakthrough potential is reached.

Another mechanism is supposed by Misawa and Tanabe (1996) who observed the presence of nitrate ions $NO_3^-$ and no ammonia ions $NH_4^+$ in austenitic nitrogen steel 316LN at the stage of pit initiation and described the favourable effect of nitrogen in terms of a combination of nitrogen and oxygen ions in the passive film improving the stability or self-repair of the passive film in presence of hydrated chloride ions.

***Pit growth.*** Once being formed a pit grows with an extremely high dissolution rate corresponding to high local current densities. The composition of the pit electrolyte during the dissolution of metal is changed so that the value of pH is decreased, i.e. the solution becomes locally more acidic. A possible role of nitrogen amounts to the formation of $NH_4^+$ species which take hydrogen ions out of the solution raising pH immediately at the metal surface, which assists repassivation of growing pits.

***Repassivation*** occurs if the aggressive anions are replaced by a passive layer. In this case the diffusion of aggressive anions from the surface of a pit to the bulk of the electrolyte controls the rate of repassivation. A positive effect of nitrogen has to be the same as was suggested for pit nucleation, i.e., the $Cl^-$ anions can be removed from the free metal surface due to repulsive interaction with the segregated $N^{x-}$ ions.

Nitrogen-enhanced repassivation was observed by Lu et al. (1991 I) in the steel Cr25Ni20Mo6N0.2. At the same time a higher concentration of chromium was found at the pit surface suggesting that nitrogen assists the surface enrichment in chromium and thereby inhibits the anodic dissolution during the pit development.

### *3.2.2.3*
### *Crevice corrosion*

Crevice corrosion can be described as a kind of pitting corrosion enhanced by the stagnant state of the electrolyte enclosed within the crevice. The hydrolysis of metal ions leads to a decrease of pH in the closed volume of the solution, i.e. it increases the aggressivity. That is why crevice corrosion can occur in milder conditions than pitting corrosion does (see e.g. Suzuki and Kitamura 1972, Olsson 1995). In general, steels resistant to the pitting attack have to be also resistant to crevice corrosion with the only remark that a higher Cr+Mo+N content (see Eq. 3.6) is needed to provide the same stability as under pitting corrosion.

The following stages are observed in the course of crevice corrosion (see Azuma et al. 1996): (i) consumption of dissolved oxygen and formation of a passive film; (ii) decrease of pH and increase of chloride concentration in the electrolyte; (iii) breakdown of the passive film; (iv) propagation of corrosion due to initiation of local electrochemical cells; (v) possible repassivation by decreasing the acidity of the solution.

The potentials of depassivation and repassivation are shown to be not affected by nitrogen suggesting that nitrogen does not influence the static characteristics of crevice corrosion (Okayama et al. 1987, Azuma et al. 1996).

At the same time nitrogen produces a beneficial effect on those stages in which the dissolution kinetics at the metal/electrolyte interface are included: initiation, rate of propagation and repassivation.

The ammonium content in the electrolyte was found to be increased with increasing nitrogen content in austenitic steels. As the ammonium formation consumes hydrogen ions, the acidity of the solution decreases, and, therefore, nitrogen suppresses the acidification in the crevice thereby increasing the time before crevice corrosion is initiated.

It was also shown (Azuma et al. 1996) that nitrogen decreases the penetration depth in the crevice sites, i.e. prevents the dissolution of metal during the propagation of crevice corrosion, which can also be due to the ammonium formation.

A decrease in the acidity of electrolyte reduces the concentration of hydrogen ions at the dissolving metal, which assists also the precipitation of metal hydroxide accelerating the formation of the passive oxide film. The anodic current for the repassivation was observed to be decreased by nitrogen (Newman et al. 1984,

Azuma et al. 1996), which means that nitrogen kinetically accelerates repassivation.

An increase of the critical crevice corrosion temperature by nitrogen was proven for high alloy austenitic CrMnMo steels and used for the development of body-friendly nickel-free corrosion-resistant steel grades (Menzel et al. 1996, Uggowitzer et al. 1996, see also Sect. 5.2.4).

### 3.2.2.4 Stress corrosion cracking

Stress corrosion cracking (SCC) in water and, in particular, in hot chloride solutions led to many service failures. Widely used austenitic stainless steels like AISI 304 (Cr18Ni9) are not stable to this kind of corrosion attack. This is demonstrated by the so-called „Copson-curve" which shows that the time to SCC failure versus the nickel content $x$ in steels with18 to 20 mass % Cr (Copson 1959) has a minimum just at $x = 8$ mass % of nickel. Nickel-free ferritic chromium steels and austenitic steels with high nickel content (about 30% Ni) are stable against SCC in the most aggressive environments.

The effect of alloying elements on SCC in austenitic steels is not thoroughly studied. An increase in the chromium content from 18 to 25 mass % did not improve the resistance of austenitic steels to SCC while molybdenum additions rendered a strong beneficial effect (Speidel 1981). Molybdenum is shown not only to retard SCC by preventing the initiation of local corrosion but also to prevent the growth of preexisting cracks under stress in aggressive environments.

A correlation between the SCC resistance and the stacking fault energy was proposed (see e.g. Swann 1963, Thompson and Bernstein 1980) and ascribed to a detrimental effect of planar slip that occurs in materials of low stacking fault energy. This approach explains the low SCC resistance of metastable austenitic steels with 8 mass % Ni having a low SFE and the improved stability against the SCC with increasing nickel content because nickel, as $\gamma$-stabiliser, increases the SFE in austenitic steels. The above mentioned absence of a positive effect of chromium in austenitic steels is also consistent with this explanation as chromium decreases the stacking fault energy of $\gamma$-iron solid solutions. In contrast, chromium, as a ferrite-forming element, increases the stacking fault energy of $\alpha(\delta)$-iron solid solutions and provides the excellent SCC resistance of ferritic 18%Cr steels.

However, it is not so easy to attribute the beneficial effect of molybdenum to a correlation between SCC and SFE because molybdenum is a ferrite-former and can hardly increase the stacking fault energy of austenitic steels. From this point of view, nitrogen has also no chance to be favourable for SCC resistance as nitrogen in most cases decreases the value of stacking fault energy (see Sect. 1.1.5.1) and always promotes planar slip in austenitic steels (Sect. 1.1.4.2). In contrast, carbon would be expected to improve the SCC resistance because it increases the stacking fault energy and assists wavy slip in iron austenite.

In fact, the influence of nitrogen on stress corrosion cracking does not follow such a simple prediction while carbon is definitely detrimental for the SCC resistance of austenitic steels.

Another hypothesis suggests a possible role of hydrogen in stress corrosion cracking. A similarity between fracture surfaces produced by stress corrosion cracking and hydrogen embrittlement is often observed (see e.g. Eliezer et al. 1979, Liu et al. 1980). The dependence of the crack growth rate on the stress intensity $K_I$ shows three stages of which the second one develops at a nearly constant rate. This steady state stage reveals some features of hydrogen embrittlement (e.g. serrated growth of the crack, decrease in the crack growth rate by factors preventing hydrogen permeability etc.), i.e. it is probably controlled by the transport of hydrogen to the tip of the crack. Within the frame of this approach, Mummert et al. (1996) explained the difference between the influence of nickel on hydrogen embrittlement and stress corrosion cracking of iron-nickel alloys by the nickel-enhanced migration of hydrogen atoms in the $\gamma$-iron. Nickel increases the mobility of hydrogen atoms in austenitic steels (see e.g. Gavriljuk et al. 1996 I). That is why it eases the transport of hydrogen atoms to the zones of lattice dilatation during a hydrogen attack, assists hydrogen embrittlement and promotes the migration of hydrogen atoms from the surface into the bulk during stress corrosion cracking, thereby lowering the accumulation of hydrogen at the tip of the propagating crack.

Following this approach, one can ascribe the absence of a positive effect of chromium on the SCC resistance in austenitic steels to the chromium-decreased mobility of hydrogen atoms. Such a suggestion can be made based on the high values of the hydrogen migration enthalpy in austenitic CrNi steels (0.52 to 0.67 eV, according to Quick and Johnson 1979) and its low values in pure nickel (0.42-0.44 according to Brass et al. 1990) and pure iron. So far there is no experimental data about the effect of nitrogen on the hydrogen migration.

***Nitrogen austenitic steels.*** Nitrogen was found to enhance transgranular stress corrosion cracking of AISI 304 type steels in boiling $MgCl_2$ solutions (see e.g. Longinow and Bates 1969, Eckel and Clevinger 1970). At the same time, their resistance to intergranular SCC in the sensitised state is significantly increased by the addition of nitrogen up to 0.16 mass % (Mozhi et al. 1986 I), which was linked to a retarded sensitisation by inhibiting the growth of chromium carbides. Relatively low nitrogen contents may be useful for the resistance to intergranular SCC failure in high alloy austenitic CrNi and CrNiMo steels because of preventing the precipitation of chromium carbides and intermetallic phases (Jargelius-Pettersson 1992)

The nickel-free steel Cr18Mn18 with nitrogen reveals an excellent resistance to stress corrosion cracking in hot water and nitrate solutions, whereas the carbon steel of Mn18Cr4 chemical composition is liable to the SCC failure (see Fig. 3.40 after Magdowski and Speidel 1989). No effect of a nitrogen concentration above 0.2 mass % or of cold work on the SCC resistance was observed in this study.

More aggressive environments, such as chlorides, cause stress corrosion cracking of steel Cr18Mn18N (Pedrazzoli and Speidel 1990, Tsai et al. 1993).

Pedrazzoli and Speidel (1990) studied the influence of the nitrogen and carbon concentration on stress corrosion cracking in Cr18Mn18 steels and concluded that the lowering of the carbon content below 0.1 mass % is the only reason for the improved SCC behaviour while the role of nitrogen amounts to an increase in the yield strength without deteriorating the SCC resistance. This conclusion was made based on the tests in the hot water. However, Tsai et al. (1993) tested two Cr18Mn18 steels with 0.58 and 0.9 mass % of nitrogen in the 1% NaCl solution and have shown that an increase in the content of nitrogen results in a shift from 500 to 650 mV of the potential above which SCC occurs. Thus, the SCC resistance increases with increasing nitrogen content in environments more aggressive than hot water. In both cases SCC occurred in the potential range where pitting is observed, in other words, a positive effect of nitrogen could be due to the retarded initiation of local corrosion sites.

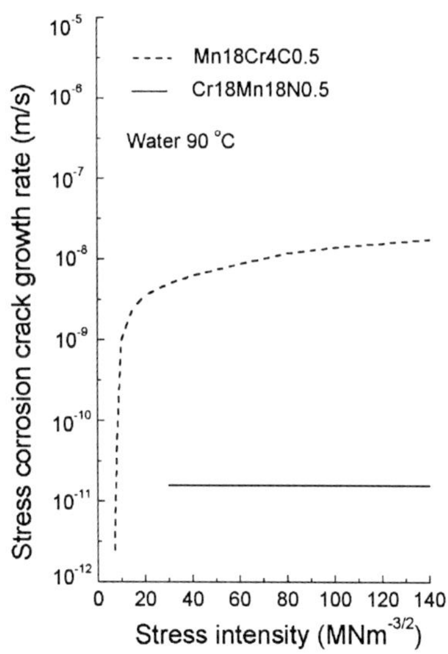


**Fig. 3.40.** Stress corrosion crack growth rate of austenitic nitrogen and carbon steels with 18 mass % Mn in hot water (after Magdowski and Speidel 1989)

A positive combined effect of nitrogen and molybdenum on the SCC resistance of steel Ni25Cr20 in boiling LiCl and $CaCl_2$ solutions was observed by Jargelius-Pettersson (1995). An explanation based on the influence of these elements on the stages of depassivation, dissolution and repassivation was proposed. A negative effect of nitrogen on SCC in $MgCl_2$ solutions was obtained by Swann (1963) and ascribed to the nitrogen-caused slip planarity, which can be hardly accepted on account of the complicated influence of nitrogen on all stages of the development of localised corrosion. A negative effect of both nitrogen and molybdenum on SCC in $MgCl_2$ was observed in some studies (e.g. Longinow and Bates 1969,

Adachi 1991). A change in the balance between effects of these elements on propagation and repassivation is thought to be a reason for such a scatter in the experimental data (Jargelius-Pettersson 1995).

***Nitrogen duplex steels*** reveal a higher resistance to SCC than austenitic steels of similar chromium and molybdenum content (see e.g. review articles by Bernhardsson 1991, Nilsson 1992). Even the low alloyed steel Cr23Ni4N0.1 (SAF 2304) is proven to be superior to the austenitic steel Cr18Ni14. The SCC behaviour of duplex steels is much improved by increasing chromium and molybdenum contents. The steel Cr25Ni7Mo4N0.27 (SAF 2507) is fully resistant to SCC in chloride solutions with concentrations of $Cl^-$ anions up to 10 mass % at a testing time of 1000 h and applied stresses equal to the yield stress at temperatures up to 300 °C (Bernhardsson 1991). However, this superiority of duplex steels is mainly due to the mictrostructure. The SCC resistance of duplex steels is increased by an increasing fraction of ferrite (Flowers et al. 1963) because transgranular cracks in the austenite are blocked at the grain boundaries of the harder ferrite.

### 3.2.3 Hydrogen embrittlement

Practically all metals and alloys are "transparent" for hydrogen atoms and the permeation of hydrogen under stress in aggressive environments causes brittle fracture that reveals some special features. One of them is that there is no clear correlation between the hydrogen-caused brittleness and the yield strength so that pure iron and high strength steels are embrittled nearly to the same extent under hydrogen attack (see e.g. Bernstein 1996). To provide a longer time to fracture is the aim pursued at the development of new hydrogen-resistant engineering materials. That is why the measurement of the hydrogen permeation is so important in order to predict and control the service time of any material intended for service in hydrogenating environments.

**Table 3.4.** Plasticity of steel Cr19Mn10Ni7N steel tested in vacuum and under a hydrogen pressure of 350 bar (after Gavriljuk and Yephimenko 1990)

| N content (mass %) | medium | elongation (%) | reduction in area (%) |
|---|---|---|---|
| 0.12 | vacuum | 40 | 62 |
| | hydrogen | 26 | 33 |
| 0.31 | vacuum | 57 | 66 |
| | hydrogen | 54 | 62 |

Nitrogen increases the resistance of austenitic steels to hydrogen embrittlement by high pressure hydrogen gas or by electrolytically charged hydrogen. We illusthis effect by two examples of mechanical properties under gaseous (Table 3.4) and electrolytic (Fig. 3.41) hydrogen saturation.

With increasing nitrogen content, the austenitic steel Cr19Mn10Ni7 acquires a resistance to hydrogen attack under applied hydrogen pressure. Of course, a further increase in the hydrogen pressure would deteriorate plasticity of this steel.

Electrolytic charging can provide a hydrogen content in austenitic solid solutions comparable with extremely high pressures of gaseous hydrogen. Rozenak (1990) studied the effect of nitrogen on the mechanical properties of hydrogen-charged austenitic steels Cr17Ni13Mn2Mo2 (≈AISI 316), Cr18Ni11Mn2 (≈AISI 321) and Cr16Ni11Mn2 (≈AISI 347). The tensile test curves are shown in Fig. 3.41 for steels 316 and 347.

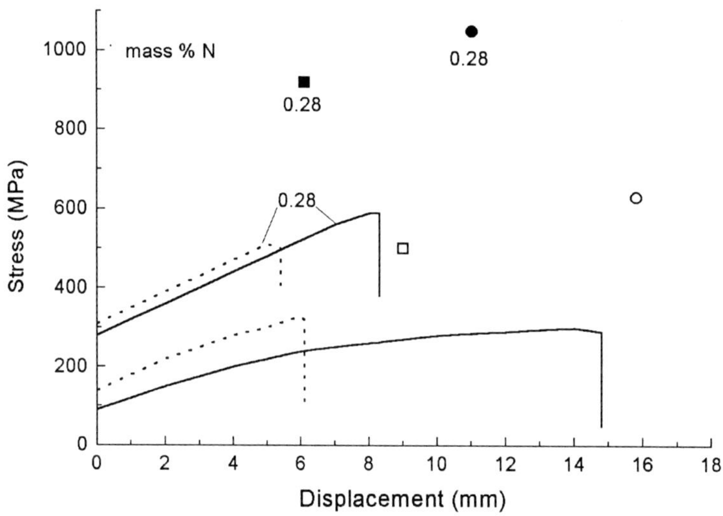


**Fig. 3.41.** Stress-displacement curves of nitrogen-free and nitrogen-containing steel AISI 316 tested in air (solid lines) or under hydrogen electrolytic charging (dotted lines). Points correspond to the ultimate strength of the nitrogen-free (open) and nitrogen-containing (solid) steel AISI 347 tested in air (circles) or under hydrogen charging (squares), after Rozenak (1990)

Hydrogen charging resulted in a reduction of elongation by 58% in nitrogen-free and by 33% in nitrogen-containing steel AISI 316. The values for AISI 321 and AISI 347 are 46 and 36%, 43 and 39%, respectively.

Before discussing possible mechanisms of the effect of nitrogen, we will briefly describe theories proposed for hydrogen embrittlement. Three main hypotheses explain the harmful effect of hydrogen on the mechanical properties: hydrogen-caused decohesion (Steigerwald et al. 1960, Oriani and Josephic 1974 etc.), hydrogen-induced brittle pseudo-hydrides (Holzworth and Loutan 1968, Szumer and Janko 1979, Narita et al. 1982, Ulmer and Altstetter 1993 etc.), hydrogen-enhanced localised plasticity (Beachem 1972, Birnbaum 1990, Rozenak et al. 1990 etc.). The first one attributes hydrogen embrittlement to a decrease in the interatomic bond that eases the opening and growth of a crack under hydrogen attack. According to the second hypothesis, hydrogen induces phase transformations resulting in the formation of brittle hydride-like phases. The third hypothesis considers hydrogen embrittlement as a macro-brittle phenomenon which develops from hydrogen-assisted microplasticity: a local decrease in the stress required for the generation and motion of dislocations.

The hydrogen-caused decrease of interatomic bonding was often claimed but never experimentally proven. Mössbauer measurements of the hydrogen effect on the probability of the absorption of γ-quanta in austenitic steels have shown that hydrogen increases the Debye temperature (Gavriljuk and Shivanyuk 1999), i.e. enhances interatomic bonding in the Debye approximation. According to electron spin resonance measurements, hydrogen significantly increases the concentration of free electrons in austenitic steels (Shanina et al. 1999), i.e. assists metallic interatomic bonding. The latter is weaker than the covalent bonding, however, it cannot ease the opening or propagation of cracks during hydrogen embrittlement, as usually presented in the hydrogen decohesion theory (see e.g. Knott 1996), because metallic bonds are not broken during the movement of dislocations and, moreover, assist the plastic relaxation of stresses in the zone of crack nucleation and at the tip of the propagating crack.

The formation of an fcc pseudo-hydride (γ* phase) as a reason for hydrogen embrittlement was suggested based on X-Ray diffraction studies of hydrogen-charged austenitic steels (see e.g. Narita et al. 1982, Ulmer and Altstetter 1993). However, as shown in low temperature X-Ray diffraction studies (Gavriljuk et al. 1994), diffraction reflections ascribed to the γ* phase belong, in fact, to the hydrogen-rich solid solution. Moreover, Ieki and Asano (1994) have proven in the studies of hydrogen-charged Fe-Ni-Mn alloys that hydrogen embrittlement occurs in absence of any hydrogen-induced phase transformations, i.e. it is the brittleness of the hydrogen γ solid solution not that caused by pseudo-hydrides.

The hypothesis of hydrogen-enhanced localised plasticity was supported by *in situ* TEM observations of a decrease in the stress required to move dislocations in front of a crack and to propagate cracks in hydrogenated steel AISI 316 (Rozenak et al. 1990). At the tip of the cracks ε and α′ phases were also observed and the stress required for their formation was decreased by hydrogen.

A decrease in the starting stress for the generation of dislocations and an increased velocity of dislocations moving under the applied stress were observed by Gavriljuk and Shivanyuk (1999 I) in hydrogen-charged steels Cr25Ni20 and Cr18Ni16Mn10 by means of amplitude-dependent internal friction.

On account of the above mentioned hypotheses and experimental data, one can state that the beneficial effect of nitrogen on the mechanical behaviour of austenitic steels under hydrogen attack is not concerned with interatomic bonding. Nitrogen is shown to increase the concentration of free electrons in austenitic steels (see Sect. 1.1.1.2) exactly as does hydrogen.

The effect of nitrogen on hydrogen-induced phase transformation was studied using X-Ray diffraction (see Gavriljuk et al. 1993 I, 1994, 1995). It was shown that nitrogen increases the concentration of hydrogen in the γ solid solution and prevents a hydrogen-induced γ→ε transformation (see Fig. 3.42).

Thus, if the hydrogen-induced phase $\varepsilon_H$ really promotes brittle fracture, the positive effect of nitrogen on the plasticity can be understood.

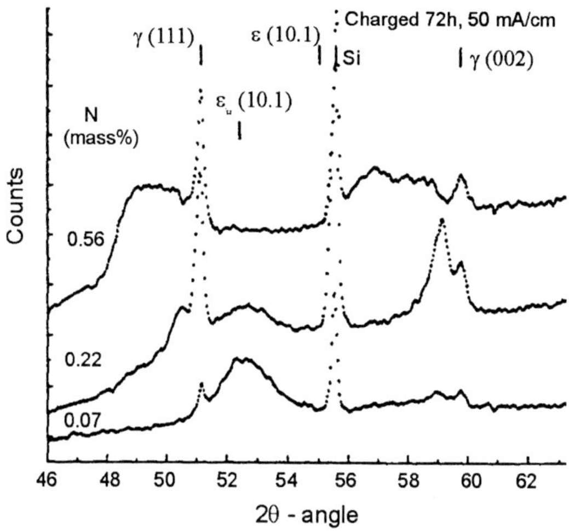


**Fig. 3.42.** Segment of the diffraction pattern of the hydrogen-charged steel Cr18Ni16Mn10 having different contents of nitrogen (after Gavriljuk et al. 1993 I). Electrolytic hydrogen charging results in a broadening and a shift of the γ reflections (111) and (002) to smaller angles (higher nitrogen contents in the solid solution). The intensity of low-angle components of reflections is increased with increasing content of nitrogen. The reflection (10.1) of the hydrogen-induced $\varepsilon_H$ phase (the most intensive reflection in the hcp lattice) is suppressed by nitrogen

As hydrogen brittleness occurs in absence of hydrogen-induced phases (Ieki and Asano 1994), the hypothesis of hydrogen-enhanced localised plasticity seems to be more realistic. An increase in the concentration of free electrons due to hydrogen charging, i.e. promotion of the metallic component of interatomic bonding, can be a physical reason for a decrease in the starting stress of dislocation sources and the enhanced mobility of dislocations. From this point of view, a broadening and a shift of γ reflections to smaller angles and an increase in the intensity of the shifted components of the reflections (see Fig. 3.42) is an important evidence of the nitrogen-increased concentration of hydrogen in the γ solid solution. This means the retarded mobility of hydrogen atoms in the fcc iron lattice. Thus, nitrogen can limit the hydrogen transport to the zones of lattice dilatation under applied stress and, in such a way, prevent a critical hydrogen concentration needed for starting local microplastic deformation as well as for the generation of microcracks and their growth. Experimental data concerning the effect of nitrogen on the migration enthalpy of hydrogen atoms in austenitic steels could be useful for further discussion of the role of nitrogen in hydrogen embrittlement.

# 4 Manufacturing

Nitrogen is a volatile element and causes manufacturing problems to HNS not encountered in comparable carbon grades. One is to get the nitrogen into the steel, another to keep it there during high temperature processing. Therefore new procedures have been developed or adjusted to make, shape and treat HNS.

One must also bear in mind that HNS are on the high alloy side. Thus, their behaviour during manufacturing is different from low alloy or plain carbon steels and comes closer to e.g. conventional stainless steels.

## 4.1 Uptake of nitrogen

The equilibrium conditions of the nitrogen solubility given in Sect. 2.1 are briefly summarized to derive Eq. 4.7 which allows to quantify the different possibilities of enhancing the poor solubility of pure iron. The equilibrium content of nitrogen in steel is depending on three variables of state: temperature *T*, nitrogen pressure $p_{N_2}$ and alloy concentration of the steel. Due to the interatomic spacing the N-solubility is lowest in ferrite as compared to austenite and the melt. In a nitrogen atmosphere the dissolution of nitrogen in the steel is governed by the reaction

$$\tfrac{1}{2} N_2 = [N] \tag{4.1}$$

yielding the equilibrium constant for a given temperature

$$K_N = \frac{a_N}{p_{N_2}{}^{1/2}} = \frac{f_N \cdot [N]}{\sqrt{p_{N_2}}} \tag{4.2}$$

which is the ratio of the nitrogen activity in solution $a_N$ and in the surrounding gas atmosphere expressed by the partial pressure of nitrogen. The activity coefficient $f_N$ relates the chemical activity of dissolved nitrogen to its concentration [*N*] in mass%. This implies the validity of Sieverts' law $[N] \sim \sqrt{p_{N2}}$ (Sieverts 1931). In a dilute solution of nitrogen in pure iron $f_N$ reaches unity. By adding an alloying element *X* the concentration of dissolved nitrogen is changed and the activity coefficient is adjusted accordingly

$$f_N^X = \frac{[N]_{Fe}}{[N]_{Fe-X}} \tag{4.3}$$

The influence of dissolved *[X]* in mass% on $a_N$ was interpreted by Wagner (1962). A Taylor series expansion leads to first and second order interaction parameters $e_N^X$ and $r_N^X$ including the effect of nitrogen on itself $e_N^N$

$$\log f_N^X = e_N^X[X] + r_N^X[X]^2 + e_N^N[N] \tag{4.4}$$

At equilibrium the rate of reaction (4.1) becomes equal in both directions and the free energy of solution $\Delta G$ is zero

$$\Delta G = \Delta G_0 + RT\ln K_N = 0 \tag{4.5}$$

The standard free energy $\Delta G_o$ of solution is then related to the standard values of enthalpy $\Delta H_o$ and entropy $\Delta S_o$

$$\Delta G_0 = -RT\ln K_N = \Delta H_0 - T\Delta S_0 \tag{4.6}$$

In combination with (4.2) it becomes evident how the nitrogen content of a steel may be raised.

$$\ln[N]_{Fe-X} = \ln\sqrt{p_{N_2}} - \ln f_N^X - \frac{\Delta H_0}{RT} + \frac{\Delta S_0}{R} \tag{4.7}$$

An increase of the partial pressure $p_{N_2}$ and the addition of an alloying element $X$ with a negative interaction parameter $e_N^X$ will support the nitrogen solubility. A higher temperature enhances the solubility if the heat of solution $\Delta H_o$ is positive. This holds true for a binary melt and δ-ferrite. In binary austenite $\Delta H_o$ is negative and the solubility decreases as the temperature is raised. Because $\Delta H_o$ is proportional to $e_N^X$ (Chipman and Corrigan 1965) a ternary high chromium alloy ($e_N^X$ negative) will show a decrease of nitrogen solubility in all three phases upon heating (Fig. 4.1).

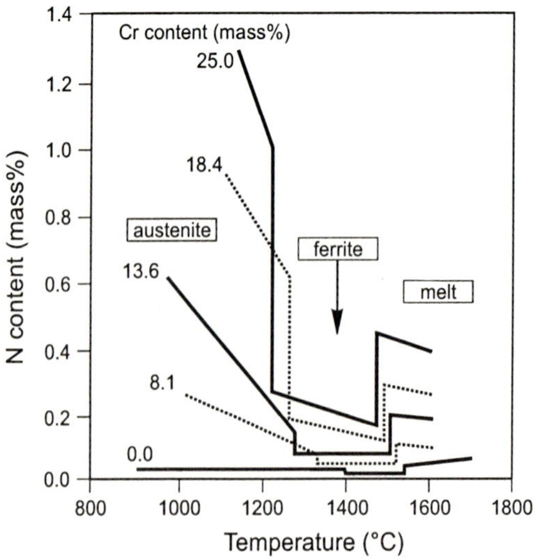


**Fig. 4.1.** Nitrogen solubility of Fe-Cr alloys in equilibrium with a nitrogen atmosphere of 1 bar (Feichtinger and Zheng 1991)

By selecting the temperature, the pressure and the alloy concentration a desired nitrogen content may be achieved in the steel melt or in the solid state. The prediction requires thermodynamic data, a collection of which is given by Satir-Kolorz et al. 1989, 1990 also Kumar et al. 1983, Zheng 1991 and in the ThermoCalc data base also providing routines of calculation (Sundman 1997). The effect of different alloying elements $X_i$ is added up and at higher concentrations interaction parameters like $r_N^{X_1X_2}$ may be introduced (Zheng 1991).

### 4.1.1 Alloying

Alloying is the easiest way of enhancing the nitrogen solubility. By adding elements with a negative interaction parameter and avoiding those of positive sign the low nitrogen content of an iron melt may be raised by more than two orders of magnitude. In the order of Cu, Co, Ni, Al, Si, C, N alloying becomes increasingly detrimental, while in the order of W, Mo, Mn, Cr, Nb, V, Ti nitrogen is more and more attracted to the melt. Thus its activity is lowered and its solubility raised (Table 2.1). However, the strongest promoters Nb, V, Ti can hardly be used, because of the low solubility of their respective nitrides in the solid state, which may lead to undissolvable embrittling precipitates. This is illustrated for example by the soluble content $X_\gamma$ of the alloying element $X$ in austenite at 1100 °C and a content of 0.5 mass% N. For X = Ti, Nb, V, Cr ThermoCalc gives $X_\gamma$ (mass%) = $1.73 \cdot 10^{-7}$, $5.57 \cdot 10^{-4}$, $3.43 \cdot 10^{-3}$, 1.6, respectively. Above 10.5 mass% Cr the austenite becomes homogeneous again (Fig. 4.2). Therefore Cr is the key alloying element followed by Mn. It seems reasonable to express the effect of alloying on the nitrogen solubility in terms of a chromium equivalent $Cr_{eq}[mass\ \%] = \Sigma\, c_N^{Xi} \cdot [X_i]$. In Table 2.1 $c_N^{Cr}$ is equal to one and the effect of other alloying elements $X_i$ relative to Cr is given by $c_N^{X_i}$ (Satir-Kolorz et al. 1989, 1990).

Except for Mn the elements enhancing the solubility of N are stabilizing ferrite. Under the above conditions of 1100 °C and 0.5 mass% N the transition from austenite to ferrite starts at [mass%] Mo = 12.4, W = 13.4 and Cr = 18.5. An increase of N will extend the austenitic regime at high but that of nitrides at lower Cr-contents (Fig. 4.2). To procure a regime of homogeneous austenite during powder compaction, hot working or solution annealing other elements may be added. Ni, Co, Mn stabilize the austenite even down to room temperature without forming nitrides, but only the latter increases the N-solubility. Because of price Co is rarely used. C like N supports the formation of austenite and hard precipitates. As it occupies the same interstitial sites as N it lowers the N-solubility.

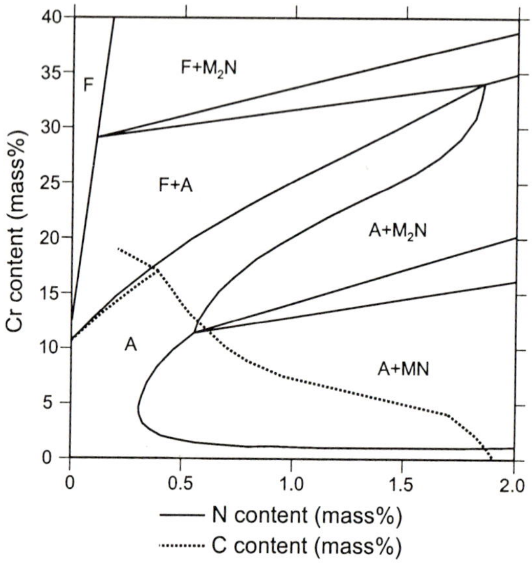


**Fig. 4.2.** Fe-Cr-N phase diagramme at 1100 °C derived by ThermoCalc.
F = δ-ferrite,
A = autenite,
MN = cubic CrFe-nitride,
$M_2N$ = hexagonal CrFe - ε nitride.
Note that the nitrogen pressure $p_{N2}$ of the system is not constant but increases roughly from the upper left to the lower right. The dotted line depicts the austenitic regime of Fe-Cr-C for comparison

## 4.1.2
## Pressure metallurgy

An increase of $p_{N_2}$ above ambient pressure requires a pressure chamber around the melting unit. This route of enhancing the N-solubility is therefore restricted to special equipment of higher cost. Thus it is generally used in combination with alloying to reach a nitrogen level above that feasible by alloying alone. According to equation (4.7) the $[N]$ content in the melt is proportional to $p_{N_2}^{0.5}$. If we assume that the cost of equipment increases with $p_{N_2}^{>1}$, the diverging pressure exponents imply an economical pressure limit.

***Melting:*** Porous ingots were one of the early problems in industrial production of HNS. Others were concerned with safety. A sudden drop of pressure would cause uncontrolled boiling of the melt (Rashev 1995). The penetration of cooling water into the melt posed a danger, which required a gas pressure somewhat above the water pressure. A solution to these problems was seen in the reduction of the liquid fraction by zone melting. Therefore a shift from induction furnaces poured under pressure to remelting under pressure was observed. Early industrial efforts were recorded in the early sixties in Japan (Okamoto et al. 1962) and Austria (Fig. 4.3a), in the seventies in Bulgaria (Fig. 4.3b) and in the eighties in Germany (Fig. 4.3c). Since a number of years a PESR electroslag remelting unit using an alloyed electrode and a $Si_3N_4$ slag under a pressure of $p \leq 42$ bar has been producing ingots of different HNS grades (Menzel and Stein 1993). Their weight is limited to 20 tons. $Si_3N_4$ is added to the slag to accelerate the uptake of N, which is rather slow from an $N_2$-atmosphere. However Si is picked up as well. Recent industrial PESR equipment is run at $p \leq 16$ bar (Bernauer et al. 1998). Plasma arc remelting (PAR) does without a slag and leads to a nitrogen pick-up in spite of a low gas pressure of ( $p_{N_2} \leq 4.5$ bar) because the plasma - assisted dissociation of $N_2$ exceeds a mere thermal one and leads to a higher concentration of N in the atmosphere. Yet degassing may occur in molten regions not covered by the arc (Rashev 1995; Torkhov et al. 1978), see also Sect. 4.3.2. Arc Slag Remelting (ASR) makes use of the higher stability of a transferred electric arc allowing for a higher ambient pressure (e.g. 10 bar) as compared to PAR. Applying a hollow electrode a high pressure $N_2$-jet may penetrate the slag, ionize within the arc and hit the melt surface. After Medovar et al. (1996) a low oxygen content is important to reach equilibrium in less than 50 seconds.

If alloying alone offers a sufficient N-level, nitrided ferro-alloys like ferrochrome are charged to open induction or electric arc furnaces to add nitrogen. The more advanced Argon Oxygen Decarburisation (AOD) may be extended to introduce N by blowing $N_2$ instead of Ar during the final step (Hooli and Kupari 1998).

An example of the additive effect of alloying and pressure is given in Fig. 4.4 for Fe-Cr melts. However, a deviation from Sieverts' law is observed as the Cr-content and pressure increase, reducing the expected nitrogen content (see also Sect. 2.1.1.3). The same was observed for Fe-Mn, Fe-Cr-Mn and Fe-Cr-Ni melts. From a study with $p_{N_2} \leq 2000$ bar the pressure dependence of the interaction coefficient was derived (Rawers and Gokcen 1993).

***Solidification:*** Usually the solubility of alloying elements is lower in the solid than in the liquid state. Thus the solute is repelled by the solid and enriched in the liquid phase of the interdendritic zone (Flemings 1974). In Fig.4.5 the situation is schematically presented for N in HNS. The partitioning coefficient $k_N = N_S/N_{max}$ is slightly raised as the solidification proceeds and the local cooling rate decreases supporting the diffusion of nitrogen. In addition the convection of the melt flushes some of the enriched alloy content from the interdendritic zone into the core thereby spreading the microsegregation within the zone to a macrosegregation within the cross-section.

The lower solubility of N in ferrite as compared to austenite entails smaller values of $k_N$ for all HNS which solidify primarily ferritic, i.e. ferritic, martensitic, duplex and some austenitic steels. The equilibrium between melt and δ-ferrite rich in chromium leads e.g. to $k_N = N_S/N_L \approx 0.4$ (Fig. 4.1). The real value depends on the growth kinetics, though. Increasing the local cooling rate will diminish the diffusion exchange of nitrogen and raise $N_L$ to $N_{max}$, which is concurrently lowered by more convection. Because of the rapid N-diffusion a solidifying ingot is closer to equilibrium than a weld pool. In contrast to non-gaseous alloying elements, which are fully transferred from the liquid to the solid state, some of the nitrogen may escape during solidification in form of bubbles. Those caught between dendrites will remain as pores in the material. This has to be suppressed by raising the pressure $p_a$ of the ambient atmosphere.

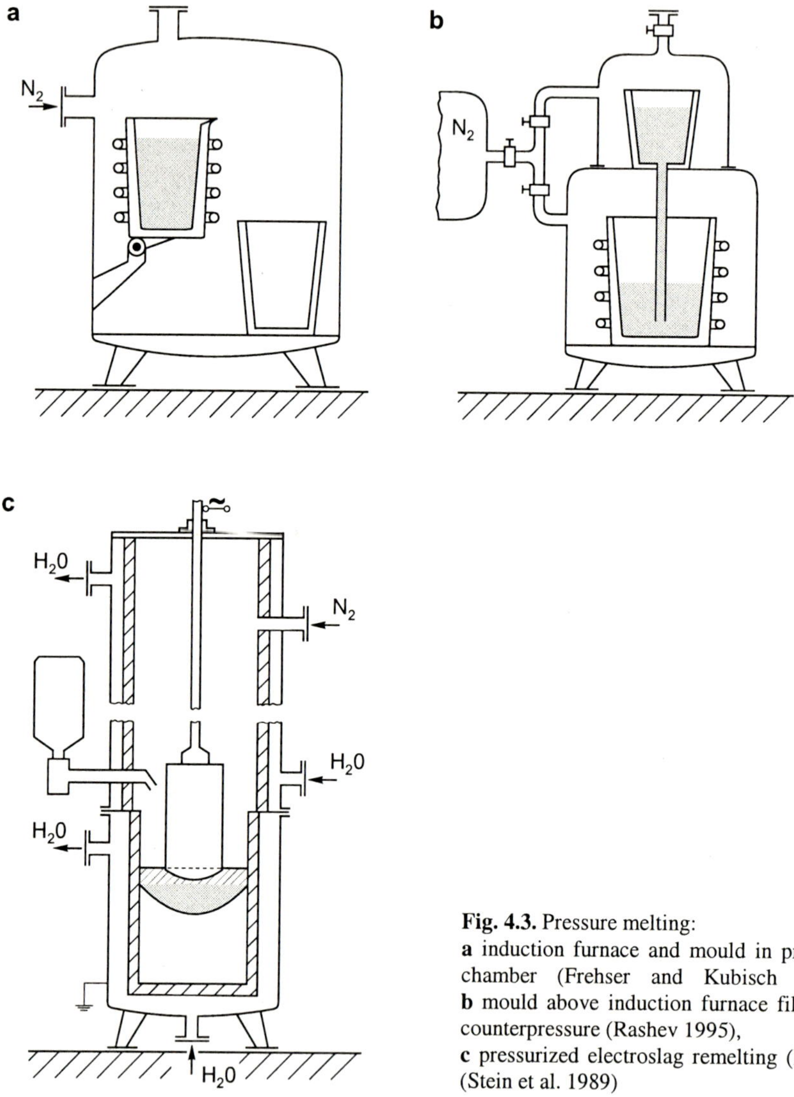


**Fig. 4.3.** Pressure melting:
**a** induction furnace and mould in pressure chamber (Frehser and Kubisch 1963),
**b** mould above induction furnace filled by counterpressure (Rashev 1995),
**c** pressurized electroslag remelting (PESR) (Stein et al. 1989)

During melting $p_{N_2}$ is applied to gain a desired N-content according to Eq. (4.7). To overcome the solubility gap of δ-ferrite $p_a$ may be raised by adding Ar leaving the partial pressure $p_{N_2}$ unchanged. The gas pressure within a bubble $p_b$ is counteracted by $p_a$, the ferrostatic pressure of the melt $p_m$ and a combination of surface stress $\sigma$ and bubble radius $r$ (Flemings 1974). To avoid bubbles the following rule has to be obeyed locally for $N_{max}$:

$$p_b < p_a + p_m + 2\sigma / r \tag{4.8}$$

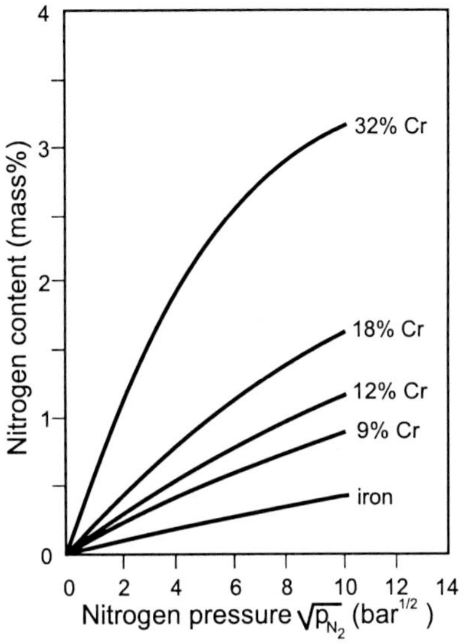


**Fig. 4.4.** Influence of partial pressure of nitrogen on its solubility in Fe- and Fe-Cr-melts at 1600 °C (Satir-Kolorz et al. 1989)

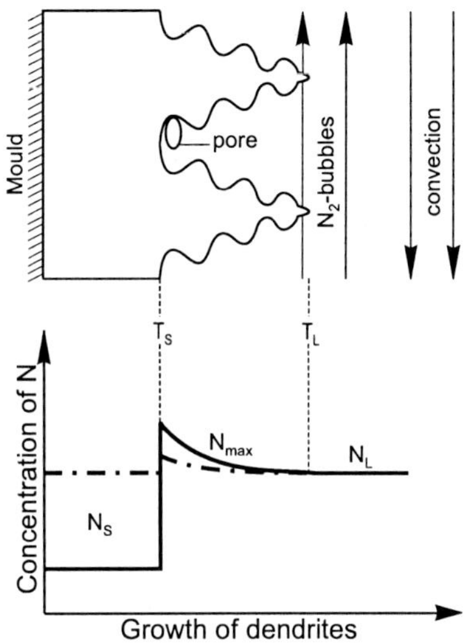


**Fig. 4.5.** Schematic representation of an HNS melt solidifying to δ-ferrite;
full lines: $p_a \approx p_{N_2}$ ($N_2$ escape);
dash-dotted lines: $p_a >> p_{N_2}$ (no bubbles)

Surface active impurities have a strong effect on $\sigma$ (Svyazhin et al. 1998). A change of alloy concentration from the lower to the upper bound of the chemical analysis defined for a certain HNS grade may influence the fraction of δ-ferrite in a liquid + γ + δ regime having an impact on the required $p_a$. An increase of the Si-content during PESR will shift the system to more δ-ferrite. The situation is illustrated in Fig. 4.6 for a martensitic stainless HNS solidifying primarily ferritic. At $p_{N_2} = 5$ bar about 0.4 mass% N are dissolved in the melt at 1450 °C. As δ-ferrite appears degassing of the system to about 0.3 mass% N occurs. As soon as austenite appears the N-solubility is raised again, but nitrogen will not be replenished. Using $k_N = 0.42$ between melt and δ-ferrite Feichtinger and Stein (1998) calculated that for this alloy $p_a + p_m + \sigma/r$ has to be raised to 37 bar to avoid the formation of bubbles and pores thus transferring all the nitrogen of the melt to the solid state. As one cannot count on $p_m$ and as 2 σ/r amounts to a minor contribution the operating pressure $p_a$ has to be increased well above 30 bar. This is in agreement with practical experience and means that $p_a \approx 7\ p_{N_2}$. Starting with $p_{N_2} \approx 12$ bar would result in a primarily austenitic solidification but would change the steel composition to about 0.6 mass% N.

The solubility of nitrogen in austenite may be higher than in the melt. A primarily austenitic solidification is therefore accompanied by an equilibrium $k_N$ close to or even above unity. In this case $p_a \approx p_{N_2}$ is usually sufficient to suppress pores.

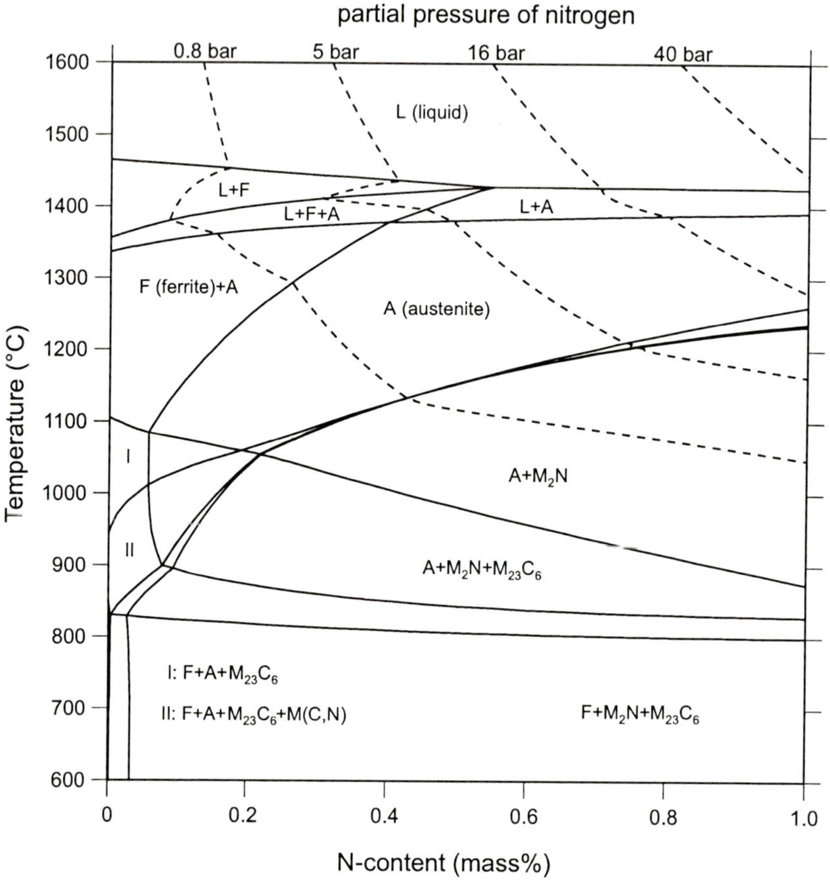


**Fig. 4.6.** Phase diagram for Cr16Mo1Si0.7Mn0.5C0.3N and partial pressure of nitrogen $p_{N_2}$ in the system derived by ThermoCalc

***Segregation:*** Microsegregation is mainly caused by insufficient diffusion in the solid state adjacent to the interdendritic melt. As the diffusion coefficient of N is orders of magnitude above that of substitutional elements this interstitial element is not prone to generate segregation. But as the substitutional elements segregate nitrogen will be attracted to areas of higher Cr, Mn, Mo content and repelled by those of higher Ni, Co, C content in accordance with the sign of $e_N^x$. Convection will flush some of the enriched interdendritic melt into the liquid core of the cross section giving rise to macrosegregation, which however remains small for nitrogen.

After hot-working the microsegregation is stretched to a banded structure. A severe case is shown in Fig 4.7 for a stainless martensitic HNS grade. After etching bands of different alloy content are revealed in the hardened condition (Fig. 4.7a). The bright grain boundaries are not decorated with precipitates

(Fig. 4.7b) but point to grain boundary precipitation of nitrides and carbides during slow cooling after forging. These precipitates are rich in Cr, which is not as readily redistributed as N or C after their dissolution at hardening temperature. Orlich et al. (1973) called this structure “inhomogeneous austenite”. After prolonged holding at hardening temperature the distribution of substitutional elements becomes homogeneous. After fracturing the bands are apparent in the fracture surface (Fig. 4.7c).

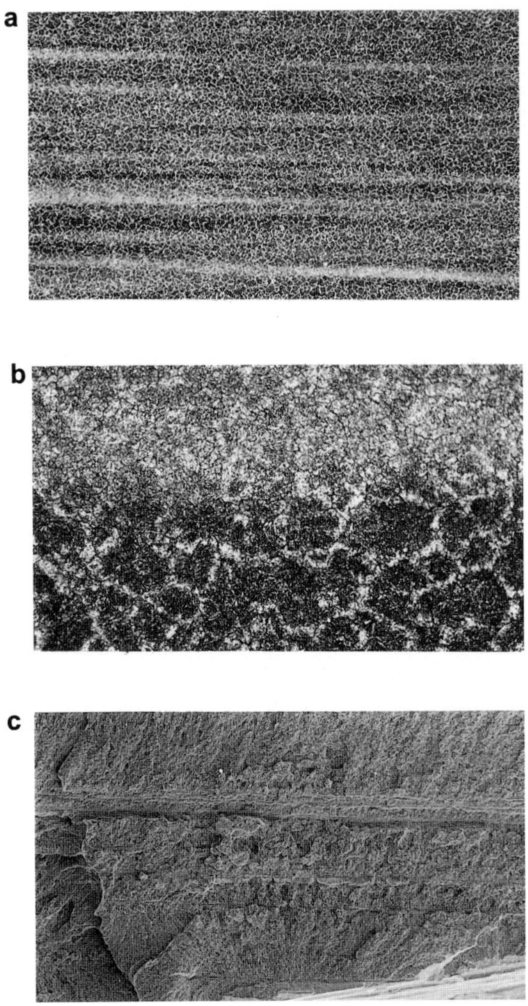


**Fig. 4.7.** Microsegregation in Cr15.6Mo1C0.3N0.42 PESR steel forged to 60 mm ∅ and hardened to 720 HV30,
**a** segregation lines, **b** transition from bright area (680 HV0.05) to dark area (780 HV0.05),
**c** fracture surface

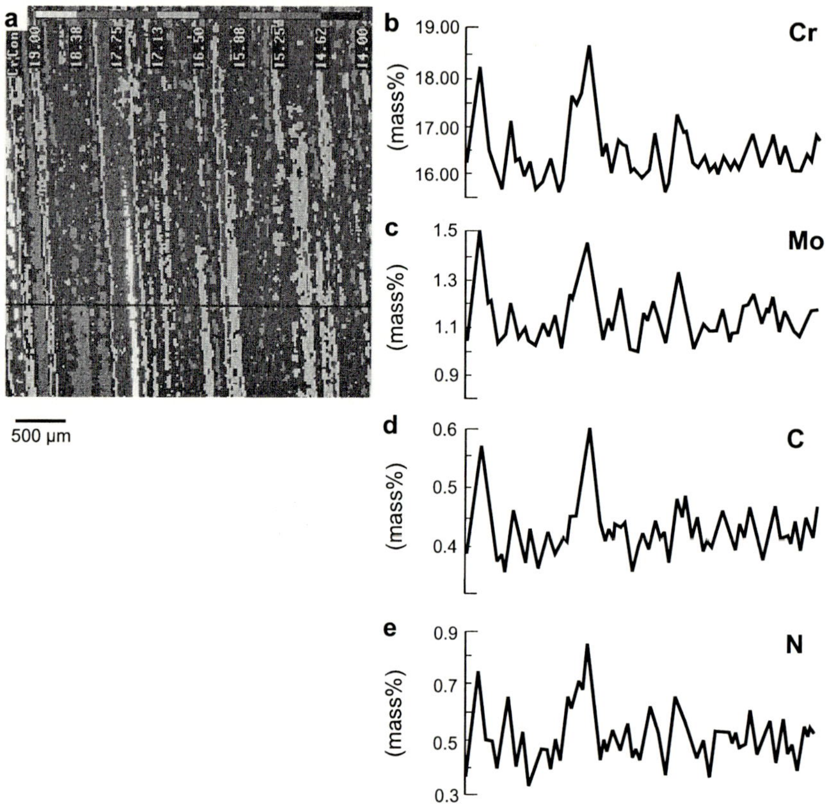


**Fig. 4.8.** WDX analysis of microsegregation depicted in Fig. 4.7, **a** mapping of Cr, **b-e** line scans along the line shown in **a**

Mapping by wave length dispersive spectroscopy (WDX) is shown for Cr in Fig. 4.8a. The results for Mo, C, N are similar as can be derived from the line scan (Fig. 4.8b to e). All four elements are attracted to the same areas. Calculating the phase composition for the maximum and minimum concentrations of the line scans reveals a distinct difference in solidification and precipitation of these two areas (Fig. 4.9).

Beside the alloy content segregation depends on the local cooling rate during solidification, i.e. on the size of the cross-section. Atomized powders are quenched rapidly, sand castings take time. Remelting in a water cooled chill is situated somewhere in-between and the rate of cooling drops with growing pool depth. The slower the solidification the wider is the dendrite spacing and the distance from maximum to minimum concentration. Thus an exchange by diffusion annealing or during soaking before hot-working is impeded.

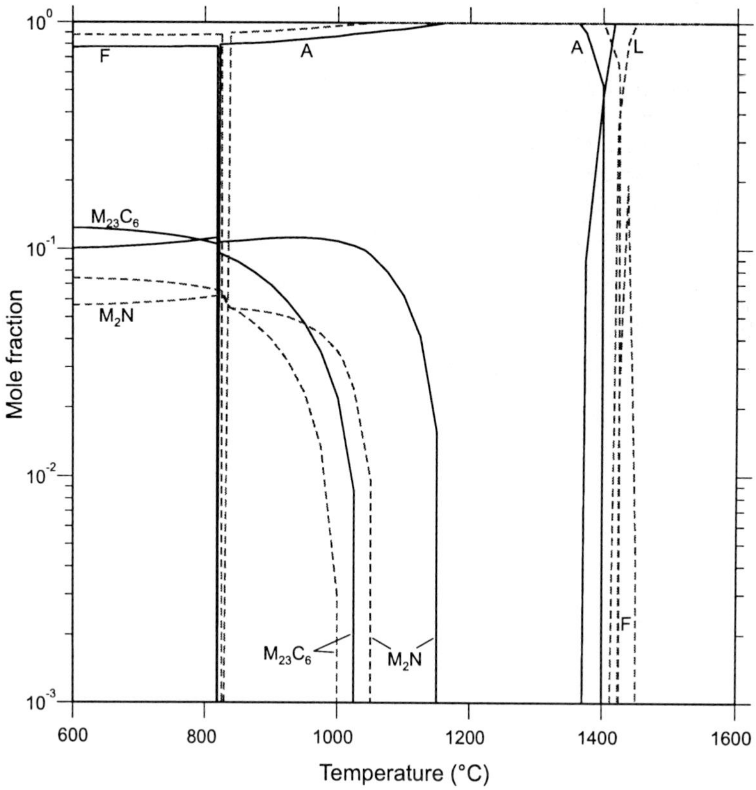


**Fig. 4.9.** Equilibrium mole fraction of constituents in dependance of temperature as derived by ThermoCalc for areas of highest and lowest alloy content in the CrMoCN co-segregation of Fig. 4.8

| (mass%) | Cr | Mo | C | N |
|---|---|---|---|---|
| —— high | 18.8 | 1.36 | 0.61 | 0.85 |
| ---- low | 15.5 | 0.96 | 0.35 | 0.41 |

## 4.1.3 Powder metallurgy

Instead of adding nitrogen to a melt, it may be introduced into a steel powder thermally by diffusion or mechanically by milling. Subsequently the powder is compacted to give a pre-shape or a near-net shape of even nitrogen content. A number of nitriding procedures have been reported and they are briefly introduced from (a) to (i).

(a) The end of a rotating alloy steel rod is melted by an $Ar/N_2$-plasma-arc so that droplets enriched with nitrogen spin off and solidify (Gammal et al. 1996). Solid state diffusion seems to be more common, though. (b) A high chromium steel powder is nitrided in ammonia at 590 °C in the ferritic regime by means of a slowly rotating furnace to avoid sintering. Nitrided powder of known nitrogen content is mixed with untreated powder to give a desired nitrogen concentration

(Berns and Wang 1990) (Fig. 4.10a). The powder mix is filled into cans which are subsequently evacuated and sealed. During hot isostatic pressing (HIP) at ≈ 1200 °C in the austenitic regime the nitride layer around each powder grain redissolves and nitrogen is fully exchanged between nitrided and untreated grains. (c) Also a fluidized bed is used for nitriding steel powder (Romu et al. 1996). After only 10 min in a bed of ammonia and austenitic steel powder AISI 316 L at 570 °C a nitrogen content of 0.4 mass% was achieved. Instead of two consecutive steps, namely nitriding and compacting, a combined method may be employed. One is (d) the sintering of cold compacted shapes in a nitrogen atmosphere which does not yield full density (Nakamura and Takaki 1996). The other therefore makes use of pressure (e). The powder is canned and the evacuated can is filled with nitrogen and hipped (Feichtinger 1990). For (d) and (e) the nitrogen content in the austenitic regime is controlled by $p_{N_2}$ (see equation 4.7). This equilibrium has to be reached before the porosity is disconnected by sintering. Gaseous nitriding is not the only means of solid state diffusion. (f) A chromium steel powder is mixed with chromium nitride powder CrN. During HIP the reaction $2CrN \rightarrow Cr_2N + N$ provides nitrogen to saturate the austenite. Upon hardening a mixture of $CrN/Cr_2N$ particles are embedded in a martensitic matrix to give a stainless wear resistant material (Berns and Wang 1990, Fig. 4.10b). (g) Metal injection moulding (MIM) is suited for mass production of near-net shape parts. Granules of a powder/binder–mix are fed to an extruder and injected into a mould. After debinding nitrogen is picked up from the $N_2$ atmosphere during sintering, which leads to an austenitic HNS of about 97 % density (Uggowitzer et al. 1998). Green compacts of high speed steel are sintered in $N_2$ to promote the density (see Sect. 5.1.3).

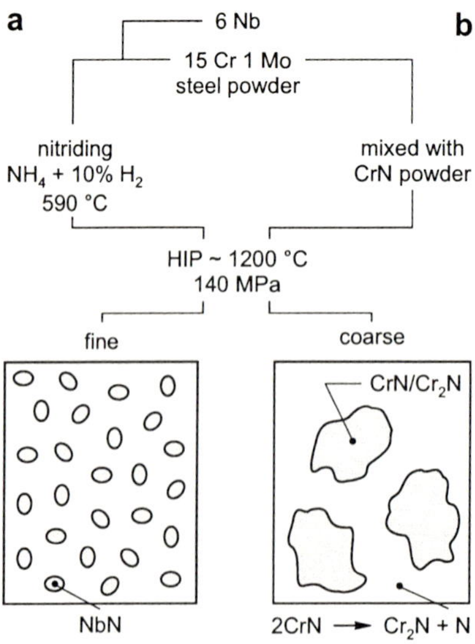


**Fig. 4.10.** PM-production **a** by nitriding of alloy steel powder and hot isostatic pressing of evacuated cans to give a dispersion of fine nitrides if the solubility limit is exceeded, **b** by mixing steel and nitride powder and compact as before to give a dispersion of coarse nitrides in a steel matrix

Mechanical alloying is brought about by high energy impacts between the hard balls of an ATTRITOR mill and the powder grains, which are heavily deformed, welded together and fractured continuously. (h) Milling in a nitrogen atmosphere leads to adsorption and dissociation of $N_2$ and finally to absorption of N (Rawers et al. 1996). (i) Another nitrogen source is a nitride powder which is disintegrated and absorbed by the metal powder (Rawers et al. 1998). Instead of a prealloyed steel powder those of iron and alloying metals may be milled (Miura et al. 1996). The microstructure of mechanically alloyed HNS powder is characterized by extremely fine grains down to the nano-crystalline regime and even by amorphous constituents. Refinement and amorphization are enhanced by elements with a negative interaction parameter like V, Nb, Ta given in Table 2.1 (Ogawa et al. 1998). To retain this structure to some extent explosive compacting may be employed or HIP at moderate temperatures making use of the higher sintering activity of a milled powder as compared to an atomized one.

The thermal PM route starts from prealloyed, gas-atomized powder of spherical shape and requires a compacting pressure to reach full density. Cans have been hipped or hot extruded (Isomoto et al. 1998). In the first case the can is heated in a pressure chamber under a gas pressure of 1 to 2 kbar, while in the second case the pressure of degassing nitrogen in the can may exceed the ambient pressure. Some sintering before extrusion appears to be beneficial. The contamination of the vast powder surface by oxygen is enhanced by the main alloying elements, i.e. Cr and especially Mn. It is accompanied by a loss in toughness which speaks against the production of near-net shape parts. After hot working the oxides are more dispersed but may form a banded structure evoking an anisotropy of toughness, thus counteracting an advantage of PM material which lies in the absence of segregation. An incentive to follow the costly thermal PM route is seen in the feasibility of nitrogen contents way above those by pressure metallurgy. This is less attractive for austenitic steels, the nitrogen content of which is - for other reasons - usually kept below 1 mass%. Stainless wear resistant martensitic steels with up to 3 mass% of nitrogen may profit, though.

As a sideline of the thermal PM route the nitrogen content of a spray coat is raised during plasma spraying of alloy steel powder using a mixture of Ar and $N_2$. In the low pressure process the $N_2$-fraction is limited to impede corrosion of the tungsten electrode (Eckardt 1996).

The mechanical PM route is even more costly and therefore restricted to parts of lower volume and very high requirements. Superplasticity, superior strength and toughness as well as corrosion resistance and magnetic behaviour are some of the properties in view.

### 4.1.4 Solution nitriding

The previous sections dealt with the uptake of an even nitrogen content throughout the cross-section of HNS. In the following the diffusion of nitrogen into a surface zone of a near-net shape part is considered forming a high nitrogen case (HNC) around a low nitrogen core. Conventional nitriding is carried out in the ferritic regime below $Ac_1$ temperature. Due to the low nitrogen solubility of the bcc lattice nitrides are precipitated to give a hard yet brittle case of ≤ 0.5 mm in thickness. Ion implantation may dissolve N in the lattice, but in a depth of usually < 1 µm. Pulsed plasma immersion was introduced as a version of lower cost for a case depth of < 15 µm (Richter and Günzel 1997). As the intensity increases precipitates are formed (Kliauga 1997; Cordier-Robert et al. 1998). In contrast solution nitriding is a high temperature process done in the austenitic regime of high nitrogen solubility (Berns and Siebert 1993, 1996). The surface content of dissolved nitrogen is in equilibrium with the $N_2$ gas atmosphere in the furnace along Eq. (4.1), (Fig. 4.11). According to Eq. (4.7) it is depending on $T$, $p_{N_2}$ and the concentration of alloying elements $X_i$. Alloying with Cr, Mn, Mo will keep the pressure at a moderate level. Because of a negative heat of solution the nitrogen content increases as the temperature is lowered. However, both measures lower the coefficient of nitrogen diffusion in austenite (Fig. 4.12) and impede the inward diffusion of nitrogen along the second law of Fick thus prolonging the solution nitriding time $t$ to reach a required case depth $s$ where $s \sim \sqrt{t}$ . An optimization of the process parameters has led to the following ranges: Cr > 13 mass%, 1100°C< T < 1150 °C, 0.1 bar < $p_{N_2}$ < 3 bar, t < 24h, s < 3 mm (Berns and Juse 1997).

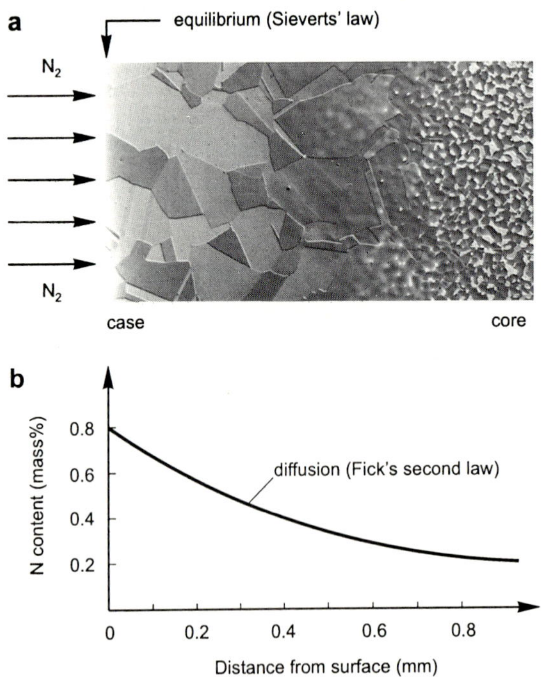


**Fig. 4.11.** Enriching the case of near-net shape, stainless steel parts with nitrogen by solution nitriding at 1150 °C in $N_2$-gas, **a** example of an austenitic case on a duplex core of steel Cr22Ni5Mo3N0.2, the surface content of which is controlled by $p_{N_2}$ , **b** inward diffusion of nitrogen

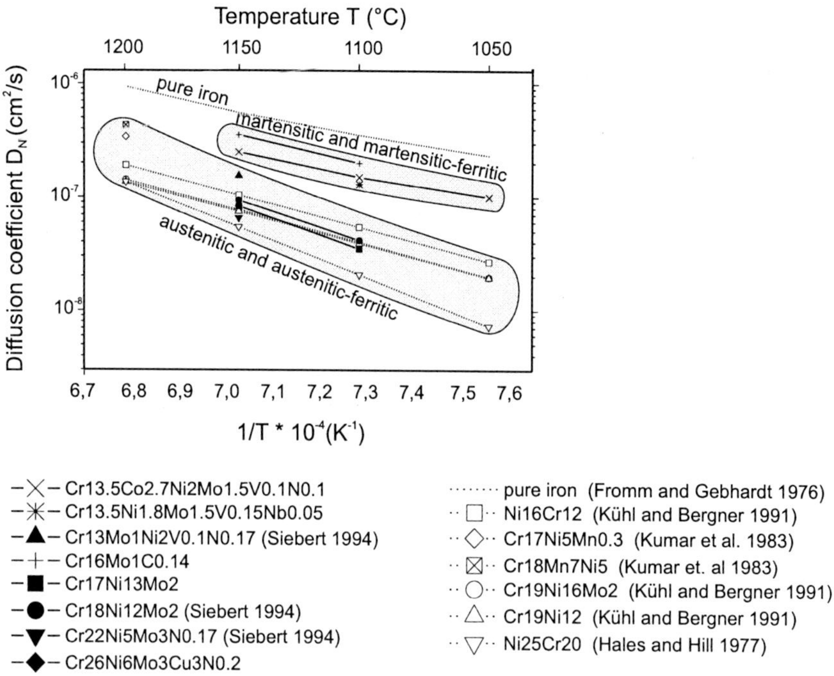


**Fig. 4.12.** Effect of temperature T on the diffusion coefficient of nitrogen $D_N$ in pure iron and stainless steels

After solution nitriding the parts are quenched to keep the nitrogen in solution. Depending on the alloy content and the resulting phase stability of the case, the high temperature austenite is retained during cooling or transforms to martensite. We therefore discern between a high strength, yet ductile austenitic HNC and a hard martensitic HNC which may contain retained austenite. The latter treatment is also called case hardening with nitrogen (Fig. 4.13). Compared to case hardening after carburizing in a $CO/CO_2/H_2$ atmosphere a number of advantages are evident. The gas atmosphere is neither explosive nor toxic. No gas flux is required as Eq. (4.1) at equilibrium works either way meaning that $N_2$ dissociates and recombines without a drop in activity. This is beneficial for parts with deep cavities like blind holes (Berns et al. 1999). An energy loss due to gas flux is avoided. No embrittling inner oxidation along boundaries of surface grains will occur. No oxygen probe is required to control the equilibrium: At a given steel composition and temperature the surface content of N is set by $p_{N_2}$. Last not least the solubility of N in high chromium steels is higher than that of C (Fig. 4.2)

Solution nitriding may also be used to enrich thin products with nitrogen. This was discussed for strip by Feichtinger (1990, 1991), for wire by Rayaprolu and Hendry (1988) as well as by Paul et al. (1996) and for tubes by Rabinovich et al. (1993).

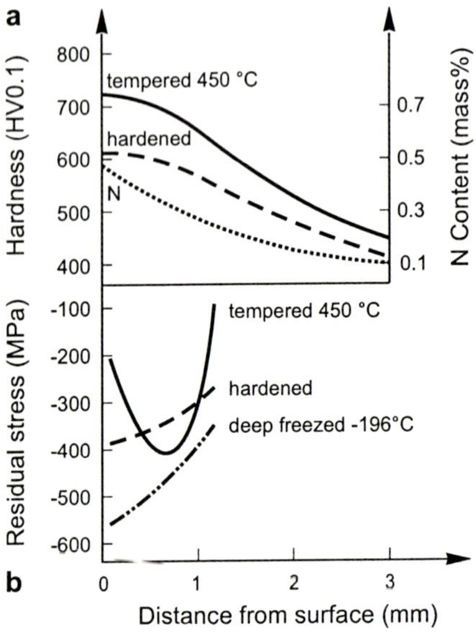


**Fig. 4.13.** Solution nitriding: a stainless CrMo steel of low (C+N) content is case hardened with nitrogen to give a hard martensitic HNC around a tough core, **a** nitriding profile, **b** residual stresses

Conventional cold wall vacuum furnaces with high pressure gas-quenching are used for solution nitriding. For conventional vacuum treatment they are filled with nitrogen up to 800 °C to make use of convection. Then vacuum is drawn and radiation takes over. For solution nitriding the furnace is loaded with gas up to 1150 °C and heat is transferred from the graphite chamber to the water cooled wall of the pressure chamber. Therefore a well sealed graphite chamber and an effective water-cooling system are prerequisites. Hot wall furnaces are cheaper but usually built for a pressure of one bar allowing for a partial $p_{N_2} \leq 1$ bar in $N_2$/Ar mixtures. At $p_{N_2} > 1$ bar a creep resistant chamber is required. Hot wall furnaces offer a more rapid liquid quench as compared to the gas quench in vacuum furnaces.

It was also proposed to split the process into two steps. A nitriding one at a lower temperature forming a nitride layer and a diffusional step at a higher temperature to dissolve the nitrides in the austenite. Plasma nitriding at 550 °C was used by Pinedo et al. (1998) in the first step and gas nitriding at 890 °C was employed by Berg et al. (1995).

## 4.2 Hot working

As shown in Sect. 4.1 alloying is an essential prerequisite of HNS. The resulting solid solution strengthening of high alloy HNS calls for stronger forging and rolling equipment capable of exerting the required stress. Because of their low nitrogen solubility ferritic HNS are not in use. Martensitic steels are austenitic at hot working temperature, but in general of lower substitutional alloy content than austenitic steels. Of the latter the conventional ones are usually of low interstitial content, while austenitic HNS contain (N+C) $\leq 1$ mass% and therefore exhibit about the highest hot strength of all steels. Kaputkina et al. (1998) showed that dynamic recrystallization of steel Cr18Ni6Mn5N during hot rolling between 1100 and 900 °C was distinctly retarded by increasing the N content from 0.13 to 0.38 mass%.

A more severe problem of HNS is the susceptibility to cracking, which is enhanced by second phases, segregation, a high level of hot strength and a high triaxiality of stress. The second phase may consist of hard particles like nitrides or near surface oxides as well as of soft δ-ferrite.

## 4.2.1 Process

Temperature $T$, true logarithmic strain $\varphi$ and strain rate $\dot{\varphi}$ lead to a true flow stress $R$ of the material which has to be overcome by the van Mises resolved shear stress $\sigma_v$ induced by the tools. Isothermal uniaxial tensile or upsetting tests reveal $R$ as a function of $\dot{\varphi}$. Due to necking or friction, respectively, $\varphi$ is rather limited and may be increased in torsion tests on tubes. In the order of compression, torsion and tension the hydrostatic stress $\sigma_h$ is raised, which may not reflect the actual stress state in a work piece. As the triaxiality expressed as $\sigma_h/\sigma_v$ is increased the probability of cracking grows, because in this respect tension is more detrimental than shear.

A hot working strategy proposed by Kopp and Holl (1990) starts from upsetting tests to measure true stress-strain-curves and defines a forming window of crack-free operation (Fig. 4.14). This basic information is then implemented into a finite element programme, which calculates the stress, strain and temperature distribution in a real work piece indicating when and where cracks are to be anticipated. Some of the high alloy austenitic HNS suffer from a rather narrow window and require frequent reheating during forging. If cracks appear flame cutting is difficult and may cause local degassing. Cooling and grinding is the alternative of higher costs.

The susceptibility to cracking is highest at the beginning of hot working. Therefore a hydraulic press is recommended for the first part of reduction. Later on many grades tolerate the higher $\dot{\varphi}$ of rolling or hammer forging. A square forging under flat tools is more evenly strained than a round one but tends to cool faster along the edges where cracks may be initiated. Pressure and powder metallurgy favour round or polygonal ingots which are forged in V-shape tools to begin with, because compressive forces are induced at four instead of two points. Even more suited are radial forging machines with four actuators. Especially the hydraulic type allows a large $\varphi$ because $\dot{\varphi}$ may be increased as hot working proceeds to give an almost isothermal operation.

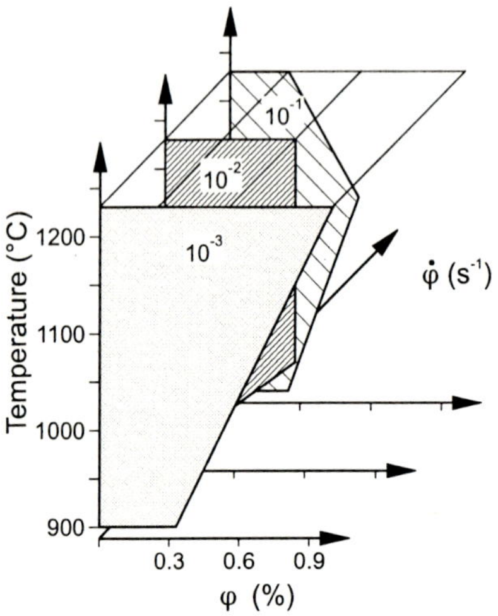


**Fig. 4.14.** Hot forming window for crack-free forging of the austenitic steel Cr20Ni8Mn6Mo3N0.85, $\varphi, \dot{\varphi}$ = true strain and strain rate in upsetting (Kopp and Holl 1990)

## 4.2.2 Microstructure

As shown in the examples of Fig. 4.15 the regime of homogeneous austenite is encased by δ-ferrite to the left and nitrides to the right. Both of these second phases affect the workability. The high alloy content of HNS promotes segregation resulting in local deviations from the calculated phase compositions. The segregation depends on the rate of solidification which increases from sand casting to remelting and powder metallurgy leading to smaller grains and precipitates.

The hot strength of bcc ferrite is distinctly lower than that of close-packed fcc austenite. In addition the higher nitrogen solubility of austenite leads to an instant partitioning of interstitial atoms which further increases the resistance to deformation of this phase. If the starting temperature of hot working exceeds the borderline to the A + F regime, some δ-ferrite is formed. Above ~ 10 vol% the individual ferrite grains start to unite in a net around the austenitic grains. This is harmful because the deformation is localized in the softer ferrite net. As its deformability is exhausted cracks may appear. At a high content of δ-ferrite like in duplex steels the localization is less severe. As the A/A+F borderlines in Fig. 4.15 are rather steep, small changes in the starting temperature of forging may have a considerable effect on the content of δ-ferrite.

During soaking at forging temperature within the A+F phase field a partitioning of substitutional elements occurs: Cr, Mo.... are enriched in the ferrite and Ni, Mn.... in the austenite. Subsequent austenitizing at a lower temperature may not fully reverse this distribution except for the interstitial elements. Based on the initial microsegregation the deviation from average composition is re-sharpened during soaking in the two-phase field.

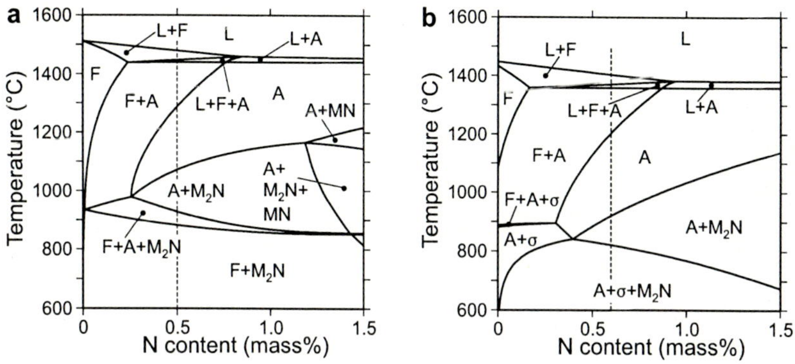


**Fig. 4.15.** Examples of phase diagrams: **a** martensitic HNS Cr15Mo1N0.5, **b** austenitic HNS Cr18Mn18N0.6, derived by ThermoCalc

Most HNS are composed so that during hot working most of the nitrides go into solution. Exceptions are e.g. austenitic valve steels with additions of Nb or martensitic PM tool steels with up to ~ 3 mass% of nitrogen. In the first case the NbN particles are precipitated from the melt. They are rather coarse (10 μm) but low in content (~ 1 vol%). In the second case the content of nitrides is high (~ 20 vol%) but their size is small (~ 1 μm) and hot working still feasible.

But even in grades of moderate N content nitrides are accumulated along primary grain boundaries. $Cr_2N$ dissolves more readily than VN and NbN, which may be of eutectic composition, thus limiting the starting temperature of hot working. Undissolved nitrides in segregated areas diminish the strain to cracking. At the lower end of the temperature interval of hot working nitrides start to precipitate again within the grains and on grain boundaries. However, the grains and the nitrides are refined, which leads directly to strengthening and indirectly to a reduction of ductility. Yet a low finish temperature of hot working followed by accelerated cooling to ~ 600 °C is worth to strive for, because the distribution of nitrides is shifted from a grain boundary net to an even dispersion and because the precipitates are refined and more readily dissolved during later heat treatment.

## 4.2.3 Surface effects

All HNS are high in Cr and the PESR melts contain some Si due to the $Si_3N_4$ slag. Thus heavy scaling during heating and hot working in air is prevented, but inner oxidation of a surface zone is liable to occur. Grain boundaries decorated with oxides may crack during hot working. The resulting fissures are arrested or extend depending on the processing and the microstructure below.

During heating the nitrogen solubility of austenite decreases (Fig. 4.1) and degassing is to be expected if the soluble content is surpassed. Even a formation of blisters was observed at the surface (Ullakko et al. 1993). In moderate form degassing would - like decarburization - not pose a serious problem unless it was

**Fig. 4.16.** Grain boundary cracking at the surface of steel Cr15Mo2Ni1V0.2N0.5 during forging after soaking at 1250 °C for 8 hours in a gas-fired furnace and cooling in air

aggravated by the appearance of δ-ferrite. Instead of degassing nitriding was observed in some cases after soaking in a gas furnacc at ≥ 1200 °C resulting in a surface zone enriched with nitrides (Fig. 4.16). Work of Wagemann (1995) and Graat (1998) points to an accumulation of nitrogen below the oxide layer due to a displacement by oxygen. During hot working cracks may be initiated at the nitrided surface zone.

To demonstrate the effect of scale ground specimens of three HNS were annealed at 1200 °C in air (1 bar) or nitrogen (0.8 bar). The results are summarized in Table 4.1. For the martensitic steel (1) the calculated equilibrium content of nitrogen at $p_{N_2}$ = 0.8 bar corresponds to the alloy content and remains unchanged by annealing except for some enrichment only 0.12 mm in depth. Annealing in air almost doubles the surface content of nitrogen and the depth of penetration is 2.5 mm. About the same holds true for the austenitic steel (3). Steel

**Table 4.1.** Change of nitrogen content at the surface of two martensitic (1, 2) and one austenitic stainless steel (3) by annealing at 1200 °C in dry air or in nitrogen of respective partial pressure.

| steel | nitrogen content (mass% ) | | |
|---|---|---|---|
| | ThermoCalc | measured at the surface after 8 h | |
| | $p_{N_2}$ = 0.8 bar | nitrogen (0.8 bar) | air |
| (1) Cr15Mo1.5V0.4C0.2N0.39 | 0.4 [α+γ+MN] | 0.57 (0.12) | 0.73 (2.5) |
| (2) Cr16Mo2V0.2N0.66 | 0.08 [α+MN] | 0.4 (2.0) | 1.13 (2.5) |
| (3) Mn17Cr15Mo4N0.89 | 0.7 [α+γ] | 0.99 (0.17) | 1.26 (3.0) |

in round brackets the distance from the surface is given in mm, where the core content is met (depth of penetration)
in square brackets the predicted phases at 1200 °C are listed

(2) is calculated to be fully ferritic at 1200 °C. Thus, in nitrogen gas the surface is denitrided to a depth of 2 mm. Yet in a narrow zone of 0.04 mm in depth the nitrogen content is raised again to a surface content of 0.52 mass%. This is similar for all three steels and probably due to impurities in the furnace atmosphere. In contrast the nitrogen content at the surface of steel (2) after annealing in air is almost twice as high as in the core and about three times above that encountered without oxygen.

# 4.3 Welding

Pressure welding is done in the solid state (except for partial melting in arc-assisted process) and the exchange of nitrogen with the atmosphere is negligible. However, during fusion welding such an exchange is most liable to occur causing a change in properties and e.g. pores.

## 4.3.1 Pressure welding

The high alloy content of HNS lowers their cold formability. Therefore elevated temperatures are required for pressure welding. Local heating is usually brought about by electric resistance or friction. The contact pressure prevents scaling and a possible escape of nitrogen. Heating by flame is not recommended for high chromium steels. An example of friction welding is shown in Fig. 4.17. The nitrogen content is unchanged throughout the weld. Pressure welding processes like flash-butt welding and friction welding are considered to be unproblematic in respect to nitrogen (Stein and Trösken 1990). However, the usual precautions of high alloy material have to be taken into account as e.g. controlled cooling to avoid excessive precipitation as well as cracking.

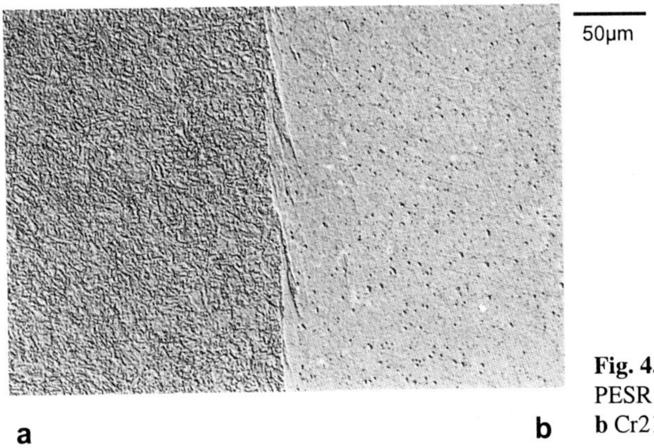


**Fig. 4.17.** Friction welding of two PESR steels: **a** Cr15Mo2V0.1N0.6, **b** Cr21Mn9.5Ni6.5W2Mo1Nb1N0.9

## 4.3.2 Fusion welding

Electric arc, plasma and laser are common heat sources for welding high alloy steels in a shielding atmosphere. The first two are usually accompanied by a filler metal, while the last one mostly goes without.

***Weld pool:*** As an example tungsten inert gas (TIG) welding is schematically shown in Fig. 4.18. Like in plasma and laser welding the heat source is much hotter than the weld pool which is cooled by convection. Therefore a hot spot is created just below the heat source. Nitrogen in the shielding gas will dissociate and even ionize in the arc, so that its content in the adjacent melt is no longer controlled by $p_{N_2}$ alone but by contributions of $p_N$ and $p_N^+$ in the plasma. In a flame of ~ 3000 °C the thermal dissociation of $N_2$ to 2 N is rather low. In arc welding with covered electrode 5000 °C are exceeded and the dissociation of $N_2$ is considerable. It is about complete in an arc of $Ar/N_2$ shielding gas at ~ 10 000 °C, shifting from N to $N^+$ as He is added and the temperature climbs to > 15 000 °C (Kuwana et al. 1989).

Additions of up to 10 % $N_2$ to the Ar shielding gas suggested that the nitriding capacity of TIG plasma is increased by a factor of 4 to 5 as compared to the respective molecular gas mixtures (Hertzman et al. 1996). Modelling the process gave a best fit at a factor of 10 (Hertzman and Wessman 1998). This implies that a take-up of nitrogen in the hot spot is opposed by a loss outside of it. The weld pool is therefore in contact with two different atmospheres, the nitriding capacity of which may differ by an order of magnitude. The more the weld pool is covered by the arc the more nitrogen may be contained. It is therefore recommended to keep

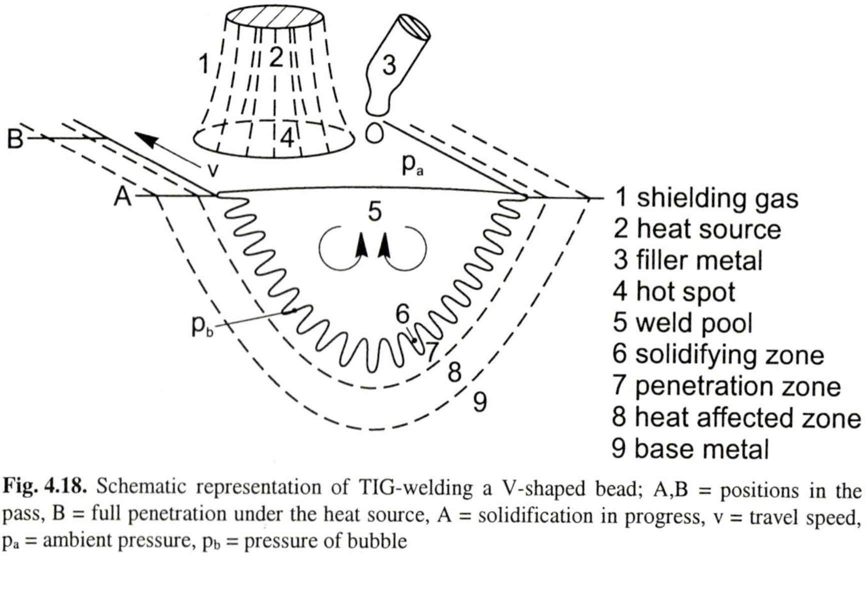


**Fig. 4.18.** Schematic representation of TIG-welding a V-shaped bead; A,B = positions in the pass, B = full penetration under the heat source, A = solidification in progress, v = travel speed, $p_a$ = ambient pressure, $p_b$ = pressure of bubble

the weld pool small (Stein and Trösken 1990; Hertzman et al. 1996; Beyer et al. 1990). Surface active elements like O and S occupying surface sites not available for N-atoms impede the nitrogen transfer to the pool (Katz and King 1989). Adding hydrogen to the shielding gas counteracts the formation of an oxide film on the pool and eases the escape of nitrogen outside the hot spot (Hertzman et al. 1996).

Nitrogen contained in the penetration zone and maybe some of the heat affected zone (HAZ) will be flushed into the weld pool, the composition of which is adjusted by the alloy content of the filler metal. A somewhat higher Cr, Mn, Mo content will reduce the effusion of nitrogen.

Summarizing the situation for TIG welding at location B in Fig 4.18 it becomes evident that the weld pool interacts with the base metal and two different atmospheres, i.e. inside and outside the hot spot. A nitrogen flux is to be expected between 1 and 2, 2 and 4, 1 and 5, 5 and 7 in the direction of lower nitrogen activity (Hertzman et al. 1996). That of the weld pool is depending on its chemical composition, which is made up of the molten penetration zone and the filler metal. Strong convection provides for rapid mixing. It seems possible to maintain a nitrogen content in the weld pool which is even above the equilibrium value in air of normal pressure.

Nitrogen in the shielding gas of TIG welding leads to corrosion of the W-electrode and its content is therefore limited. The same holds true for plasma sources. Metal inert gas (MIG) welding is not sensitive in this respect, but the arc temperature is changed by a larger nitrogen addition. The nitrogen content in the atmosphere of submerged arc welding and welding with covered electrode is close to that in air and the slag reduces the effusion of nitrogen. Laser welding may be done in pure nitrogen. Too little is known though about the effect of numerous welding procedures and parameters on the nitrogen content in the weld pool.

***Solidification:*** Behind the heat source travelling at speed v the weld pool starts to solidify (Fig. 4.18, location A). We now go back to section 4.1.2 and recall the remarks on "Solidification". The solubility gap of δ-ferrite will enhance the pressure $p_b$ in bubbles (see equation 4.8). The counteracting atmospheric pressure $p_a$ cannot be raised, though, in normal welding practice and bubbles are prone to grow. Thus some of the nitrogen dissolved at B will be lost at A and pores may remain in the weld metal. Increasing v tends to suppress the nucleation of bubbles because of rapid solidification. For a primarily austenitic solidification the partitioning coefficient of nitrogen $k_N$ is closer to unity. A loss of nitrogen and the formation of pores is less liable (Groß et al. 1998). However, hot tearing is more likely to occur than in case of a primarily ferritic solidification connected with a lower thermal contraction.

For specific welding operations a pressure chamber is feasible. Here the nitrogen content of the weld pool at full penetration may be influenced by the partial pressure $p_{N_2}$ and the addition of Ar will give a sufficient $p_a$ to suppress bubbles and pores during solidification. Metal inert gas (MIG) welding of a CrNiMo austenitic stainless steel in a pressure chamber filled with nitrogen of up to 6 bar increased the nitrogen content of the weld metal to ~ 0.6 mass% (Kikuchi et al. 1996). Starting from this nitrogen level in the base metal the respective content of the weld metal was further raised to ~ 0.8 mass%. Nitride precipitates were observed but no defects.

***Weld seam:*** Nitrogen promotes austenite, strength and corrosion resistance. It is therefore important to at least maintain the nitrogen content of the base metal throughout the HAZ and the weld metal. In some cases an increase may be beneficial, e.g. for suppressing δ-ferrite. This initial nitrogen content is accumulated in the weld pool at full penetration. Reaching a sufficient nitrogen level to start with is the first obstacle to overcome.

The second one is to keep the nitrogen in the steel during subsequent solidification without forming pores. Bubbling is equal to a loss of nitrogen and tends to result in pores unless the bubbles are fully liberated from the melt. They are caused by a lack of $p_a$. A high travel speed as in laser welding may increase the solidification rate above the nucleation rate of bubbles. In all, the second obstacle appears to be the more serious one.

As a compromise a lower nitrogen content is adjusted in the bead by shallow penetration and a low-nitrogen filler. The expected losses in properties are made up by alloying via the filler metal. HNS produced by pressure metallurgy are difficult to handle, because the respective $p_a$ is not available during welding. But even grades melted in air may cause problems, if e.g. a small loss of nitrogen shifts the solidification to δ-ferrite. Yet an optimizing strategy has led to encouraging results in arc and laser welding of HNS.

***Weld deposit:*** Fusion welding may be used to increase the nitrogen content of e.g. a stainless weld deposit on a low alloy steel substrate thus improving strength and corrosion resistance. A shallow penetration is essential to keep the CrMnMo-content in the pool high. In plasma-powder-welding nitrogen is picked up from the $Ar/N_2$ atmosphere. In laser-powder-welding Cr25Ni25 and Cr25Mn25 powders were mixed with 2 mass% $Si_3N_4$ which led to about 0.4 respectively 0.8 mass% N in the deposit (Huang et al. 1996).

## 4.4 Heat treatment

HNS require about the same heat treatments as comparable high alloy grades without nitrogen or with carbon. Effusion of nitrogen makes the difference. It is accelerated by an increasing temperature. Therefore treatments like stress relief annealing, soft annealing and tempering are less liable to provoke degassing. Problems are more likely to occur at the temperature of hardening or solution annealing usually ranging above 1000 °C. Also different from conventional steels of low interstitial content is the higher susceptibility to nitride precipitation during cooling after austenitization.

## 4.4.1
## Holding at high temperature

The high Cr content of HNS yields an oxide layer upon heating in air which seems to impede the escape of nitrogen. In contrast a comparable high carbon grade develops a decarburized surface zone, because carbon is oxidized at the scale/steel interface and the larger $CO/CO_2$ molecules provoke cracks on their way out. After treatment in a vacuum furnace the situation is reversed. The HNS shows a denitrided surface zone while the respective carbon grade goes unharmed.

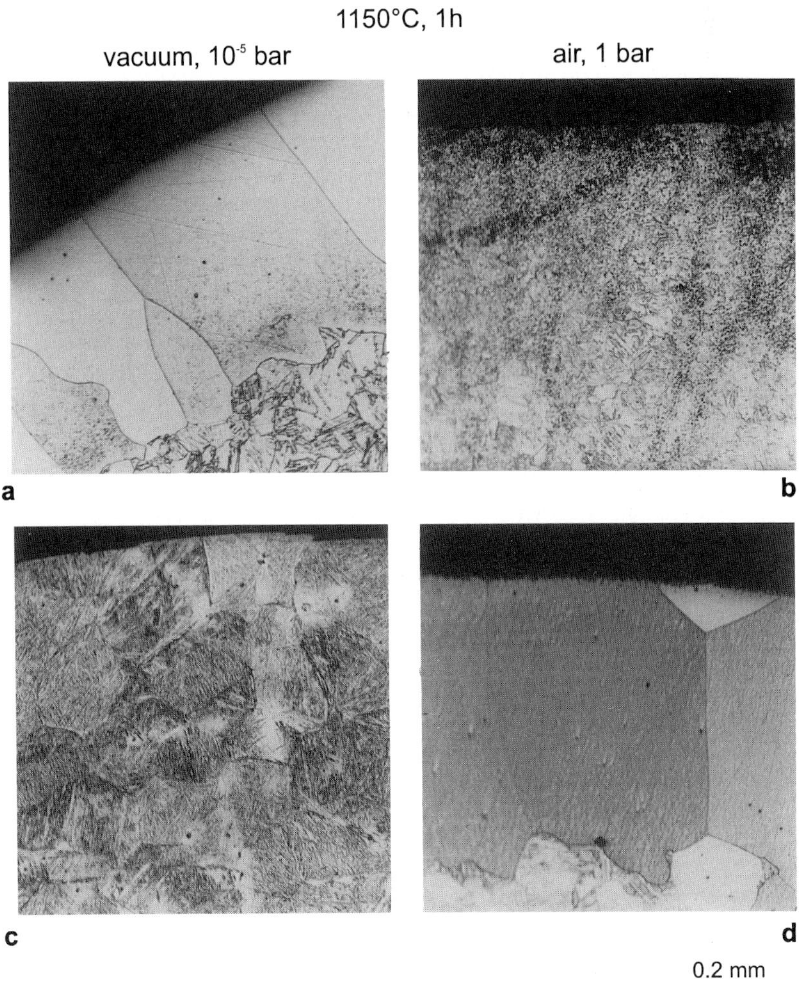


**Fig. 4.19.** Effect of austenitizing (1150 °C, 1 h) **a, c** in vacuum and **b,d** in air on the surface zone of **a,b** steel Cr15Mo1N0.5 and **c,d** steel Cr15Mo1C0.5 after quenching; **a** denitriding in vacuum and **d** decarburizing in air was observed

The example given in Fig. 4.19 does not reveal the whole story, though. The partial pressure of nitrogen in air may be sufficient to suppress effusion if the CrMnMo content is high. As described in section 4.2.2 the nitrogen activity at the scale/steel interface may differ from that of the gas. Therefore the oxide layer is not a tight seal on HNS heated in air. On the other hand do industrial vacuum furnaces contain some oxygen resulting in a thin scale and less nitrogen effusion. Although the equilibrium conditions of nitrogen in the presence of oxygen at the steel surface are not clear, hardening and solution annealing of HNS has been carried out in air or "poor" vacuum without excessive denitriding. A safer way of avoiding denitriding consists of a treatment in an equilibrium nitrogen atmosphere as described in Sect. 4.1.4.

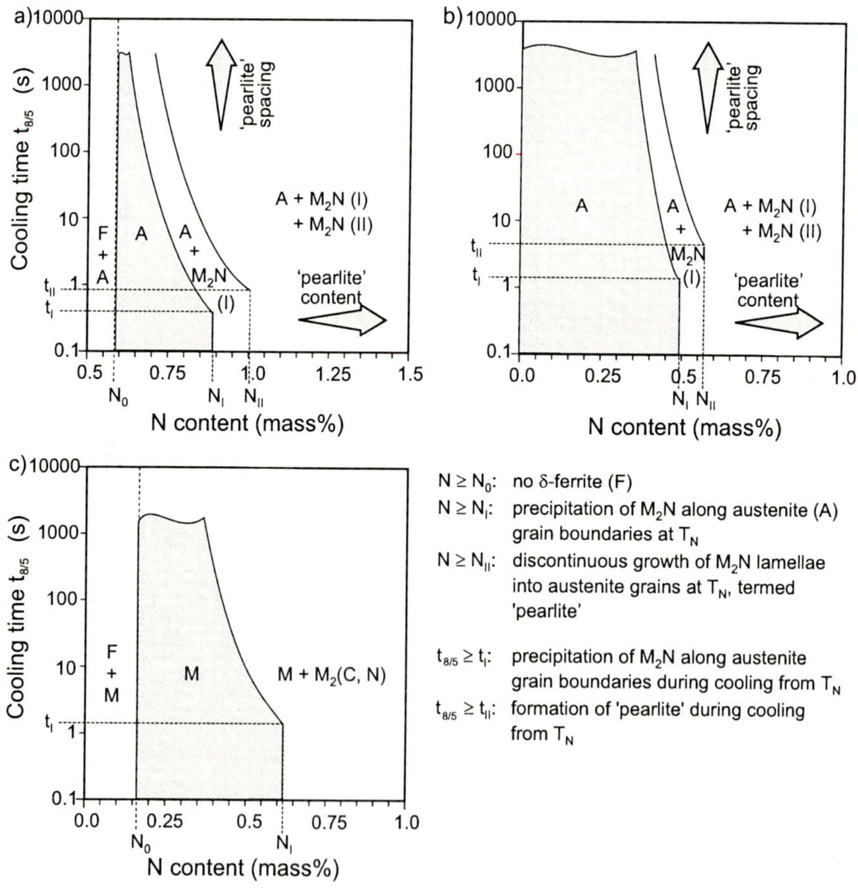


**Fig. 4.20.** Influence of nitrogen content and cooling time on nitride precipitation during quenching after solution nitriding at 1150 °C: **a** stainless duplex steel Cr22Ni5Mo3N0.2, **b** stainless austenitic steel Cr17Ni13Mo2, **c** martensitic steel Cr16Mo0.5C0.15 (Juse 1999)

### 4.4.2
### Cooling

According to equilibrium diagrams of most HNS (see Fig. 4.15) a precipitation of nitrides like $Cr_2N$ will occur during slow cooling from austenitizing temperature. As their nucleation is promoted by grain boundaries an intercrystalline net of hard particles is formed followed by a discontinuous pearlite-like precipitation within the grains. Thus the toughness is impaired and the matrix is depleted of Cr lowering the corrosion resistance. This is well known from hardenable stainless steels of high carbon content behaving quite similar to respective HNS. However, the interstitial content of conventional austenitic steels is low and the transition to austenitic HNS is accompanied by a distinct change in cooling behaviour. Certain cooling rates are required to suppress the precipitation of nitrides within the grains and along the grain boundaries. The respective cooling time from 800 to 500 °C ($t_{8/5}$) has to be reduced as the nitrogen content dissolved before cooling increases (Fig. 4.20).
As $t_{8/5}$ depends on the quenchant the cross-section of thicker parts may require a liquid quench to avoid precipitates in the core. This is feasible in combination with a hot wall chamber furnace usually run at ambient pressure. Parts of smaller cross-section may do with a gas quench at elevated pressure which is offered by cold wall vacuum furnaces. They allow an equilibrium treatment in a nitrogen atmosphere, even at $p_{N_2} > 1$ bar (see Sect. 4.1.4).

As a first approximation the liability of HNS to distortion and cracking during heat treatment is similar to that of respective carbon grades.

## 4.5 Machining

Beside materialproperties of the workpiece metal cutting depends on process parameters like geometry, cutting speed, cooling, lubrication and type of tool bit. The evaluation of cutting tests has to consider the volume removed in time, the surface and sub-surface conditions of the workpiece, the shape of chip and the tool life. It therefore takes a lot of experimental effort to gain reliable cutting parameters for new materials.

### 4.5.1
### Experimental conditions

In a first step a cutting pendulum described by Berns et al. (1996) was used to acquire some information on the cuttability of HNS. It consisted of a modified 300 J impact pendulum bearing a WC-Co hardmetal tool bit of 10 mm in diameter on the arm. A specimen of $w = 2.5$ mm in width and 40 mm in length was mounted below the swinging arm on a double wedge to allow the adjustment of the cutting depth h. The tangential and normal forces acting on the specimen, i.e. the cutting force $F_c$ and the passive force $F_p$, were measured by piezo transducers. Due to the orthogonal cutting conditions a feed force $F_f$ did not occur. After

**Table 4.2.** Austenitic (A), hard martensitic (MH) and soft annealed martensitic (MS) specimens alloyed with carbon (index C) or nitrogen (index N) for cutting tests, after Schellewald (1999)

| | steel | heat treatment | hardness H (HV 30) |
|---|---|---|---|
| A | Cr17Ni12Mo2 | 1050 °C/water | 197 |
| $A_N$ | Cr21Mn9Ni6W2Mo1N1 | 1180 °C/water | 297 |
| $MH_C$ | Cr14Mo0.5VC0.48 | 1040 °C/oil | 661 |
| $MH_N$ | Cr15Mo2VN0.47 | 1140 °C/oil + 450 °C | 683 |
| $MS_C$ | Cr14Mo0.5VC0.48 | soft annealed | 280 |
| $MS_N$ | Cr15Mo2VN0.47 | soft annealed | 220 |

pre-cutting to pre-shape the specimens, Schellewald (1999) carried out dry cutting test on martensitic and austenitic HNS and reference steels according to Table 4.2 at an initial speed $v_c$ of 214, 300, and 400 m/min, a rake angle $\gamma$ of +6°, 0°, and -6° and $h \leq 0.5$ mm. The arithmetic mean roughness $R_a$ of the as-cut specimens was measured as well as the contact length l between rake face and chip. After cutting selected specimens and chips were sectioned in cutting direction and observed by light and scanning electron microscopy (SEM). A quick-stop device allowed to investigate the chip root.

## 4.5.2 Results

The extensive use of the cutting pendulum (Segtrop 1995, Liu 1996, Theisen 1997) for testing various steels and hard alloys revealed that, in spite of the short cutting length $s \approx 40$ mm, conditions similar to continuous turning are achieved in the workpiece. However, an assessment of tool wear is not possible.

The cutting forces $F_c$ and $F_p$ increase and the specific forces $k_c$ and $k_p$ related to the area of cut $(w \cdot h)$ decrease as $h$ is raised. While $k_c$ is only moderately increased with hardness $H$, $k_p$ is raised by a factor of 5. The as-cut surface becomes rougher as $h$ increases and more so for the martensitic states. A higher $v_c$ and a negativ $\gamma$ tend to reduce $R_a$. The oscillations of $R_a$ over $s$ coincide with those of $F_p$. The softer the material the thicker the deformed sub-surface zone reaching up to 40 μm. Segmented chips prevail at small $h$ and low $v_c$ while in the opposite direction discontinuous chips are observed.

The above observations are related to the generation and dissipation of heat in the process zone and occur in nitrogen and carbon grades. However, some differences between HNS and conventional steel are apparent. The high nitrogen austenitic state $A_N$ requires a higher cutting energy compared to the standard state A (Fig. 4.21a). The reduced contact length $l$ of $A_N$ points to a more difficult chip deformation which may entail a higher tool wear on the rake face. Compared to A

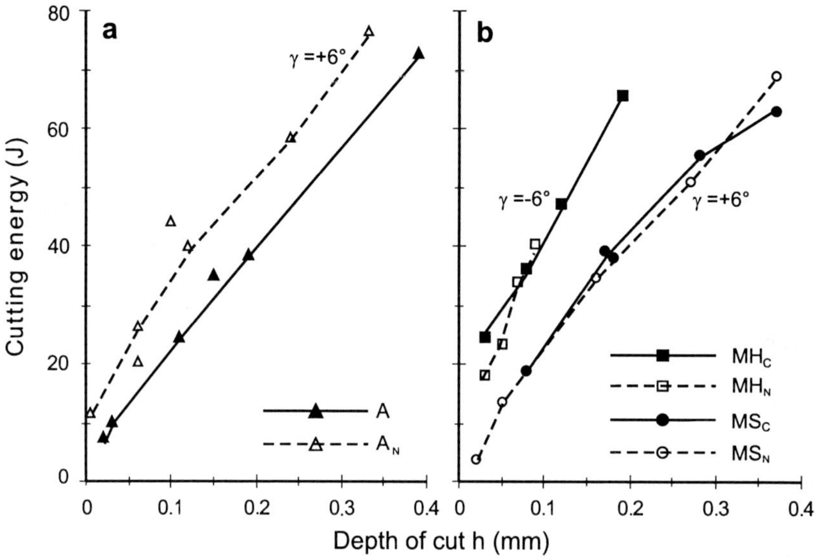


**Fig. 4.21** Cutting energy of HNS (index N) compared to conventional grades given in Table 4.2 at a cutting speed $v_c$ = 400 m/min, after Schellewald (1999)

the passive force $F_p$ of $A_N$ is about twice as high and is expected to have a detrimental effect on the tool wear of the flank face. The HNS shows an embrittlement by a more pronounced formation of discontinuous chips, which is also connected with a somewhat higher $R_a$.
The differences between A and $A_N$ are due to the difference in the interstitial content and are to be expected for N and for C as well. The difficult machining of Hadfield steel Mn12C1.2 supports this view. The martensitic grades contain a similar amount of N or C and little difference is found in the cutting energy (Fig. 4.21b) and the other criteria. But new martensite is found in a surface zone of the hard martensitic HNS. At $v_c$ = 300 m/min and $h$ = 0.085 mm a continuous zone of a few microns in thickness is formed on the workpiece while at $h = 0.21$ mm a wavy zone of 5 to 25 µm in thickness appears which follows the chatter marks. An SEM study showed a dense population of undissolved secondary carbides in steel $MH_C$ while $MH_N$ was practically free of detectable precipitates. The latter is therefore readily reaustenitized if the process temperature exceeds $Ac_1$. Rehardening the surface of $MH_N$ but not of $MH_C$ is prone to cause differences in residual stresses, fatigue resistance and even corrosion resistance which are not indicated by the above mentioned measured quantities.

## 4.6 Products

The previous sections have revealed two trends in general: (1) The volatility of nitrogen calls for specific manufacturing procedures. (2) A high nitrogen content impairs the workability of steels. Both of these trends are cost effective. An increase of cost has to be justified by superior performance. Numerous solutions are feasible but only few will pay off. The favoured ones depend on the required volume as well as on size and shape. New steel grades usually start from a small-scale production route, which may have to be changed as the volume increases. With these premises in mind one may try to judge the prospects of semi-finished HNS products.

***Castings:*** As melting and pouring under pressure is costly this type of near-net shape product obtains nitrogen just by alloying. Its content in the melt is limited by the solubility of the solidifying phase in air at normal pressure of which ferrite dissolves less than austenite. Obeying this limit the formation of bubbles and pores during solidification may be avoided.

***Bars:*** Flat-die forging and rolling are employed, often in consecutive order. Parts are produced mainly by machining and are rarely fusion welded. The production of bars is therefore suited for most HNS grades which stem from open melted, pressure melted or powder metallurgical ingots. The higher the nitrogen content the smaller the window for hot forming is to be expected, though.

***Die-forgings:*** This product mostly starts from forged or rolled billets. The near-net shape parts are further machined and rarely fusion welded. Like bar stock this type of product is suited for most HNS grades but maybe limited to less complex shapes.

***Sheet:*** Strip and plates are mainly used in welded structures and what was said for castings applies to fusion weldings, too. A controlled liberation of $N_2$ bubbles from grades of higher nitrogen content is risky in respect to pores. However, in non-welded applications like cold formed parts, grades above the normal nitrogen solubility may be employed.

***Tubes:*** Fusion welding is quite common, either for the seam or for joining even seamless tubes to each other or to flanges and sheet. Seamless tubes not carrying fluids but mechanical loads or those sectioned into rings usually go without fusion welding, which allows for a higher nitrogen content. Pressure welding is feasible. Hot working is impeded by a high nitrogen level, though.

***Rings:*** Larger thick-walled rings are made by forging. Thin-walled ones may be produced by ring rolling. Smaller ones are e.g. manufactured from tube, bar or sheet. Welding is usually not involved except e.g. for flanges.

***Wire:*** Starting from forged or rolled billets wire is gained by rolling. For parts made thereof fusion welding remains an exception and pressure welding seems feasible.

***PM-parts:*** Hot worked shapes from hipped PM-ingots may be further machined to PM-parts, However, their near-net shape production directly from powder is hindered by the low sinter activity of the rather coarse atomized alloy powder. MIM and sinter-forging are means of reaching a high but not full density. However, individual parts may be manufactured from pre-shaped cans filled with powder, which are nitrided and hipped like ingots.

***Cold worked products:*** The resistance to cold deformation increases with the nitrogen content and intricate shapes are more difficult to achieve. Some cold forging operations may require more stress than the tools will hold. Compared to cold forging wire is easier to be drawn, which means that bolts and fixtures may be harder to manufacture than cables and coil springs. Looking at strip, cold rolling of high-strength flats e.g. for springs seems easier to be handled than deep drawing.

***Nitrided products:*** To circumvent some of the manufacturing problems caused by nitrogen a nitriding heat treatment in the austenitic regime is applicable in two ways. Thin products like wire and strip are enriched with nitrogen all the way through. Thicker, near-net shape parts receive a high-nitrogen case (HNC).

In summarizing one may conclude that weldable HNS of moderate nitrogen content are open to a wide range of products. A really high content of nitrogen above the solubility at normal pressure implies some restrictions with respect to castings, welded structures and the forming of intricate shapes. Here most of the volume is seen in hot worked stock machined to final shape. For some applications HNC offers a low-cost alternative to HNS.

# 5 Steels and applications

In HNS nitrogen is effective in solid solution or in precipitates. Except for some wear-resistant grades fine precipitates prevail, which are obtained by ageing a prior solid solution. Therefore solution annealing is usually the first step of heat treatment followed by a rapid quench to suppress precipitation. The resulting microstructure is of key interest and allows to assign a steel grade to a field in the non-equilibrium Schaeffler diagram, which was originally developed for weld-metal, i.e. rapidly quenched melts (Schaeffler 1949). Elements stabilising austenite are combined to a Ni equivalent and plotted over the ferrite forming elements represented by the Cr equivalent. A revision by Delong (1974) was also concerned with welding, which usually deals with a low level of interstitials. The Ni equivalent factor of 30 for (C+N) decreases, though, as the concentration is raised. For HNS a revised diagram was proposed (Fig. 5.1). Speidel and Uggowitzer (1993) proposed a smaller $Ni_{eq}$ factor for N than for C which is at variance with the higher stabilization of austenite by N demonstrated in Sect. 2.5. However, an increased concentration of segregated carbon is to be expected at grain boundaries (see Sect. 1.1.4.3), i.e. in the areas of ferrite nucleation. Thus an apparently stronger stabilization of austenite by carbon is observed at low N and C contents where the effect of segregation is decisive.

Schaeffler-type diagrams reveal fields for as-quenched martensitic, austenitic or ferritic steels. In addition steels of mixed microstructure are feasible like ferritic-martensitic (e.g. dual-phase) steels, ferritic-austenitic (e.g. duplex) steels and austenitic-martensitic ones. All of these groups are suited for HNS, except for merely ferritic steels because of their low nitrogen solubility. At least part of the microstructure has to be austenitic at solution annealing temperature to enhance the solubility of nitrogen. The austenite may be stable to room temperature or transform to martensite during quenching. After tempering some martensitic steels are called ferritic, which is misleading in the present context and will be avoided. The different groups of HNS are designated according to their microstructure in the as-quenched state irrespective of subsequent aging treatments. Other ways of designation are based on steel properties like resistance to corrosion, wear and creep or on applications like tool steels, bearing steels and valve steels. Last not least the manufacturing route and the type of product are helpful in characterizing a steel. We expect some difference in properties between PESR and PM steel or between castings, sheet and wire (see Chap. 4).

Chromium is the key alloying element of HNS because it strongly enhances the solubility of nitrogen, yet allows for a sufficient solubility product of nitrides during solution annealing (see Sect. 4.1.1). From about 10 to 15 mass% Cr martensitic HNS are located in Fig. 5.1, i.e. creep resistant grades at the lower and

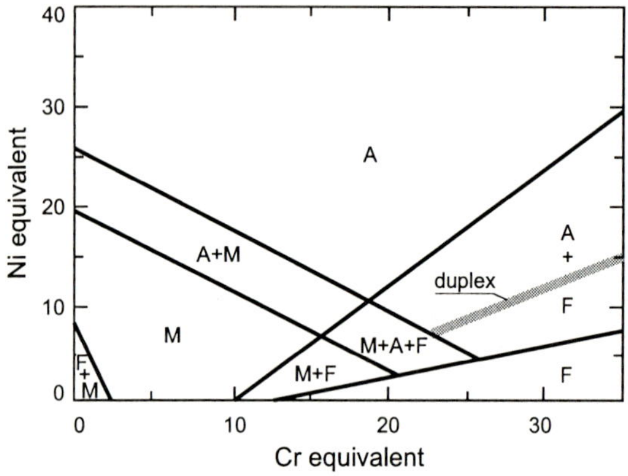


**Fig. 5.1.** Schaeffler diagram, revised after Speidel and Uggowitzer (1993) with (mass%) $Cr_{eq}$=Cr+1.5Mo+1.5W+0.48Si+2.3V+1.75Nb+2.5Al and $Ni_{eq}$=Ni+Co+0.1Mn−0.01$Mn^2$+18N+30C, *A* = austenite, *M* = martensite, *F* = ferrite

stainless grades at the upper end. Above this level austenitic steels are centred around 18 mass% of Cr providing the lowest Ni equivalent. This is composed of nitrogen, of manganese which supports the nitrogen solubility and of nickel which does not. Substitutional elements are required to stabilize the austenite because nitrogen alone would amount to ≥ 0.6 mass%, which are difficult to introduce to the steel and hard to keep in solution during quenching. Stabilizing the austenite to subzero temperatures calls for an even higher alloy content. The amount of CrMnNiNC offers the basis for designing HNS of martensitic, austenitic and mixed microstructure. Additions of Mo, W, V, Nb are used to improve certain properties like corrosion resistance or hot strength. As Cr the latter are nitride formers and some promote intermetallic phases like σ-FeCr and Laves $Fe_2$(MoW). For details see Sect. 2.4.1. The content of Si is usually kept low because it decreases the nitrogen solubility and stabilizes ferrite. Remelting under pressure and an $Si_3N_4$ slag tends to raise the Si content, though.

## 5.1 Martensitic steels

The martensitic regime of the Schaeffler diagram is valid at room temperature and restricted to the matrix composition. Deep freezing after hardening shifts the border M/M+A away from the origin. This treatment is therefore frequently used to transform retained austenite of high alloy steels, thus extending the martensitic regime. At a hardening temperature of e.g. 1100 °C the dissolution of at least 0.3 mass% N is necessary to fully austenitize alloy Cr15Mo1N0.5 (Fig. 4.15). Above 0.6 mass% N nitrides are precipitated from the austenite depleting the matrix of alloying elements. Besides steels of homogeneous martensite those of heterogeneous microstructure consisting of martensite and nitrides may be obtained at a high level of nitrogen. The hard $M_2N$ and MN particles bear on the tribological properties. At the CrMo content given PM manufacturing is about the only way to add one or more mass% of nitrogen to the alloy. It is therefore not possible to discern between secondary and eutectic nitrides.

As revealed in Sect. 1.1.2.6 nitrogen promotes ordering of Cr-atoms in an austenitic solid solution, while carbon supports thcir clustcring. IIowever, a combination of N and C yields the highest degree of ordering. A more homogeneous atomic distribution improves the stability of austenite with respect to transformation, i.e. precipitation and martensite. The latter does inherit the ordered structure from the parent austenite and offers a better corrosion resistance. The nucleation of precipitates during tempering is refined which has a beneficial effect on the mechanical properties.

As shown in Fig. 4.2 the borderline austenite/austenite + carbide of martensitic CrC-grades is of hyperbolic shape, i.e. $Cr{\sim}C^{-1}$, which is characteristic of a constant solubility product. In contrast the austenitic regime of martensitic HNS extends about linearly to high alloy contents, i.e. Cr~N. This result of ordered interaction of Cr and N is the first advantage of martensitic HNS as compared to carbon grades. It allows to increase the interstitial content of high chromium steels at hardening temperature without provoking precipitates and to enhance the hardness of martensite. For binary iron-alloys Mittemeijer (1983) proved that equal atomic concentrations of C and N entail about the same hardness increase in martensite. Hardening is limited, though, by the growing stability of austenite which lowers the start and finish temperatures of martensitic transformation ($M_s$, $M_f$) and leads to retained austenite. This is reflected by the as-quenched hardness (Table 2.6). Nitrogen enhances secondary hardening during tempering which is the second advantage.

In contrast to carbon, dissolved nitrogen improves the resistance to aqueous corrosion and especially to pitting corrosion (see Sect. 3.2.2.2). In carbon grades undissolved carbides, rich in Cr, impair the corrosion resistance. During tempering secondary hardening is brought about by $\zeta$ $(Fe,Cr)_2N$ nitrides, which are not enriched in Cr so that the matrix is not depleted. This is the third advantage of martensitic HNS. The respective carbides consume more Cr (see Sect. 2.5) and tempering beyond secondary hardening leads to $M_7C_3$ and $M_{23}C_6$ carbides binding more Cr atoms per interstitial atom than the respective nitrides MN and $M_2N$. In all there are several reasons to prefer nitrogen in martensitic stainless steels.

In the following we differentiate between low-tempered and high-tempered martensitic HNS. The first group is tempered up to the regime of secondary hardening (< 500 °C). The aim is to produce hard steels with ≥ 0.5 mass% N+C in solution, the wear-resistance of which may be further enhanced by nitride precipitates. A high Cr content is necessary to provide a sufficient solubility of nitrogen at hardening temperature resulting e.g. in stainless HNS. Steels of the second group contain less than 0.5 mass% (N+C) and are tempered well above 500°C. Of these nickel-martensitic grades, quench-and-temper steels as well as valve steels are stainless, while creep resistant HNS do with less Cr.

## 5.1.1
## Hard stainless steels

***Development:*** This group of HNS is aimed at a martensitic microstructure of maximum hardness, which is nearly free of undissolved nitrides or carbonitrides. The hardness of martensite depends on its nitrogen content, which stems from the prior austenite and may be raised at a given hardening temperature by increasing the Cr equivalent (Fig 4.2). The higher $Cr_{eq}$+N the lower $M_s$ and the more austenite is retained after quenching (Fig. 5.1). Work of Lueg (1990) showed that the steel Cr15Mo1N0.5 reached a hardness of 590 HV30 after quenching from 1100 °C in oil. The $M_s$ temperature was 255 °C and the content of retained austenite below 1%. After tempering in the regime of secondary hardening the highest hardness was 650 HV30 measured after 450 °C. The weight of Mo is 1.5 in the Cr equivalent but 3.3 in the pitting resistance equivalent number (PREN) (see Sect. 3.2.2.2). This explains the part of Mo in the Cr equivalent and also reveals the potential of N to reduce corrosion in chloride solutions (Speidel 1989, Lueg 1990) because k in Eq. 3.6 is found between 10 and 30 which is often averaged by k = 16.

In applications like tools and bearings a hardness of 60 HRC (700 HV30) is often required. This called for a higher interstitial content and an improvement of secondary hardening. To stay within the pressure limits of an existing PESR unit up to 0.3 mass% of carbon was introduced beside 0.3 to 0.4 mass % of nitrogen. The resulting increase in retained austenite was counteracted by deep freezing and tempering. Berns and Trojahn (1993) added small amounts of V and Nb to enhance secondary hardening, having in mind the low solubility product mentioned in Sect. 4.1.1. A hardness increase of about 1 HRC was gained by 0.28 mass%V. The addition of 0.1 mass% Nb was effective only after hardening from ≥ 1150 °C, which is usually considered too high a temperature in view of distortion and grain growth.

Today a family of PESR-CRONIDUR steels are available which contain about (mass%) 15Cr, 1Mo, 0.4 to 0.7(N+C), 0 to 0.4 V and some Ni at the lower end of (N+C) to avoid δ-ferrite (VSG 1995). After hardening, deep freezing to - 80 °C and tempering at 450 to 500 °C a hardness of 57 to 61 HRC (650 to 720 HV30) is reached. While comparable steels without nitrogen loose most of their corrosion resistance after tempering in the range of secondary hardening, the HNS grades deteriorate only slightly and current density/potential-curves still show a passive regime in diluted aqueous $H_2SO_4$ or NaCl solutions. Salt spray tests after DIN 50021 did not attack Cr15Mo1N0.4C0.15 but severely stained Cr17Mo0.5C1 (AISI 440 C, see Lueg 1990, Lueg and Berns 1991)

Precipitation strengthening during secondary hardening offers temper resistance and hot hardness. Steel Cr15Mo1N0.4C0.15 hardened from 1075 °C in oil and tempered three times at 450 °C for 1 h to a hardness of 58.5 HRC (665HV30) remained at this level after tempering for 1000 h at 400 °C. Thus, HNS offer a combination of high hardness and corrosion resistance at elevated service temperature, which up to now has not been available. This is important for stainless components which intermittently serve at room and elevated temperature.

At high hardness the bending strength and the fatigue strength of steels depend strongly on the size and frequency of hard inclusions. Hard stainless carbon grades like Cr13C0.65 and Cr17Mo0.5C1 are located outside the area of homogeneous austenite at hardening temperature (Fig. 4.2) and contain $M_7C_3$-carbides, which -supported by microsegregation- are in part eutectic. Precipitation from the melt increases the size by at least an order of magnitude compared to secondary precipitation from a solid solution. Eutectic $M_7C_3$-carbides of e.g. 10μm in size are more frequent than oxides and impair the mechanical properties of hard stainless carbon grades. In comparison hard stainless PESR-HNS stay near the border of the austenitic regime at hardening temperature (Fig. 4.6) and only some secondary nitrides and carbides of ≤ 1μm in size were found even in extreme cases of segregation (Fig. 4.7 to 4.9). The higher degree of short-range atomic ordering discussed in Chap. 1 and the resulting increase of interstitial solubility in HNS-austenite are a prerequisite for the success of the CRONIDUR steels.

The higher cost of PESR has spurred the development of conventionally melted steels with a moderate N content. Starting from Cr17Mo0.5C1 Tomasello et al. (1998) decreased the Cr, Mo, and C content to reduce the carbide size. Steel Cr14.5C0.65N0.1 showed some carbide refinement and improved corrosion resistance. Pichard et al. (1998) increased the Cr content of Cr13C0.65 and added Mo and V to enhance the nitrogen solubility. About one third of the C content was replaced by N which was added to the ladle by gaseous stirring and nitrided ferro-alloys. After ESR at atmospheric pressure steel Cr16Mo1.7V0.3C0.4N0.2 was obtained. It contained less and smaller eutectic carbides. The steel showed a hardness of 59HRC (675HV30) and a better corrosion resistance. Horovits et al. (1996) injected gaseous nitrogen into steel melts with 13 to 16 mass% Cr and 0.11 to 0.16 mass% of carbon to obtain 0.16 to 0.19 mass% N. At a maximum of 0.32 mass% of interstitials the hardness remained below 600 HV100 which is too low for some applications.

Besides melting PM was employed to manufacture hard stainless HNS, starting from nitrided steel powder to follow route (a) of Fig. 4.10. Berns and Wang (1990) described PM-HNS Cr15Mo1N0.5 which after quenching from 1100 °C and tempering at 450 °C reached a hardness of 661 HV30 (58.5 HRC) and was much more resistant to corrosion in aqueous solutions of $H_2SO_4$ or NaCl than respective carbon grades. Jargelius et al. (1990) used about the same steel powder but introduced the nitrogen by high-temperature nitriding in the austenitic regime. They also emphazise the superior mechanical and chemical properties.

***Application:*** Hard HNS are especially suited for stainless tools and bearings. Other applications are seen e.g. in the field of stainless and temper-resistant gears as well as wear resistant parts for processing of food, polymers, pharmaceutics, paper and pigments.

Today stainless knives and scissors are less hard than plain carbon grades. HNS like Cr15Mo1C0.3N0.35 offer a higher hardness than respective stainless carbon grades used for cutlery like Cr15Mo0.5C0.48. In addition HNS are much more resistant to pitting corrosion induced by $Cl^-$ in dish-washers. The same holds true for medical instruments like scissors and chisels which need a high hardness and tend to corrode during cleaning. Ehrhardt (1995) proved that even in the quenched and tempered condition HNS are tougher and more corrosion resistant than respective stainless carbon grades. This allows to combine a tough, die-forged HNS body with an induction or laser hardened cutting edge.

In plastic moulding some polymers corrode parts of the extruder and the mould. Polyvinyl chloride (PVC) for instance emits some HCl upon heating. During cooling remainders of PVC in the extrusion channel react with moisture to form traces of hydrochloric acid, which attack even stainless steels like Cr17Mo1C0.35. The results of Lueg (1990) gave a clear indication of the higher resistance to $Cl^-$ induced pitting corrosion of PESR-HNS as compared to the stainless carbon grade. In addition the higher resistance of HNS to tempering is an advantage in processing of polymers that require an extrusion temperature above 250 °C especially if they contain abrading filler materials, which call for hard tools. Small polymer parts are usually shaped in a multiple mould which is demanding in respect to precision of the matching halves of the tool. Size stability of the tool steel becomes an important issue. No size change by precipitation or by transformation of retained austenite is allowed at service temperature. Secondary hardening anticipates microstructural changes and is now available in HNS combined with high hardness and corrosion resistance. Ernst und Rasche (1992) reported a substantial increase of tool life by using HNS Cr16Mo1N0.34C0.17 instead of standard steel Cr17Mo1C0.35 because of superior corrosion resistance. However, moulds are usually manufactured along a tight schedule, which is asking for a sufficient size range of forged stock. This will only grow with an emerging market and delays the introduction of any new tool steel.

Stainless HNS bearings dwell on the lack or small size of nitrides and carbides and the superior resistance to corrosion and tempering. The PESR process does not only provide a sufficient nitrogen content, but also reduces the size and frequency of non-metallic inclusions like in ordinary ESR. An immediate field of application are miniature bearings, which do not allow for hard particles of ≥ 10 μm in size as encountered in the conventional stainless bearing steel Cr17Mo0.5C1 (AISI 440C). The life of larger angular bearings was measured for different steels and test conditions. The inner ring with an inner diameter of 25 mm consisted of the material tested while the balls and the outer ring were made of Cr1.5C1 (SAE52100). A series of equal bearings was tested and the time to 10% failure probability (L10-life) was derived from a Weibull plot. A calculation according to DIN/ISO 281 yielded L10=50h which was well in line with the stainless carbon grade. The result of HNS was more than 40times above that level (Fig. 5.2a). Intermittent salt spraying brought L10 of the stainless carbon

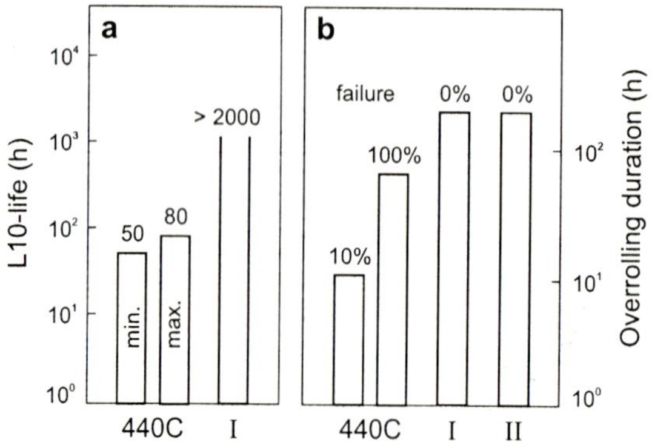


**Fig. 5.2.** Bearing life under a Hertzian pressure of 2800 MPa and poor lubrication at a rotating speed of 12 000 rpm, **a** life at 10% failure probability in a continuous run, **b** interrupted run (4 h salt spray + 24 h run), materials:
AISI 440 C = Cr17Mo0.5C1,
**I** Cr15Mo1C0.15N0.38,
**II** Cr16Mo1C0.33N0.33
(Berns and Trojahn 1993)

grade down to 10 h and all bearings had failed after 80 h but none of the HNS rings after 240 h (Fig. 5.2b). The poor lubrication in these experiments allowed some metal to metal contact, which does not occur under full elasto-hydrodynamic (EHD) lubrication. For EHD conditions L10=1300h was considered a good result for steel Cr1.5C1. No failures had occurred among the HNS bearings of Fig. 5.2 after 2000 h continuous EHD run. The wear of these bearings was derived in modified tests without lubrication. HNS was superior to high speed steel M50 especially in combination with $Si_3N_4$ balls (Fig. 5.3). In practical application the contamination of bearings with abrading particles leads to indentations in the contact zone which reduce the lifetime. To copy this effect a pre-damage was introduced by Rockwell C indentations on the inner ring. The results of subsequent runs are given in Fig. 5.4 and indicate that L10 of HNS is about 7 times longer than that of standard ball bearing steel and about twice as long as that of M 50.

This superior performance of HNS bearings has led to advanced applications e.g. in aviation (Hucklenbroich et al. 1998). One makes use of their high resistance to pitting corrosion. During landing aircraft passengers may observe part of the wing being pushed backward. The actuators for the wing flap movement consist of ball screws, which are exposed to humidity and even salt spray in winter time or near the sea. To prolong the service life of screw shafts and nuts two manufacturers of civil airplanes are equipping new models with PESR-HNS which is induction hardened in the contact zone. The balls are made of $Si_3N_4$.

Another application is based on the good combination of resistance to stress corrosion, liquid oxygen compatibility and endurance. The fuel pumps of the space shuttle are driven by gas turbines, which are cooled by the liquefied gas. The rotor bearings run under high Hertzian pressure and poor lubrication in contact with liquid $O_2$ or $H_2$. Between missions condensation causes corrosion (Chin et al. 1994). Previously the bearings were renewed after each flight. Since the introduction of PESR-HNS the first inspection is scheduled after ten flights, because simulating experiments have demonstrated a high life expectancy (Trojahn 1998).

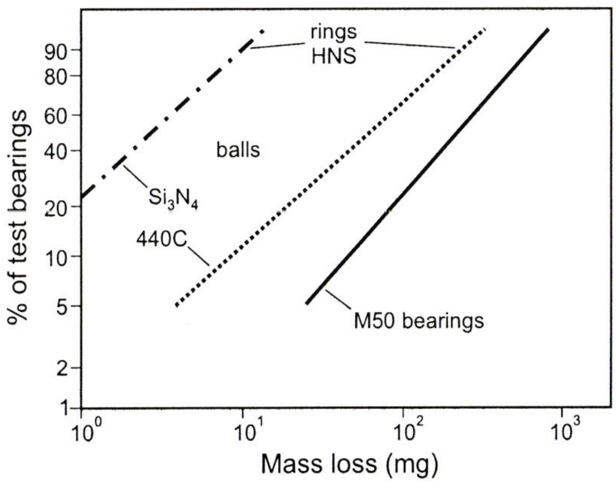


**Fig. 5.3.** Mass loss of ball bearings after 50 h of dry operation under a Hertzian pressure of 865 MPa at a rotating speed of 500 rpm. HNS = Cr15Mo1C0.15N0.38, 440C = Cr17Mo0.5C1, M50 = Cr4Mo4V1C0.8 (Trojahn 1992)

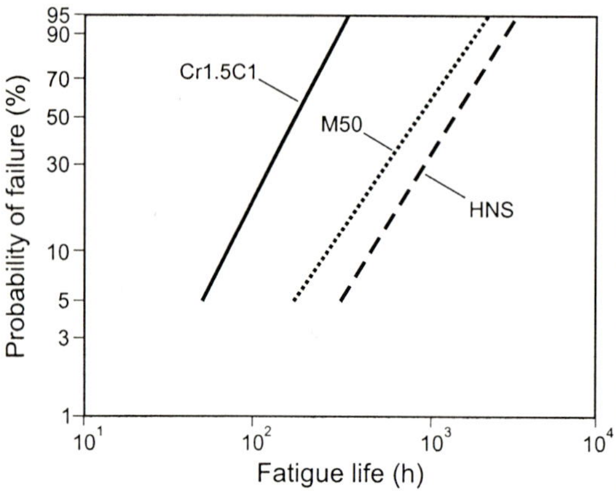


**Fig. 5.4.** Bearing life under a Hertzian pressure of 2500 MPa and EHD lubrication at a rotating speed of 12 000 rpm, the inner rings were made of materials as indicated and pre-damaged by Rockwell C indentations. Standard steel Cr1.5C1 was used for outer rings and balls. HNS = Cr15Mo1C0.3N0.33, M50 = Cr4Mo4V1C0.8 (courtesy FAG,Schweinfurt)

The main bearings of aircraft turbines are air-cooled in flight but after landing residual heat may raise their temperature to about 300 °C. During overnight stops cooling to ambient temperature entails condensation of flue gas which causes corrosion. At present large HNS-bearings of e.g. 550 mm outer diameter are undergoing simulating tests.

In hydraulic systems fire resistant fluids along ISO 6743/4 are used for safety reasons. Their high content of water causes a deterioration of bearing life so that stainless steels are used for bearings submerged in these fluids. Werries (1995) compared the performance of grooved ball bearings in class C hydraulic fluid (HFC) constisting of polyglycol with 40 % water. While steel Cr17Mo0.5C1 (AISI 440C) failed after 13 % of the calculated L10, HNS Cr15Mo1C0.3N0.33 (CRONIDUR 30) had reached 140 % unharmed when the overrolling run was stopped after 4000 h. In axial piston pumps used in coal mining bevel ball bearings were surrounded by HFC fluid. Bearings made of steel Cr1.5C1 failed at 45 °C after 81 % of the calculated L10 while HNS Cr15Mo1C0.3N0.33 remained unharmed after 133 % even at 65 °C (Reichel 1993, Lösche et al. 1997).

According to recent informations by FAG (1999) CRONIDUR 30 is gaining a foothold in other applications of mechanical engineering because of prolonged bearing life. Examples are bearings for high-performance chainsaws, spindle bearings for machine tools and large bearings of 7 Mg mass for rolling mills

## 5.1.2
## Wear resistant stainless steels

The previous section dealt with HNS almost free of nitrides and carbides in the as-hardened condition. Increasing the nitrogen content leads to precipitates which are not dissolved at hardening temperature. The resulting microstructure consists ideally of fine globular hard particles embedded in martensite and is expected to enhance the tribological properties.

***Development:*** One way to achieve a higher N content was discussed in Sect. 4.1.2 (Solidification). As shown in Fig. 4.6 $p_{N2}$ >10bar offers an austenitic solidification of steel Cr16Mo1N0.6C0.3 with a reduced liability of pore generation. During cooling the precipitation of nitrides occurs preferentially along grain boundaries and inhibits crack-free hot working. To retain the corrosion resistance of the matrix the Cr content has to be raised or nitride formers of higher N affinity like V or Nb have to be added. This reduces the size of the austenitic regime. Melting in existing PESR units seems feasible but has not been tried for the above reasons. However, a steel with an N/C ratio of 1/2 instead of 2/1 i.e. Cr16C0.6N0.3 was investigated by Shimada et al. (1994) which did not form coarse eutectic $M_7C_3$ carbides found in carbon grades like Cr17Mo0.5C1 (AISI 440C). Instead fine secondary precipitates such as $M_{23}C_6$ and $M_2N$ were detected which improved the toughness and corrosion resistance.

PM production poses an alternative route towards high N-contents (see Sect. 4.1.3). As depicted in Fig. 4.10a atomized steel powder was gas nitrided in a slowly rotating furnace to form a nitride layer on each powder grain without sintering. The basic composition of the powder was Cr15Mo1 (Wang 1992). Nb or V was added at high N levels to keep enough Cr in the matrix and provide corrosion resistance (Berns and Wang 1993). In this respect Ti is even more effective but its high affinity to oxygen makes melting more difficult. During HIP of evacuated cans nitrogen was evenly distributed and formed fine globular nitrides as shown in Fig. 5.5. Some properties of selected, hardened and tempered PM-HNS are given in Table 5.1. In respect to hardness and wear resistance they are in line with stainless carbon grades, but their corrosion resistance is quite superior (Fig. 5.6). A can of steel C15Nb5Mo1V0.7N1.6 with a diameter of 130 mm was forged to a bar of 60 mm diameter without any problems. This good workability is similar to conventional PM tool steels and is brought about by the small size and even distribution of the hard particles in the steel. Even more important than the resistance to abrasion is the reduced tendency to adhesion as compared to the HNS grades of the previous section. In deep-drawing of steel sheet or cold forging of wire hypoeutectoid tool steels are prone to adhesion, which is less pronounced for hypereutectoid or hypoeutectic tool steels containing

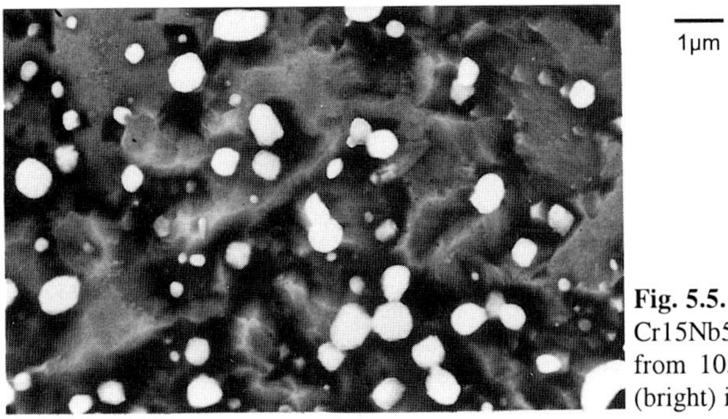


**Fig. 5.5.** Microstructure of PM steel Cr15Nb5Mo1V0.7N1.6 hardened from 1050 °C in oil, MN nitrides (bright) in martensitic matrix (dark)

**Table 5.1.** Wear resistant stainless PM steels (except 5), 1 to 3: HNS, 4: HNS-MMC, 5 and 6: stainless carbon grades for reference (Berns and Wang 1993)

| steels | hardness (HV30) quenched | hardness (HV30) tempered 450 °C | nitrides content (vol%) | nitrides mean size (μm) | [2] wear resistance $10^4$ quenched | [2] wear resistance $10^4$ tempered |
|---|---|---|---|---|---|---|
| (1) Cr15Mo1N0.8 | 543 | 671 | 2.3 | 0.35 | 8.7 | 7.4 |
| (2) Cr15Nb5Mo1V0.7N1.6 | 618 | 705 | 15 | 1 | 9.5 | 9.1 |
| (3) Cr15V6.5Mo1N3.2 | 648 | 698 | 22 | 1 | 8.1 | 7.1 |
| (4) Cr15Mo1 + 10CrN | 605 | 705 | 10.1 | 12 | 17.3 | 16.2 |
| | | | 10.8 | 0.35 | | |
| (5) Cr17Mo1C0.35 | 600 | 545 | ~0 | - | 6.3 | 5.3 |
| (6) Cr17V5.8Mo0.5C2.2 | 736 | 700 | 17[1] | 2 | 23 | 13 |

[1] carbides, [2] against flint paper of 220 mesh size

carbides. It seems justified to deduce a similar effect of the nitrides in alloy 2 or 3 of Table 5.1 independent of their application as tools, bearings or wear parts.

As the groove width of abrasive wear increases larger hard particles are required in the steel to protect the matrix. For this type of attack metal matrix composites were developed. These HNS-MMC were composed of atomized Cr15Mo1 powder and crushed CrN powder as in no (4). During HIP a partial transformation to $Cr_2N$ furnished sufficient nitrogen to saturate the austenitic matrix and provide hardenability (Fig. 4.10b). The coarser nitrides with a hardness of 1670 HV0.025 are more effective in enhancing the wear resistance than the fine particles (Table 5.1). The corrosion resistance of the HNS-MMC is close to the standard carbon grade Cr17Mo1C0.35, although about 11 vol% of fine nitrides have been precipitated consuming some of the CrMo content of the matrix. This requires future adjustments of the alloy content and HIP parameters.

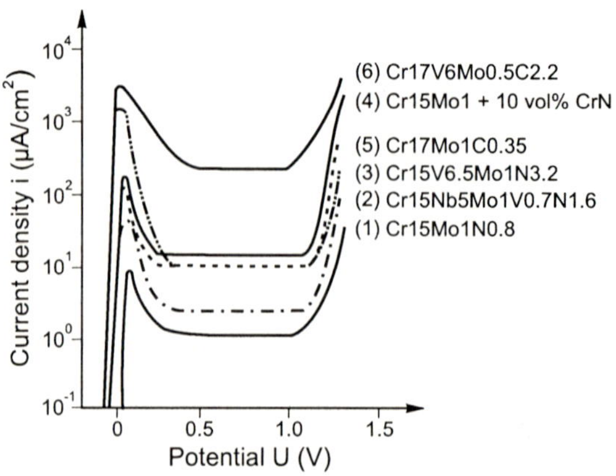


**Fig. 5.6.** Passivation of as-hardened martensitic stainless steels in 1-n $H_2SO_4$ at room temperature, for manufacturing see Table 5.1

Retained austenite usually improves the resistance to abrasion at least as long as it transforms to martensite in the wear surface. Its content and stability is governed by the temperature of hardening, quenching and tempering and may be adjusted accordingly. Tempering in the range of secondary hardening may yield the highest hardness but not the best wear resistance because the retained austenite is transformed or stabilized.

*Application:* Modern melting procedures have considerably raised the cleanliness of atomized steel powders. Yet there is still some risk that accidental remainders of slag are atomized resulting in non-metallic inclusions of a size equal to that of the metal powder grains. Therefore PM steel is generally accepted in the tooling but not in the bearing industry. An increasing amount of hard particles lowers the sinter activity of steel powders and HIP is essential to reach full density. Depending on handling the large powder surface may be contaminated by oxygen. Also surface segregation may occur. Both lead to changes in the properties of powder-grain-boundaries after HIP. Tensile tests on hot work PM tool steels revealed localized deformation and fracture along these boundaries in dependence of temperature (Wendl 1985). This embrittlement was not observed after hot working which changes the grain structure and allows for diffusion. As hardness increases brittle transcrystalline cleavage takes over and failure along powder grains is suppressed. Hot-working is expected to have a beneficial effect on PM-HNS but not on PM-HNS-MMC. The large hard particles of the latter are prone to fracture resulting in some loss of wear-resistance. At a higher fraction of coarse hard particles hot-working is impaired by cracking.

Products in view of PM-HNS are hot-worked bars and die-forgings of simple shape. In contrast PM-HNS-MMC start from the can which may be pre-shaped according to the final part. Applications are seen in the field of stainless tools and wear parts. Processing of aggressive, fiber or particle reinforced polymers calls for the superior combination of wear resistance and corrosion resistance offered by HNS. The same holds true for wear parts in the chemical industry or the pulp and paper industry. HIP cladding may be an intermediate step to introduce the new HNS. It is a means of applying a thick coating to a pre-machined part, a casting or a die-forging resulting in excellent bonding. We see some of this protection in polymer extruders. Currently iron- or cobalt-alloys enriched with carbides are used. HNS are superior in corrosion resistance to the former and cheaper than the latter.

### 5.1.3 High-speed steels

*Development:* The first approach to HNS-HSS was the addition of nitrogen to standard HSS produced by pressure or powder metallurgy. Lueg (1990) examined a WMoV grade with 2 mass% V and one with about 3 mass%V each with and without nitrogen. Their response to heat treatment is shown in Fig. 5.7. Some properties are given in Table 5.2. The differences between the standard steels and the HNS are not significant and in part may be explained by the higher sum of

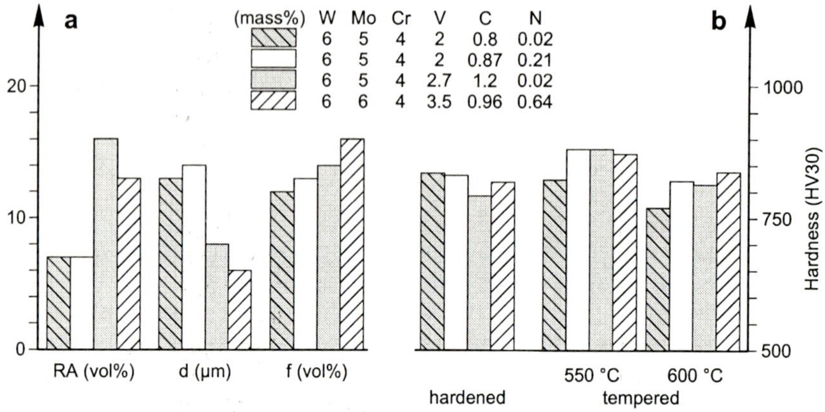


**Fig. 5.7.** Influence of nitrogen on high speed steels manufactured as in Table 5.2 and hardened from 1220 °C in oil, **a** content of retained austenite RA, grain size of austenite d, content of carbides/nitrides f, **b** hardness as-quenched and tempered

C+N. Element mapping by WDX analysis revealed that about all the nitrogen was contained in the coarse M (C,N) vanadium-carbonitrides. Thus N was hardly available for secondary hardening of the matrix. The solubility product of MX is lower for N than for C which leads to the observed accumulation of N. The resulting increase in the fraction of coarse precipitates might also be achieved by raising the carbon content instead of adding nitrogen.

These results were supported by extensive work of Rasheva et al. (1993 I). They found that 90% of the nitrogen content is trapped in the M(C,N) and only 3% in the matrix. However, they recommended the addition of 0.2 mass% N because of superior secondary hardness as well as improved properties and tool life (Rashev et al. 1993). Earlier work by Rasheva and Rashev (1990) pointed to a refinement of the as-cast eutectic structure and to less segregation by nitrogen, which enabled the production of larger ingots. The investigations cited above and the work of Stoichev et al. (1993) did not reveal what the equivalent rise of carbon instead of nitrogen would have amounted to.

To reach a nitrogen-based secondary hardening carbon has to be excluded. In this case a nitrogen content of ≈ 1 mass% is required, which may be added by powder metallurgy. A project supported by the German Research Foundation (DFG) dealt with the development of a HNS-HSS free of carbon. Thermodynamic calculations by Golczewski and Fischmeister (1996) showed that the solubility of nitrogen in austenite at hardening temperature was too low in steel W6Mo5Cr4V2N. They proposed to raise the Cr level to 8 mass% and limit the sum of WMoV to 8 mass%. Experimental verification of the calculations suggested a further increase in Cr and led to a maximum of secondary hardening at 475 °C which is about 75 °C below that of W6Mo5V2C0.9. This may be explained by the solubility product of VN at hardening temperature being much

**Table 5.2.** Influence of nitrogen on some properties of high-speed steels (Lueg 1990)

| steels[1)] | C + N (mass%) | tempered [2)] (°C) | $R_{c\,0.01}$ [3)] $R_{cf}$ (MPa) | $R_{b\,0.01}$ [3)] $R_{bf}$ (MPa) | $\varepsilon_{cf}$ [3)] $\varepsilon_{bf}$ (%) | $W_{ab}$ $10^{-6}$ [4)] flank wear | $v_c$ [5)] (m/min) |
|---|---|---|---|---|---|---|---|
| (1) W6Mo5Cr4V2C0.8 | 0.82 | 564 | 2058 | 2019 | 2.83 | 6.5 | 153 |
| | | | 3188 | 3437 | 0.23 | 87 | |
| (2) W6Mo5Cr4V2C0.87N0.21 | 1.08 | 595 | 2348 | 1948 | 2.72 | 4.9 | 150 |
| | | | 3296 | 3367 | 0.18 | 93 | |
| (3) W6Mo5Cr4V2.7C1.2 | 1.22 | 595 | 2217 | 2115 | 2.25 | 5.4 | 160 |
| | | | 3308 | 3588 | 0.25 | 72 | |
| (4) W6Mo6Cr4V3.5C0.96N0.64 | 1.60 | 606 | 2229 | 2078 | 2.77 | 9.6 | 152 |
| | | | 3375 | 3144 | 0.13 | 59 | |

(1) 100 kg ESR melt, (2) 100 kg PESR melt, (3) conventional stock, (4) PM steel
[1)] hotworked and hardened from 1200 °C, [2)] to a hardness of 64 HRC (800 HV30)
[3)] strength R, plastic deformation ε in compression (c) or bending (b), at fracture (f)
[4)] $W_{ab}$ = abrasive wear against flint paper of 220 mesh size, flank wear (μm) in slow cutting of carbon steel
[5)] $v_c$ = critical cutting speed at which failure of tool occurs after constant acceleration.

lower than that of VC. The precipitation during tempering is therefore shifted from V-based to Cr-based particles, i.e. to the ζ-phase $(FeCr)_2N$, which reaches its optimal strengthening effect at a lower tempering temperature than VC (Berns 1998). This is not a promising outlook: A high Cr content required to enhance the content of interstitial nitrogen in martensite is lowering the temper-resistance, which is an essential property of HSS.

The above development focussed on nitrogen as an alloying element influencing the properties in service. However, nitrogen is also used as a sintering aid raising the density of PM-HSS parts. A comparative study on the fatigue behaviour of hipped, vacuum sintered and conventionally cast and forged HSS showed, that the residual porosity of vacuum sintered material led to a severe reduction of the fatigue life (Berns et al. 1987). Palma et al. (1992) observed that sintering in an atmosphere of (vol%) $90N_2$, $9H_2$, $1CH_4$ at about 1200 °C entailed a transformation of coarse MC to fine MX precipitates in carbon based HSS. This reaction attracted V and released C which raised the volume fraction and lowered the V content of $M_6C$ (Urrutibeaskoa et al. 1993). As a result the uptake of about 0.5 to 0.8 mass% N during sintering led to full density at a lower temperature especially in an HSS of high V content (4.7 mass%). The sintering window was enlarged avoiding liquation and the formation of an embrittling net-like eutectic. Upon hardening a high content of retained austenite was found which decomposed during tempering, though. The hardness and temperature of the secondary hardening peak was raised as compared to vacuum sintering. The average fracture toughness $K_{Ic}$ after atmospheric and vacuum sintering dropped from 34 MPa $\sqrt{m}$ at a tempered hardness of 500 HV50 to 12.5 MPa $\sqrt{m}$ at 1000 HV50 although the scatter at 900 HV50 is about 2 to 1. These results spurred the development of dense high vanadium HSS with 6 to 11 mass% V and 1.6 to 2.7 mass% C which absorbed up to 2.4 mass% N during sintering in $N_2$ with 7 vol% $H_2$ (Wright et al. 1998). Full density (> 99 %) was reached at a temperature as low as 1130 °C. However, at too high a C/V ratio $M_3C$ appeared promoting liquation, a net-like eutectic and even distortion. The exploration of the pseudo-binary phase diagram of HSS by thermodynamic calculations visualized the effects of alloying on the sintering window (Wronski et al. 1998). The improved sintering behaviour is brought about by a high content of interstitials (C+N < 4 mass%) which entails a content of hard precipitates far above the usual level of HSS. Giminez et al. (1998) therefore started with nitriding green compacts at 930 °C under a pressure of 2 and 4 bar reaching a nitrogen content of 0.86 and 1.07 mass% respectively. During subsequent sintering at 1210 °C in a vacuum furnace with a nitrogen pressure of 5 mbar most of the absorbed nitrogen was extracted again and yet full density was reached.

***Application*:** Two limitations impede the use of high-nitrogen HSS. (a) Pressure metallurgy allows for the addition of N to a carbon-based HSS but appears to be expensive with respect to the gain in properties. (b) A PM-HNS based on Cr and N is afflicted with a lack of hot hardness. This seems to restrict the use of nitrogen to a level soluble at normal pressure, which still may influence solidification. Alloying with 0.2 mass% N is reported to refine the eutectic and reduce microsegregation (Rasheva and Rashev 1990; Lueg 1993). However, at equal cross-section of small ESR ingots remelted at ambient and high $N_2$-pressure no refinement of the eutectic carbides was observed (Lueg 1990). Finer eutectic carbides were found in bars of cold-work tool steel with 12 mass% of Cr containing nitrogen (Lueg 1993). It is not clear, though, if the conditions of solidification and hot working were equal. If there is a refinement of the eutectic in ledeburitic tool steels due to nitrogen it could be based on primary M(C,N) acting as nuclei. This type of inoculation might even work at a nitrogen level of normal pressure because of the low solubility product of VN. In the 1960ties the inoculation of conventional HSS by Ti on a production scale was not successful because of coarse TiC. The precipitation of VN is expected to reduce the liquidus/solidus – interval compared to TiC leading to nuclei with less time to grow. These considerations need further work but may be useful not only with respect to tool performance but also to ingot size and hot-workability.

The use of nitrogen as a sintering aid is aimed at the low-cost production of dense near-net shape tools of complex geometry. It is yet to be seen, if the resulting increase of hard precipitates provides a superior combination of wear-resistance and toughness. Hopefully the limited market of vacuum sintered HSS parts will be revived by the higher density of the new sintering technique.

## 5.1.4 Nickel - martensitic stainless steels

***Development:*** In this group of low-carbon steels the austenite is stabilized at hardening temperature by Ni to avoid hard grain-boundary precipitates and to enhance weldability and toughness. Commercial grades are Cr15Mo1Ni5 and Cr13Ni4. The latter was microalloyed with N,V and Nb by Berns and Xie (1992) to enhance secondary hardening. At 0.025 mass% C about 0.06 mass% N were dissolved by conventional melting together with 0.12 mass% V and 0.06 mass% Nb. After hardening from 1050 °C secondary hardening during tempering was considerably improved. After tempering at 570 °C the tensile strength was raised by about 200 MPa compared to standard Cr13Ni4 (Table 5.3). At equal strength of about 950 MPa microalloying allowed for an increase of the tempering temperature by about 100 °C and a decrease of the ductile-to-brittle transition temperature (DBTT). To reach a high strength level the tempering temperature of the standard steel is sometimes lowered to < 570 °C, where the temper-carbides are encased by a Cr-depleted zone which impairs the corrosion resistance. This problem does not arise in the micro-alloyed grade which also offers an increase of the $Ac_{1b}$-temperature by 30 to 40 °C. Thus reaustenitization during tempering may be suppressed and the formation of new martensite during cooling is avoided. Although the nitrogen content is low the effect on properties is high.

***Application:*** The martensitic transformation is used to provoke grain refinement and a disperse precipitation during tempering. Compared to stainless ferritic steels the yield strength of the modified grade is three times as high. Yet ductility and impact toughness remain at a high level and the DBTT may be shifted to the sub-zero range. Therefore microalloyed grades are best suited for high-strength

**Table 5.3.** Properties of nickel-martensitic stainless steels,
A = standard steel Cr13Ni4.8C0.025N0.025,
B = microalloyed steel Cr13.1Ni4.7V0.12Nb0.06C0.025N0.06,
hardening temperature 1050 °C (Berns and Xie 1992)

| Steel | A | B |
|---|---|---|
| $Ac_{1b}$(°C) | 565 | 600 |
| $Ac_{1e}$ (°C) | 710 | 750 |
| $M_s$ (°C) | 250 | 230 |
| $M_f$ (°C) | 110 | 125 |
| tempered at 570 °C | | |
| $R_{p0.2}$ (MPa) | 788 | 1009 |
| $R_m$ (MPa) | 836 | 1074 |
| $A_5$ (%) | 15.8 | 13.9 |
| Z (%) | 73 | 65 |
| tempered to 315 HV30 | | |
| TT (°C)[a] | 504 | 623 |
| toughness at 25 °C (J)[b] | 150 | 145 |
| DBTT (°C)[c] | | |
| at 90 J | - 15 | - 35 |
| at 27 J | - 25 | - 65 |

[a] tempering temperature to give equal hardness, [b] ISO-V impact tests,
[c] ductile-to-brittle transition temperature of ISO-V impact tests

stainless applications in the machine industry. Parts may be produced as castings, from hot-worked stock or as welded components. Nickel-martensitic stainless steels are widely used in hydroelectric power generation for turbines and armatures. A higher strength at good toughness allows to save weight. This is even more pronounced in rotors of radial compressors. The increase of pressure is related to the speed of rotation which is limited by the strength. If it is raised by lowering the tempering temperature embrittlement and corrosion are liable to rise. Microalloying offers strength without these disadvantages. Also new martensite in segregated areas is less prone to appear after tempering, which may be an advantage in case of $H_2S$ in the compressed gas.

## 5.1.5 Stainless quench-and-temper steels

***Development:*** The standard steels Cr17Ni2C0.2 and Cr17Mo1C0.35 are characteristic for this group. The first one contains some Ni to suppress δ-ferrite while the second relies on C alone. Both are hypereutectoid and tend to precipitate grain boundary carbides after solidification or hot working which cause an

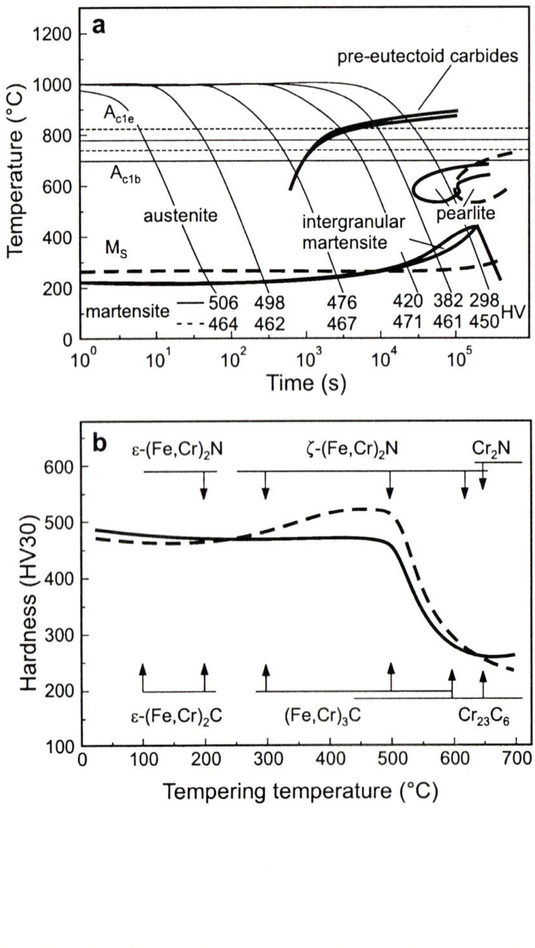


**Fig. 5.8.** Response to heat treatment of stainless quench-and-temper steels Cr15Ni2Mo1C0.2 (solid lines) and Cr15Ni2Mo1N0.2 (dashed lines) , **a** TTT diagram after austenitisation at 1000 °C, **b** hardness and precipitates after tempering of specimens quenched from 1000 °C in oil

embrittlement especially in larger cross-sections. After hardening these steels are usually tempered above 600 °C and carbon is predominantly contained in carbides.

Ehrhardt (1995) explored the exchange of carbon by nitrogen in a range of 0.1 to 0.5 mass% of the interstitial content. Up to 0.2 mass% N melting at normal pressure is sufficient while above this level pressure or powder metallurgy is required making the HNS expensive. The lower the nitrogen the higher the nickel content to suppress δ-ferrite and the higher the cost. Therefore 0.2 mass% seems to be an optimal content with respect to price. Part of Cr may be replaced by Mo to improve the resistance to pitting corrosion. These considerations led to the HNS Cr17Ni2N0.2 and Cr15Ni2Mo1N0.2 which were compared with respective carbon grades (Berns and Ehrhardt 1996). The TTT-diagramme (Fig. 5.8a) shows that the HNS in contrast to the carbon grade is free of pre-eutectoid precipitation up to long $t_{8/5}$ cooling times (see Sect. 4.4.2). Thus an embrittling grain-boundary precipitation is avoided. This reflects the observation that N-atoms do not tend to segregate into grain boundaries while C-atoms distinctly do (Petrov 1993). It is supposed that the increase of free electrons by dissolved N compensates the elastic lattice distortion driving interstitial atoms towards a surface (Petrov et al. 1999, see also Sect. 1.1.4.3). During tempering the $M_3C$ and $M_{23}C_6$ carbides grow much faster than the respective nitrides, i.e. $(Fe,Cr)_2N$ and equilibrium $(Cr,Fe)_2N$. This is reflected in the as-tempered hardness (Fig. 5.8b). In the usual tempering range of 600 to 670 °C there is little difference in yield strength between N- and C-grades. However, the impact toughness is raised by about 40% as a result of soft

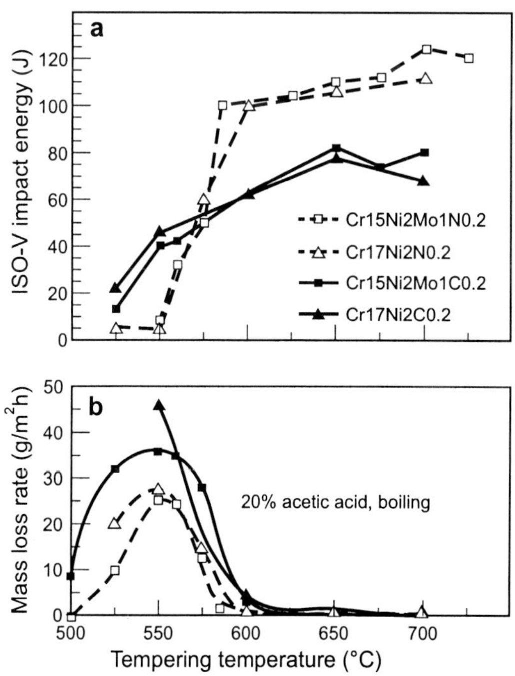


**Fig. 5.9.** Properties of stainless quench-and-temper steels with carbon or nitrogen depending on the tempering temperature **a** impact toughness at about equal 0.2% proof strength, **b** corrosion resistance (Ehrhardt 1995)

grain boundaries and smaller precipitates in HNS (Fig. 5.9a). The influence of N on the corrosion resistance is different from low-tempered HNS (see Sect. 5.1.1) because nitrogen is no longer dissolved in the matrix but almost fully contained in the nitrides. Therefore it cannot participate in the formation of the passive layer (see Sect. 3.2.2.2). Instead the number of Cr atoms consumed per interstitial atom becomes important. Looking at the equilibrium precipitates the ratio is about 4/1 for $Cr_{23}C_6$ but only 2/1 for $Cr_2N$. In addition the Cr content of $M_{23}C_6$ was found to be twice as high as that of $M_2N$ after tempering Cr15Mo1C0.6 and Cr15Mo1N0.6 at 650 °C (Table 2.5, see also Gavriljuk and Berns 1998). As a result the pitting potential of HNS is clearly above that of the respective carbon grade after tempering above 575 °C (Ehrhardt 1995). Below this temperature the difference in corrosion resistance becomes even more pronounced because the Cr diffusion is retarded and depleted zones are formed around the precipitates. The amount of precipitates, their size and their Cr concentration is higher for carbon grades aggravating anodic dissolution e.g. in boiling acetic acid (Fig. 5.9b). After tempering at 600 °C the Cr depletion around the finer nitrides is evened out but not around the coarser carbides. Therefore HNS allow for a lower tempering temperature and offer a better combination of corrosion resistance and high strength – both at a superior level of toughness.

A nitrogen content of 0.2 mass% is not only a good choice with respect to cost but also with respect to properties, which is documented in Fig. 5.10 for a given tensile strength of 900 MPa. At 0.1 mass% of interstitials the tempering

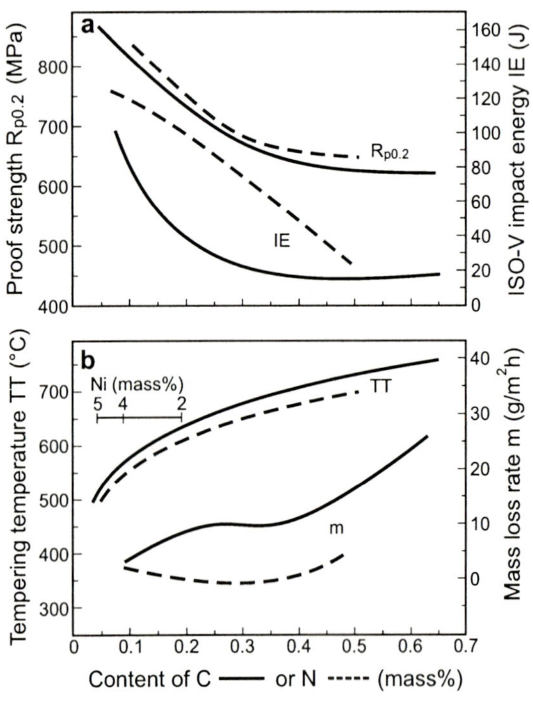


**Fig. 5.10.** Properties of stainless quench-and-temper steels with mass% 15Cr, 1Mo, ≤ 5Ni (to suppress δ-ferrite) depending on the N or C content at an equal strength level of $R_m$ = 900 MPa, **a** strength and toughness, **b** required tempering temperature to reach $R_m$ and corrosion rate in boiling 20% acetic acid (Ehrhardt 1995)

temperature is too low to differentiate between C and N alloys. At 0.5 mass% the mechanical properties of C and N alloys are approaching each other, but the lower Cr depletion of HNS is still visible in the lower mass loss.

***Application:*** The new HNS-grades Cr15Ni2Mo1N0.2 and Cr17Ni2N0.2 are superior to conventional Cr17Ni2C0.2 in many aspects. One is the reduction of grain-boundary precipitates, as they increase the risc of cracking after cooling from forging temperature. All stainless martensitic steels possess a high hardenability and a lower $M_s$ temperature compared to low-alloy grades. Through-hardening in air after forging causes positive residual stresses at the surface when the core transforms. Especially in larger cross-sections this may lead to cracking which is supported by brittle grain boundaries. HNS promise to defuse this manufacturing problem.

Like low-alloy quench-and-temper steels the stainless grades are used for a large variety of parts usually machined from forged or rolled stock, like e.g. shafts, spindles, valves, bolts and devices which serve in a moderately corrosive environment and require strength and toughness. Because nitrogen is narrowing the temperature range of Cr-depletion around precipitates it allows to apply a lower tempering temperature to gain a higher strength level. This offers a reduction of weight or e.g. an increase of rotational speed like in turbo machinery. For as-cast or welded components nickel-martensitic steels are preferred (see Sect. 5.1.4)

## 5.1.6 Steels for inlet valves

The inlet valves of automotive engines have to withstand cyclic mechanical, chemical and tribological loads. Although cooled by the incoming air/gas-mix the service temperature in the head rises to about 500 °C. During longer stops the engine cools down and condensation of the exhaust leads to corrosion. The compression causes bending of the head and small fretting movements between valve and ring of the seat resulting in wear.

Cr9Si3C0.45 is the standard steel for inlet valves. However, in case of higher sulphur and chloride contents of some heavy oiles for big Diesel engines a stainless valve steel like Cr18Mo2V0.4C0.85 is used. The steels are quenched and tempered to a tensile strength of 1000 to 1350 MPa and require a strength of > 600 MPa at 500 °C. The stem end is subsequently induction hardened to > 50 HRC to bear the load of the rocker arm. Because of the high Cr and C content coarse eutectic $M_7C_3$ carbides appear in the microstructure of the stainless grade entailing a lower ductility and fatigue strength than for the CrSi grade, which contains only smaller secondary carbides. In addition tool wear during hot extrusion of the valves is increased by the eutectic carbides.

***Development:*** As shown in Figs. 2.18 and 4.2 the solubility of austenite enriched with Cr is considerably higher for N than for C. This opens the possibility of creating a stainless grade with sufficient hardenability but without coarse

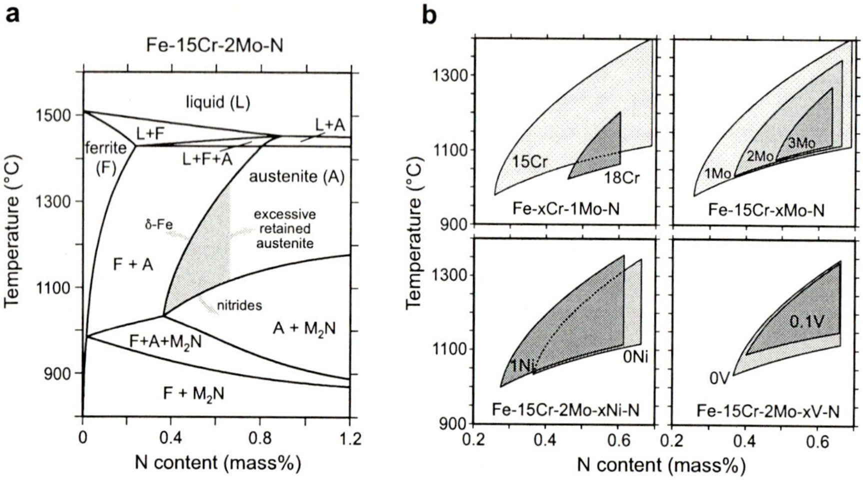


**Fig. 5.11.** Constitution of prospective HNS for inlet valves derived by ThermoCalc **a** shaded area offers homogeneous austenite for forging and hardening, **b** change of this area by variations in alloy content (Escher 1999)

precipitates (see Sect. 5.1.1). The previous section revealed that nitrogen offers advantages not only in the as-quenched but also in the highly tempered state, because the precipitates are finer and consume less Cr compared to respective carbon grades. The constitution of a stainless HNS for inlet valves was explored by ThermoCalc. Fig. 5.11a points out that the austenitic regime required for hot working and hardening is encased by δ-ferrite to the left, excessive retained austenite after quenching to the right and $M_2N$ precipitation below. Fig. 5.11b hints to changes of the austenitic phase field induced by variations of the alloying content. As a result the HNS Cr15Mo2V0.1N0.6 was produced by PESR and tested in comparison with the two carbon grades mentioned above, all hot extruded, quenched and tempered to a strength of 1100 to1200 MPa (Berns et al. 1998 I, II).

A brief summary of the test results is given in Table 5.4. They reveal that the ductility and fatigue strength of HNS at service temperature were distinctly above those of the stainless carbon grade and close to Cr9Si3C0.45 which is, however, not corrosion resistant. In this respect the HNS and the stainless carbon grade were close to each other and their passivation in 1-n $H_2SO_4$ is almost equally good. For wear tests in a Diesel engine only a slightly different HNS was available, which was compared to standard Cr9Si3C0.45 in different states against two different types of rings. The HNS showed a superior wear resistance.

***Application:*** The continuing reduction of the valve mass enables a light-weight valve drive but increases the cyclic bending stresses in the valve head. HNS may replace stainless grades, if a higher fatigue resistance is required and corrosion is a major issue. The high hardness required at the stem end calls for a nitrogen content which affords a manufacture by pressure or powder metallurgy and raises

**Table 5.4.** Properties of steels for inlet valves (Escher 1999)

| properties | | Cr9Si3C0.45 | Cr18Mo2V0.4C0.85 | Cr15Mo2V0.1N0.6 |
|---|---|---|---|---|
| hardness (HRC) | | | | |
| as-quenched, | | 59 | 54 | 56 |
| after tempering at 500 °C | | 56 | 49 | 59 |
| tempering temperature (°C)[1] | | 710 | 760 | 660 |
| results of tensile tests at 20 °C / 500 °C | | | | |
| strength $R_m$ (MPa) | | 1210 / 650 | 1060 / 775 | 1220 / 820 |
| reduction of area (%) | | 46 / 80 | 32 / 42 | 46 / 70 |
| fatigue strength (MPa)[2] | | 588 / 390 | 419 / 348 | - / 435 |
| passive current density in 1-n $H_2SO_4$, 20°C ($\mu A/cm^2$) | | 260 | 6 | 5 |
| break-through-potential in 3% NaCl, 20°C (mV) | | - 240 | 125 | 385 |
| relative wear rate (%) in 400 h of service[3] | | | | Cr15Mo1.5V0.35N0.39C0.2 |
| valve[4] | ring | | | |
| hardened steel | as-cast steel | 148 | - | 97 |
| tempered | | 100[5] | - | 102 |
| hardened | sinter steel | 207 | - | 109 |
| tempered | | 167 | - | 159 |
| coated | | 102 | - | - |

[1] applied before the following tests, [2] rotating bending, 50 Hz, $10^7$ cycles, [3] 6 cylinder Diesel engine without blow by and with increased compression, [4] seat induction hardened, tempered or weld coated with a Co-base hard alloy, [5] reference condition

the costs. Therefore the application of HNS will probably be confined to niche engines. For large valves in heavy Diesel engines driving a ship or large stationary machines the lack of eutectic nitrides and the reduced tendency to grain boundary precipitation may be an advantage in producing toughness in thicker cross-sections.

### 5.1.7 Hot work tool steels

***Development:*** The enhancement of secondary hardening by nitrogen (Fig.2.27) was considered to be useful for hot work tools. In a first step carbon in standard steel Cr5Mo1V1C0.4 (~ H 13) was partially replaced by nitrogen to an extent which was feasible at the limited Cr content and pressure of the PESR unit. The lower solubility product of VN compared to VC in austenite decreased the hardenability of Cr5Mo1V1N0.22C0.17 in comparison to the standard grade. In addition undissolved carbo-nitrides lowered the toughness. Therefore V was replaced by up to 3 mass% of Mo. Key results of 4 HNS and the standard grade are given in Table 5.5.

The chemical composition is described in part (a). In (b) the HNS reveal a lower $Ac_{1e}$-$Ac_{1b}$ interval, a higher $M_s$ temperature and a lower critical cooling time from 800 to 500 °C ($tc_{8/5}$) to avoid pearlite. This is based on a reduced alloy content of the austenite due to undissolved precipitates which also act as nuclei. The temperature of secondary hardening (c) was not reduced by nitrogen like in high speed steels. A reason for this is given in Sect. 2.5. In carbon-free HNS ζ-$(Fe,Cr)_2N$ is responsible for secondary hardening, while cubic (Cr,Fe)N is found in (N+C)-alloys. The hot strength was considerably improved by N (e). Not so the creep properties (f) at 550 °C. This was assigned to a different growth rate of the precipitates. Hede and Aronsson (1969) found the growth of carbides proportional to $t^{1/3}$ and that of nitrides proportional to $t^{1/4}$. Reppich (1982) demonstrated that

**Table 5.5.** Properties of hot work tool steels with 5 mass% Cr, 100 kg ESR (1) or PESR (2 to 5) ingots forged to bars (Lueg 1990)

| steel | 1 | 2 | 3 | 4 | 5[1)] |
|---|---|---|---|---|---|
| (a) Mo/V (mass%) | 1.34/0.95 | 1.16/0.91 | 1.45/0.30 | 1.41/0.09 | 3.23/0.11 |
| C/N (mass%) | 0.40/- | 0.17/0.22 | 0.14/0.30 | 0.14/0.30 | 0.05/0.27 |
| (b) $Ac_{1b}$ (°C) | 850 | 835 | 840 | 850 | 830 |
| $Ac_{1e}$ (°C) | 905 | 870 | 870 | 895 | 865 |
| $M_s$ (°C)[2)] | 280 | 350 | 340 | 338 | 295 |
| $tc_{8/5}$ (s)[2)] | 3970 | 1240 | 1420 | 1480 | 1990 |
| (c) hardness (HV30)[5)] hardened 1050/1150 °C | 627/666 | 490/555 | 551/630 | 560/575 | 543/- |
| (d) DBTT (°C)[3)] | 15 | - 95 | - 150 | - 35 | - 75 |
| $K_{Ic}$ (MPa $\sqrt{m}$ )[4)] | 49.8 | | | 44.6 | |
| (e) $R_m$ (MPa) | 1055/708 | 1308/1045 | 1235/805 | 1245/915 | 1168/971 |
| $R_{p0.2}$ (MPa) | 970/657 | 1215/964 | 1121/786 | 1149/867 | 1082/928 |
| Z (%) at 500/600 °C[4)] | 64.3/82.5 | 49.3/51.2 | 53.5/64.5 | 49.1/54.2 | 52.4/58.7 |
| (f) $t_f$ (h) | 51/167 | 72/120 | 58/115 | 44/115 | 54/216 |
| $\dot{\varepsilon}_{min}$ $10^{-2}$· (%/h) | 2.4/0.66 | 1.9/1.4 | 1.4/1.8 | 3.8/1.7 | 3.9/1.9 |
| $\varepsilon$ ( $\dot{\varepsilon}_{min}$ ) stress 600/500 MPa at 550 °C[4)] | 1.20/1.19 | 0.83/0.53 | 1.24/1.01 | 0.62/0.84 | 0.81/0.75 |

[1)] plus 3.44 mass% Mn to enhance N solubility, [2)] austenized 1050 °C, [3)] ductile-to-brittle transition temperature of unnotched 7 x 10 x 55 (mm) impact specimens hardened from 1150 °C and tempered to a strength of 1500 MPa, [4)] hardened from 1050 °C and tempered to a strength of 1500 MPa, [5)] after tempering at 500 °C smaller particles become effective as the stress and strain rate increase. This may account for the higher hot strength of HNS reaching the effective size sooner than the standard carbon grade. At a given room temperature strength level the ductility at elevated temperature – as given by the reduction of area Z – was reduced due to enhanced precepitation strengthening. The fracture toughness $K_{Ic}$ was lowered by (V+N) yet N shifted the brittle to ductile transition temperature DBTT of smooth 7 x 10 x 55 mm impact specimens to the sub-zero range. This was considered as an indication of superior resistance to crack initiation due to a refined microstructure. Ernst and Rasche (1992) found a higher resistance to thermal fatigue for an HNS similar to steel 3 of Table 5.5.

***Application:*** An 8 ton ingot of HNS Cr5.4Mo1.3V0.3N0.25C0.18 was produced by PESR and subsequently forged to bars of suitable cross-section. From these, dies for drop forging, hot extrusion and die casting were manufactured and tested in the field. The results were encouraging. In drop forging of turbine blades 2.6 times as many INCONEL sockets were produced in the first step and 1.3 times as many Cr13C0.2 blades in the final step compared to standard Cr5Mo1V1C0.4 (Ernst and Rasche 1992). However, these authors found some surface porosity after electro-discharge machining due to effusion of nitrogen which required a final pass with low current (EDM-facing). In addition they were not able to completely suppress pores during TIG welding. This seems to be a severe drawback because the design of parts is frequently changed and the dies have to be adjusted accordingly which is done by welding and machining. Welding is also used to repair worn tools.

Therefore the mere exchange of carbon by nitrogen is apparently not the ideal way to create a high nitrogen hot work tool steel. Instead one would have to refrain from pressure melting and spur the nitrogen solubility by alloying with Cr as described in section 5.1.5. The interstitial content of the stainless quench-and-temper steel Cr17Ni2N0.2 would have to be raised, though, to meet the strength level of hot work tools thus lowering the Ni-content required to suppress δ-ferrite. Berns and Wendl (1986) proposed to lower the carbon content of Cr5Mo1V1C0.4 to 0.28 mass% to enhance toughness and creep resistance, because less CrMoV is spent on coarse secondary carbides and more is available for secondary hardening. Steel Cr16Mo1V0.3Ni1N0.16C0.12 resembles a tentative composition. A high chromium steel was used for aluminium die-casting dies before. Sheet of the standard creep resistant steel Cr12Mo1V0.3C0.2 was successfully employed to build up the ribs of dies for the production of air-cooled cylinder heads. A steel with a high Cr and N content would be targeted at tools of complex shape calling for a high ductility. However, one has to bear in mind, that a high alloy content impairs the thermal conductivity and therefore supports heat checking.

### 5.1.8 Creep resistant low-alloy steel

***Development:*** The carbon content of the standard steel Cr1Mo0.4C0.13 which is often above the mean level of 0.13 mass% was partially exchanged by nitrogen via PESR (Table 5.6). The addition of 0.26 mass% V led to undissolvable precipitates and poor hardenability. Manganese additions, meant to enhance the nitrogen solubility, raised the DBTT because Mn supports covalent interatomic bonds in the iron matrix (see Sect. 1.1.1.2). Therefore only three steel compositions are compared (part a). Nitrogen shifts the austenitization to higher temperatures and the critical cooling time $tc_{8/5}$ to avoid pearlite is reduced due to undissolved precipitates (b). Secondary hardening is improved by N especially at a high austenitizing temperature (c) and the hot yield strength is considerably raised, e.g. by 75 % at 650 °C (d). Impact test with ISO V-notch specimens in longitudinal direction revealed a rather low DBTT for all grades. Short time creep tests at 150 MPa and temperatures between 550 and 650 °C resulted in a superior performance of the nitrogen grades. Deformation induced strengthening combined with precipitation reduced the minimum creep rate $\dot{\varepsilon}_{min}$ of the nitrogen grades and also the creep $\varepsilon$ ($\dot{\varepsilon}_{min}$) to reach the lowest creep rate (Fig. 5.12). Transmission electron microscopy (TEM) revealed that the precipitates in steel 3 tempered at 720 °C were still 5 times smaller than those of steel 1 tempered at 670 °C. In addition more N than C seems to be in solid solution to be precipitated during creep.

***Application:*** The results show that even in low-alloy quench- and -temper steel N has a beneficial effect on the mechanical properties at elevated temperature. However, alloying with nitrogen requires pressure melting and is probably more expensive than increasing the CrMoV-level of carbon grades. Also weldability is required as these steels are used for tubes, boilers and the like. But after melting under pressure pores are liable to remain in the weld metal (see Sect. 4.3.2). Therefore this investigation was dedicated more to the potential of nitrogen than to real application.

**Table 5.6.** Properties of creep resistant low-alloy steels; 1 = hot rolled stock; 2,3 = 100 kg PESR ingots, forged (Berns and Wang 1989)

| | steel | 1 | 2 | 3 |
|---|---|---|---|---|
| (a) | Cr/Mo (mass%) | 0.92/0.42 | 1.05/0.52 | 1.01/0.45 |
| | C/N (mass%) | 0.19/.007 | 0.05/0.09 | 0.04/0.15 |
| (b) | $Ac_1$ (°C) | 753 | 756 | 834 |
| | $Ac_3$ (°C) | 860 | 895 | 937 |
| | $tc_{8/5}$ (s) | 36 | 13 | <13 |
| | cooled from (°C) | 930 | 950 | 1000 |
| (c)[1)] | hardness (HV 30) | 260 | 305/340 | 350/440 |
| | hardened at (°C) | 930 | 950/1050 | 950/1050 |
| (d)[2)] | $R_{p0.2}$ (MPa) | 494/229 | 516/383 | 563/400 |
| | $A_5$ (%) | 28/30 | 15/19 | 15/17 |
| | at 550/650 °C | | | |
| | DBTT (°C) | - 130 | - 145 | - 90 |
| | (Charpy-V) | | | |

[1)] after tempering at 500 °C, [2)] tempered at 680 °C to a hardness of about 220 HV30 ($R_m \approx 700$ MPa)

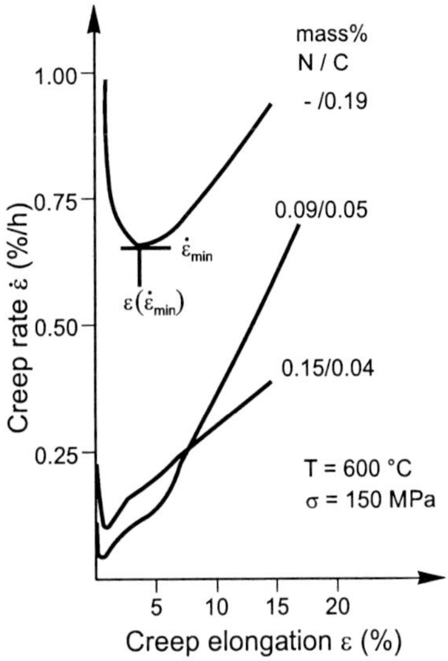


**Fig. 5.12.** Creep rate of steels with ~ 1 mass % Cr, ~ 0.5 mass% Mo and N/C as indicated after hardening and tempering to a hardness of ~ 220 HV30, after Berns and Wang (1988)

## 5.1.9
## Creep resistant high alloy steels

***Development:*** The Cr content of the standard grade Cr12Mo1V0.3C0.2 was reduced to 9 mass% to lower the content of carbides and enrich them with Mo and V thus raising their stability without enhancing δ-ferrite or scaling. Making use of nitrogen within the solubility limit at normal pressure and of microalloying with Nb a very stable array of fine precipitates was reached in the quenched and tempered state which lowered the creep rate. The commercial steel Cr9Mo1V0.2Nb0.07C0.1N0.05 (P/T91) is an example of this development.

As PESR became industrially available the effect of higher N contents was investigated. Berns und Krafft (1987) stepwise replaced C by N in steel Cr11.5Mo1V0.3CN. They found that the dissolution of precipitates upon austenitization between 1000 and 1200 °C was impeded by nitrogen and that the austenite grain size was reduced. The highest toughness was reported for the alloy with 0.12 mass % N and 0.07 mass % C (Fig. 5.13). However, the hot strength increased up to 0.25 mass % of N if the hardening temperature was sufficiently high. At 0.33 mass % N the dissolution of nitrides remained so incomplete even at 1150 °C that not enough temper nitrides were available to promote hot strength. In short time creep tests similar results were obtained. The extrapolation of creep data at a stress of 105 MPa and temperatures between 620 and 680 °C pointed to a fracture life at 600 °C of $10^4$ h for the steel with 0.25 mass% N and of $10^3$ h for the standard steel with 0.21 mass% C. This strengthening effect of nitrogen was assigned to much finer and more evenly distributed nitrides compared to respective carbides (Berns and Krafft 1989). Anthamatten et al. (1987) partially replaced V by Nb and the superior creep resistance of steel Cr10Mo1.5V0.22Nb0.07C0.06N0.16 was explained by the precipitation of fine

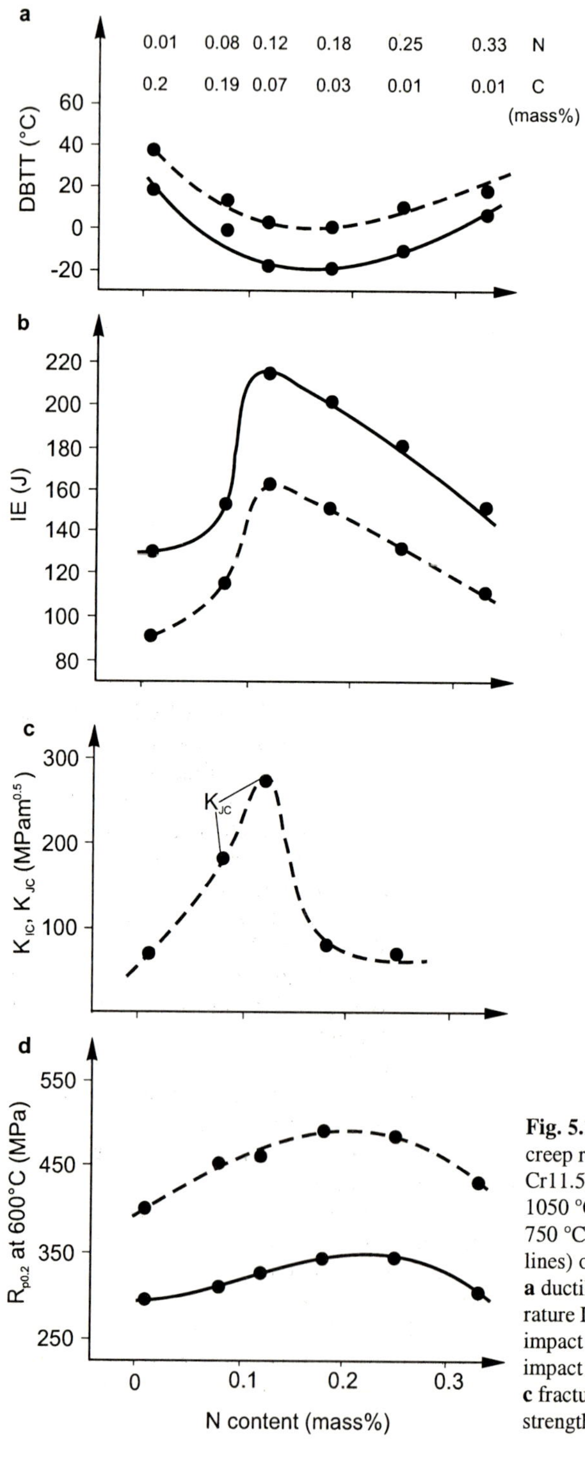


**Fig. 5.13.** Mechanical properties of creep resistant steel Cr11.5Mo1V0.3CN hardened from 1050 °C and tempered at 650 °C or 750 °C to $R_m \approx 1000$ MPa (dashed lines) or $R_m \approx 800$ MPa (full lines) **a** ductile-to-brittle transition temperature DBTT derived by ISO-V impact tests, **b** upper shelf ISO-V impact energy IE at 100 °C, **c** fracture toughness at 20 °C, **d** yield strength $R_{p0.2}$ at 600 °C (Krafft 1991)

$M_2X$ and MX which was more pronounced after hardening from 1200 °C compared to 1050 °C. An evaluation of creep rupture data derived from different HNS alloyed with 9 to 12 mass% Cr and Mo,V,Nb led to the conclusion that the target of 100 MPa creep strength at 600 °C after $10^5$ h could be met (Anthamatten et al. 1989). However, a high hardening temperature entails grain growth which is detrimental to toughness. Therefore Berns and Krafft (1990) combined a high temperature and a fine grain by means of a thermo-mechanical treatment (TMT). After austenitizing at 1250 °C the steels were cooled to 1000 °C and subsequently forged down to 850 °C while reducing the cross-section to one third, followed by hardening and tempering. This TMT led to a reduction of the DBTT compared to conventional heat treatment and to an increase of creep rupture life. Table 5.7 contains some results of the development described above. Column (a) describes the beneficial effect of N on toughness and strength. From (b) we conclude that Nb strengthens HNS more than a carbon grade. Dissolving the precipitates requires a hardening temperature which causes embrittlement. This is overcome by TMT in (c).

Göcmen et al. (1996) investigated steel Cr9Mn3Mo1V0.7Nb0.05N0.16 hardened from 1200 °C in air. They found that during tempering at 708 °C for 28 h the initially uniform precipitation changed into a decoration of subgrain boundaries within the recovered substructure. To reach a uniform and stable distribution of precipitates the continuous cooling from hardening temperature was interupted below $Ac_1$ enabling a pre-precipitation in the metastable austenite before martensitic transformation and tempering (Göcmen 1997). Due to the coherency between austenite and fcc MN this ausaging treatment produced a uniform precepitation which remained stable during tempering if initially overaged. The MN precipitates are rich in V and Nb and the precipitation of

**Table 5.7.** Toughness and strength of high-alloy creep resistant steels with (mass%) 10 to 12 Cr, 0.9 to 1.5 Mo, 0.2 to 0.3 V and C, N, Nb as given below, tempered at 750 °C, (Krafft 1991)

| | | | (a) | | (b) | | (c) | |
|---|---|---|---|---|---|---|---|---|
| | N | 0.01 | 0.12 | 0.25 | 0.01 | 0.14 | 0.20 | |
| (mass%) | C | 0.20 | 0.08 | 0.01 | 0.19 | 0.08 | 0.08 | |
| | Nb | - | - | - | 0.05 | 0.07 | 0.04 | |
| hardened from (°C) | | 1050 | | 1050 | 1250 | 1250 | 1150 | 1250 +TMT |
| DBTT (°C) | | 10 | - 60 | - 25 | 50 | 50 | - 18 | - 30 |
| $R_{p0.2}$ at 20°C (MPa) | | 597 | 599 | 650 | 666 | 744 | 615 | 807 |
| 650 °C (MPa) | | 296 | 333 | 344 | 386 | 476 | 355 | 400 |
| $t_f$ (h) at | | | | | | | | |
| 105 MPa, 640 °C | | 500 | 1000 | 1700 | - | - | - | - |
| 100 MPa, 675 °C | | - | - | - | 240 | 1078 | 762 | 856 |

*DBTT* ductile-to-brittle transition temperature of ISO-V impact tests,
$t_f$ time to fracture of creep tests in tension, *TMT* thermomechanical teatment

rapidly growing $M_2N$, rich in Cr, is suppressed in the presence of Nb. Ausaging of creep resistant steels offers new possibilities of precipitation control and microstructural design. A high hardening temperature is required, though, to dissolve enough nitrides for the intended re-precipitation. Göcmen et al. (1998) stressed the importance of microalloying with Nb and Ti to control the grain size of austenite. First results on the influence of ausaging on the hot strength are quite promising (Fig. 5.14) but the optimal microstructure for long-time service has yet to be evaluated.

Masuyama et al. (1996) investigated the oxidation behaviour of steel Cr9W2Mo0.5V0.2NbN with 0.045 or 0.168 mass% of N in the range of 500 to 900 °C and up to $10^4$ h. Nitrogen considerably reduced the weight gain by oxidation in air which is probably due to less Cr-depletion at the metal/oxide interface. Thus scaling is subdued by N and no problems are exspected for HNS in this respect.

***Application:*** The development of high-alloy creep resistant steel is aimed at a higher efficiency of fossile power plants by increasing the steam temperature to 600 °C or even above using bcc steel of low thermal expansion compared to fcc steel. Turbine blades are a possible application of HNS. Welded structures such as tubes and armatures are usually seam-less but are joined by circumferential weldings. Using nitrogen-free consumables part of the nitrogen set free by the penetration may be dissolved in the weld metal. If one succeeds to weld seams free of pores, they may contain less nitrogen but also bear less load, because the longitudinal stress of hollow cylinders under inner pressure amounts to only half

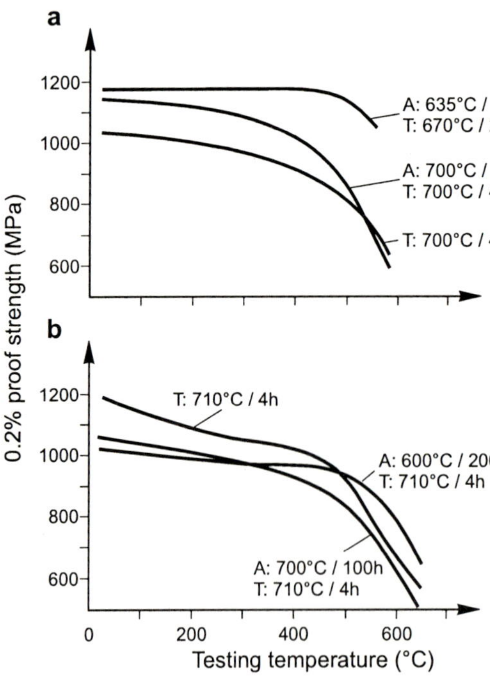


**Fig. 5.14.** Effect of ausaging (A) on hot strength of high-alloy creep resistant steels subsequently tempered (T), **a** Cr12Co10Mn2Mo1.5V0.7NbTiN0.16, **b** Co13Cr12Mo4.5V0.7NbTiN0.16 (Göcmen et al. 1998)

of the hoop stress. Other possible applications make use of the high toughness of HNS. Discs for gas turbines are an example. The requirement of $R_{p0.2} \geq 800$ MPa at 450 °C and Charpy impact energy $\geq 50$ J at room temperature can be met by HNS (Krafft 1991). Rotors of steam turbines are in some cases composed of forged rings joined by welding to avoid the interior degradation of very large forgings. The size limitation for PESR ingots (see Sect. 4.1.2) allows for the production of rings, though. Gerdes and Redecker (1990) reported that a creep resistant steel with about 12 mass% Cr and 0.18 mass% N was successfully welded with an N-free filler of similar composition. The mechanical properties of the weld were somewhat better than for a comparable carbon grade. The improved creep resistance of HNS rings may therefore be used to cope with the centrifugal forces while rotating bending is met by a welding of usual quality.

## 5.2 Austenitic steels

The Schaeffler diagram in Fig. 5.1 reveals that a Cr equivalent of about 18 requires the lowest Ni equivalent to give a stainless austenitic structure after solution annealing and quenching. Such a lean composition is still close to a martensitic transformation, which is to be expected during deep freezing or cold working (see Sect. 5.3.2). A further increase of $Cr_{eq}$ to enhance e.g. the corrosion resistance calls for about an equal rise of $Ni_{eq}$ to stay within the austenitic phase field, as the slope of the A/A+F borderline amounts to nearly 1. An addition of Mo requires more $Ni_{eq}$ than Cr. But at equal consumption of $Ni_{eq}$ a rise of Mo is more effective with respect to the pitting resistance (see Eq. 3.6). The rich alloys are more resistant to martensitic transformation. Most of the rich austenitic steels are located near the borderline to δ-ferrite because of minimizing the alloying costs and because of a full or partial solidification to δ-ferrite, which lowers the shrink strains compared to an austenitic solidification and helps to avoid hot tearing e.g. in castings and weldings. Based on microsegregation the solid state transformation to austenite may be incomplete and yield remnants of δ-ferrite, usually below 10 vol%. If the alloy content increases from lean to rich, intermetallic phases are more liable to precipitate between 1000 and 600 °C (see Sect. 2.4.1), which is usually detrimental to toughness and corrosion resistance.

In conventional stainless austenitic steels Ni contributes most to $Ni_{eq}$. Carbon is kept low because of an embrittling carbide precipitation along grain boundaries and a susceptibility to intercrystalline corrosion e.g. after welding. However, a high carbon content is used to stabilize the austenite in wear resistant steels like Hadfield manganese steel Mn12C1.2 or in non-magnetic steels like Mn18Cr4C0.5. Manganese is added instead of Ni because cost and stacking fault energy are lowered and the solubility for interstitial elements is raised.

How does nitrogen come into this picture of austenitic steels? The advantages of this interstitial element over carbon are discussed in the Chaps. 1 to 3. Some key words are: Enhancement of atomic ordering and solubility, higher stability of austenite to precipitation and martensitic transformation, higher strength, ductility and work hardening also at sub-zero and elevated temperatures, distinctly improved corrosion resistance, especially to localized corrosion. But the superior mechanical and chemical properties of austenitic HNS are limited to about 1 mass% N because of brittle, cleavage-like failure at room temperature and especially below (see e.g. Uggowitzer et al. 1992 and Fig. 3.27). To achieve this N level by alloying alone a high content of Cr and Mn is required to enhance the N solubility (see Sect. 4.1.1). The nickel-free stainless steel Mn23Cr21Mo0.7N1 is an example (Carpenter 1998). The uptake of N is also favoured by a fully austenitic solidification and by the lower solidification temperature of high alloy grades. If the substitutional part of $Ni_{eq}$ is made up mainly by Ni instead of Mn, pressure or powder metallurgy (Sects. 4.1.2 and 4.1.3) are usually required to approach 1 mass% of N. However, most austenitic HNS do with less nitrogen and are conventionally melted unter atmospheric pressure. In CrNi(Mo) steels we see N contents between 0.15 and 0.4, in CrNiMn(Mo) grades between 0.2 and 0.5, and in CrMn(Mo) alloys between 0.5 and 1 mass%. Due to Cr being a key element for the N solubility, all grades are stainless although sometimes used in high strength, non-magnetic and other applications.

Compared to conventional stainless austenitic steels based on substitutional elements and a low interstitial content, stainless HNS depend to a considerable extent on N to stabilize the austenite. This entails some changes in manufacturing properties. As the diffusion coefficient of N is orders of magnitude above that of substitutional elements, the partitioning of alloying elements in the ferritic-austenitic region after solidification is more readily recovered by the former upon further cooling to the austenitic region. Thus a local delay of the ferrite to austenite transformation is less liable, which helps to avoid a precipitation of $\sigma$- or $\chi$-phase in the ferrite during welding. On the other hand a precipitation of chromium nitride $M_2N$ is the more likely to occur the higher the N-content (see Fig. 2.22). These precipitates usually start at grain boundaries during cooling from casting, hot working, solution annealing or welding and between 900 and 800 °C the incubation time is shorter for CrNi than for CrMn steels and often below one minute. Upon slow cooling, e.g. in heavy cross-sections, the precipitation extends into the grains in a discontinuous manner forming a lamellar array of austenite and nitride. This microstructure is called cellular, pseudo or nitrogen pearlite or just "pearlite". Compared to low-interstitial austenitic steels the properties of some HNS are more size-dependant with respect to this type of deteriorating precipitation.

As shown in Sect. 2.4.3 nitrogen retards the precipitation of intermetallic phases. This is used to increase the CrMo content and raise the corrosion resistance leading to superaustenitic CrNiMnMo steels with PREN > 40 (see Eq. 3.6). Melting under pressure allows to further explore this field but at the expense of hot-workability. The extremely high hot strength of such rich alloys combined with segregation, element partitioning in the A+F region and ready precipitation impedes manufacturing and limits some developments. The higher yield and flow strength of HNS makes cold working and machining more difficult. In cold forging the required stresses may exceed the strength of tools.

Nakamura et al (1998) searched for stainless austenitic steels without a substitutional part of $Ni_{eq}$ using only Cr and N. The latter was introduced by solid state diffusion. At 23 mass% Cr and 0.9 to 1 mass% N a rather stable austenitic structure was obtained after quenching from 1200 °C in water which revealed an 0.2% proof strength of 700 MPa and about 50% fracture elongation (Takaki et al.

1998). However, the incubation time for $M_2N$ precipitation at 900 °C was down to about 10s but increased with the addition of 5 mass% Mn to about 50 s. Some assistance by substitutional elements like Mn and Ni therefore seems to be crucial in stabilizing the austenite in thicker products. This is corrobborated by Ustinovshikov et al. (1996) who investigated steels with 18 mass% Cr and 0.6 to 1.3 mass% N. They found that ≥ 0.9 mass% N has to be dissolved to stabilize the austenite down to room temperature requiring a solution temperature of 1200 °C. A lower N content leads to a structure of austenite with martensite lamellae. After aging for about 2 min at 600 °C a distinct change of hardness and lattice parameter is already observed depending on the as-quenched microstructure. The high solution temperature and the low stability of austenite does not speak for an application of CrN austenite. As shown in Sect. 2.5 ordering and austenite stability are enhanced, if N and C are simultaneously dissolved in steel Cr15Mo1N0.3C0.3. Maybe this result of a martensitic steel can be transferred to an austenitic CrNC grade.

The previous remarks dealt mainly with austenitic HNS containing N in solid solution for use in the range of climatic temperatures. However, we also encounter N intentionally precipitated as nitrides in steels for service at elevated temperature. This development is aimed at precipitates of higher stability compared to standard steels and at a contribution of N in solid solution.

In the following sections different fields of austenitic HNS will be discussed starting with standard stainless CrNi(Mo)N grades. Next come low-cost CrMnN steels, high strength grades and Ni-free CrMnMoN alloys for body friendly applications. Other areas of interest are alloys of highest corrosion resistance, non-magnetic steels for cryogenic service, steels for outlet valves in combustion engines and creep resistant HNS. Finally Ni alloys of high nitrogen content will be considered. Although not steels, those materials are of respective fcc "austenitic" structure.

## 5.2.1 Standard stainless steels

Standard austenitic CrNi and CrNiMo steels contain about 0.05 mass% C which is a compromise of contrasting aims. Interstitial carbon enhances the yield strength and the stability of austenite. However, upon heating, e.g. during stress relief annealing, hot straightening or welding, a grain boundary precipitation of $M_{23}C_6$ carbides rich in Cr may lead to a Cr depletion of the adjacent matrix causing intercrystalline corrosion (IC) in service (Sect. 3.2.2.1). One way to avoid this is an extra low carbon (ELC) level which is connected with higher melting costs and a loss of yield strength. This is where N comes in, which does not tend to segregate into grain boundaries thus retarding the precipitation of chromium nitrides. At about 0.15 mass% N the strength is distinctly increased without IC impending. This development is demonstrated in Fig. 5.15 for two standard grades, one without Mo and the other with about 2.2 mass% Mo. It is evident that a change of interstitial content is bearing much more on the 0.2% proof strength than the variation in Mo and Ni of which the latter has to be adjusted to subdue δ-ferrite.

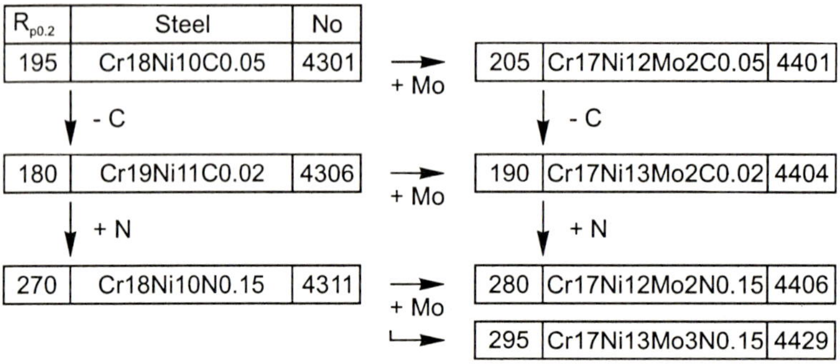


**Fig. 5.15.** Schematic representation of minimum proof strength $R_{p0.2}$ (MPa), steel grade and grade number according to DIN standard 17440 showing the effect of an extra low carbon (ELC) level and the addition of N in Mo-free and Mo-containing steels (after Gümpel et al. 1988)

Lowering the test temperature to liquid He increases $R_{p0.2}$ of Cr18Ni10N0.15 about 2.5 times as much as that of Cr18Ni10C0.05. This stronger temperature dependence of the nitrogen grade extends to higher temperatures (see also Fig. 3.3) and diminishes the advantage of this steel with respect to strength at elevated design temperature (Gümpel et al. 1988). Although the strength at room temperature is raised by N the ductility remains at a high, constant level (Fig. 5.16a). The notch impact toughness is high, too, but less relevant for sheet. Cold working provokes a partial transformation from austenite to martensite which is impeded by N. At a given content of martensite the strength is considerably raised by nitrogen, though, underlining the solid solution hardening of both phases (Fig. 5.16.b). The intentional strengthening of lean unstable austenite is discussed in more detail in Sect. 5.3.2. The unintentional implications affect the formability of standard steels especially during deep drawing and stretch forming of sheet.

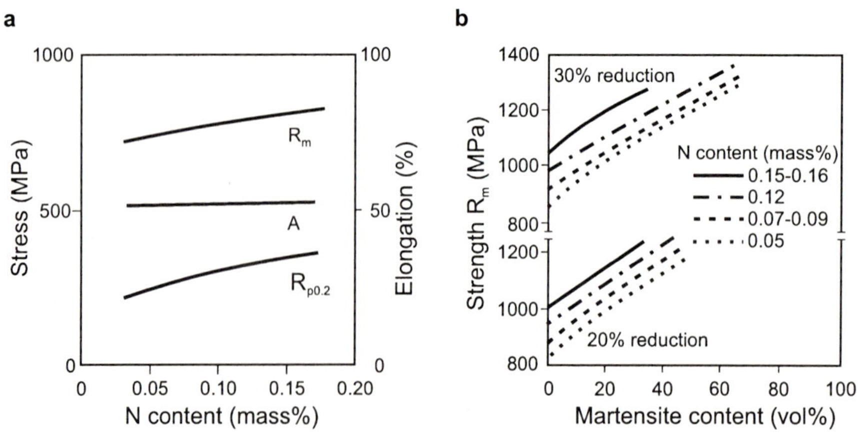


**Fig. 5.16.** Effect of nitrogen on the ultimate tensile strength $R_m$, the proof strength $R_{p0.2}$ and the fracture elongation A of standard austenitic CrNi steel, **a** solution annealed, **b** cold rolled at 70 °C (Tanaka et al. 1983)

A gradual formation of martensite during cold forming supports the plasticity of austenitic steels, because necking is suppressed as well as the formation and extension of microcracks (Becker et al. 1986). The formation of martensite is likely to occur at local tensile stress concentrations which are relieved by the volume increase. A low ratio of $R_{p0.2}/R_m$ is accompanied by a high uniform elongation which raises the stretch formability of unstable austenitic steels above that of stable ones. As nitrogen promotes the phase stability of austenite the forming parameters have to be adjusted accordingly. A lower deformation rate and less friction reduce the forming temperature and thereby the phase stability thus counteracting some of the nitrogen effects. Another problem are cracks appearing some time after deep drawing. Less martensite, by the addition of nitrogen, is bound to reduce this risk. But a higher interstitial content entails a rise of hardness which is much more pronounced in the martensite and likely to increase the risk of delayed cracking.

As far as the corrosion resistance of standard austenitic steels is concerned the distinctly reduced sensitivity of N alloyed steels to IC as compared to carbon grades is turned into higher strength and easier manufacturing. Pitting, the other frequent type of localized corrosion, is considerably retarded by N, the PREN factor of which (Eq. 3.6) is close to 30 at the N level of the present steels. This and the improvement of strength has even led to an elevated N content in some standard steels which do not disclose this addition in their chemical analysis.

***Application:*** The production volume of standard steels with nitrogen (Fig.5.15) is far behind that of grades without. The reason is that standard austenitic steels are often applied because of better availability, weldability and toughness compared to ferritic steels. In these cases their corrosion resistance and strength are not fully exploited and the advantages offered by N are required only in specific application. Furthermore strength is often derived by deformation, e.g. in deep drawings, bolts and other cold forged parts. This is different from a pressure vessel in the chemical industry, where the higher yield strength of solution annealed plate caused by N is quite welcome in meeting the design stresses. Flat parts stamped out of sheet are often embossed to give stability which reduces the importance of additional strength by nitrogen. In contrast castings, hot forgings and machined parts rely solely on the yield strength of the solution annealed state, which is improved by nitrogen. In all, nitrogen plays its part in standard austenitic steel, but not as exclusively as in grades of high strength (Sect. 5.2.3) or high corrosion resistance (Sect. 5.2.5) where it cannot be substituted by appropriate design or manufacturing.

### 5.2.2 Low cost stainless steels

***Development:*** To lower the alloying costs Ni is replaced by N and Mn which are part of the $Ni_{eq}$ in Fig. 5.1. In addition Mn enhances the stability of austenite indirectly because it raises the solubility of N quite in contrast to Ni (see Table 2.1). To keep the production costs low melting and solidification in air of normal pressure is a prerequisite which also improves the weldability. In an early attempt to save Ni Rapatz (1941) proposed the steels Cr18Mn10Ni1.5N0.2C0.12, Cr20Ni5.5N0.2C0.07 and similar grades with Mo and/or Nb.

A look at the FeCrMnN phase diagram (see Cotton et al. 1998 I) reveals a large phase field of austenite at the temperature of solution annealing and 0.5 mass% N (Fig. 5.17a). The solubility of this nitrogen content in the melt at 1500 °C under an $N_2$ pressure of 1 bar is given by the dashed line. Alloys to the lower left of this line are not capable to pick up the desired N content of 0.5 mass% in the melt at atmospheric pressure. Alloys to the upper right, though, offer a sufficient solubility. After conventional melting the steels 1 and 2 contain close to 0.5 mass% N and are fully austenitic after solution annealing. Both make use of a high Cr content to enhance the N solubility and the corrosion resistance but stay clear of δ-ferrite.

By raising the N content to 1 mass% the field of $M_2N$ precipitation in austenite is quite extended (Fig. 5.17b). As the appearance of δ-ferrite is shifted to a higher Cr level a good-size austenitic field still remains for solution annealing. But a much higher CrMn level is required to dissolve the doubled N content in the melt. Steel 3 is positioned so that 1 mass% N is dissolved during melting and homogeneous austenite without δ-ferrite or nitrides appears after solution annealing.

The three steels in Fig. 5.17 demonstrate that the N content of the melt is retained in the solid state, although a ferritic-austenitic solidification is observed for 1 and 2 (Fig. 4.15b). In Sects. 2.3.3 and 4.1.2 the reduction of N solubility connected with the appearance of δ-ferrite was discussed leading to a liberation of $N_2$ bubbles if the counteracting pressure is too low (Eq. 4.8). The solubility during solidification of the present steels is shown in Fig. 5.18. The nitrogen content of the melt is reached by blowing $N_2$ or by adding nitrided ferro-chromium and is equilibrated by the partial $N_2$ pressure of 0.8 bar in air under a slag. During

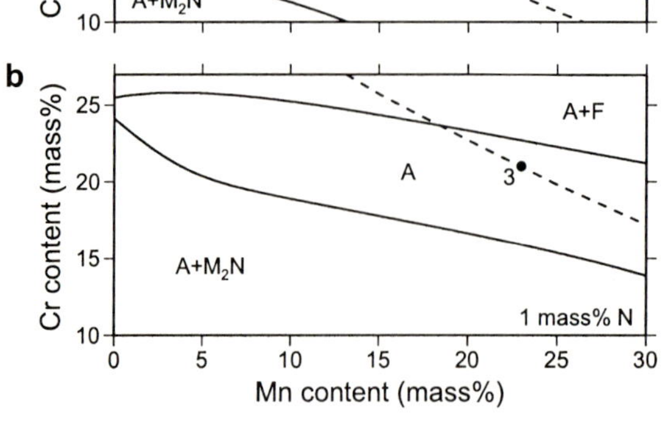


**Fig. 5.17.** Sections of the FeCrMnN phase diagram at 1050 °C derived by ThermoCalc, **a** for 0.5 mass% N, **b** for 1 mass% N. The dashed lines indicate the solubility of N at 1500 °C under an $N_2$ pressure of 1 bar. The numbers designate the steels 1 = Cr19Mn10N0.5, 2 = Cr18Mn18N0.6, 3 = Mn23Cr21N1

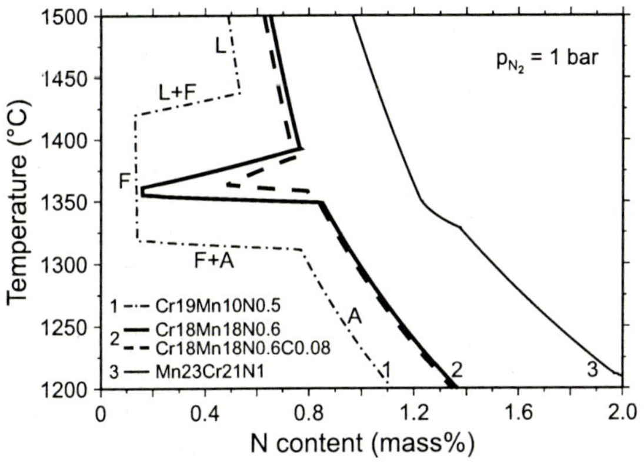


**Fig. 5.18.** Nitrogen solubility during solidification derived by ThermoCalc, steels 1 and 2 reveal a solubility gap caused by the formation of ferrite while steel 3 solidifies to austenite without a gap

solidification the atmospheric pressure is 1 bar and the ferrostatic pressure 0.2 bar assuming a feeder 250 mm high. Apparently there is a sufficient contribution by the term 2 $\sigma$/r of Eq. 4.8 because little nitrogen is lost during solidification and pores are rare.A carbon impurity helps to bridge the gap. Steel 3 shows an austenitic solidification without a gap. In contrast to a martensitic steel with (mass%) 16 Cr and 1 Mo in Fig. 4.6 the high alloy content of the present austenitic grades makes it possible to start with an equilibrium $N_2$ pressure which is more than 6 times lower and get by the gap without raising the atmospheric pressure (see also Feichtinger and Stein 1998).

The mechanical properties of low cost CrMnN steels are quite impressive. The 0.2% proof strength is 2.5 to 3.5 times as high as of standard steel Cr18Ni10 and yet the ductility is hardly impaired (Table 5.8). A high ISO-V impact toughness may be obtained at room temperature (Rennhard 1998). Cooling to -80°C reduces the toughness of steel 1 from about 280 to 70 J. Steel 2 is less sensitive to deep freezing (Fig. 3.27). Compared to Ni, manganese has a detrimental effect on the resistance to localized corrosion, which may be more than compensated by the higher N content soluble in CrMn steels (Jargelius - Pettersson 1998). Critical pitting and crevice temperatures above those of steel Cr22Ni13Mn5 are reported for steel Mn23Cr21Mo0.7N1 (Carpenter 1998).

**Table 5.8.** Approximate mechanical properties of three solution annealed CrMnN steels with less than 0.08 mass% C,
steel 1 Zapp (1997), steel 2 VSG (1996), steel 3 Carpenter (1998)

| | steel | $R_{p0.2}$ (MPa) | $R_m$ (MPa) | A (%) | Z (%) |
|---|---|---|---|---|---|
| 1 | Cr19Mn10N0.5 | 550 | 900 | 50 | 65 |
| 2 | Cr18Mn18N0.6 | 550 | 900 | 50 | 65 |
| 3 | Mn23Cr21Mo0.7N1 | 600 | 950 | 50 | 65 |

***Application:*** The high work hardening capacity of CrMnN steels is used to further raise the strength, which shall be discussed in the next section. By addition of Mo the corrosion resistance is enhanced to the level of Ni-free, body friendly applications (see Sect 5.2.4). The present section deals with work hardening in the surface of wear parts. Impact and sliding, e.g. of mineral particles, causes plastic deformation in a surface zone. The higher the resulting work hardening rate, the higher the surface hardness. In this respect N is superior to C because it lowers the stacking fault energy (see Sect. 1.1.5.1 and Fig. 1.33). An interesting aspect is the self-renewal of the strengthened layer on a tough substrate during service. This is an advantage over through-hardened martensitic steel of equal hardness. The latter is often not achieved because of weldability limiting the C content and thereby the initial hardness of wear resistant martensitic steels which are not prone to work harden.

A self-hardening surface is successfully used in wear resistant, austenitic Hadfield steel Mn12C1.2 and similar alloys like Mn17Cr2C1.4 or Mn6Mo1C1.4 which are not corrosion resistant. It is, however, well known that mineral wear is higher under wet conditions (Uetz and Sommer 1986). In numerous wear situations we do not only deal with humidity or water but with acidity, $Cl^-$ ions and elevated temperatures like in deep pit mines. Here, corrosion during service and stand still contributes considerably to the mass loss and may impair the function of machinery.

This is where stainless CrMnN steels come in. Compared to CrNiN grades more N may be dissolved and the work hardening effect of this element is more pronounced in combination with Mn (Sect. 3.1.3). The use of carbon for solid solution strengthening like in Hadfield steel is rather limited in the presence of a stainless Cr level because of clustering and carbide precipitation (see e.g. Fig. 4.2). Atomic ordering improves the solubility of N, though, which is another advantage of CrMnN steels. The lower the alloy content and the higher the deformation in the wear surface, the more martensite may be formed, the hardness of which is expected above 60 HRC at $N \geq 0.5$ mass %. Solid solution hardening of bcc martensite is more pronounced than that of the fcc austenite and accompanied by an increase of volume, which entails compressive residual stresses at the surface. They are believed to contribute to the abrasion resistance because the transformation of retained austenite in the whole cross-section by deep-freezing before wear does not yield as high a wear resistance as a wear-induced transformation in the surface (Berns 1998).

Wear parts are often made of sheet or produced as castings. They are used e.g. in mineral extraction and processing, in the mining and cement industry, in road construction and the food industry. Sheet is part of bunkers, buckets, chutes, chains and the load area of lorries. Cold formability and weldability are requirements, which are usually met by low cost stainless CrMnN steels like Cr19Mn10N0.5 (Zapp 1997). Cold forming of sheet may require a somewhat higher force and welding is carried out with a low heat input to avoid nitride precipitation. In some of these applications work hardening of the wear surface is not fully developed and the inherent wear resistance is not exploited. However, the corrosion resistance is usually quite sufficient and distinctly above the low-cost ferritic stainless steel Cr11 which is sometimes applied if corrosion prevents the use of weldable high-strength-low-alloy (HSLA) steels.

Castings serve as crushing tools like jaws, rollers, rings and hammers. Rennhard (1998) reported a hardness increase from 25 to > 60HRC in the surface of shredder hammers with (mass%) 24Cr, 20 Mn, < 1.2 N and alternative additions of Ni and Mo to adjust the corrosion resistance. Up to 1 mass% of V and Nb was added by Cotton et al. (1998 II) to increase the N solubility and to form MN precipitates. These rotating tools are used to desintegrate cars and shred bodies, engines, gears and axles to hand-size bits. The severe impacts result in an optimal work hardening of the wear surface. Conventionally a shredder hammer is made of a steel with e.g. (mass%) 1.7 Ni, 1.2 Si, 0.7 Cr, 0.3 Mo, 0.1 V and 0.55 C, which is fully hardened and partially tempered at the bore end to avoid brittle failure in service that would damage the shredder severely. A high strength austenitic steel is capable of bearing the stress concentration at the bore avoiding its expansion and yet offers superior toughness. The self-hardening tool edge allows a high wear depth.

### 5.2.3 High strength stainless steels

Strength and toughness are related to size which is demonstrated by two examples: Microalloyed pearlitic steel cord, reinforcing tires, reaches 4000 MPa after cold drawing to 0.15 mm in diameter. Maraging steels designed to obtain 2000 MPa just by heat treatment, even in heavy cross-sections, lose toughness by segregation as the size increases. The size relation applies to stainless steels as well. In the two previous sections nitrogen was used to raise the yield strength of the solution annealed state and a limitation of size arises out of nitride precipitation which is more likely to occur at the lower cooling rate of a larger cross-section, the incubation time according to Harzenmoser (1990) being about an order of magnitude shorter in CrNiN steels as compared to CrMnN ones.

***Development:*** Work hardening was recognised to be especially effective in the wear surface of CrMnN steels. Therefore cold working will now be considered to raise the strength level of products, the cross-section of which becomes the smaller the higher the cold reduction is considered. A fine grain size (Fig. 3.12) and aging (Fig. 3.25) are employed to lift the strength further. A partial transformation to martensite is not intended here but in Sect. 5.3.2. Although contributing to strength martensite raises the ductile-to-brittle transition temperature (DBTT) and promotes hydrogen embrittlement as well as stress corrosion cracking. As explained in Sect. 3.1.3 CrMn steels are particularly suited for cold worked, high strength applications because of superior N-solubility, intense solid solution hardening and a low stacking fault energy which enhances the work hardening rate by texturing and twinning.

The low-cost steel Cr18Mn18N0.6 introduced in the previous section was selected by Uggowitzer and Speidel (1991) to demonstrate the additive effect of the different strengthening mechanisms. Grain refinement by deformation and recrystallisation plus subsequent cold drawing and aging led to wire with a yield strength of about 3400 MPa which is within the range of pearlitic steel wire but

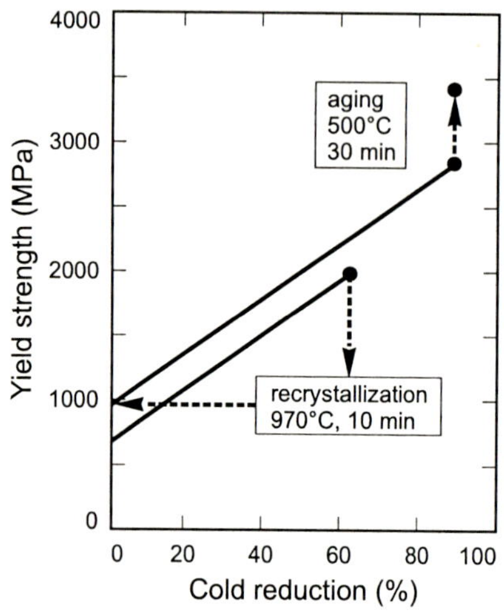


**Fig. 5.19.** Strengthening of steel Cr18Mn18N0.6 by wire drawing. Initial grain size 30 μm, after first reduction and recrystallization down to 8 μm. Final reduction and aging, after Uggowitzer and Speidel (1991)

offers corrosion resistance as an additional property (Fig. 5.19). The well known creep of austenitic steels at room or slightly elevated temperature and high stress leading to an elongation ε in time t along $\varepsilon = a \log t + b$ (Gümpel et al. 1988) is effectively reduced by cold working of CrNi steels (Schmidt et al. 1986) and CrMn steels, the creep resistance of which is improved by N already in the solution annealed state (Uggowitzer 1993).

In thicker cross-sections cold working is limited to a lower level and strength is supported by a higher N content which is introduced by pressure melting. The simultaneous effect of N content and cold working is shown in Fig. 5.20 (see also Sect. 3.1.3 and Fig. 3.14). In a first step steel Cr18Mn18N0.6 was turned into

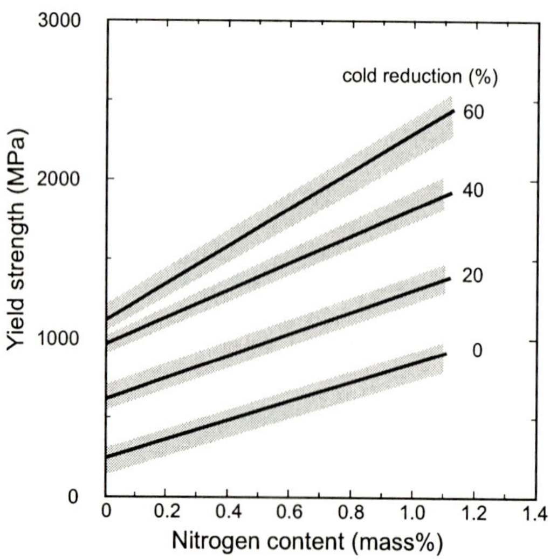


**Fig. 5.20.** Effect of nitrogen and cold working on the yield strength of austenitic CrMnN and CrNiN steels with a grain size of 50 to 150 μm after solution annealing, after Uggowitzer and Speidel (1991)

Cr18Mn18N0.9 by PESR (see Sect. 4.1.2). Next the resistance to pitting corrosion was improved by adding Mo which required the high N content provided by pressure melting to prevent δ-ferrite. This led to steel Cr18Mn18Mo2N0.9 and further to steel Cr16Mn14Mo3N0.9 (VSG 1996).

Although the yield strength of solution annealed austenitic steels is raised by a factor of more than 3 if the N content increases from about zero to 0.9 mass%, the fracture toughness remains unchanged at a high level of $\geq 500\,\text{MPa}\sqrt{\text{m}}$ (Fig. 3.26). After cold deforming steel Cr18Mn18N0.6 by 40 % this value drops to $\approx 200\,\text{MPa}\sqrt{\text{m}}$ which is an extremely good toughness at a yield strength of $\approx 1350$ MPa (Speidel 1989). After cold working Cr18Mn18N0.6 and Cr18Mn18N0.7 to a yield strength of about 1100 MPa no stress corrosion cracking (SCC) was found by Speidel (1981), neither in aerated water nor in chloride or nitrate solutions up to 90°C. Even after raising the N content to 1 mass% and the cold worked yield strength to ≈ 1550 MPa no SCC was observed (Magdowski and Speidel 1989). However, at temperatures above 90°C a crack growth rate increasing with temperature was measured up to 288 °C (Pedrazzoli and Speidel 1990).

As discussed in Sect. 3.1.4 cleavage along non-active slip planes is found in stable austenitic CrMnN steels below room temperature, the DBTT being the higher the greater the N content (Fig. 3.27). For notched parts at winterly temperatures the higher strength of PESR steels cannot be fully exploited. A partial substitution of N by Ni improves the sub-zero toughness at the expense of strength. A higher Cr content helps to make up for the loss of N solubility by Ni and to get along without pressure metallurgy. Steels for cryogenic temperatures like Cr24Ni15Mn4N0.35 are treated in Sect. 5.2.6.

Ikegami and Nemoto (1996) combined grain refinement and work hardening by controlled thermomechanical hot working to a low finish temperature followed by accelerated cooling. They forged a billet of 140 mm square between 1100 and 850 °C to 80 mm square and then to 85 mm round ending below 800 °C. Compared to the solution annealed state the 0.2% proof strength changed from 540 to 1180 MPa, the fracture elongation from 51 to 22% and the ISO-V impact energy from 288 to 43.2 J. This work reveals that thermomechanical treatment may be a way to circumvent the size restrictions of cold working mentioned above.

***Application:*** High cold reduction and strength is achieved in the products e.g. by cold drawing of wire or tube or cold rolling of strip. Stainless leaf and coil springs, reinforcing wire of rubber composites like belts or high pressure hoses, wire and tube for medical applications, supporting wire in over-ground power transmission, membranes and flexible tubes are some of the possible applications of CrMnN steels, the corrosion resistance of which may be further enhanced by Mo.

Fastening elements are an interesting field of applications e.g. in the building industry. Air pollution and the continuous change of weather from dry to wet to dry promotes the formation of a crust on the steel surface which accumulates aggressive ions, so that standard CrNiMo steels are required to cope with general corrosion as well as with crevice and pitting corrosion. However, these steels are of low yield strength and not resistant to SCC. This is where the superior combination of high strength and toughness of HNS in the range of climatic temperatures comes in, which is accompanied by resistance to SCC. Heavy plates of concrete or natural stone are attached to buildings and the steel fasteners like bolts, suspensions and supports are destined to safe, long time service. This needs more attention. The manufacturing of bolts by cold forging of HNS wire may overstress the tools and requires an appropriate design. Some shapes may need additional machining.

Electric generators in power plants rotate at 1500 to 3600 rpm depending on the frequency and the number of poles. To keep the wiring in place, a retaining ring is shrink fitted to each end of the shaft. These rings bear a high centrifugal force entailing a tremendous hoop stress, which can only be met by a high specific strength $R_{p0.2}/\rho$ of the ring material. Because of elevated service temperatures it is difficult to reduce the density $\rho$ e.g. by taking to Al alloys or reinforced polymers. Therefore high strength steel has been used which has to be non-magnetic to avoid induction heating and a loss of strength and shrinkage. The solution is a stable austenitic steel strenghtened by cold expansion to $R_{p0.2} > 1000$ MPa without formation of martensite. Steel Mn18Cr4C0.5 was widely used until failures occurred due to SCC destroying whole plants. This initiated the development of steel Cr18Mn18N0.6 and of the PESR process to reach an even higher N content like in steel Cr18Mn18N0.9.

The capacity of power plants is limited by the size of the rotor and the strength of the retaining rings, which measure up to 2 m in diameter at a length and wall thickness sometimes exceeding 1 and 0.1 m respectively. Steel Cr18Mn18N0.6 is cold worked to $R_{p0.2} \leq 1350$ MPa requiring up to about 50 % expansion (Fig. 5.21). A higher deformation is risky because of fracture during manufacturing. With the higher N content of steel Cr18Mn18N0.9 the same proof strength is reached at about 20% less expansion which facilitates $R_{p0.2} > 1350$ MPa. At a shrink temperature of ≈ 400 °C recovery and aging determine the final strength, although the cold reduction is not as high as in Fig. 5.19. The circumferential

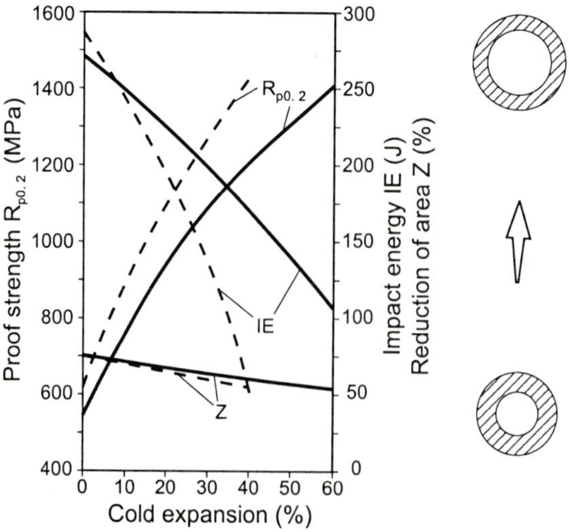


**Fig. 5.21.** Cold expansion of retaining rings made of steel Cr18Mn18N0.6 (full lines) or Cr18Mn18N0.9 (dashed lines) and resulting properties (after Stein 1987)

elongation of the rings during expansion entails anisotropic properties. As to the ratio of radial/tangential properties e.g. at $R_{p0.2}$ = 1200 MPa Stein (1987) gives 0.7 to 0.85 for $R_{p0.2}$, 0.65 to 1.1 for fracture elongation $A_5$ and 0.4 to 0.75 for ISO-V impact energy. A similar anisotropy was reported for $R_{p0.2}$ by Orita et al (1990). The fracture toughness $K_{Ic}$, however, remained in a common scatter band almost independent of radial or circumferential crack extension. At $R_{p0.2} \approx 900$ MPa the $K_{Ic}$ of retaining rings, derived from $J_{Ic}$, amounted to about 350 and at $R_{p0.2} \approx 1400$ to about 200 MPa $\sqrt{m}$ .

In recent years the temperature resistance of insulating materials for the electric wiring was raised to about 130 °C which permits to increase the service temperature of rotor and retaining ring. This calls for a higher resistance to pitting corrosion, because the stress concentration at pits promotes SCC. Steel Cr16Mn14Mo3N0.9 is a promising candidate (Stein et al. 1998).

## 5.2.4 Body friendly steels

The previous sections dealt with Ni-free CrMnN steels which are also the basis of body friendly alloys. About 10% of the human population are allergic to Ni, which is transferred from Ni coatings like on spectacle frames, from Ni alloys like coins and bracelets or from stainless austenitic CrNi steels used e.g. for wrist watches. Although the latter are corrosion resistant, enough Ni ions may be dissolved by sweat or body fluid to cause local inflammations of the skin or tissue. According to Menzel et al (1996) medical requirements call for a Ni content of less than 0.2 mass%. Uggowitzer et al. (1996) cite a European directive of 1994 in which the Ni content of inserts for pierced parts of the body is limited to 0.05 mass%.

***Development:*** Of the Ni-free stainless bcc steels the standard martensitic ones lack corrosion resistance, while the ferritic as well as the martensitic ones may not offer a sufficient toughness. In addition a non-magnetic material is preferred in some applications. Therefore CrMnN austenites were adapted to body friendly applications along the following lines: Nitrogen is used as the major substitute of Ni but is limited to 1 mass% to avoid cleavage-like, brittle failures (see Sect. 3.1.4). The intense solid solution strengthening by N has a beneficial effect on the fatigue resistance. The resistance to pitting and crevice corrosion is considerably increased by nitrogen. The PREN (Eq. 3.6) is also supported by Cr and more so by Mo, the latter being especially effective in relation to $Ni_{eq}$ required to prevent δ-ferrite (Fig. 5.1). However, Cr raises the N solubility more than Mo (Table 2.1). Thus high Cr levels are encountered in steels melted under normal pressure, while high Mo contents prevail in PESR grades. Manganese is added to promote the N solubility and to retard the precipitation of $M_2N$ (Takaki et al. 1998). Jargelius-Pettersson (1998 I) showed that about 10 mass% Mn impaired the pitting resistance of stainless austenitic CrNiMo steels with up to 0.25 mass% N leading to a term of −1.6Mn in the PREN. However, this trend seemed to be reversed by raising the N content to 0.5 mass%. She concluded that the detrimental effect of Mn may be more than compensated by the higher N contents attainable in Mn-alloyed steels.

An example of the first group of steels melted under normal pressure is Mn23Cr21Mo0.7N1. Pressure melted steels aiming at increasing PREN by Mo are Cr18Mn18Mo2N0.9 and Cr16Mn14Mo3N0.9 (Menzel et al. 1996). A PREN of > 45 is achieved by (mass%) 15 to 18 Cr, 3 to 6 Mo,10 to 12 Mn and about 0.9N (Uggowitzer et al. 1996). The proof strength of these solution annealed steels is in the range of 600 to 750 MPa. The impact toughness at room temperature is high but deteriorates to the cooler side. Further strengthening is reached by cold working as shown in the previous section.

Body friendly steels belong to the low cost grades because they do without expensive Ni and more so if they are melted under normal pressure. But in contrast to some applications in Sect. 5.2.2 a high corrosion resistance is a prerequisite in most body friendly applications. The present steels are also part of the high strength group discussed in Sect. 5.2.3 because their CrMnN base offers an exceptional rate of work hardening. Ni-free stainless steels were reviewed by Speidel (1998), who also discussed duplex grades (see Sect. 5.3.3).

***Application:*** Stainless steel parts may be in contact with the skin of the human body or with the tissue inside. The duration and frequency of contact vary from short and rare to permanent. A handrail in a store does not call for a Ni-free grade. But zippers, rivets or costume jewellery touch an area of the skin for a prolonged time. Wrist watches or rings may stay in place almost permanently. Dental braces are attached for an extended period of time. They mark the transition from outside to inside of the body. Springs of cold drawn wire provide the required force to align the teeth. Splints and screws usually stay in the body for a limited time but plates and stents mostly remain there permanently. As allergies are on the rise a growing market for body friendly alloys is in view. The PESR steels Cr18Mn18Mo2N0.9 and Cr16Mn14Mo3N0.9 are currently applied to some of the above areas (Stein et al. 1998).

## 5.2.5 Steels of high corrosion resistance

In Sect. 5.2.2 to 5.2.4 a CrMn base composition was used to achieve low cost, high strength and body friendly behaviour. Due to the negative effect of Mn on the corrosion resistance we now come back to a CrNi base composition and tie on to Sect. 5.2.1.

***Development:*** To increase the resistance to pitting and crevice corrosion a high CrMo content is required, the latter element being more effective (Eq. 5.1). The resulting increase of $Cr_{eq}$ in Fig. 5.1 calls for about an equal rise of $Ni_{eq}$. In this respect N offers several advantages over Ni: (a) Nitrogen shifts the formation of embrittling σ-phase to lower temperatures and longer times thus diminishing the collateral effect of higher CrMo contents far more than Ni. Uggowitzer et al. (1994) saw the upper bound of σ in austenitic CrNi steels at slightly below 950 °C and calculated the shift by 1 mass% of N, Cu, Ni, Mn, Cr, Mo at -320, -35, -20, -15, +50, +70 °C, respectively. (b) Nitrogen distinctly enhances the resistance to pitting and crevice corrosion (Eq. 3.6). (c) Solid solution strengthening by N is quite pronounced (Sect. 3.1.1). (d) Costly Ni is replaced by N. Acknowledging the benefits of N the solubility of this element is supported by increasing the Cr content far above the 17 mass% of standard steels and by a moderate addition of ≤ 10 mass% Mn. With respect to corrosion resistance Cr and Mn are opposing each other, while they are in line in the σ-phase, although Mn is less critical. As the N content is raised to subdue the formation of intermetallic phases like sigma, chi and Laves, the precipitation of another brittle phase is promoted, i.e. chromium nitride $M_2N$ (Sect. 2.4.3). It starts to appear on grain boundaries and may grow discontinuously into the grains during cooling after hot working or solution annealing if the cross-section is too large to allow for a sufficient cooling rate.

Table 5.9 contains a few examples of steels that characterize the development of recent years. Number 1 denotes a standard grade mentioned in Sect. 5.2.1 which serves as a basis. In steel 2 the N content is enhanced by Cr and Mn. The content of Ni is raised to suppress δ-ferrite. In the next step the Mo content is lifted from 3 to 4.5 mass% (steel 3). A further increase of Mo entails a reduction of Cr (steel 4) unless the assistance of N is fully exploited (steels 5 and 6). However, the latter requires pressure melting and the resulting N content of about 0.8 mass% afflicts costs, hot working, and welding. Comparing the first six examples one recognizes that by simultaneous increase of Cr, Mo and N balanced by Ni and Mn the PREN and the proof strength are doubled. Steels 3 and 5 in relation to 4 emphasize the key part of N. The steels of high CrMo content with a PREN of more than about 40 are often called superaustenitic.

Steel 7 in Table 5.9 stands for a different concept. Since Mo is known to accelerate the precipitation of intermetallic phases it is kept at a moderate level. Manganese is omitted because it impairs the corrosion resistance. Instead Cr is raised to 33 mass% providing a high corrosion resistance and nitrogen solubility, the latter being attenuated by 31 mass% Ni added to suppress δ-ferrite. Embrittling precipitation is shifted to a lower temperature range and retarded.

**Table 5.9.** Austenitic steels of high corrosion resistance, solution annealed at 1100 to 1150 °C. Number 6 is a PESR steel proposed by Uggowitzer et al. (1994), the remainder are commercial alloys melted under normal pressure

| Steel No. | Approximate composition (mass%) | | | | | PREN (Eq. 3.6) | $R_{p0.2}$ [a] (MPa) |
|---|---|---|---|---|---|---|---|
| | Cr | Ni | Mn | Mo | N | k = 16 | |
| 1 | 17 | 13 | - | 3 | 0.15 | 29 | 295 |
| 2 | 23 | 17 | 6 | 3 | 0.4 | 39 | 420 |
| 3 | 24 | 18 | 6 | 4.5 | 0.4 | 45 | 440 |
| 4 | 20 | 18 | - | 6 | 0.2 | 43 | 300 |
| 5 | 24 | 22 | 3 | 7 | 0.5 | 55 | 450 |
| 6 | 24 | 20 | 7 | 7 | 0.8 | 60 | 600 |
| 7[b] | 33 | 31 | - | 1.6 | 0.4 | 45 | 380 |

[a] approximate lower bound, [b] plus 0.6 mass% Cu

This is depicted in Fig. 5.22. If steel Cr20Ni18Mo6N0.2 is low in nitrogen (0.16 mass%) and moderate in silicon (0.17 mass%) annealing at 1000 to 950 °C entails an immediate drop of notch impact energy from > 300 J in the solution annealed state to 250 J in the annealed state. After about 15 minutes the notch impact energy is down 150 J. Avoiding Si at all almost triples this duration which delineates the strong accelerating effect of this element on precipitation. Already a minor increase of N to 0.19 mass% shifts the 150 J line to about one hour although 0.22 mass% Si are present. The high content of Ni and N as well as the low content of Mo contained in steel Cr33Ni31Mo1.6Cu0.6N0.4 distinctly lower the temperature range of precipitation which is also delayed.
The deterioration of toughness between 1000 and 700 °C is accompanied by a sensitization to corrosion e.g. in boiling azeotropic nitric acid (Huey test). Gümpel and Michel (1986) found a mass loss of about 0.15 $g/m^2h$ for steel Ni24Cr19Mo5.5Cu1N0.15 in the solution annealed state. After annealing at 900 °C for 5 h the mass loss remained constant if the steel contained 0.08 mass% Si but was raised to about 7 $g/m^2h$ at 0.49 mass% Si. The mass loss of steel Cr33Ni31Mo1.6Cu0.6N0.4 was unchanged at 0.04 to 0.05 $g/m^2h$ after annealing between 1000 and 600 °C for up to 10h (Köhler et al. 1996). Adding the results of Jargelius-Pettersson (1998 I) it is apparent that nitrogen becomes the more important in suppressing sensitization the higher the content of Mo and Si. Thus some nitrogen is a prerequisite for good properties of weldments. If the nitrogen level becomes too high sensitization may increase again in part by nitride precipitation. Jargelius-Pettersson (1996) found the highest critical pitting temperature (CPT) after sensitization at e.g. 900 °C, 30 min for an intermediate nitrogen content, which was about 0.2 mass% for steel Cr20Ni18Mn5Mo4.5N and almost 0.4 mass% for steel Cr20Ni18Mn10Mo4.5 N of higher N solubility, which probably impeded $M_2N$ precipitation. A depletion of the matrix in respect to CrMo or N around precipitates enriched in these elements is thought to supplement the sensitization at grain boundaries. Truman and Lomax (1989) tested steel Cr21Ni9Mn4Mo3Nb0.3N0.4 near the surface and in the center of cross-sections up to 480 mm of thickness after solution annealing at 1100 °C followed by

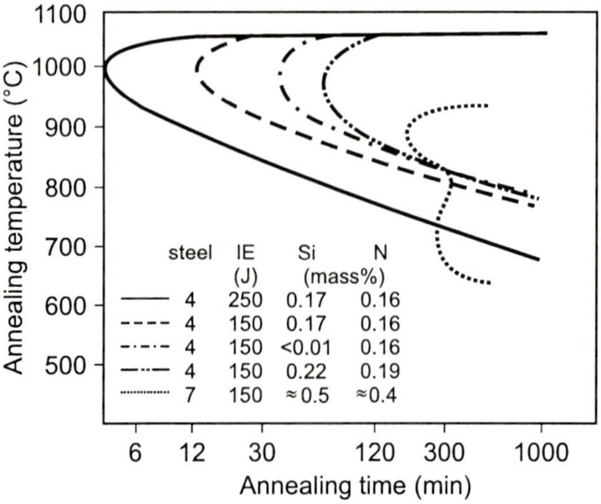


**Fig. 5.22.** Embrittlement of steel 4 = Cr20Ni18Mo6N0.2 and steel 7 = Cr33Ni31Mo1.6Cu0.6N0.4 of Table 5.9 upon annealing between 1000 and 700 °C made visible by lines of constant notch impact energy (IE) after Gümpel and Michel (1986, steel 4) and Köhler et al. (1995, steel 7)

cooling in air. Ductility and toughness were lower in transverse specimens compared to longitudinal ones and the difference increased with size. But besides this regular anisotropy hardly any deterioration was observed in the specimens of longitudinal taking. The authors conclude that little nitride precipitation occurred and N stayed mainly in solid solution. Simmons et al (1996) investigated the PESR steel Cr19Ni5Mn5Mo3N0.7 after aging at 800 °C. The precipitation of $Cr_2N$ started within minutes followed by Laves phase after about 2.5 h. At 25h sensitization reached a maximum followed by healing up to 300 h.

Between corrosion resistance and strength procured by N in solid solution on one side and embrittlement or sensitization caused by nitrides on the other, an N content of 0.4 +/-0.1 mass% currently seems to be a compromise.

***Application:*** The high resistance to corrosion, especially localized corrosion determines the use of the present steels although increased strength is welcome to meet the design stresses as in components like tubes and vessels under inner pressure. The stability of austenite induced by Ni and N does not permit the formation of martensite during cold forming or at winterly service conditions. In general a high resistance to stress corrosion cracking is encountered. Paulus et al. (1993) showed for steel Cr24Ni18Mn6Mo4.5N0.4 that a further increase of yield strength is feasible without a loss of corrosion resistance.

In flue gas desulphurization (FGD) plants energy is recovered by condensation of the flue gas causing severe general corrosion on the condenser. In the scrubber sulphurdioxide is washed from the flue gas at elevated temperature and the acid environment, contaminated by $Cl^-$ ions, causes pitting corrosion especially in the lower part of the plant, i.e. the absorber vessel and quencher. Grades of high Mo content like steel 5 in table 5.9 are a match for the Ni alloy 625 with (mass%) 22Cr and 9Mo (Hertzman 1995).

Marine environments often require superior resistance to pitting and crevice corrosion as for instance in desalination plants and condensers of power plants on-shore or reinjection piping systems and fire fighting systems on offshore platforms. The pulp and paper industry also deals with high chloride concentrations and severe general corrosion like in digesters, reheaters, coolers, scrubbers and bleaching equipment. In these fields of application the superaustenitic steels have progressed.

The chemical industry is another wide area of application for HNS of high corrosion resistance especially in those processes where strong acidity and halogen concentrations come together (Rockel et al. 1990). Steel 7 of high Cr content proved e.g. highly resistant to $H_2SO_4$ (97.5 to 99%) in the temperature range of 150 to 250 °C which is interesting with respect to heat recovery during the production of this acid. Because of its convincing resistance to mixtures of $HNO_3$ and HF this alloy is also applied to pickling plants in the metal industry (Köhler et al. 1996).

## 5.2.6
## Cryogenic non-magnetic steels

Austenitic MnC and CrNi steels are used in the range of climatic temperatures if disturbing interactions with magnetic field lines are to be avoided. Nitrogen comes in, if a higher strength is required or if service occurs at a lower temperature. An example of the former was given in Sect. 5.2.3 describing a high strength, non-magnetic retaining ring forged from stainless CrMnN steel and subsequently cold expanded. Let us now consider non-magnetic components within a strong magnetic field at cryogenic temperatures like in superconduction. The lower the magnetization of the steel, the less heating is induced by eddy-currents, the less harm is done to the field and the smaller are the mechanical forces induced by the field.

***Development:*** To begin with non-magnetic steels have to be free of δ-ferrite, even in segregated areas or after welding. Next a transformation to bcc martensite during cooling or plastic deformation is not permissible. The latter is a collateral effect of bending and blanking sheet or of machining components. Localized plastic deformation at notches during cryogenic service combines both prerequisites for a martensitic transformation. Here, nitrogen is most helpful in stabilizing the austenite, thus avoiding δ-ferrite as well as martensite, which are both ferromagnetic, i.e. easily magnetized. As delineated in Fig. 5.23 even the standard steel Cr18Ni10C0.05 may be cold worked and subsequently deep-freezed without a sign of martensite. However, deformation at low temperatures entails martensitic transformation and it takes a higher alloy content to suppress this ferromagnetic phase.

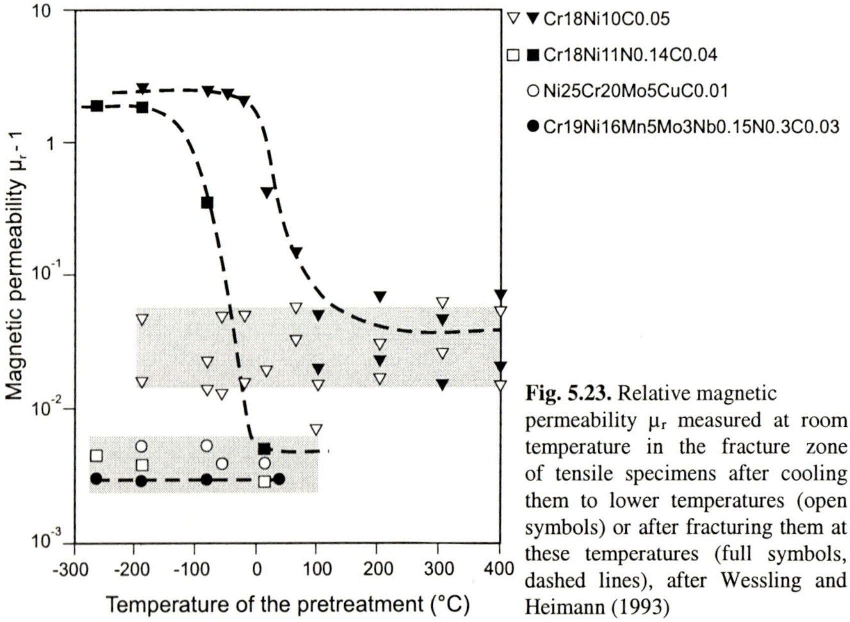


**Fig. 5.23.** Relative magnetic permeability $\mu_r$ measured at room temperature in the fracture zone of tensile specimens after cooling them to lower temperatures (open symbols) or after fracturing them at these temperatures (full symbols, dashed lines), after Wessling and Heimann (1993)

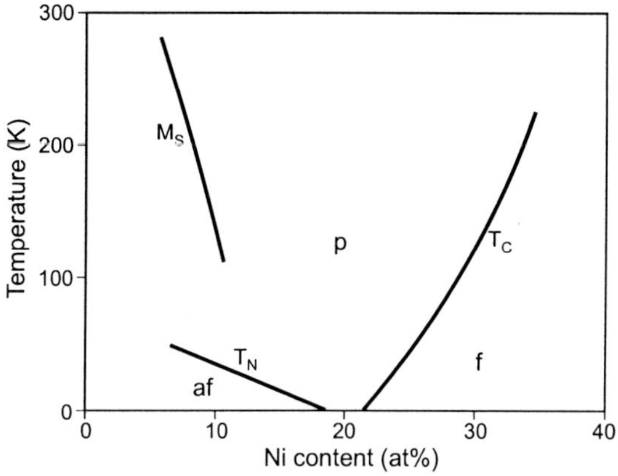


**Fig. 5.24.** Magnetic transformation of paramagnetic (p) to antiferromagnetic (af) or ferromagnetic (f) austenite at the Néel temperature $T_N$ or Curie temperature $T_C$ upon cooling a steel with 20 at% Cr and an increasing content of Ni. $M_s$ denotes the start of martensite formation, after Pepperhoff (1992)

The elements Ni, Mn, C and N are most effective in stabilizing the austenite, but are supported by the ferrite forming elements Cr and Mo to a lesser extent as can be deduced from the Schaeffler diagram (Fig. 5.1). Since most applications require stainless grades, the elements Cr, Ni, Mn, and Mo form the substitutional backbone of non-magnetic steels. Of the interstitial elements carbon is hardly used because of intercrystalline corrosion that may arise after welding. If it comes to very low service temperatures the paramagnetic austenite may transform to antiferromagnetic or ferromagnetic austenite depending mainly on the Ni content. As depicted in Fig. 5.24 it takes about 10 at% Ni in a stainless steel with 20 at% Cr to avoid martensite. The transition from paramagnetic to antiferromagnetic austenite at the Néel temperature $T_N$ decreases as the content of Ni is raised. Above 20 at% Ni the Curie transition temperature $T_C$ from paramagnetic to ferromagnetic austenite increases with the Ni content. This transformation is accompanied by an increase of the specific volume. Interstitial N expands the lattice and eases the transformation thereby increasing $T_C$, as was shown for Invar alloy Ni36N by Rochegude et al. (1993). The paramagnetic and antiferromagnetic states are termed non-magnetic, which means that they are hardly magnetizable in technically feasible fields. In spite of a high magnetization field strength H the magnetic polarization J of a non-magnetic steel with nitrogen remains at some 10mT (Fig. 5.25). In comparison $\alpha$-iron is spontaneously magnetized already by a weak field and J reaches a saturation of more than 2 T (Pepperhoff 1992). The magnetic susceptibility $\kappa$ per unit volume serves as a measure of the magnetization rate of non-magnetic steels. At room temperature the relationship between J and H is linear, i.e. $J = \kappa H$. As the temperature is lowered $\kappa$ increases until $T_N$ is reached, below which $\kappa$ declines again. Schlump (1975) pointed out that reducing the Ni content of a stainless steel from 18 to 12.8 mass% raises $T_N$ from 12 to 37 K thereby cutting the maximum value of $\kappa$ between 100 and 0 K from 7 to 2 $cm^3/10^4g$. According to Sumitomo and Nakatuka (1991) the relative permeability $\mu_r = 1 + 4\ \pi\kappa$ at 4 K is inversely proportional to $T_N$ (Fig. 5.26). In this figure $T_N$ was calculated after Jones et al. (1987). An increasing field strength lowers $T_N$ which according to Heimann et al. accounts for the curvature in Fig. 5.25 at cryogenic temperatures.

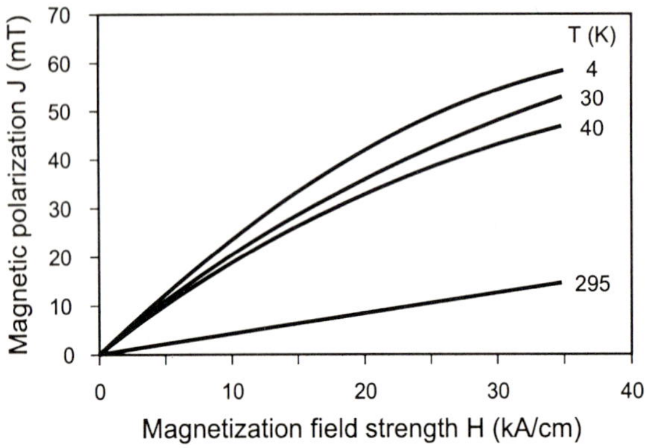


**Fig. 5.25.** Magnetic polarization of non-magnetic steel Cr19Ni16Mn5Mo3Nb0.15N0.3 solution annealed at 1100 °C and measured at room and cryogenic temperatures T, after Heimann et al. (1979)

Nickel is a key stabilizer of austenite, but beyond about 20 at % an unwanted ferromagnetic order is impending (Fig. 5.24). A lower content seems to have a beneficial effect on the susceptibility, which is reduced via an antiferro-magnetic transition. However, a low Ni content calls for other means of preventing deformation martensite in cryogenic service. In this situation nitrogen alloying offers a solution, which also enhances the yield strength. According to the equation in Fig. 5.26, N lowers $T_N$ about 10 times as much as Ni. However, with respect to the stability of austenite the factor is 15 to 30 times (Fig. 5.1). Thus stabilizing the austenite by N entails a higher $T_N$ as compared to Ni. The N solubility is raised by increasing the Cr content or by addition of Mn which raises $T_N$. The steels Cr19Ni16Mn5Mo3N0.3 (Wessling and Heimann 1993), Cr25Ni15Mn4N0.35 (Nakajima et al. 1990; Nyilas et al. 1993) and Cr19Mn10Ni6N0.35 (Sumitomo and Nakutuka 1991) are examples of this

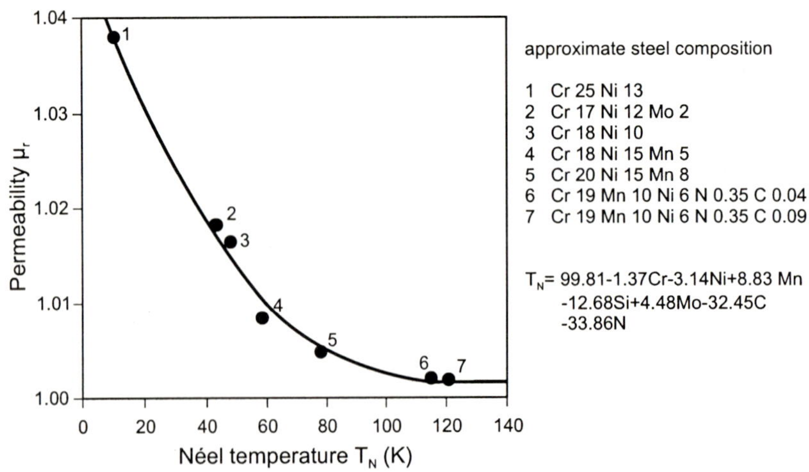


**Fig. 5.26.** Relation between relative permeability and calculated Néel temperature, after Summitomo and Nakatuka (1991)

development. In contrast to Cr18Ni11N0.14 the new steels are not prone to develop sub-zero deformation martensite (Fig. 5.23).

In Sect. 3.1.1.2 the intense strengthening of austenite by N at low temperatures was discussed. As shown in Fig. 3.3 the increase of yield strength is the stronger the lower the temperature. In comparison, C is about equal to N with respect to strengthening at room temperature but quite inferior at cryogenic temperatures (Fig. 3.8). The favourable influence of N on strength and austenite stability is limited by toughness, though. This was discussed in Sect. 3.1.4 and a ductile-to brittle transition temperature (DBTT) is encountered in some stable austenitic steels with ≥ 0.5 mass% N (Fig. 3.27). Today the N content of non-magnetic steels for cryogenic service ranges between 0.3 and 0.4 mass%, which is seen as a compromise of magnetic properties, strength and toughness.

***Application:*** According to Dietrich (1964) nitrogen was used in stainless non-magnetic steel Cr18Mn10Ni4N0.17 as early as 1957. Later N was adopted to strengthen standard steels for cryogenic service like in transport and storage of liquefied gas. The combination of both, cryogenic and non-magnetic suitability became important when superconduction entered application. The construction of electric generator rotors with super conducting windings were considered (Thien 1979) as well as superconducting magnets for magnetic levitation transport systems, for colliders in plasma physics and for nuclear fusion reactors to harness the hot plasma (Nakajima et al. 1990). In addition to a low magnetic susceptibility at cryogenic temperature a high strength is required to cope with the strong magnetic forces. These cannot be borne by the superconducting material itself which consequently has to be reinforced or framed by strong and tough non-magnetic steel components. The temperature depends on the coolant and great efforts are being made to raise the critical temperature of superconductors. The goal is a change from liquid He (4K) and $H_2$ (20 K ) to $N_2$ (78 K), which not only would save costs but also ease the requirements for non-magnetic steels. Nakajima et al. (1990) quote the following requirements for coil cases of 50 to more than 100 mm wall thickness at 4 K: yield strength $R_p > 1200$ MPa, fracture toughness $K_{Ic} > 200$ MPa$\sqrt{m}$ , ISO-V impact energy IE > 100 J, corrosion resistance = stainless (non-rusting), good workability and weldability. For a field of 12 T they calculated only a minor advantage of a non-magnetic over a saturated ferromagnetic steel and therefore the permeability was not specified. At 4 K steel Cr25Ni15Mn4N0.35 reached $R_p = 1403$ MPa, IE = 183 J and $K_{Ic} = 201$ MPa$\sqrt{m}$. These results were corroborated by Nyilas et al. (1993). By thermomechanical hot rolling with a low finish temperature and controlled cooling Ikeda et al. (1990) produced a 70 mm thick plate of steel Mn22Cr13Ni5N0.23 which showed an increase in 0.2% proof strength at 4 K of about 100 MPa compared to the solution annealed state. A fracture toughness of 194 MPa$\sqrt{m}$ and an 0.2% proof strength of 1240 MPa were measured at 4 K. To minimize radioactivity in non-magnetic cryogenic components of fusion reactors and improve waste management after end of service, Nyilas and Obst (1989) exchanged Ni by Mn, because the half life of Ni-63 amounts to about 100 years compared to 303 days for Mn-54. Steel Mn34Cr13.5N0.32 revealed a yield strength of 1547 MPa but only 10% fracture elongation and 13% reduction of area at 4K. Ductility was improved only after 5 mass% Ni were added. Sumitomo and Nakatuka (1991) considered the low thermal expansion of high Mn steels in the sub-zero range ($\approx 7 \cdot 10^{-6}$ K $^{-1}$) compared to CrNi and CrMnNi steels ($\approx 12 \cdot 10^{-6}$ $K^{-1}$) as a disadvantage, because the superconductors receive less clamping from the non-magnetic collars or frames upon deep-freezing. Shimada (1990) addressed the problem that $Nb_3Sn$ superconductors are formed from Nb and Sn by annealing at 630 to 730 °C for more than 50 h which is sometimes done within the non-magnetic frames. In this case a high N content tends to precipitate embrittling $Cr_2N$ nitrides and in addition sensitization by σ-phase is impending (Fig. 5.22 and Sect. 2.4.3).

## 5.2.7
## Steels for exhaust valves

The combined mechanical, chemical, thermal and tribological loading of valves in combustion engines were briefly outlined in Sect. 5.1.6. Unlike inlet valves the head of exhaust valves is not cooled by the incoming air/gas mix and therefore reaches temperatures of about 850 °C, which calls for close-packed steels of respective hot strength and resistance to high temperature corrosion and wear. During stand-still wet corrosion has to be avoided.

***Development:*** An example of early austenitic valve steel was Cr18Ni9Si2.5W1C0.45 which was aged to precipitate secondary carbides. According to Müller and Weintz (1998) the introduction of HNS began 1952, when steel Cr21Mn9Ni4C0.5N0.5 appeared, which is still in use. This alloy was developed to improve the resistance to corrosion by lead oxide but was also accompanied by a better stability of the austenite to the formation of ferrite and σ - phase. However, long time exposure to the service temperature caused growth of $M_{23}C_6$ along grain boundaries and of lamellar $M_2N$ into the grains. The resulting embrittlement was seen less harmful than the deterioration of corrosion resistance. In a next step the microstructural stability in service was improved by adding Nb to give steel Cr21Mn9Ni4Nb2W1C0.5N0.45. After hot working and solution annealing at 1140 °C and aging at 760 °C a dispersion of primary and eutectic M(N,C) particles is embedded in a stable austenitic matrix. These precipitates subtract some of the grain boundary decoration in service and provoke a higher Cr content in the matrix without promoting σ - phase. The formation of lamellar $M_2N$ "pearlite" within the grains was subdued. To further improve strength the Nb content was split into about 1 mass% of Nb, V and Mo each. Because of its affinity to sulphur in the fuel Ni was omitted but the corrosion resistance of steel Cr21Mn10MoVNbC0.6N0.5 was impaired by Mo and V at high temperatures.

The three austenitic HNS are widely used (Müller and Weintz 1998). They emerged from a number of grades most of which contained N beside carbon (Kocis and Matlock 1978; Wessling and Ulm 1993). In contrast to N the influence of C on the corrosion resistance is regarded as detrimental, though. Therefore Escher (1999) replaced C by N leading to PESR steels with 0.9 to 1 mass% N. He investigated several grades with about (mass%) 21 to 24 Cr, 9 Mn, 6 to 10Ni, 0.9 to 1.5 Si, ≤ 2 W, ≤ 1 Mo, ≤ 1 Nb. Aging at e.g. 800 °C led to an intense precipitation of lamellar $M_2N$ "pearlite" which lowered the ductility and thereby the high temperature fatigue life. Silicon raised the lamellar spacing and aggravated the embrittlement. To alter the precipitation pattern Nb was added. Small amounts like 0.1 mass % to avoid large precipitates from the melt proved insufficient to suppress the "pearlitic" reaction. However, at 1 mass% Nb no "pearlite" occurred after aging at 800 °C and the MN particles were smaller compared to the standard grade with C, N and 2 mass% Nb. The new steel Cr21Mn9Ni6W2Mo1Nb1N0.9 showed superior mechanical properties at 700 °C (see e.g. Fig. 3.35), but at longer duration or higher temperature the standard steel Cr21Mn9Ni4Nb2W1C0.5N0.45 prevailed (Fig. 3.32 and 3.34). A microstructural inspection revealed "thick" grain boundaries after 200 h of aging at 800 °C, which were identified as $\sigma$ - phase and amounted to about 10 vol%. This was corroborated by thermodynamic calculation using a recent up-date of ThermoCalc: While the standard steel precipitates $\sigma$ - phase at $\leq$ 750 °C the new steel contains this phase up to about 870°C equilibrium temperature. One reason for this unfavourable development is the difference in Cr, Mo, W, Si content of the matrix derived by microprobe analysis after aging for 200 h at 800 °C. The other is discussed in Sects. 2.5, 3.1.5.1 and 3.1.6 and based on the higher degree of ordering found in austenite containing C + N instead of N alone. Atomic ordering enhances the stability of austenite to transformation thus retarding the formation of $\sigma$ - phase.

***Application:*** To summarize N has been and still is a key alloying element of austenitic valve steels, which today are only surpassed by Ni alloys of higher cost like NiCr20Ti2.5Al1. Prospective materials are titanium aluminides and $Si_3N_4$ allowing a service temperature of about 1000 °C, but their costs are still too high for the present market needs. The colder stem is made of a martensitic steel to endure the load of the rocker arm and joined to the austenitic head part by friction welding. To lower the massforces and improve the control of the combustion cycle the number of valves per cylinder was increased from 2 up to 3 to 5. In an extensive overview on valve materials Müller and Weintz (1998) estimated the number of inlet and outlet valves required worldwide per year at 750 million. The stem diameter ranges from about 5 mm in some passenger cars to several cm in large stationary or ship engines. The stem of the smaller valves is usually hot extruded and the head is die forged in the same stroke. Electric resistance heating of a rod end is used to compress a bulb from which the head of a larger valve is die forged. The seat of valve heads is often weld-clad with hardfacing CoCrC alloys to reduce wear and avoid channelling of the exhaust. Valves in Diesel engines are usually exposed to a higher pressure within the combustion chamber but remain somewhat lower in temperature compared to those in gasoline engines. Standard high nitrogen austenitic valve steels comply with these manufacturing procedures and service loads.

Nitrogen exerts not only a favourable effect on hot strength as shown in Fig. 5.27 but also improves the resistance to wet corrosion by $SO_4^{2-}$ or $Cl^-$ ions and in exhaust condensate as shown by Escher (1999). He also found that a renunciation of C enhanced the resistance to oxidation at 850 °C and a test duration of 500 h. The higher cost of PESR production was not accompanied by respective advantages of high N over (N+C) grades, though.

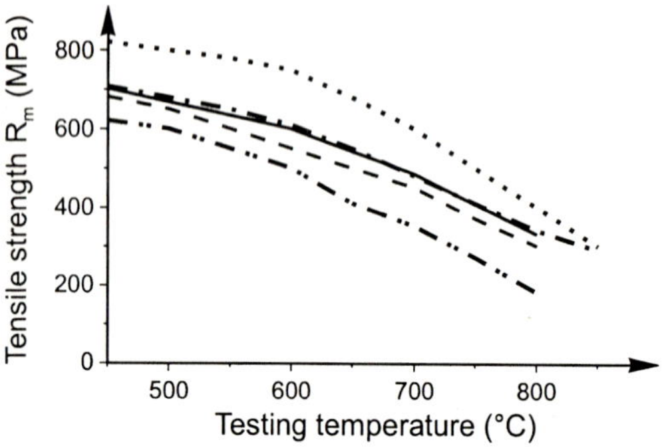


| steel | $R_m$ at 20°C (MPa) |
|---|---|
| ····· Cr21Mn10Mo1V1Nb1C0.6N0.5 | 1000 |
| —— Cr21Mn9Ni6W2Mo1Nb1N0.9[a] | 995 |
| —·— Cr21Mn9Ni4Nb2W1C0.5N0.45 | 950 |
| – – – Cr21Mn9Ni4C0.5N0.5 | 950 |
| —··— Cr18Ni9Si2.5W1C0.45 | 800 |

**Fig. 5.27.** Hot tensile strength $R_m$ of austenitic valve steels after DIN 17480 respectively [a] Escher (1999)

## 5.2.8 Creep resistant steels

In most room temperature applications of austenitic HNS nitrogen stays in solid solution. At elevated temperature N tends to precipitate and the resulting nitrides may be used to enhance the creep resistance in addition to solid solution strengthening. According to an early review on nitrogen in stainless austenitic steels by Grützner and Schüller (1967) the investigation of nitrogen in creep resistant steels goes back to the fifties and early sixties but has had only a modest impact on standardization since. Maybe the creep strength of HNS at > $10^5$ h was not as convincing as that at < $10^4$ h.

***Development:*** The favourable influence of 0.19 mass% N on the creep strength of a low carbon CrNiMo steel tested up to about $10^4$ h at 650 °C is shown in Fig. 5.28. The addition of 0.3 mass% Nb improved the steel without N but did not change the creep strength of the nitrogen grade (Jesper et al. 1966). The

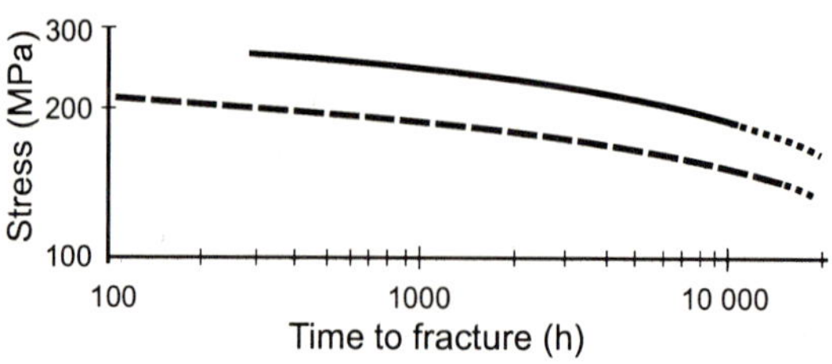


**Fig. 5.28.** Creep strength of the steels Cr18Ni14Mo3 (dashed line) and Cr18Ni14Mo3N0.2 (full line) containing 0.03 mass% C, tested in tension at 650 °C, after Jesper et al. (1966)

contributions of solid solution and precipitation strengthening were not separated. Nakazawa et al. (1989) looked into creep of steel Cr17Ni13Mo2.5 containing ≤ 0.09 mass% C and ≤ 0.12 mass% N for about $10^4$ h at 550 °C. They found that the creep strength is increased more by N than by C and that N does not impair creep ductility as C does (see also Sect. 3.1.6). This is consistent with grain boundary segregation of C which is not shown by N (see Sect. 1.1.4.3). To obtain a strong yet ductile steel of superior resistance to creep fatigue they proposed a moderate addition of 0.07 mass% N at 0.01 mass% C. Matsuo et al. (1989) investigated steels with about (mass%) 25 Cr, 28 Ni, < 0.07 C and < 0.26 N at temperatures between 700 and 1000 °C up to about 2000 h. While C led to an intergranular precipitation of $M_{23}C_6$ the distinct reduction of the minimum creep rate depicted in Fig. 5.29 was mainly assigned to solid solution strengthening by N. The transition from a more uniform distribution of dislocations to subgrains was shifted by N to higher stresses leading to a more pronounced strengthening by N at elevated stress, respectively shorter exposure (Matsuo et al. 1990).

The nitrogen contents discussed up to now were limited by the solibility at normal pressure. As pressure melted steels became available Pant et al. (1985) investigated PESR grades of a much higher N level at 600 and 800 °C up to about 50 000 h. At 800 °C under constant load they derived a 1 % creep strength after $10^4$ h of 30 MPa for steel Cr22Mn5Ni4Mo3VNbN0.77 and of 40 MPa for steel Cr19Ni16Mn5Mo3NbN0.66 which is in the range of Ni base alloys and appears to be high compared to e.g. steels for exhaust valves containing about 0.9 mass% of N or (C+N). These are expected to reach the same 1 % creep strength within a few

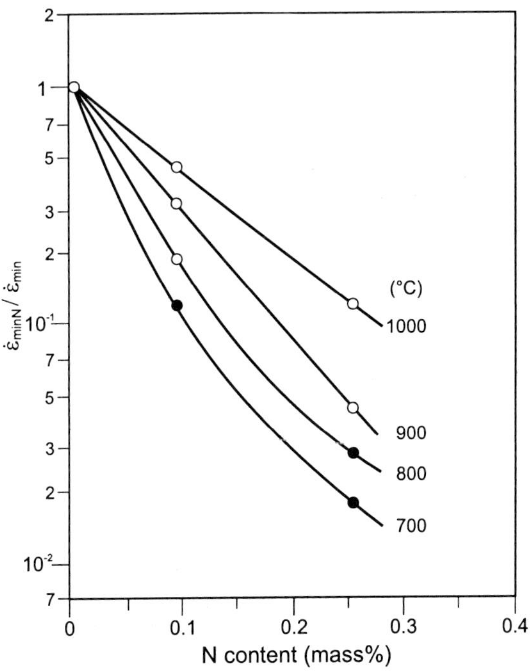


**Fig. 5.29.** Minimum creep rate of an austenitic steel with nitrogen $\dot{\varepsilon}_{min\,N}$ related to $\dot{\varepsilon}_{min}$ of the same grade without N at a constant load of 49 MPa, approximate composition (mass%) 25 Cr, 28 Ni, ≤ 0.01 C. Filled symbols indicate $M_2N$ precipitate observed after testing (Matsuo et al. 1989)

hundred hours (see Escher 1999 and Fig. 3.34). Pant and co-authors found a reduction of the as-quenched lattice parameter already after 10 h at 800 °C which came to an end within 100 h. The contraction corresponded to the precipitation of $M_2N$ binding 0.54 of the 0.77 mass% N in the low Ni grade. The saturated nitrogen austenite transformed extensively to "pearlite" composed of lamellar $M_2N$ and depleted austenite (see Sect. 2.4). The authors assigned a strengthening effect to the nitride lamellae. This is at some variance with results of Escher (1999) who found a higher creep resistance and ductility as the content of "pearlite" and its lamellar spacing decreased (Table 5.10). However, there is also a difference in solid solution strengthening of the three alloys by Si, W, Mo and a contribution by Nb which is used to change the lamellar nitride morphology into a globular one. Pant et al. (1985) as well as Escher (1999) report that about one third of the nitrogen remains in solid solution after aging which is reduced to just below one quarter after creep (Escher 1999). In both investigation of PESR-HNS the suppression of intermetallic phases by N is considered a prerequisite of superior creep resistance.

***Application:*** The development described above reveals two nitrogen levels. At the lower one, up to about 0.25 mass% N is kept mainly in solution during creep. The upper limit of N depends on the service temperature as well as on the Cr content and is expected to be higher for CrMn than for CrNi steels. The solid solution strengthening by N decreases as the duration of service is prolonged. In contrast to carbon embrittling precipitates along grain boundaries are avoided.

At the high level up to 1 mass% N leads to a considerable amount of nitrides, the shape and size of which affect the strength and damage evolution during creep. The remainder of dissolved N is comparable to that of low N steels. Pressure melting favours the application of forgings or parts machined from stock, while as-cast or welded housings are less feasible.

**Table 5.10.** Results of short time creep tests at 800 °C under a constant stress of 80 MPa. The steels contained about (mass%) 21 Cr, 9 Mn, 6 Ni, 0.95 N and other elements as indicated and were solution annealed and aged at 750 °C (steel 1 and 2), respectively 800 °C (steel 3), after Escher (1999)

| | alloy<br>content (mass%) | "pearlite"<br>content (vol%)<br>spacing (μm) | time (h) to<br>1 % creep,<br>fracture | minimum creep rate ($s^{-1}$)<br>fracture elongation (%) |
|---|---|---|---|---|
| 1 | 2 Si | 42<br>1 | 1.4<br>29 | $13 \cdot 10^{-7}$<br>33 |
| 2 | 2 W, 1 Mo<br>0.1 Nb | 34<br>0.5 | 9.3<br>> 178 | $2.5 \cdot 10^{-7}$<br>> 48 |
| 3 | 2 W, 1 Mo<br>1 Nb | < 5<br>- | 19.2<br>> 196 | $1.15 \cdot 10^{-7}$<br>> 48 |

Long-time creep properties of austenitic HNS are scarce and so are the applications. One is concerned with non-magnetic steels in fusion reactors, which are less activated by radiation, if the Ni content is low. This was already described for cryogenic components in Sect. 5.2.6 but applies to creep resistant ones as well. Bott et al. (1986) outlined the respective remodeling of creep resistant CrNi steels in which Ni is replaced by Mn and N. Miyahara et al. (1996) looked into steel Mn15Cr11W2Ni1 alloyed with (mass%) 0.13 to 0.25 N, ≤ 0.2 C, ≤ 0.5 V, ≤ 0.55 Ta and ≤ 0.2 Ti. Aging and creep tests at 600 °C up to about 8000 h revealed that carbon entailed an embrittling precipitation of $M_{23}C_6$ and had to be limited to 0.02 mass%. A grade with (mass%) 0.23 N, 0.25 V and about 0.15 Ti and Ta each lived up to the $10^4$ h creep rupture strength of the standard steel Cr17Ni12Mo2 (AISI 316). After aging the ISO-V impact toughness of the nitrogen steel was somewhat lower but remained above 80 J at a testing temperature between –50 and +100 °C.

Sekiguchi et al. (1991) reported the application of steel Cr18Mn15N0.35 for non-magnetic gaskets and springs in automotive engines to avoid an interaction with electronic devices at exhaust temperatures of ≤ 750 °C. At this highest temperature the hot strength of the solution annealed state was about twice that of steel Cr18Ni10 (AISI 304). After cold rolling and aging up to 700 °C the hardness remained far above that of steel Cr18Ni7 (AISI 301) and stayed above 400 HV after prolonged aging at 750 °C for 16 h.

Another special application of HNS in engines was presented in the previous section on exhaust valves.

## 5.2.9 High temperature nickel alloys

At high service temperatures austenitic steels are often replaced by Ni alloys of the same close-packed fcc structure. Until recently N was not considered a suitable alloying element for these high temperature materials because of the low N solubility of Ni. However, the uptake of N is quite enhanced by Cr, just like in austenitic steels. Grade 7 of table 5.9 may serve as an example. This corrosion resistant steel consists of about one third of Fe, Cr, and Ni each and attains 0.4 mass% N. If Fe is replaced by Ni and pressure melting is employed, a Ni alloy of comparable N level seems to be feasible.

***Development:*** Nickel alloys with (mass%) 30 Cr, ≤ 2.3 Si, ≤ 0.86 N, and ≤ 0.23 Y or Hf were PESR melted and investigated as to their microstructure and properties by Brill (1998). After solidification homogeneous austenite was obtained up to 0.1 mass% N, above which $Cr_2N$ appeared that turned into a eutectic morphology above 0.55 mass% N. After hot and cold rolling, recrystallization at 1050 °C, solution annealing at 1260 to 1300 °C, and aging between 750 and 1100 °C a precipitation of π nitride $Cr_{13}Ni_7N_4$ was observed, which had formed by a peritectic transformation from austenite and $Cr_2N$ (see also Sect. 2.3.1). Most strikingly the shortest incubation time was encountered at about 1000 °C which is about 250 °C above the respective envelope of γ' $Ni_3(Al,Ti)$ precipitation (Fig. 5.30). This indicates the potential of π nitrides to strengthen Ni alloys at temperatures above 800 °C.

During creep of solution annealed specimens π nitrides were formed and their amount increased with the Si content up to about 12 mass% in the alloy NiCr30Si2N0.4. At 1000 °C under a load of 10 MPa the time to creep fracture was extended with the content of π phase being formed and especially so between 1 and 5 mass% π. The effect of N on the creep resistance is shown in Fig. 5.31. The high and low cycle fatigue life at 1000 °C was improved by Si as well.
Cyclic oxidation at 1000 °C stayed below 0.1 g/m$^2$ h which is quite satisfactory. The same holds true for the resistance to nitriding in $N_2$ at 1000 °C. The high nitrogen alloys proved to be superior to e.g. NiCr22Co12Mo9Al1Ti. This reference alloy also revealed a distinctly lower resistance to sulfidation at 850 °C compared to the nitrogen grades. At a Si content of only 1.1 mass% and a temperature of 850 °C the mass increase of a nitrogen alloy in a caburizing atmosphere was about twice as high as in the above reference alloy. A higher Si content is expected to impede carburization.

According to Brill (1998) future work is aimed at raising the content of π phase to the level of γ' phase in Ni super alloys, i.e. up to 50 mass%. In tentative experiments with Ni alloys containing up to about (mass%) 40 Cr or 10 W, 10 Mo, 4 Si up to 35 vol% of precipitates were obtained. However, the effect of these additions on manufacturing has to be considered, too. Melting point, hot workability and weldability are among the crucial manufacturing properties, which were demonstrated to be quite satisfactory for NiCr30Si2N0.4 during the pilot production of a 2 Mg slab.

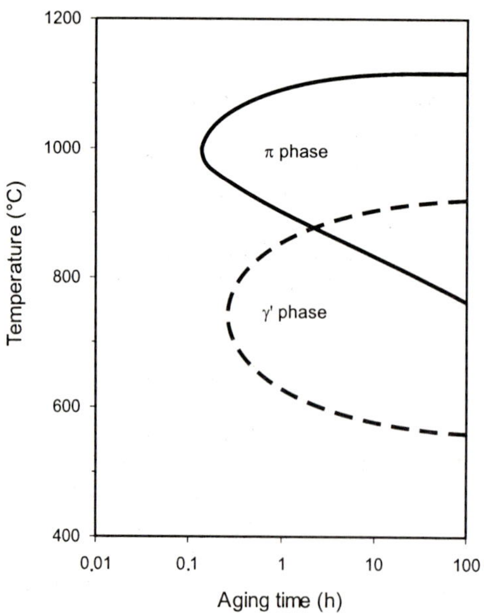


**Fig. 5.30.** Preliminary time-temperature-precipitation diagram of Ni alloy NiCr30Si2.3N0.4 precipitating π nitrides (full line) compared to the precipitation of γ' phase in Ni superalloys (dashed line), after Brill (1998)

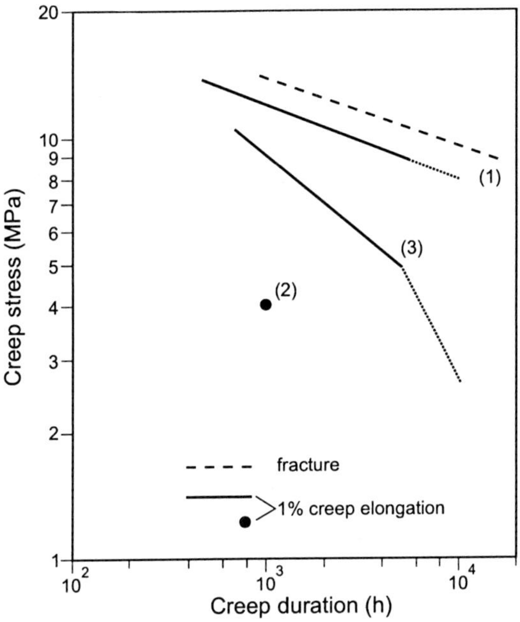


**Fig. 5.31.** Creep resistance at 1000 °C under constant load of Ni alloys with 30 mass% Cr and about 0.4 mass% N (1) or without N (2) compared to the Co alloy 188 (CoNi22Cr22W15) (3), after Brill (1998)

## 5.3 Steels of mixed microstructure

As shown in the Schaeffler diagram (Fig. 5.1) the phase field of martensite, austenite and ferrite are separated by two-phase zones and a three-phase region. In these the austenitic (A), respectively martensitic phase (M) contains more dissolved nitrogen than the ferrite (F). Of the low-alloy ferritic-martensitic steels the dual-phase grades with about 10 to 25 vol% of martensite are used as sheet for deep drawing. A stainless dual-phase steel with nitrogen was developed recently. Austenitic steels of lean alloy content partially transform to martensite upon deep-freezing or cold working. In these austenitic-martensitic grades nitrogen is used to improve strength and corrosion resistance. Stainless ferritic-austenitic duplex steels with about 50 vol% of each phase were first introduced in the 1930ies because of their high strength and corrosion resistance. In recent years up to about 0.4 mass% of nitrogen have been added to enhance their properties.

A common feature of steels with a mixed microstructure is the presence of phase boundaries besides grain boundaries. A duplex microstructure is e.g. characterised by an equal concentration of F/F and A/A grain boundaries and contains F/A phase boundaries. One phase is percolating the other. In contrast M/M boundaries are ideally avoided in dual-phase steels so that the martensitic grains are dispersed in a ferritic matrix. Deformation induced martensite in austenitic-martensitic grades is again distributed quite differently. The size and distribution of the metal phases have a strong influence on the mechanical properties. Their even distribution is disturbed by segregation leading e.g. to a net-like arrangement around primary dendrites or to a banded structure after hot working resulting in anisotropic behaviour. The higher thermal expansion of the fcc phase as compared to the bcc ones leads to microstresses and precipitation is promoted in the bcc phases because their diffusion coefficient is about two orders of magnitude above that of the fcc phase. Adding the lower solubility of the bcc phase for interstitials it becomes obvious that a heterogeneous microstructure of quenched fcc and bcc metal phases tends to develop an uneven array of precipitates upon heating. A recent study summarizes some results on HNS consisting of a mixture of metal phases (Berns 1997).

## 5.3.1
## Stainless ferritic-martensitic dual-phase steels

***Development:*** A dual-phase (DP) microstructure is obtained by intercritical hot working or annealing in the ferritic-austenitic regime followed by quenching to transform the fraction of austenite to martensite. To assure a good corrosion resistance and a sufficient N solubility the Cr content was set at 17 mass% and the Mo content at 1 mass%. The content of austenite respectively martensite was adjusted by alloying (mass%)1Ni and 0.054 or 0.09 N as well as by choosing the intercritical temperature $T_i$. As derived by ThermoCalc austenite starts to form just below 900 °C during heating and its content increases as the nitrides dissolve. To avoid grain growth $T_i$ is confined to the lower part of the F+A regime where the volume ratio of both phases is insensitive to temperature. To further impede grain growth 0.1 mass % V was added which entailed an increase of $T_i$ , though. The steels Cr17Mo1Ni1N0.1 and Cr17Mo1Ni1VN0.05 contained between 28 and 12 vol% of martensite which is more than twice as hard as the ferrite. This is caused mainly by the partitioning of elements between ferrite and austenite, as derived by WDX microprobe analysis (Table 5.11). It reveals that nitrogen is concentrated in the martensite which contains less CrMo.

This provides the key to the development of stainless DP steels with nitrogen instead of carbon: The martensitic areas which are less corrosion resistant because of the lower CrMo content and a higher concentration of lattice imperfections are protected by N enriched in solid solution. In contrast C would decrease the corrosion resistance and the more so because of insufficient solubility in the austenite leading to carbides which consume Cr. In spite of the "low" overall N content of the steels the local content in the martensite is "high" which justifies to classify these stainless DP grades as HNS.

The above DP steels were investigated by Kleff (1996) and compared to similar HNS and respective carbon grades. It is well known that strength and ductility of DP steels are enhanced by a fine grain size. Just as Mn influences the nucleation of austenitic grains in low alloy DP steels, the enrichment of Ni in the martensite impedes the formation of austenite upon heating so that austenite is preferentially nucleated at the dispersed nitrides in the ferrite leading to an even distribution of martensite after quenching (Berns et al. 1996). In tensile tests continuous yielding

**Table 5.11.** Properties of stainless ferritic (F) – martensitic (M) dual-phase steels (Kleff 1996)

| $T_i$ (°C) | Cr17Mo1Ni1N0.1 annealed 950 | Cr17Mo1Ni1N0.1 extruded 960 | Cr17Mo1Ni1VN0.05 annealed 1000 | Cr17Mo1Ni1VN0.05 extruded 1000 |
|---|---|---|---|---|
| chemical compos. F/M (mass%) | | | | |
| Cr | 17.93/16.63 | | | |
| Mo | 1.18/0.78 | | | |
| Ni | 0.76/1.02 | | | |
| N | 0.01/0.28 | | | |
| M content (vol%) | 21 | 28 | 16 | 12 |
| F grain size (μm) | 90 | 25 | | |
| hardness F/M (HV 0.05, HV 0.02*) | 197/471 | 213*/533* | 202/409 | 199*/500* |
| $R_{p0.2}$ (MPa) | 404 | 470 | | 378 |
| $R_m$ (MPa) | 761 | 839 | | 614 |
| $R_{p0.2} / R_m$ | 0.53 | 0.56 | | 0.62 |
| $A_5$ (%) | 25 | 22 | | 27 |
| Z (%) | 64 | 52 | | 65 |
| DBTT (°C) | > 20 | < -80 | >20 | < -80 |
| $N_B \cdot 10^3$ $\sigma = 500$ MPa | 265 | 330 | | 141 |
| mass loss (g/m²) 10% $H_2SO_4$ | 191 | | 235 | |

$T_i$ intercritical temperature, *extruded* 58% extension, *DBTT* ductile-to-brittle transition temperature of unnotched impact specimens 7 x 10 x 55 mm, $N_B$ cycles to fracture in rotating bending at 30 Hz as an average of two smooth specimens, *M* grain size < 10 μm

was observed as well as a low ratio of $R_{p0.2}/R_m$ and a much higher initial work-hardening rate at small strains as compared to subcritically annealed states. This is proof of a typical DP behaviour (Fig. 5.32). The yield strength increases with the content of martensite and the grain refinement which is brought about by hot extrusion (Table 5.11). A maximum of $R_{p0.2} = 470$ MPa is reached at > 20% elongation and > 50% reduction of area. Only the grain refined extruded state provided a satisfactory impact toughness.

Fatigue starts with the formation of slip bands in the ferrite which initiate microcracks. These are stopped by the martensitic islands or bands and are forced to deviate along F/M phase boundaries. Thus crack growth is impeded and fatigue life is extended. The small number of tentative fatigue tests did not allow to predict a probability of survival. But two tests with each of the extruded states, run at a bending stress of $\sigma = 400$ MPa, had not failed after $N_B = 10^7$ (Berns and Kleff 1997).

Immersion tests in 10% aqueous $H_2SO_4$ revealed an initial weight loss before passivation occurred. Corrosion started at the martensite but the weight loss was lower for the steel of higher martensite and nitrogen content (Table 5.11).

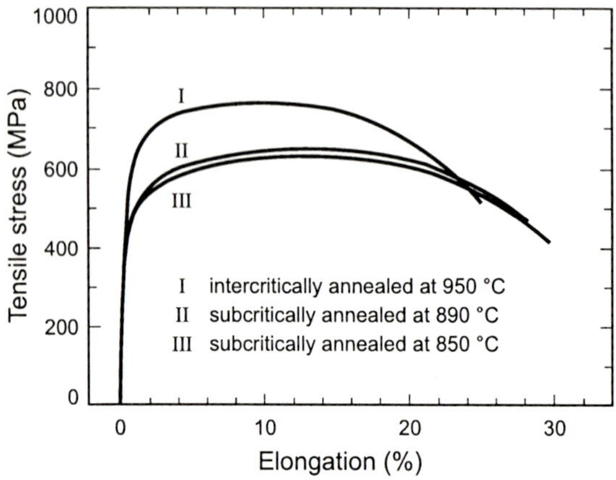


**Fig. 5.32.** Engineering stress/strain curves of steel Cr17Mo1Ni1VN0.05 with **I** ferritic-martensitic dual-phase microstructure **II, III** ferritic microstructure (ferrite + annealed former martensite), after Kleff (1996)

***Application:*** Low-alloy, high strength DP steels are produced as sheet for deep-drawing application. Intercritical rolling provides an extremely fine grain, which is required for stainless DP as well to ensure sufficient toughness. The limited thickness of deep-drawing sheet allows a rapid quench thus suppressing nitride precipitation. High strength, yet good formability and a superior corrosion resistance are attractive for many applications, especially in vehicles and marine equipment.

The respective stainless carbon grade Cr15Ni2MoTiC0.05 of low free interstitial content and soft martensite was successfully used for beer barrels, chain segments and for body parts as well as for the skin of suburban rail cars. To enhance the corrosion resistance nitrogen is needed in solid solution, which also favours tribological applications, because the hard martensitic inclusions promise a higher resistance to abrasive wear, as was shown for low-alloy DP steels (Hornbogen et al. 1984). This may be used for the load area of tipper trucks, for chutes and containers hauling wet and aggressive minerals. Welding seems possible, but will locally change the microstructure and properties.

Continuous intercritical rolling requires a larger quantity of sheet to be produced at a time. This is an obstacle which so far has prevented the transfer of the above laboratory results to industrial application.

## 5.3.2 Stainless austenitic-martensitic steels

In principle there are four ways to obtain an austenitic-martensitic microstructure. The first two start from lean austenite along the A/A+M borderline in Fig. 5.1 which is either deep-freezed or cold worked to initiate a partial transformation to martensite. The third way is based on an alloy content within the A+M field leading to the desired microstructure after quenching to room temperature. The fourth way requires e.g. a nickel-martensitic composition within the M field, which is partially reaustenitized during tempering.

Industrially deep-freezing is used to lower the content of retained austenite e.g. in bearing steels or to enhance the formation of martensite during cold rolling of austenitic steels. The latter was called zerolling and resembled a combination of cooling and cold working. In research deep-freezing was employed to gain different austenite/martensite mixtures in iron with 32 mass% of Ni and to study their microstructure-property relation (Eckstein and Guimaraes 1984). Cold working alone is applied to lean austenitic steels of thin cross-section like strip and wire or to the surface of thicker parts by shot peening, surface rolling or explosives. Wear or cavitation induced martensite may be observed in a near-surface zone of austenitic steels after service. In addition unintentional martensite may impair the cold forming of e.g. standard steel Cr18Ni10, especially during deep-drawing and stretch forming. As-quenched austenitic-martensitic steels with nitrogen were e.g. investigated by Haddick et al. (1978) but have not entered the market, probably because of difficulties in controlling the amount, size and distribution of the phases. Reaustenitization during tempering looks promising, though. Of the four ways to produce the said microstructure only cold working of thin products and reaustenitization of thicker parts will be discussed, because nitrogen is used to enhance their properties.

***Cold working:*** The manufacturing of high strength strip and wire is of major technical importance but requires a brief introduction before the effect of nitrogen is discussed. To characterize the resistance of austenitic steels to martensitic transformation, $M_d$ is used besides $M_s$. While the latter designates the start temperature during cooling, $M_{d30}$ stands for the temperature at which, after a plastic elongation of $\varepsilon = \Delta\, l/l_0 = 30\%$, just 50 vol% of martensite have been formed. In a review on the martensitic transformation of stainless austenitic steels Becker et al. (1986 I) presented a collection of empiric formulas predicting $M_s$ and $M_{d30}$ in dependence of the chemical composition. In seven of them N is set equal to C, in two the stabilising effect of N is rated 52, respectively 59 % higher. After Schumann and Fircks (1969) hcp $\varepsilon$ - martensite constitutes an intermediate step in the $\gamma \rightarrow \varepsilon \rightarrow \alpha'$ transformation especially at a high Mn/Ni ratio. But even in standard steel Cr18Ni10 some 10 vol% of $\varepsilon$ - martensite was found with $M_{s\varepsilon} = 40$ °C and $M_{s\alpha'} = -110$ °C. After cold working little $\varepsilon$-martensite is left in this steel. While some elements promote and others prevent bcc $\delta$ -ferrite, almost all elements suppress $M_s$ and $M_d$, because they increase the shear strength of fcc austenite requiring more supercooling or deformation to build up the energy necessary for the formation of $\alpha'$-martensite.

Moreover the amount of $\alpha'$ - martensite is influenced by the production parameters. A coarse initial grain size leads to more martensite during cold forming (Küppers 1982 I). The same holds true for an increase of hydrostatic stress e.g. from rolling to stretch forming and a change in temperature from summer to winter. A slight heating of sheet and tool to ≤ 100 °C subdues unwanted martensite during deep drawing, while cooling enhances the transformation during rolling. Raising the deformation rate brings the system closer to adiabatic conditions and the resulting elevation of temperature lowers the content of martensite (Küppers 1984).

The influence of martensite on the mechanical properties of steel Cr18Ni9 was investigated by Küppers (1982 II) in tensile tests between +50 and –50 °C covering the range from a stable to an unstable behaviour with little retained austenite after uniform elongation (Fig. 5.33a). A maximum of elongation was obtained, if about 20 vol% of martensite formed before necking started (Fig 5.33b). As the temperature was lowered the ultimate strength was raised much more than the proof strength, because of martensite emerging in-between (Fig. 5.33c). Thus the ratio $R_p/R_m$ grows with temperature. True stress/strain-plots showed that martensite supports the strain hardening of austenite until the major part is transformed and the lower strain hardening capacity of martensite prevails (Fig. 5.33d). Under a constant load near or above yielding austenitic steels suffer from logarithmic creep $\varepsilon = a \log t + b$ during time t which is considerably reduced by cold working steels of different stability (Schmidt et al. 1986, Gümpel et al. 1988).

A growing interstitial content augments the stability of austenite to martensitic transformation. As the temperature of solution annealing is increased more carbides or nitrides are dissolved so that the stability may be adjusted to a specific need. Austenite and martensite are of similar interstitial content leading to quite superior strengthening of the latter phase. Therefore the development of AM-HNS is confined to moderate N contents to avoid an embrittlement. Due to the large

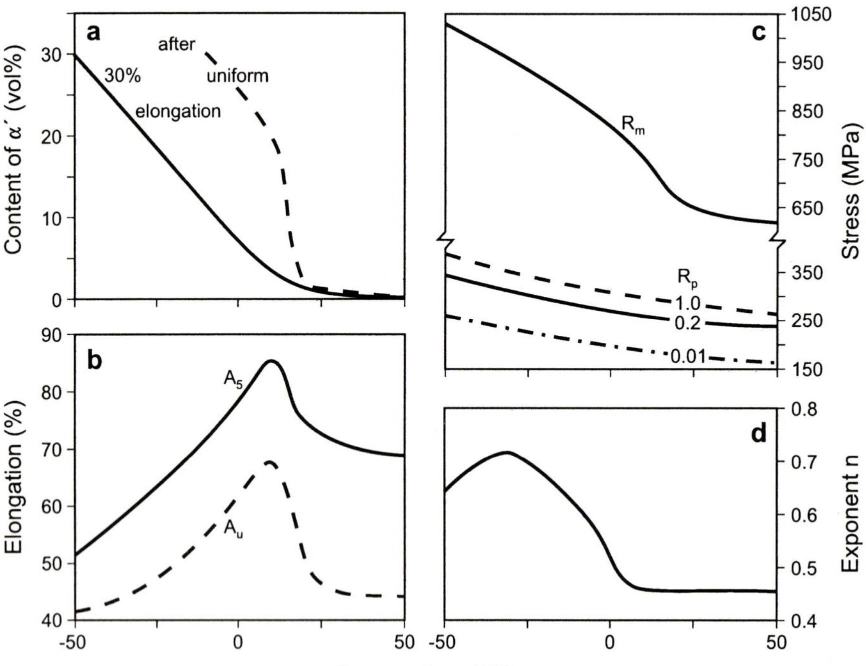


**Fig. 5.33.** Tensile properties of steel Cr18Ni9 depending on the testing temperature **a** content of martensite **b** uniform and fracture elongation $A_u$, $A_5$, **c** ultimate and proof strength $R_m$ and $R_p$, **d** mean strain hardening exponent n (Küppers 1982 II)

difference in the diffusion rate each phase responds differently to subsequent tempering, which may be applied to impede further transformation in service.

Cold working of unstable austenitic steels of moderate interstitial content yields high strength. This was demonstrated for cold rolling of the lean austenitic steel Cr17Ni7C0.12 (Becker et al. 1986 II). At 70 % reduction a tensile strength $R_m$ of about 2000 MPa was obtained. An exchange of C by N in this steel would enhance strength and corrosion resistance. Tanaka et al. (1983) showed that an increase of nitrogen from 0.05 to 0.16 mass% raised the tensile strength $R_m$ from 1000 to 1200 MPa after cold rolling at 70 °C to 30 % reduction and 20 vol% of α'-martensite. The low nitrogen grade required about 50 vol% of deformation martensite to reach the higher strength level. Takemoto et al. (1990) found that an increase of N from 0.1 to 0.21 mass% in steel Cr18Ni9Mn2N raised the hardness after cold rolling and lowered the martensite content, which demonstrates the strengthening of both phases by nitrogen (Fig. 5.34). To lower the cost of decarburisation, some carbon may be left in the melt. Jointly N + C promote ordering and the stability of austenite most, allowing for some decrease of Ni. This type of development has not yet been carried out but seems to promise improved cold rolled strip for stainless springs.

Holmberg et al. (1990) developed a low cost, non-magnetic stainless steel by making use of ε - martensite to strengthen the austenite during cold rolling without much further transformation to α'-martensite. After 75 % of reduction and aging at 450 °C steel Cr18.5Ni7Mn4N0.27C0.1 reached a strength level of $R_{p0.05}$ = 1412 MPa, $R_{p0.2}$ = 1660 MPa and $R_m$ = 1840 MPa at a relative magnetic permeability of 1.048. Aging was especially beneficial for the 0.05% proof strength which was raised by 30%. The material seems suitable for the production of non-magnetic stainless springs.

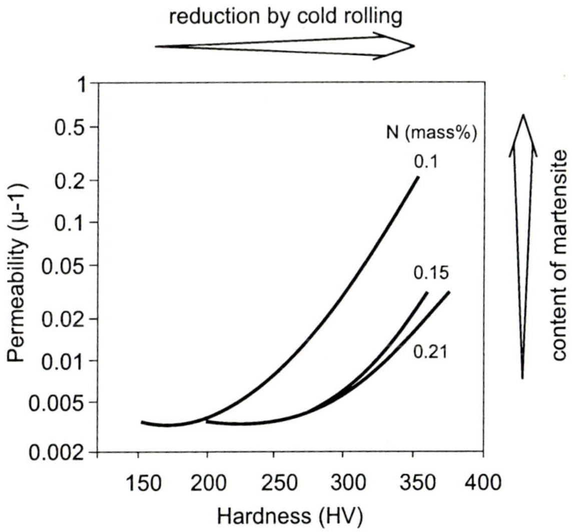


**Fig. 5.34.** Relation between hardness and permeability of steel Cr18Ni9Mn2N (Takemoto et al. 1990)

The other product made of cold worked austenitic steel is cold drawn, high strength wire for stainless ropes. Gümpel and Strom (1988) investigated three steels with increasing N-content: (a) Cr18Ni13Mo2.6Mn2N0.17, (b) Cr19Ni15Mn5Mo3NbN0.27 and (c) Cr21Ni16Mn6Mo3NbN0.33. To reach the desired strength of $R_m \approx 1500$ MPa the steels required a cold reduction of about 50, 38 and 34%, respectively. This reflects a growing strain hardening of the austenite by planar slip and twinning. Martensite was not observed in these high-alloy grades. Steel (c) showed the best resistance to pitting corrosion, crevice corrosion and stress corrosion. All three grades performed well in reversed bending tests and in tensile tests with knotted specimens. A third technological test, relevant for the production of wire ropes, is the torsion test, in which the number of revolutions to fracture $N_t$ under a tiny tensile stress is counted. With mean values of 2 to 6 the three stable austenitic HNS performed poorly because of localised shear. Shear bands and low energy shear fracture was also discussed by Paulus et al. (1993) for heavily cold worked, high nitrogen superaustenitic steels.

The thin plates of cementite in lower pearlite of plain carbon steels with 0.7 to 0.8 mass% C are turned into the direction of the wire axis during cold drawing. They seem to resemble effective obstacles for a localisation of shear resulting in $N_t$ values between 25 and 35 (Gridnev et al. 1974). Based on this observation Berns et al. (1997) studied the influence of a textured second phase on the plasticity of cold-drawn wire produced from stainless austenitic steel. Strain-induced martensite and/or δ-ferrite were chosen as second phases. A series of unstable austenitic steels with (mass%) 19Cr, 10Mn or 5Ni+4Mn, 0.2 to 0.3 N without or with 2Mo, 2 Cu and 1.5 Si were compared to a stable austenitic steel and a duplex steel. Wire of 2.4 mm in diameter contained up to 50 vol% of thin ferritic bands and – after a reduction of $\varepsilon \leq 86\%$ by cold drawing – up to 90 vol% of martensite, the initial microsegregation being the cause of the banded structure. Transmission electron microscopy revealed that the dislocation structure in the bands of ferrite formed cells and even some dynamic recrystallization occurred as the cold reduction increased. In the austenitic bands the dislocations split, forming stacking faults followed by twinning. At the highest deformation some dynamic recovery of austenite was observed near martensitic plates. Arrays of fine plates of austenite and martensite with a thickness of about 100 nm were identified by convergent beam electron diffraction.

Mechanical properties of 5 selected alloys are depicted in Fig. 5.35. Alloy A represents a stable austenitic steel, alloy B an unstable one as can be derived from the insert in Fig. 5.35a. The alloys C to E contain from 15 to about 50 vol% of ferrite and the remaining austenite behaves unstable in C and D and rather stable in the duplex steel E. At 28 vol% the strength $R_m$ is impaired but the reduction of area Z is improved. After a cold reduction of 75% additional aging up to 550 °C revealed a maximum of strength increase at 400 °C. For steel C the gain amounted to $\Delta R_{p0.2} \approx 400$ MPa and $\Delta R_m \approx 300$ MPa.

In a stable austenite (A) the $N_t$-value deteriorates with growing cold reduction. In the unstable austenite (B) $N_t$ is improved as the content of martensite is increased by drawing. Unstable austenite with some ferrite (C,D) gives the highest $N_t$. A cold-worked duplex structure (E) with a rather stable austenite is more ductile in torsion than the stable austenite A but far below C and D. A summary of

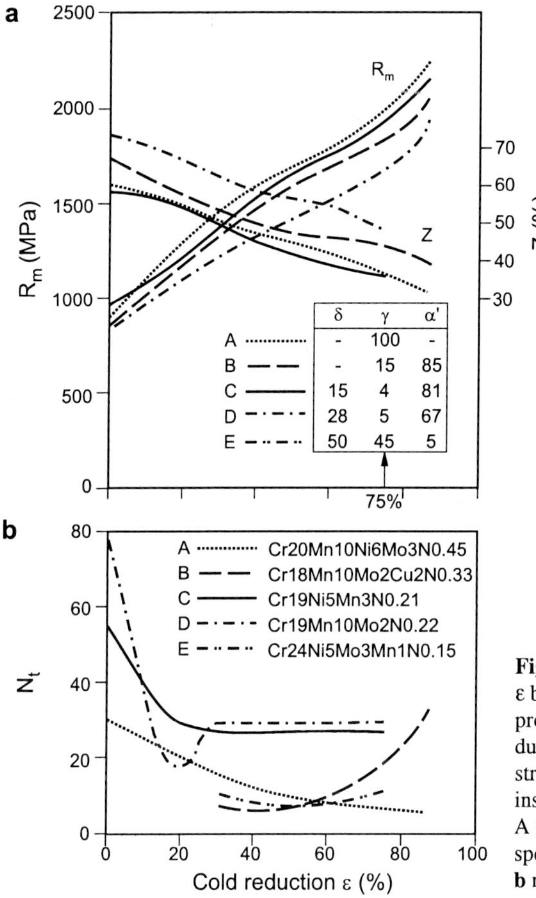


**Fig. 5.35.** Influence of cold reduction ε by wire drawing on the mechanical properties of stainless austenitic and duplex steels, **a** ultimate tensile strength $R_m$ and reduction of area Z, inset: vol% of phases after ε = 75 %, A to D derived by Mössbauer spectroscopy, E approximate values, **b** number of torsions to fracture $N_t$

all test results revealed that a banded structure of austenite and > 50 vol% of martensite increased the ductility in torsion. The addition of ferritic bands extended the regime of high $N_t$ to lower cold reduction. The highest product of $R_m \cdot N_t$ was obtained by a microstructure of about 15 vol% of δ-ferrite embedded in a mixture of austenite and about 50 vol% of martensite. The concept of avoiding localised shear by a banded structure of austenite and second phases proved to be valid.

After cold drawing (ε = 75%) the resistance to aqueous corrosion in 1-n $H_2SO_4$ and 3% NaCl was considerably improved by Mo, Cu and even Si but so was the austenite stability leading to lower $N_t$ values. All three alloying elements impede the formation of martensite but only Cu reduces the content of δ-ferrite. Therefore further iteration is required to simultaneously optimise the mechanical and chemical properties. The use of nitrogen has a beneficial effect on both while carbon would only enhance the mechanical properties.

Beside cold rolled strip and cold drawn wire thin-walled tubes may be strengthened by drawing or roll-forming making use of deformation martensite and of nitrogen.

***Reaustinitization:*** Stainless nickel-martensitic steels with (mass%) 13 to 17 Cr, 4 to 6 Ni, up to 1.5 Mo and a low interstitial content are widely used in industry, e.g. for hydro-turbines or radial air compressors, because of their high yield strength, low DBTT, and good weldability. These steels are hardened and tempered at 550 to 650 °C or even below, if they contain up to 3mass% Cu for precipitation hardening the martensite. In his review Brezina (1983) described the evolution of microstructure and properties during tempering (Fig. 5.36). Reaustenitization starts below 500 °C and part of the austenite transforms to new martensite upon cooling to room temperature leading to a maximum of about 50 vol% of austenite besides tempered and new martensite. He showed that a fine distribution of austenite in martensite with a spacing in the submicron range considerably raised the toughness at a moderate loss of strength. The austenite remained stable during deep-freezing in liquid nitrogen.
Based on these findings we extended our interest in high strength nickel-martensite steels, micro-alloyed with N (Sect. 5.1.4), to high toughness reaustenitized states. As shown in Fig. 5.37 the stainless steel Cr15Ni4Mo1 of low interstitial content usually contains a little δ-ferrite which vanishes upon adding 0.05 to 0.1 mass% of N. During tempering of the as-quenched martensite, dispersed austenitic nuclei are formed, which are stabilised by the solution of N and by an increase of Ni (Svejcar et al. 1987). As the tempering temperature is raised the reaustenitization proceeds and the amount of austenite increases leading to a dilution of stabilizing elements so that some transformation to new martensite will occur upon cooling, which may be subsequently tempered at a lower temperature. Nitrogen enhances the strength of austenite and temper nitrides have

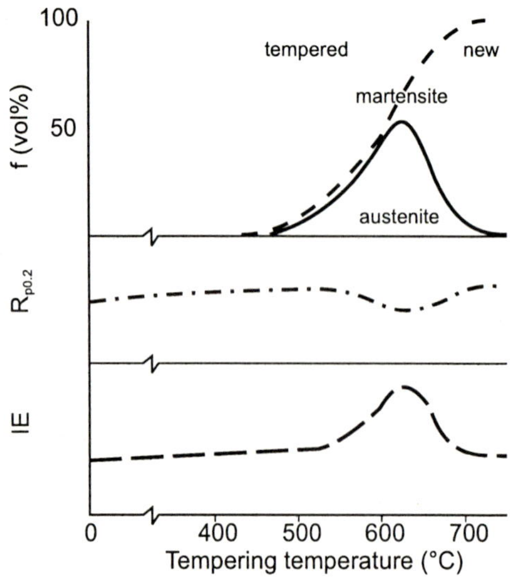


**Fig. 5.36.** Schematic representation of the effect of tempering on the fraction f of microstructural constituents, the proof strength $R_{p0.2}$ and the ISO-V impact energy IE derived for a stainless nickel-martensitic steel (after Brezina 1983)

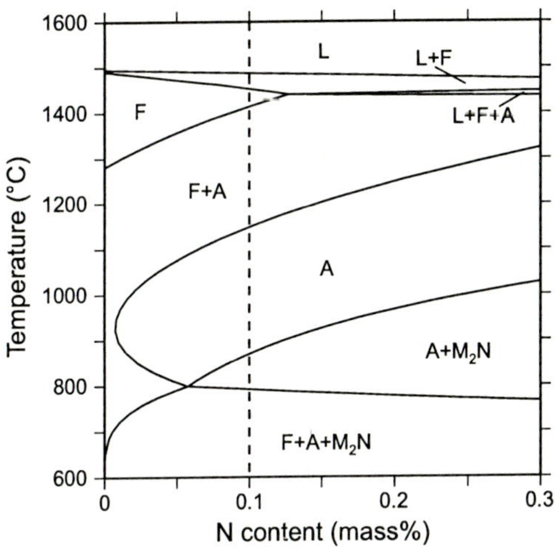


**Fig. 5.37.** Constitution of steel Cr15Ni4Mo1N0.1 as derived by ThermoCalc

a strengthening effect on the martensite. Therefore the drop in 0.01 % proof strength, indicating reaustenitization, is less pronounced in the nitrogen grade and does not impair the toughness (Fig. 5.38). In contrast to the carbon grade no intercrystalline corrosive attack is found in the nitrogen grade at ≥ 575 °C and no mass loss in boiling, 20% acetic acid.

Compared to ferritic-martensitic stainless steels the austenitic-martensitic ones offer a smaller size and a finer distribution of the phases due to the lower temperature of formation. The fine grain size, the fcc fraction of the microstructure and the high Ni content provide a lower ductile-to-brittle transition temperature than in dual-phase steels. Brezina (1983) also mentioned Co and Mn

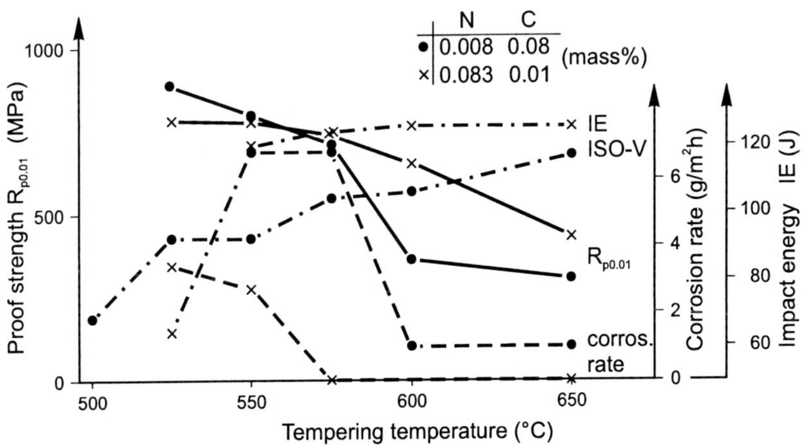


**Fig. 5.38.** Effect of tempering and reaustenitization on some properties of nickel-martensitic stainless steel Cr15Ni4Mo1 with 0.09 mass% (N+C) hardened at 1000 °C, 30 min/oil (Ehrhardt 1995)

as useful elements in controlling the phase transformation. The latter increases the N solubility and was employed by Klotz et al. (1998) to partially replace Ni in a creep resistant steel with 10 to 13 mass% of Cr and up to 0.1 mass% of N besides some Mo, V, and C. The microstructure of martensite with 20 to 40 vol% austenite is aimed at a creep strength of 250 MPa for $10^5$ h at 550 °C.

### 5.3.3
### Stainless ferritic-austenitic duplex steels

***Development:*** Steels with about 50 % of ferrite and austenite each emerged in the 1930ies to raise strength and corrosion resistance at lower cost (Charles 1991 I). The strength was enhanced by phase boundaries and the corrosion resistance by a higher Cr content at the expense of costly Ni as expressed in the Schaeffler diagram (Fig. 5.1). In castings up to about 0.4 mass% C was added to promote the wear resistance by a eutectic precipitation of carbides, however, at a severe loss of ductility during manufacturing and application. It was not until modern melting practice allowed to reduce C to nearly nothing and replace it by N staying in solid solution, that duplex stainless steels (DSS) were able to demonstrate their potential in the market.

During solidification ferrite is formed primarily, limiting the soluble content of nitrogen at ambient pressure. This is made up for by a high Cr, Mo, Mn and a low Ni, C content (Table 2.1). On a small scale DSS with up to about 0.5 mass% of N have been produced (Lardon et al. 1989, Horvath et al. 1997, Wang et al. 1999). There does not seem to be a need of further raising this level by pressure melting, because Foct and Akdut (1993) found that N is enriched in the solid solution of austenite leading to a brittle cleavage-like fracture. The same was observed for fully austenitic steels by Uggowitzer et al. (1992) if the N content approached 1 mass% (see Sect.3.1.4). The partitioning of elements in DSS was investigated e.g. by Wahlberg and Dunlop (1987). For steel Cr22Ni5Mo3Mn1.5N0.2 they found the following mass ratios of elements in ferrite/austenite: Cr1.10, Mo1.24, Ni0.65, Mn0.81, N0.15. Each element is attracted by the phase it stabilizes, especially nitrogen. Jomard and Perdereau (1991) reported ratios for steel Cr25Ni6Mo4N0.24: Ni0.75,Mo1.44, N0.14. Charles (1991 I) published respective values for 14 different alloys. This demonstrates that the N content in austenite may be nearly twice as high as in the steel. In view of brittle fracture the latter should stay below 0.5 mass%. The renunciation of pressure melting eases casting and welding and lowers the cost.

In large scale production of DSS the margin of 0.5 mass% N is not exhausted. The N content usually ranges from 0.1 to 0.4 mass% and the numerous grades of wrought steel may be devided into 4 groups, each of which is characterized by an example: (I) Cr23Ni4N0.1 low cost, Mo-free DSS with PREN ≈ 25, (II) Cr22Ni5Mo3N0.2 with PREN ≈ 35, (III) Cr25Ni7Mo4N0.27, superduplex steel with PREN ≈ 42, (IV) Cr26Mn6Ni4Mo2N0.34 with PREN ≈ 39, increase of N solubility by adding Mn at the expence of Ni. Further variations are obtained by the addition of Cu, Si, W especially in the field of castings (Nilsson 1992, Charles 1991 I). DSS are solution annealed and quenched in water to suppress a reprecipitation especially in the ferrite enriched by CrMo in which the diffusion rate is about two orders of magnitude faster than in austenite. Depending on the chemical composition intermetallic phases like σ, χ, R and nitrides like $M_2N$, MN and π ($M_{20}N_4$) were observed in DSS between 1000 and 600 °C which may entail an embrittlement if the cooling rate is too slow. Also secondary austenite $\gamma_2$ may form by different mechanisms (Josefsson et al. 1991). Another regime of potential embrittlement is encountered between 550 and 350 °C. It is called "475 °C-embrittlement" and caused by spinodal decomposition or nucleation of Cr-rich α'. In respective DSS a precipitation of Cu occurs in about the same temperature range. Compared to the rapid precipitation at e.g. 850 °C the reaction at e.g. 450 °C is quite retarded. Most of the precipitates are rich in CrMo depleting the matrix which impairs the corrosion resistance.

In super DSS with about 4 mass% of Mo the χ-phase is precipitated more rapidly than the σ-phase. To create a super DSS with a PREN of 42 but without χ precipitation Nilsson et al. (1998) raised the Cr,N content and lowered the Mo level. The steel Cr29Ni7Mo2N0.38 requires a cooling rate of 17.5 °C/min after solution annealing or welding to restrict the formation of σ-phase to 1 vol%. This corresponds well to the cooling rate reported by Charles (1991 I). At 5 vol% of σ-phase the ISO-V impact toughness was down to 27 J while the hardness was still unchanged.

As the temperature of solution annealing increases the formation of ferrite is promoted. In high nitrogen DSS this is counteracted by the dissolution of nitrides since the resulting interstitial nitrogen stabilizes the austenite. This seems to be a key advantage of HNS during welding because too high a ferrite content in the superheated and rapidly cooled HAZ is attenuated by nitrogen which is equilibrated between ferrite and austenite much more rapidly than the substitutional elements (Charles 1991 II). As to the precipitation of intermetallic phases during further cooling a retardation due to nitrogen was shown experimentally by Thier (1967) and based on thermodynamics by Hertzman (1995) for austenite. However, the ferrite may profit less because of element partitioning resulting in low N yet higher CrMo contents. Anyway heat input and interpass temperature have to be limited to avoid $M_2N$ precipitation in the HAZ. The formation of pores in the weld metal was suppressed by selecting the appropriate filler metal, flux and shelding gas (Nassau 1991). The development of weldability for DSS and super DSS (PREN ≥ 40) was a prerequisite for the success of these steels in the market and was supported by alloying with nitrogen.

Wahlberg and Dunlop (1987) showed that the austenite/ferrite hardness ratio of steel Cr22Mo3NiN increased from 0.9 to 1.0 and 1.1 as the nitrogen content was raised from 0.05 to 0.13 and 0.20 mass%. The initially softer austenite was strengthened more than the ferrite because of element partitioning and an accumulation of N in the austenite. In tensile tests and wire drawing nitrogen enhanced work hardening accompanied by planar glide, twinning and some deformation martensite. More of the latter (13 vol%) was surprisingly observed in the cold drawn high nitrogen grade, the austenite of which was assumed to be more stable. The authors suggest that the softer low nitrogen austenite did not develop the critical transformation stress. In all nitrogen has a beneficial effect on the tensile properties of DSS and supports work hardening of wire and strip (see Sect. 5.3.2). The Charpy-V-notch impact toughness is characterized by a ductile-to-brittle transition temperature (DBTT). Due to the fraction of austenite and the

Ni content in the ferrite the DBTT of quenched strip with moderate thickness is usually well below –50°C, which is an advantage over ferritic steels.

In cyclic loading the picture is more complex. Foct et al. (1991) demonstrated that in low cycle fatigue under a plastic push-pull strain amplitude $\Delta\varepsilon_p$ a DSS behaves like a ferritic steel with crack initiation in the ferritic grains at high $\Delta\varepsilon_p$ and more like an austentic steel with crack initiation in the austentic grains at low $\Delta\varepsilon_p$. In the latter case the strain was confined to the softer austenite while in the former both phases accomodated the strain, the austenite living up to the ferrite by work hardening. Nitrogen influences this mechanical coupling of the two phases in DSS. In reference to Degallaix et al. (1993) cyclic stress-strain curves of ferrite and austenite are drawn schematically in Fig. 5.39 for three different N-contents. For simplicity it is assumed that the nitrogen content of the ferrite hardly changes because it is accumulated in the austenite, where it raises the yield point. At a low level of N in the DSS the austenite starts to deform plastically while the ferrite still responds elastically until $\Delta\varepsilon_t/2$ has reached the yield point of ferrite, above which both are strained plastically, the austenite displaying the stronger work hardening. A high N-content reverses the sequence of yielding and increases the difference in strength between ferrite and austenite. Thus plastic strain is localized in the ferrite. At an intermediate level of nitrogen both phases start to yield simultaneously. There seems to be an optimal content of nitrogen to enhance the low-cycle fatigue resistance of DSS above which localized plasticity may become a disadvantage (Vogt et al. 1998).

Due to CrMo partitioning the ferrite is more resistant to pitting corrosion in chloride solutions. An addition of nitrogen is accumulated in the austenite and strongly improves the pitting resistance of this initially weaker phase. Charles (1991 II) showed a linear increase of the critical pitting temperature (CPT) with PREN up to 0.45 mass% N using a factor of k = 30 in Eq. 3.6. Tsuge et al. related the CPT directly to the N content in austenite as reported by Combrade and Andouard (1991) (Fig. 5.40). To impede pitting means also to avoid stress raisers and postpone crack initiation in corrosion fatigue. In reference to stress corrosion

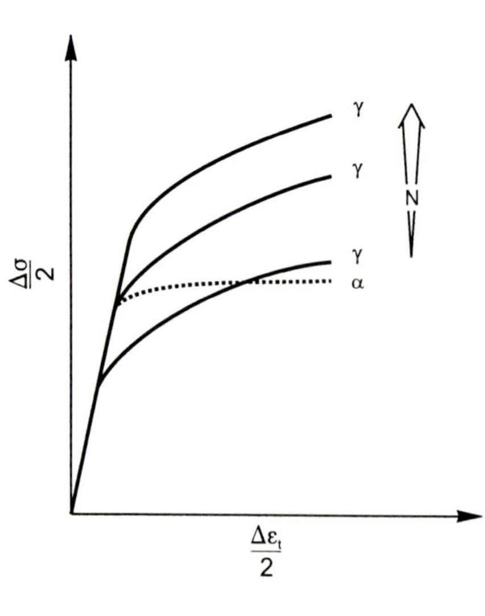


**Fig. 5.39.** Schematic representation of cyclic stress-strain curves for austenite and ferrite in three duplex steels of increasing nitrogen contents accumulated in the austenite. $\Delta\varepsilon_t/2$ = half cycle of the total strain amplitude, $\Delta\sigma/2$ = half cycle of the resulting stress amplitude

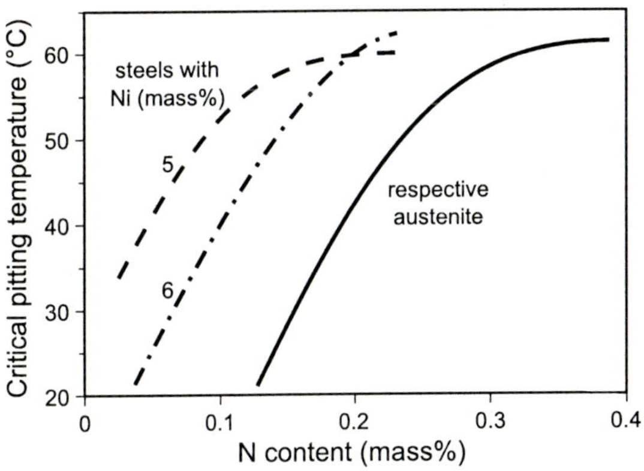


**Fig. 5.40.** Effect of steel and phase composition on the pitting resistance of Cr22Mo3NiN in 10% $FeCl_3$-$6H_2O$ - after Tsuge et al. (quoted from Combrade and Andouard 1991)

cracking (SCC) in chloride solution DSS are usually quite superior to standard austenitic steels (Bernhardsson 1991). Foct et al. (1991) revealed that in the active branch of a DSS polarization curve in acid chloride solution the softer austenite is cathodically protected by the ferrite. As the N content increases the austenite becomes stronger and more resistant to pitting. The authors discuss the superposition of an electrochemical and mechanical coupling of the two phases with respect to SCC which depends on depassivation by slip in the softer phase and the subsequent ability to repassivate. They conclude that about 0.2 mass% N is necessary to enhance the resistance to SCC and corrosion fatigue but that too high a nitrogen content may have an adverse effect because the attack is focussed on the ferrite.

To summarize, nitrogen is used today in DSS to stabilize the austenitic phase and improve the weldability in which it is superior to Ni. Pressure melting is not required to achieve a sufficient N content. This and the lower Ni content compared to austenitic steels add to the cost effectiveness of DSS, the strength and corrosion resistance of which are considerably raised by nitrogen.

***Application:*** Compared to ferritic stainless steels DSS offer a lower DBTT and higher strength, compared to austenitic steels they are less prone to stress corrosion cracking and provide a higher strength at lower cost. In application either the low cost or the high corrosion resistance is emphasized.
The low cost DSS Cr23Ni4N0.1 is still above the price of ferritic steels. This spurred the search for a Ni-free DSS in which the austenite is stabilized by Mn and N. In a diploma thesis Schug (1993) investigated two alloys for use in castings because it is easier to introduce new alloys this way than as a wrought product. In the first one the N content was a bit too low to give a good balance of ferrite and austenite (Table 5.12). The second one contained too much nitrogen resulting in some brittle cleavage. Current density-potential measurements in 1-n $H_2SO_4$ revealed about an equally good behaviour. In aqueous 3 mass% NaCl solution passivation was observed only for the grade containing Mo. Both grades embrittled rapidly between 900 and 600 °C. Cr26Mn19N0.6 did not show a 475 °C embrittlement, though. This was confirmed by Pohl et al. (1998) for a Ni-free DSS Cr26Mn17N0.5. Because of a limited amount of material, impact

tests were carried out with specimens of reduced size and a round notch. Yet the ductile-to-brittle transition occured at or slightly above room temperature (Table 5.12) reflecting a lack of Ni suppressing cleavage in the ferrite. In the final report on HNS with mixed microstructure (Berns 1997) it was concluded that a Ni-free DSS for castings might not provide a sufficiently low DBTT. This may be different for wrought steel of smaller grain size. Wang et al. (1999) investigated the DSS Cr22Mn10N0.35 and Cr20Mn10Mo3N0.45 with about 55 vol% of austenite. They found an 0.2 % proof strength of 567 and 635 MPa, respectively, at > 40 % of elongation, > 70 % of reduction in area and 260 J ISO-V-notch impact energy at room temperature. No SCC was found and especially the grade alloyed with Mo offered a high resistance to pitting and crevice corrosion. The DBTT at half of the upper shelf energy amounted to –20 °C (Wang 1998) but the proposed application as reinforcing bars in the building industry goes without sharp notches. As sheet these steels may be used in vehicles like lorries, busses and trains to cover areas of severe corrosive attack.

The application of Ni-free DSS for parts in contact with the human body could avoid a nickel allergy. Respective austenitic steels require pressure melting while DSS do without. The Cr equivalent of both types of stainless steel is in the same range (see Sect. 5.2.4).

At the high end of corrosion resistance super duplex steels with PREN above 40 rely on nitrogen as well. Steels like Cr25Ni7Mo4N0.27 are used as castings and also as wire and sheet. Impellers and housings of pumps for the chemical industry are frequent as-cast products. Digesters and bleaching equipment are built of tube and sheet exposed to high chloride concentrations. The same holds true for certain areas in flue gas desulphurization or desalination plants. High speed pump

**Table 5.12.** Microstructure, hardness and ductile-to-brittle transition temperature (DBTT) of Ni-free, high nitrogen duplex stainless steels as-cast and quenched from 1050 °C in water, respectively as indicated. The chemical composition of the phases was derived by microprobe analysis (Schug 1993)

| | Cr22Mn12Mo3N0.4 | | | Cr26Mn19N0.6 | | |
|---|---|---|---|---|---|---|
| | steel | ferrite | austenite | steel | ferrite | austenite |
| fraction (vol%) | | 67 | 33 | | 44 | 56 |
| Cr (mass%) | 22.4 | 22.5 | 21.9 | 25.8 | 29.2 | 23.1 |
| Mn (mass%) | 11.8 | 11.5 | 12.7 | 18.9 | 17.9 | 20.1 |
| Mo (mass%) | 3.1 | 3.3 | 2.5 | | | |
| N (mass%) | 0.40 | 0.17 | 1.09 | 0.63 | 0.10 | 1.11 |
| hardness | | | | | | |
| (HV30) | 253 | | | 251 | | |
| (HV0.05) | | 244 | 312 | | 191 | 313 |
| 800 °C, 24 h (HV30) | 455 | | | 520 | | |
| 475 °C, 1000 h (HV30) | 312 | | | 251 | | |
| DBTT (°C) | ≈ 40 | | | ≈ 20 | | |

impellers, ship propellers or tubes under inner pressure rely on the high strength of DSS. So do some offshore applications. The service temperature is usually limited at 250 °C to avoid long term α' embrittlement.

## 5.4 Steels for solution nitriding

By this new heat treatment described in section 4.1.4 a high nitrogen case (HNC) is formed on near-net shape parts at a temperature $T_N$ around 1100 °C in equilibrium with nitrogen of pressure $p_{N2}$. It is the aim of this process to dissolve about 0.4 to 0.9 mass% N in the austenite and obtain either a hard martensitic HNC upon quenching or a high strength yet ductile austenitic HNC around a softer core. To achieve this a high Cr content in the steel is a prerequisite. In some applications the solubility limit of nitrogen may be surpassed to precipitate hard nitrides in the case (Siebert 1994).
At a given case content of N the resistance of austenite to a martensitic transformation depends on the alloy content of the steel chosen. The interdependence of core microstructure and an either martensitic or austenitic case is depicted in Fig. 5.41 for steels of high Cr content, i.e. stainless steels. These are steels of ferritic (F), martensitic (M), austenitic (A) and mixed microstructure. Six standard stainless steels are used to explain the relation of case and core. The superferritic grade 1 suffers from excessive grain growth in the core. Steel compositions in the shaded area are therefore not recommended for solution nitriding. In the annealed

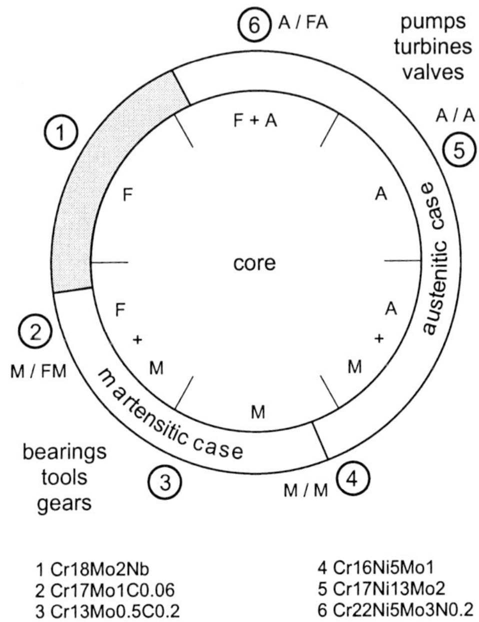


**Fig. 5.41.** Relation between the microstructure of core and case after solution nitriding (Siebert 1994). Encircled are the locations of six exemplary steels and of four combinations arranged counterclock wise from M/FM (martensitic case/ ferritic-martensitic core) to A/FA (austenitic case/ ferritic-austenitic core), see text

condition steel 2 is known as a ferritic grade but after quenching from $T_N$ the core contains ~ 15 vol% martensite pointing to a F + A microstructure at $T_N$ which impedes grain growth in the core. Steel 3 is almost austenitic at $T_N$ and therefore grain growth in the core is retarded compared to the ferritic grade 1. However, the core hardness of steel 3 is almost twice as high as that of steel 2. The case of number 2 and 3 contains martensite and retained austenite (RA) after quenching. By deep-freezing and tempering in the regime of secondary hardening at ~ 450 °C the RA is transformed to martensite. Steel 4 resembles a nickel-martensitic grade, the high Ni content of which leads to an austenitic case around a soft martensitic core. The austenitic steel 5 remains austenitic in core and case, while the ferritic-austenitic core of the duplex steel 6 bears a fully austenitic case. The application of martensitic HNC is seen in the field of stainless bearings, tools, and gears, while austenitic HNC are e.g. suitable for fluid flow components to increase the resistance to cavitation and erosion.

It is desirable to make use of standard stainless carbon grades and introduce advantages of HNS via HNC. Corrosion resistance and wear resistance are confined to the surface and are independent of the core. Here an HNC may offer a low-cost alternative to HNS. In addition the concept of "a harder case around a softer core" combined with residual stresses has advantages over HNS of uniform hardness in store that are well known from other thermo-chemical heat treatments. As to the cost of solution nitriding one has to bear in mind that stainless steels are austenitized or solution annealed at ≥ 1050 °C anyway and that it is mainly the prolonged nitriding time $t_N$ which counts.

Standard stainless steels are available and accepted by industry. However, some are not best suited for solution nitriding. Modifications have been developed which lack availability unless a major consumer appears. The following sections will demonstrate that it is justified to speak of "steels for solution nitriding". To designate the microstructure in case/core abbreviations are used like A/FA for an austenitic case around a ferritic-austenitic core.

## 5.4.1
## Martensitic case

*Martensitic core:* Low-alloy case-hardened steels are usually martensitic in case and core and about free of hard precipitates. Due to the limited C solubility of high alloy stainless Cr or CrMo steels at $T_C$ the obtainable case hardness after carburizing and quenching is unsatisfactory. Within the usual Cr range of 13 to 17 mass% the phase field of austenite in the carburized case is small and cornered by δ-ferrite and carbides (Fig. 5.42a). This draw-back is quite defused by solution nitriding instead of carburizing (Fig. 5.42b), because the atomic ordering of Cr and N atoms (see Sect. 1.1.2.6) enhances the N solubility. A further increase of Cr and N would still yield austenite at $T_N$ but the resulting solid solution strengthening impedes the martensitic transformation and leads to excessive retained austenite (RA).

The high alloy content of stainless steels adds to the core hardness. Therefore the content of interstitial elements has to be kept rather low to avoid an excessive

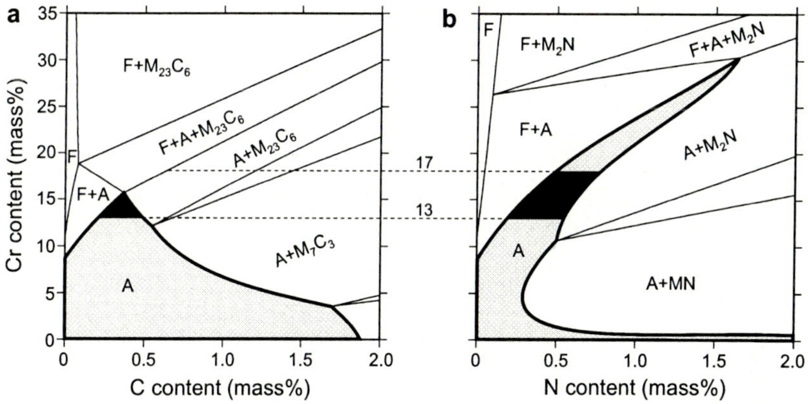


**Fig. 5.42.** Isothermal section of phase diagrams at 1100 °C derived by ThermoCalc for Fe-1 mass% Mo-Cr and additions of **a** carbon or **b** nitrogen. The dark area represent the regime of homogeneous austenite for a stainless Cr range (Juse 1999)

core hardness. This calls for the addition of Ni and Co to suppress δ-ferrite in the core which in turn supports the formation of RA in the case. From experimental results and empirical formulas Ehrhardt (1995) constructed a target field of favourable steel compositions, which promise a martensitic core of moderate hardness surrounded by a martensitic case of low RA content with a hardness of > 59 HRC (> 675 HV30) (Fig. 5.43). The field is encased to the left by an insufficient corrosion resistance and N solubility in the case, to the upper right by

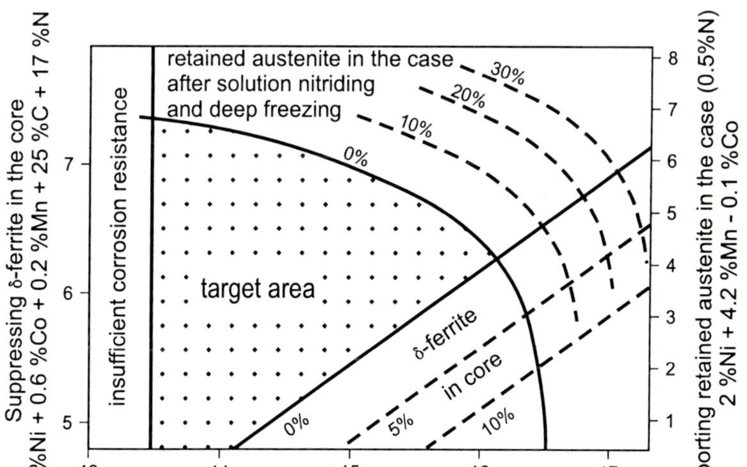


**Fig. 5.43.** Target area of stainless steel compositions to give a hard martensitic case around a softer martensitic core after solution nitriding (Ehrhardt 1995)

excessive retained austenite in the case and to the lower right by unwanted δ-ferrite in the core. Steel M/M in Fig. 5.41 stands for martensitic case/martensitic core and contains (mass%) 13.5Cr, 1.5Mo, 2Ni, 3Co, 0.07N and < 0.2(V+Nb). After deep-freezing and tempering at 450 °C the case shows compressive residual stresses (Fig. 4.13) and a hardness of ~ 700 HV30 (surface content $N_s \approx 0.5$ mass%) while the martensitic core stays at ~ 350 HV30. The ζ-phase $(FeCr)_2N$ supporting secondary hardening during tempering at 450 °C does not deplete the matrix of Cr. Therefore the corrosion resistance is hardly impaired by this tempering treatment. This advantage over stainless steels, case hardened with carbon, opens the door to applications at elevated temperatures (≤ 400 °C). V and Nb promote secondary hardening. In addition about 0.05 mass% Nb will form a dispersion of fine NbN precipitates during soft annealing after hot working which are only partially dissolved during solution nitriding. They act as nuclei for a nitride precipitation during the treatment in case the nitrogen solubility is exceeded. In this manner grain boundary precipitation may be subdued. A dispersion of MN precipitates formed by V and Nb is also effective in impeding grain growth at $T_N$ (Fig. 5.44). Another way of reducing the grain size of austenite is a second hardening from a temperature $T_H$ after intermediate soft annealing (Table 5.13).

If a core hardness of 450 to 500 HV30 is accepted the (N+C) content of the steel may be raised to 0.15 – 0.20 mass%. Cobalt is then no longer required to suppress δ-ferrite. It is well known, though, that precipitation strengthening at peak hardness is accompanied by a drop in ductility and toughness. Tempering at 450 °C, necessary to reach a sufficient case hardness, yields just this critical precipitation hardening in the core, deteriorating the reduction of area and the fracture toughness (Ehrhardt 1995). This predicament may be overcome by tempering the core above the peak of secondary hardening before the martensitic transformation in the case occurs. This is possible because of the high stability of ordered nitrogen austenite in the case and the large difference of $M_s$ temperature

**Table 5.13.** Grain growth during solution nitriding (SN) at $T_N = 1150$ °C for 24 h of steels without grain refinement by Nb (see Fig. 5.48). M, F, A stand for martensite, ferrite, austenite. $T_H$ is the temperature of subsequent hardening (Siebert 1994)

| microstructure of case/core, (steel) | ratio of mean grain diameter after/before SN | |
|---|---|---|
| | case | core |
| M/M (Cr13.5Co3Ni2Mo1.5V0.1) | 2.9 | 4.2 |
| $T_H \leq T_N$ | 2.9 | 1.2 |
| $T_H = T_N$ | 1.2 | 1.7 |
| A/M (Cr16Ni5Mo1) | 2.2 | 1.5 |
| A/A (Cr17Ni12Mo2) | 5.0 | 5.0 |
| A/FA (Cr22Ni5Mo3N0.2) | 10.9 | 3.8 |

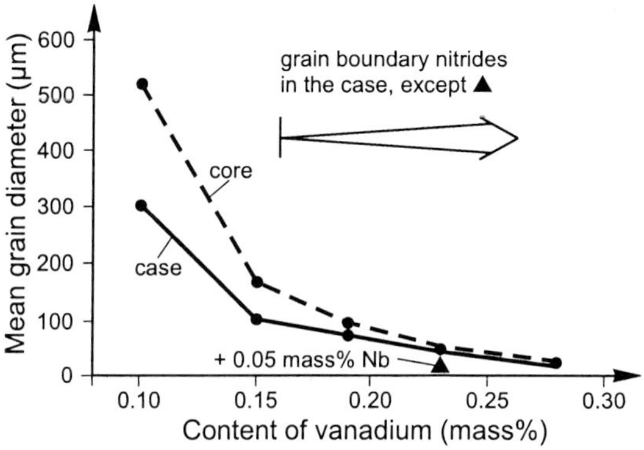


**Fig. 5.44.** Grain growth during solution nitriding of stainless martensitic steels for 10 h at 1150 °C is impeded by V and Nb (Ehrhardt 1995)

between case and core (Berns et al. 1997). At equal case hardness an intermediate tempering treatment increases the impact toughness of the core considerably (Table 5.14).

***Ferritic-martensitic core:*** To further reduce the core hardness and increase ductility a ferritic-martensitic microstructure was investigated (Fig. 5.45). Co is no longer required and (C+N) is exchanged by Ni to control the hardness and the amount of martensite. About 35 vol% of this constituent appears to be an optimum

**Table 5.14.** Effect of heat treatment on the properties of solution nitrided steel Cr13.5Mo1.5Co3Ni2V0.15N0.07 (a) hardened and tempered, (b) intermediate tempering of core before martensitic transformation of case to enhance core toughness at equal case hardness

| | (a) | (b) |
|---|---|---|
| austenitizing | 1050 °C/20 min | |
| quenching | oil, 25°C | salt bath, 160 °C |
| tempering of core | - | 570°C, 1h/air |
| deep freezing | - 80°C, 1 h | |
| tempering | 450 °C, 1 h / 3 times | |
| core: | | |
| hardness (HV30) | 465 | 397 |
| impact energy (J) | | |
| ISO-V, 20 °C | 2 | 24 |
| smooth[1)], - 80°C | 31 | >150[2)] |
| case: | | |
| hardness (HV30) | 663 | 661 |

[1)] specimen 7x10x55 (mm), [2)] no fracture

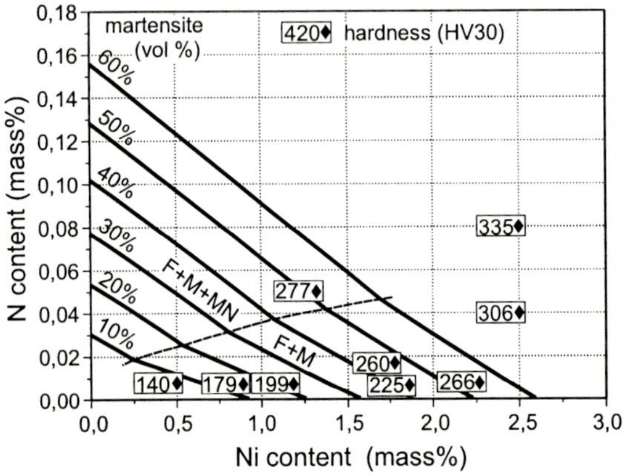


**Fig. 5.45.** Effect of Ni and N on the fraction of martensite in the ferritic (F)-martensitic (M) core of stainless steels with about 13.5 mass% Cr, < 1.5 mass% Mo and traces of V and Nb after quenching from 1100 °C as derived by ThermoCalc. The core hardness increases with the content of martensite and nitrogen (Juse 1999)

between avoiding an embrittling percolation of the harder phase and impeding grain growth at $T_N$. In addition grain growth is controlled by V and Nb as shown for steel M/M. The optimized steel M/FM (Fig. 5.41) contains (mass%) 13.5Cr, 1.5Mo, 1.8Ni, 0.15V, 0.05Nb and 0.03(C+N). After solution nitriding, deep-freezing, and tempering at 450 °C the case hardness is equal to steel M/M but the core hardness is reduced to ~ 250 HV30. The fracture toughness of the core $K_{Ic}$ reaches 65 MPa$\sqrt{m}$ as compared to 28 MPa$\sqrt{m}$ of steel M/M. The respective core microstructure of steel M/FM is shown in Fig. 5.46

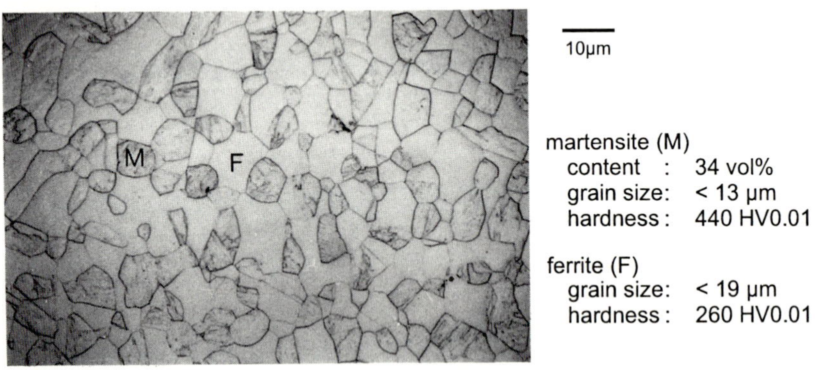


**Fig. 5.46.** Core microstructure of a stainless ferritic-martensitic steel with (mass%) 13.5 Cr, 1.8 Ni, 1.5 Mo, 0.15 V, 0.05 Nb, 0.03 (C+N) after blank hardening from 1150 °C, 30 min in oil to a macro-hardness of 250 HV 30

## 5.4.2
## Austenitic case

In the previous section some nickel was used to control the core microstructure of stainless CrMo steels. As a result RA was stabilized in the case which was subsequently transformed by deep-freezing and tempering. By further increasing the Ni content the $M_s$ temperature of the case may be lowered to sub-zero temperatures leading to an austenitic HNC. The corresponding core microstructure may either be martensitic, austenitic or ferritic-austenitic (Fig. 5.41).

***Martensitic core:*** In nickel-martensitic stainless steels δ-ferrite is suppressed by Ni instead of interstitial elements. The resulting martensite is therefore softer and tougher. After blank hardening steel Cr 13C0.2 reaches about 500 HV30 while steel Cr13Ni4 will stay below 350 HV30. Thus nickel-martensitic steels provide a high strength yet ductile core. Of the grades available steel Cr16Ni5Mo1 was solution nitrided to form an austenitic HNC around a martensitic core, which is designated by A/M (Fig. 5.47). In contrast to the state M/FM the hardness increases from case to core passing through an intermediate peak, where nitrogen strengthens the core martensite mixed with RA. About half of the HNC depth remains fully austenitic although the stability of austenite decreases with the distance from the surface. This may give rise to transformation by stress, strain or winterly temperatures affecting the size stability of thin components in service. In thicker cross-sections the size change is subdued but a rearrangement of residual

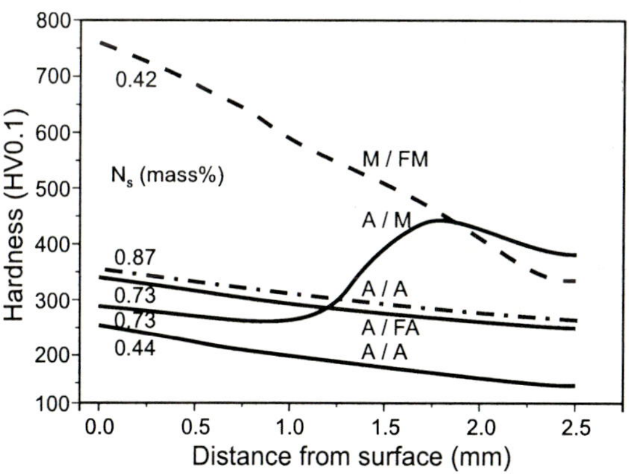


**Fig. 5.47.** Hardness penetration of different steels after solution nitriding at 1150 °C for 24 h and quenching in oil. M/FM was subsequently deep freezed and tempered at 480 °C,
M = martensite, A = austenite, F = ferrite in case/core of the steels Cr16Mo0.5C0.15 (M/FM), Cr16Ni5Mo1 (A/M), Cr17Ni13Mo2 (A/A), Cr18Mn18N0.55 (A/A, dashed line) and Cr22Ni5Mo3N0.2(A/FA),
$N_s$ = surface content of nitrogen

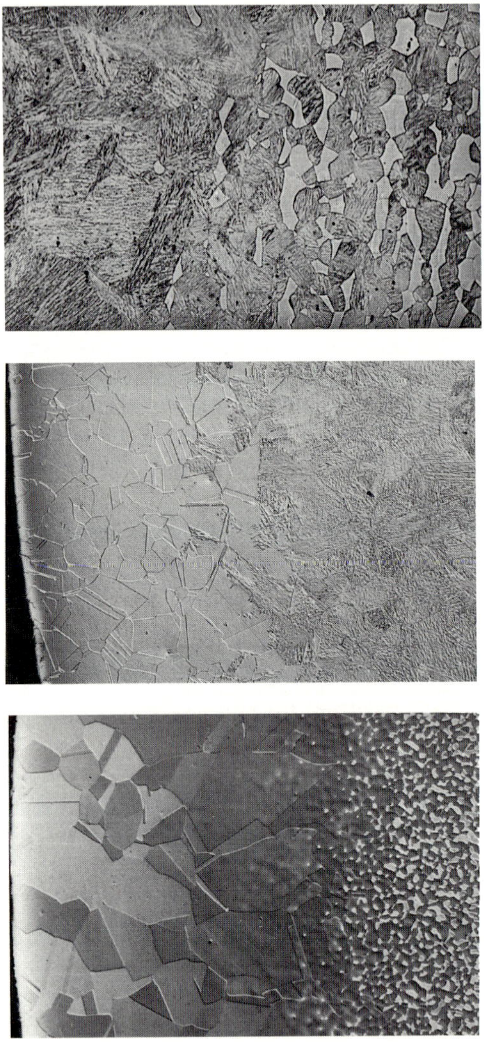


**Fig. 5.48.** Microstructure of case and core after solution nitriding A = austenite, F = ferrite, M = martensite, **a** steel Cr16Mo0.5C0.15, **b** steel Cr16Ni5Mo1, **c** steel Cr22Ni5Mo3N0.2

stresses is liable to occur. Tempering is likely to settle these changes. The microstructure of case and core is given in Fig. 5.48, the respective grain growth in Table 5.13.

***Austenitic core:*** The case of austenitic stainless steel parts is enriched with nitrogen to give a high strength austenitic HNC around an austenitic core (A/A). The solubility of nitrogen is enhanced by Cr, Mo, Mn and decreased by Ni. Therefore in the extra low carbon (ELC) steel Cr17Ni13Mo2 an $N_s$ content of about 0.45 mass% is reached while in steel Cr18Mn18N0.55 an $N_s$ of about 0.85 mass% is feasible (Figs. 5.47, 5.49). The 0.2% proof strength of an ELC grade may be raised from about 200 MPa in the core to 400 MPa in the case while the HNC of CrMn grades may top 700 MPa. Low cost CrMnNi grades have nearly vanished from the market but promise a higher $N_s$ than the ELC grades. For castings to be solution nitrided some of these grades could be revived. Todays tendency to reduce the carbon content in austenitic steels should be reconsidered in view of solution nitriding because the interstitial content of an HNC is anyway above the usual level of avoiding a sensitization to intercrystalline corrosion (IK) upon heating. Solution nitriding is a final heat treatment of near-net shape parts

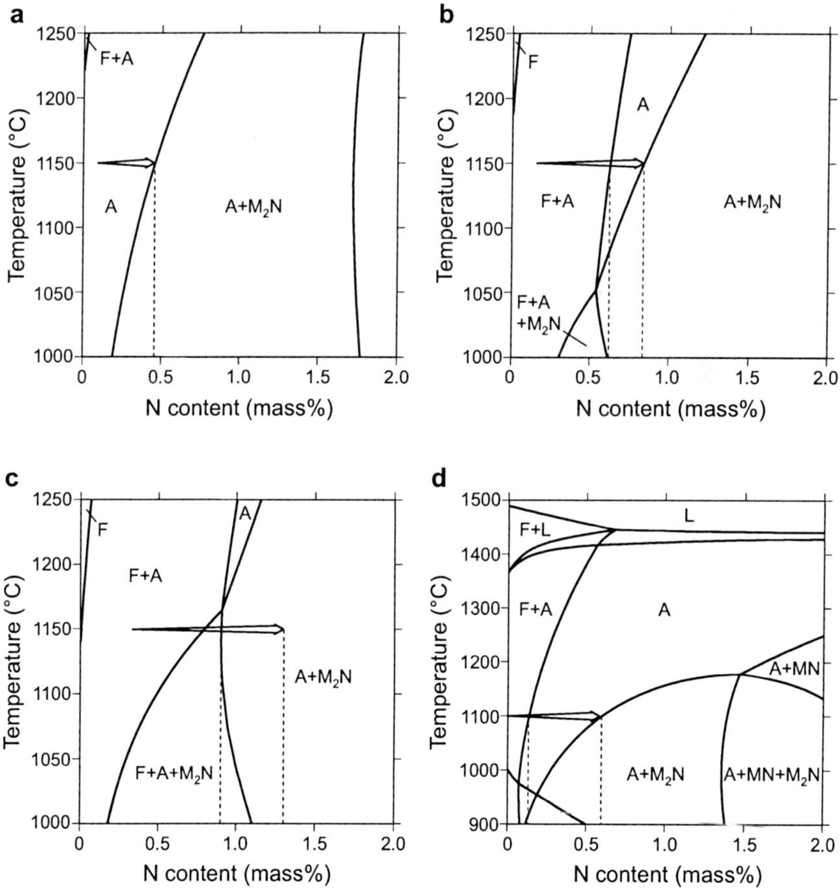


**Fig. 5.49.** Constitution of stainless steels during solution nitriding as derived by ThermoCalc, the arrows indicating the N gradient from core (left) to surface (right) of
**a** A/A, austenitic steel Cr17Ni13Mo2, **b** A/FA, duplex steel Cr22Ni5Mo3N0.2,
**c** AN/FA, duplex steel Cr26Ni6Mo3Cu3N0.2 with nitrides in the case,
**d** M/FM, ferritic-martensitic steel Cr16Mo0.5C0.15 (Juse 1999)

**Table 5.15.** Mechanical properties of stainless duplex steel Cr22Ni5Mo3N0.2 solution annealed 1050 °C/30 min or solution nitrided 1150 °C/1.2 bar/28h. Thin tensile specimens of even N content were tested (Juse and Berns 1997)

| Heat treatment | N-content (mass%) | $R_m$ (MPa) | $R_{p0.2}$ (MPa) | $A_5$ (%) | hardness (HV30) |
|---|---|---|---|---|---|
| solution annealed | 0.19 | 759 | 449 | 38 | 220 |
| solution nitrided | 0.73 | 1078 | 621 | 54 | 319 |

and heating by e.g. welding or stress relieving does not occur thereafter. Heating in service should stay below the sensitization temperature to subdue IK. Solution nitriding of stabilized steels leads to a precipitation of MN in the HNC because Nb or Ti are usually contained above the stoichiometric ratio of MC. The MN precipitates are more evenly distributed compared to the $M_2N$ which prefer grain boundaries.

***Ferritic-austenitic core:*** Stainless duplex steels with 22 to 27 mass% of Cr and only 5 to 7 mass% of Ni offer a high nitrogen solubility, i.e. $N_s \leq 0.9$ mass%. This uptake of nitrogen causes the δ-ferrite to transform to austenite so that a high strength austenitic HNC is formed around a ferritic-austenitic core (A/FA). The phase transformation within the case is accompanied by severe grain coarsing (Table 5.13), yet the mechanical properties are convincing as depicted in Table 5.15.

As shown in Fig. 5.49b the phase field of austenite broadens as the temperature is raised. At a core content of 0.2 mass% N the fully austenitic fraction of the entire HNC increases simultanously but its grain size as well. An elevation of N above the solibility limit of austenite entails the precipitation of $M_2N$ which starts at the grain boundaries followed by a discontinuous growth within the grain. The latter results in a microstructure which is termed pseudo pearlite or nitrogen pearlite although the matrix remains austenitic. The hard lamellae enhance the wear resistance, but reduce the corrosion resistance because they are enriched by Cr. For the state AN/FA (case: austenite + nitrides/core: ferrite + austenite) it is advisable to start with a higher Cr content in the steel. The $N_s$ of steel Cr26Ni6Mo3Cu3N0.2 was raised to e.g. 1.3 mass% to obtain a "pearlitic" near-surface zone Fig. 5.49c.

## 5.4.3 Application of solution nitriding

***Implementation of the process:*** To achieve a martensitic HNC the Cr content has to be confined to the lower end of the stainless range resulting in $p_{N2} > 1$ bar. A cold wall vacuum furnace with a pressurized gas quenching system has proved to be a suitable equipment. Stainless duplex steels are at the high end of the Cr range and $p_{N2} < 1$ bar is sufficient to form an austenitic HNC. These grades may also be treated in a hot wall chamber furnace of lower cost (see Sect. 4.1.4) , but up to now this type of device has only been tested in the laboratory. If it comes to large parts a chamber furnace appears to offer advantages. One is the higher cooling rate of liquid quenching compared to gas quenching in a vacuum furnace. This becomes the more important the thicker the cross-section of the solution nitrided part, i.e. the higher the $t_{8/5}$ cooling time. In Fig. 4.20 the microstructure from core to case (left to right) of three different stainless steels is described in dependence of $t_{8/5}$. The shaded area depicts the homogeneous austenite, respectively martensite and narrows as the cooling slows down. To the right grain boundary precipitates appear at $N \geq N_I$ and discontinuous precipitates within the austenitic grains at $N \geq N_{II}$ which is called "pearlite" the spacing of which increases with $t_{8/5}$. It is obvious, that in the near-surface part of the HNC, precipitation tends to start already at $t_{8/5} >$ 1 s if the surface content $N_s$ is set close to the solubility limit. To avoid embrittling grain boundary precipitates a liquid quench is desirable but usually connected with more distortion of intricate parts than a gas quench. Another means is to lower $N_s$ which diminishes the strengthening of the case.

***Martensitic case:*** The near- surface properties of this HNC are similar to those of hard stainless HNS described in Sect. 5.1.1 but are supplemented by compressive residual stresses (Fig. 4.13). Through-hardening steels of high interstitial content develop tensile stresses at the surface which sometimes initiate cracks during heat treatment, subsequent machining or service, especially in larger cross-sections. Compressive stresses in the case after solution nitriding offer an advantage. The high surface hardness of 58 to 60 HRC increases the wear resistance which may be further enhanced by nitride precipitates but at the expense of corrosion resistance. Tempering at 450 °C allows service at moderately elevated temperatures as discussed in section 5.4.1. These considerations suggest the following applications.

Stainless tools for the processing of food or polymers profit from the combination of hardness and superior corrosion resistance. In addition temperatures of ≈ 250 °C encountered in polymer extrusion do not affect the as-tempered hardness. This is important in the presence of abrading filler materials or fibres. Size stability in service is also improved by the applied tempering treatment. These advantages over stainless carbon grades are in accordance with those discussed for hard HNS in Sect. 5.1.1. If it comes to heavy tools of large cross-section, though, through-hardening stainless steels tend to embrittle because of grain boundary precipitation which is aggravated by segregation and promotes cracking. In such tooling the lower interstitial content in most of the volume of a solution nitrided part eases the heat treatment and improves the core ductility.

Large and high-speed bearings are usually made of low-alloy carburizing steel to cope with positive tangential stresses in the rotating rings and to avoid tensile residual stresses at the surface. Carburizing of stainless steels is not met by a sufficient solubility of carbon in the austenitic regime, which entails embrittling precipitates and impairs the corrosion resistance (Fig. 5.42a). This does not hold true for case hardening with nitrogen (Fig. 5.42b) but subsequent precipitation may occur during quenching (Fig. 4.20c). Overrolling tests have shown that an $M_2N$ decoration of grain boundaries deteriorates the L10 life. A liquid quench would be helpful to achieve a high cooling rate in the relatively thin case, but is not available in combination with a respective furnace to bear a $p_{N2}$ up to 3 bar.

An increase of the N solubility by a higher Cr content is limited by too much retained austenite in the case. To subtract the nitrides from the grain boundaries microalloying with Nb is essential as explained above. The requirements for bearings in air crafts call for a tight combination of surface hardness and core toughness and the development is still on its way.

Lubricated gears do not require stainless materials. However, temper resistance is needed in case of an oil leak, because an increase of friction and temperature coincides with a loss of coolant. This situation may arise e.g. in a helicopter gear and a low-alloy, case-hardened steel is not up to this loading. Case hardening of a high-alloy steel with nitrogen offers beneficial residual stresses and an improved temper resistance. Rotors in gear pumps are often exposed to aggressive fluids containing solid particles. Here solution nitriding of stainless steels may considerably reduce erosion-corrosion. The same holds true for pump impellers and valves, especially if erosion tops corrosion. In addition a casting is less liable to crack at an accidental overload if the wear resistant surface is supported by a tougher core. As a result of a European project (Kühl 1999) case hardening with nitrogen proved to be a new and effective way of providing an optimal combination of properties in case and core for parts in fluid flow machines. Steel Cr16Mo0.5C0.15 with an M/FM microstructure revealed low friction and adhesion, less hydroabrasive wear, a reduction of cavitation damage by a factor of 6 to 8 and a reduction of fretting wear by a factor of 80 as compared to the solution annealed FM state. In 0.2 molar sulphuric acid the passive current density at room temperature was reduced by 50% and the passive range was considerably extended by solution nitriding.

***Austenitic case:*** The application of this surface modification is not as much based on a high hardness increase but on a high plasticity and work hardening capacity induced by nitrogen in an austenitic solid solution. As explained in Chap. 1 nitrogen promotes interatomic metallic bonding and short range atomic ordering, lowers the stacking fault energy and develops a stronger bond to dislocations than carbon. The resulting combination of strength and toughness leads to a much higher resistance to e.g. cavitation. The scratching energy is increased as well entailing a lower erosion rate. In addition the resistance to pitting corrosion is improved considerably by the dissolution of N (Berns et al. 1998 III) . Therefore stainless austenitic HNC were extensively tested on a laboratory scale and in the field of pumps (Kühl 1999).

Solution nitriding reduced the mass loss during vibratory cavitation by a factor of 7.6, 2.3 and 10.8 for the nickel-martensitic steel Cr16Ni5Mo1, the austenitic steel Cr17Ni12Mo2 and the duplex steel Cr22Ni5Mo3N0.2, respectively (Siebert 1994). Flow induced cavitation, measured up to 200 h, was cut down by a factor of about 3 (Kühl 1999). Erosion-corrosion was tested in a bypass of a flue gas desulphurization plant (FGD). After 1000 h in an aqueous solution of 45°C and pH = 5 with 1.5 mass% $Cl^-$ and 15 mass % of suspended lime and gypsum particles the specimens rotating at a speed of 28 m/s showed the mass loss depicted in Fig. 5.50. Solution nitriding of duplex steels reduced the erosion by a factor of 1.5 to 2 and the effect is more pronounced if the HNC contains hard $M_2N$ precipitates (Berns and Siebert 1996).

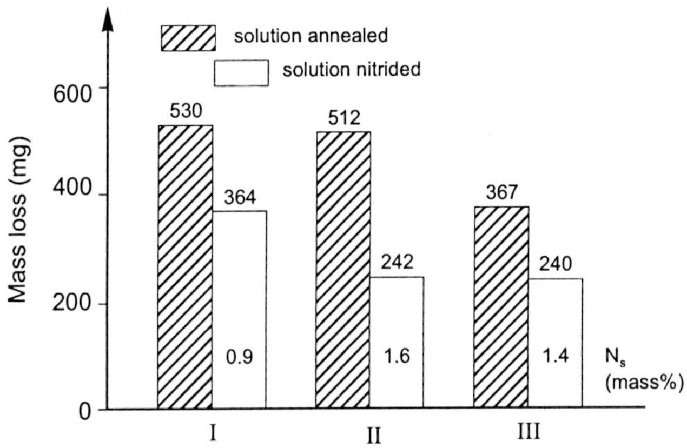


**Fig. 5.50.** Mass loss of stainless duplex steels by 1000 h erosion in an FGD bypass is reduced by solution nitriding to a surface content of nitrogen $N_s$, **I** steel Cr 22Ni5Mo3N0.2 (A/FA), **II** steel Cr27Ni6Mo2N0.15 (AN/FA), **III** steel Cr27Ni5Mo2C0.4 (ANC/FAC), A, F, N, C = austenite, ferrite, nitride, carbide in case/core, after Berns and Siebert (1996)

In a sewage plant sludge was recirculated within an anaerobic digestion system at 37 °C and 7.2 pH by two pumps working alternatively for a week at a time and altogether for 1842 h each. The sludge contained about (g/l) 2.5 $Cl^-$, 2.9 $SO_4^{2-}$, 6.5 sand (mainly quartz of ≈ 150 μm mean size) and in all 50 g/l of suspended solids. Pump A was equipped with a standard impeller made of the stainless duplex steel Cr 26Ni6Mo3Cu3N0.2, solution annealed (state FA). Pump B was run with a solution nitrided impeller of the same composition, the HNC of which consisted of austenite and "pearlitic" $M_2N$ (state AN/FA). Compared to A the mass loss of impeller B was lowered by a factor of 1.5 which resulted in the prediction of 60% more life.

Nickel-martensitic steels are widely used for turbines and other components of hydroelectric power plants because they offer a superior combination of strength and ductility at a sufficient level of corrosion resistance. The same holds true for rotors of radial compressors. Comparing the martensitic core M to the solution annealed state A/M of steel Cr16Ni5Mo1 tempered at 580 °C the vibratory cavitation was reduced by a factor of 6 and the passivation current density in 1-n $H_2SO_4$ by a factor of 12. The break-through potential in aqueous 3% NaCl was extended from 230 to 450 mV and the wear resistance against abrasive flint paper increased by a factor of 2 (Siebert 1994). In short, cavitation and wear are impeded by solution nitriding of nickel-martensitic steels providing even some gain of corrosion resistance. This is due to the high stability of the austenitic HNC, which is hardly affected by tempering at 580 °C, the usual treatment for nickel-martensitic steels. Although the improvement of properties appears to be beneficial for water power equipment, the large size of components speaks against the application of solution nitriding. Castings of e.g. 2.5 m in diameter require a respective furnace chamber for $p_{N2}$ > 1 bar. In case of e.g. welded Pelton or Francis turbines the smaller cups or blades may be solution nitrided, though, if the HNC is removed in the vicinity of the weld to avoid pores. The size of compressor wheels is mainly within the range of available furnaces. An austenitic HNC may enhance the wear resistance against dusty air or the corrosion resistance against aggressive gases or sea air. Because of a lower N solubility steel Cr13Ni4 is less suitable than Cr16Ni5Mo1 of higher cost. Micro-alloying may be employed to enhance the core properties (see section 5.1.4). An industrial application is still pending.

***General guidelines:*** There are numerous applications for low-alloy carburizing steels. As soon as corrosion resistance is added case hardening of stainless steels with N offers a solution hitherto not available. The high strength, yet ductile austenitic HNC is also a new type of surface modification but without a parallel in low-alloy grades. The hard martensitic HNC is best suited for high elastic contact stresses while the tough austenitic HNC is used if the combined stresses locally exceed the yield limit leading to work hardening and large accumulated cyclic strains. The resistance to grooving wear is proportional to the specific scratching energy which is enhanced in both types of HNC: in the martensitic one mainly by the high hardness and in the austenitic one mainly by the high toughness and work hardening rate. The corrosion resistance of austenitic HNC is usually superior to that of martensitic ones because of a higher Cr and Mo content and less lattice imperfections.

The yield strength of the core may vary in a wide range, i.e. from about 350 to 1200 MPa below a martensitic HNC and from about 200 to 800 MPa below an austenitic HNC. Support by a high core strength becomes important if the contact stresses penetrate the case. The toughness of bcc phases (M and F) in the core is diminished by an increase of grain size, hardness, hydrostatic stress, strain-rate and a lowering of Ni content and temperature. The ductile-to-brittle transition temperature (DBTT) is raised accordingly. A stable austenitic core does not fail by cleavage and no DBTT is observed unless martensite is initiated by deformation. With respect to toughness at room and sub-zero temperatures the state A/A is located at the high end and the state M/(F)M at the low end. In bending and torsion an austenitic HNC will sort of shield bcc phases in the core from cleavage, like in the states A/FA and A/M, the DBTT of which is lowered by Ni anyway. The higher DBTT of state M/(F)M is counteracted by compressive residual stresses in the case lowering the mean stress. This is an advantage of all case hardening steels during service. In an austenitic HNC the residual stresses are less pronounced because of the lower yield strength and they are less crucial because of the high ductility which promotes yielding instead of cracking.

From a manufacturing point of view the development of inner stresses during solution nitriding and subsequent quenching governs distortion, cracking and residual stresses. Most of the stresses induced by phase transformation and temperature gradient above 600 °C are decomposed by yielding, leading to distortion but rarely to cracking. Below 600 °C an austenitic HNC can usually cope with the build-up of inner stresses even in the state A/M where the martensitic transformation of the core stretches the case. In contrast the martensitic transformation of the case in state M/M or M/FM takes place at a very low temperature, i.e. between + 130 and – 70 °C. The increase of volume in the case induces triaxial tensile stresses in the core, which locally at flaws or inclusions may fall below the DBTT and fail by cleavage. Parts made of clean hot worked steel are therefore best suited to receive a martensitic HNC. For castings an austenitic HNC is preferred because it is less liable to cause cracking during production.

# 6 From structure to development

The unusual combination of properties of nitrogen steels as discussed in the previous chapters (increased strength, toughness, creep and wear resistance, excellent corrosion properties etc.) allows one to characterise them as a promising class of engineering materials. In relation to mechanical properties, a superior combination of toughness and yield strength contrasting the conventional inverse relationship between these properties is particularly important. This cannot be achieved by carbon. Moreover, attempts to increase the strength of corrosion-resistant steels using carbon were not successful because of its deteriorating influence on the corrosion properties. Although carbon dissolves easily in the liquid steel, its solubility in the austenitic state is limited by elements like Cr and Ni constituting the base of stainless steels. Precipitated chromium carbides are therefore the main reason preventing the development of stainless carbon steels. In contrast, nitrogen has a low solubility in liquid iron-based alloys, but its content in austenite can be rather high. Thus, it has to be underlined that alloying by nitrogen is expedient only at appropriate contents of elements decreasing the thermodynamic activity of nitrogen in iron, which allows to introduce nitrogen to the liquid while the alloyed nitrogen austenite is rather stable against precipitation.

Ranking beneficial effects of nitrogen in steel, one should refer first of all to the resistance to localised corrosion such as pitting, crevice and intercrystalline corrosion as a distinctive quality of nitrogen steels that cannot be procured by carbon. This is prerequisite for the successful application of nitrogen steels providing a possibility to realise other important effects concerned with the improved mechanical properties.

A clear understanding of what nitrogen does in steel and why is important to develop this class of materials. Therefore, it is a key task to clarify the nature of nitrogen-induced effects in metal science of nitrogen steels.

## 6.1 Structure

The most important feature of the crystal structure of nitrogen steels is the increase in the concentration of free electrons, i.e. a promotion of metallic interatomic bonding. In this relation nitrogen and carbon in steel differ crucially. Nitrogen makes the electron exchange between atoms less directed and the distribution of electrons in the crystal structure more homogeneous (Sect. 1.1.1).

This results first of all in the gliding of dislocations without weakening or breaking the interatomic bonds, which usually accompanies the deformation of crystals having pronounced covalent bonding. This effect of nitrogen can be compared to that of nickel and cobalt which also increase the concentration of free electrons in steels and significantly improve their ductility and toughness. Unlike these elements, nitrogen atoms occupy interstitial positions in the fcc crystal lattice causing large elastic distortions, which diminishes the favourable influence of the nitrogen-enhanced metallic character of interatomic bonding.

An important conclusion follows from the analysis of how the concentration of free electrons changes with increasing nitrogen content (Sect. 1.1.1.2). One can expect that the plastifying effect of nitrogen is diminished in austenitic steels containing more than 0.5 mass % N because the metallic component of interatomic bonding decreases. Thus, in constructional steels a content of nitrogen above 0.5 mass % seems to be unfavourable because of an increased probability of brittleness. The cleavage-like fracture observed at room temperature in austenitic steels containing about 1 mass % N and in the austenite of duplex steels alloyed by 0.6 mass % N (see Sects. 3.1.4 and 5.2.3) is consistent with this conclusion.

Another distinctive feature of nitrogen steels arising from their electronic structure is the atomic distribution. It is concerned with a close correlation between the character of interatomic bonding and the preference of nearest atomic neighbourhood in multicomponent solid solutions. As shown in Sect. 1.1.2.6, an increase in the concentration of free electrons, i.e. in the prevailing metallic interatomic bonding assists a non-identical nearest neighbourhood resulting in short range atomic ordering. If a covalent electron exchange between atoms is preferred, the interatomic bonds between identical metallic atoms are favoured, i.e. clustering occurs.

This is the most important point in which nitrogen and carbon diverge in their influence on the structure of steel. Carbon promotes clustering of substitutional atoms in iron-based solid solutions, the precipitation of chromium-rich carbides $M_{23}C_6$ being the most deteriorating result. Nitrogen suppresses $M_{23}C_6$ precipitation because of higher Cr-N interactions as compared to Cr-C ones, and, at the same time, the precipitation of chromium-rich $M_2N$ nitrides is delayed although a lower segregation of chromium atoms is needed for their formation. Like nickel and cobalt, nitrogen assists short range atomic ordering and this effect is attributed to the nitrogen-enhanced metallic character of atomic interactions. A more homogeneous distribution of alloying elements follows from the nitrogen-assisted short range atomic ordering (Sect. 1.1.2.6), which leads to an increasing stability of austenitic steels to precipitation and corrosion.

One of the peculiarities of the atomic distribution in nitrogen steels is concerned with the small affinity of nitrogen to grain boundaries (Sect. 1.1.4.3). In contrast to carbon, nitrogen does not reveal a noticeable tendency to form a grain boundary segregation, which can be explained by the increased concentration of free electrons partly compensating the nitrogen-caused elastic distortions due to some appropriate redistribution of the electric charge in the vicinity of nitrogen atoms. The latter is revealed in the experiment as a smaller value of the electric field gradient at the iron atoms in the neighbourhood of nitrogen atoms as compared to those of carbon atoms. The diminished grain boundary segregation plays an important role in the properties of nitrogen steels.

Interstitial-substitutional interactions and consequent local elastic distortions of the crystal lattice depend on the distribution of nitrogen and substitutional solute atoms in iron-based solid solutions. Both nitrogen and carbon atoms create so called *s-i* atomic complexes in the solid solution because their interaction with nitride(carbide)-forming elements is stronger than with iron. Nitrogen atoms cause a higher dilatation of the austenitic crystal lattice, which is the direct result of the increased concentration of free electrons having a higher energy than localised electrons. At the same time, because of different electronic structures of nitrogen and carbon austenites, nitrogen atoms carry a negative electric charge whereas carbon atoms are positively charged. The electrostatic interaction with a positively charged dislocation line is attractive for nitrogen and repulsive for carbon atoms. That is why nitrogen atoms in austenitic steels strongly interact with dislocations (Sect. 1.1.4.2).

The structure of dislocations is the next important characteristic of nitrogen steels. Nitrogen causes planar slip in austenitic steels, which is usually considered as a sign of a brittle mechanical behaviour of materials. But this does not occur in nitrogen steels except for the case of an extremely high nitrogen content and low temperatures where the localisation of plastic deformation in narrow slip bands leads to a considerable accumulation of planar dislocation arrays in {111} planes creating microcracks and subsequent fracture on the non-active {111} planes (see Sect. 3.1.4). In other words, it is not a typical brittle fracture but rather a pseudo-brittleness arising from nitrogen-enhanced localised plasticity.

The main reason for planar dislocation slip in austenitic nitrogen steels is short range atomic ordering. The passage of dislocations in an ordered solid solution requires an additional expenditure of energy to destroy the ordered distribution of atoms in the vicinity of the slip plane so that the localisation of slip in the once activated planes is favoured. One has to underline that short range atomic ordering, not just short order, as it is sometimes presented, is responsible for planar slip in austenitic nitrogen steels. Otherwise planar slip would be even more pronounced in austenitic carbon steels where carbon causes clustering of chromium atoms, while nitrogen delays it. The breaking of Cr-C bonds would result in a strong preference for next dislocations to follow the first one on the same slip plane. This is not observed, though.

Planar slip increases the low cycle fatigue life and the resistance of nitrogen steels to creep (see Sects. 6.3, 3.1.5 and 3.1.6)

Along with characteristics of dislocation assemblies, the properties of individual dislocations are significantly changed by nitrogen. The effect of nitrogen on the stacking fault energy of austenitic steels remains so far a complicated and controversial issue. It is generally admitted that nitrogen at least slightly decreases the value of SFE. However, as shown in Sect. 1.1.5.1, the sign of the nitrogen effect depends on the basic chemical composition of steels. Moreover, the dependence of the SFE on the nitrogen content is non-monotonous: it has a maximum in CrNiMn and a minimum in CrMn steels. Some correlation is found between SFE and the concentration of free electrons in austenite, which allows to explain the strong temperature dependence of the yield strength in austenitic nitrogen steels (see Sect. 3.1.2).

Mechanical twinning is another peculiarity of austenitic nitrogen steels that is unambiguously determined by the value of SFE: the more dislocations are split the earlier twinning starts during plastic deformation.

Unusual dislocation and twin structures play an important role in high cold work strengthening of austenitic nitrogen steels and contribute much to their resistance under mechanical loading at low and elevated temperatures.

## 6.2 Properties

As mentioned above, alloying of steels by nitrogen causes the following changes in the crystal structure:

(1) an enhanced metallic component of atomic interactions;

(2) a tendency to homogeneous atomic distribution in solid solutions and a delay of precipitation and growth of nitrides;

(3) A change in the stacking fault energy affecting the dislocation structure and twinning.

Let us summarise how this is reflected in the properties of nitrogen steels.

***Metallic character of interatomic bonding*** is a reason for improved toughness and ductility of austenitic nitrogen steels. This difference between austenitic nitrogen and carbon steels (at an interstitial contents of less than 0.5 mass %) is particularly noticeable if the mobility of dislocations is low (shock wave loading, low temperatures) and the metallic or covalent character of atomic interactions becomes decisive for the nucleation and growth of cracks.

The somewhat larger effective size of nitrogen atoms dissolved in the fcc iron lattice as compared to carbon ones arises from an increase in the concentration of free electrons and causes a higher athermal solid solution strengthening of nitrogen austenite. However, a more pronounced relation between the electronic structure and strength of nitrogen steels is revealed at low temperatures.

If the temperature decreases, the thermally activated component of the yield strength becomes significant and, in fcc crystals of low stacking fault energy (less than 100 mJ/m$^2$, which applies to austenitic steels), it is controlled by the intersection of gliding and forest dislocations. This is where the effect of nitrogen on the interatomic bonds and, indirectly, on the structure of the dislocation cores comes in. This effect is of quantum-mechanical origin and arises from the correlation between stacking fault energy and state density at the Fermi level.

Nitrogen increases the concentration of free electrons which occupy different energy states according to Fermi statistics. The Fermi level is blurred with temperature and consequently the state density is decreased if temperature increases. This causes a temperature dependence of SFE, which results in an enhanced dislocation splitting at low temperatures. As the slip of extended dislocations is accompanied by the formation of constrictions and needs higher applied stress, the yield strength of austenitic nitrogen steels increases sharply with decreasing temperature. At 4 K nitrogen is a 1.4 times more effective strengthener of austenite than carbon (Sects. 3.1.1.2 and 5.2.6).

The nitrogen-caused decrease of the stacking fault energy in most austenitic steels leads to mechanical twinning which increases the effectiveness of cold work hardening and provides a high strength of cold worked austenitic nitrogen steels (Sect. 3.1.3), an improved resistance to sliding wear (Sect. 3.1.7) etc. Twinning in austenitic high nitrogen steels occurs on intersecting planes and icludes second order twin systems creating thereby effective obstacles for plastic strain. At the same time, the role of twins in strengthening of highly deformed steels is overestimated as they blur at large strains and some other mechanisms of strengthening concerned with collective modes of deformation come in.

As mentioned above, peculiarities of the space distribution of electrons in the fcc crystal lattice are responsible for the high energy of binding between nitrogen atoms and dislocations. The effective pinning of dislocations by nitrogen atoms results in a higher athermal solid solution strengthening (Sect. 3.1.1.1) and strain ageing of austenitic nitrogen steels (Sect. 3.1.3). But a unique effect of the interaction between nitrogen atoms and dislocations is the nitrogen-enhanced grain boundary strengthening.

The efficiency of grain boundaries as obstacles for the slip transfer between adjacent grains in austenite grows with an increasing nitrogen content. That is why grain refinement is a powerful tool for the strengthening of nitrogen steels, which is so far not sufficiently used in practice. The nature of this effect lies in the increasing blocking of dislocation sources, which requires a higher applied stress to initiate slip in the adjacent grains for the transfer of deformation through the grain boundary (Sect. 3.1.2).

***Short range atomic ordering*** in nitrogen steels arises from a preferential metallic component of atomic interactions. As a result a more homogeneous distribution of substitutional solutes occurs. This leads first of all to a higher stability of austenitic stainless steels in relation to precipitation and phase transformations, which affects the properties in several aspects.

The absence of a noticeable grain boundary segregation of nitrogen contributes to the delay in the precipitation at grain boundaries. Consequently, the precipitation of chromium carbides during sensitisation treatments is shifted to longer exposures resulting in an improved toughness and corrosion resistance.

The precipitation of embrittling intermetallic phases (Sect. 2.4.3), in particular of the σ phase, is suppressed by nitrogen, which allows to avoid a deterioration of the toughness.

The creep resistance of austenitic steels is increased due to the high stability of nitrogen austenite as nitrogen atoms in solid solution being dragged by dislocations limit their mobility at elevated temperatures.

The resistance of austenitic steels to hydrogen embrittlement can be derived from the nitrogen-increased stability of austenite if one accepts hydrogen-induced phase transformations as the mechanism of embrittlement (Sect. 3.2.3). If one exploits the hypothesis of hydrogen-enhanced localised plasticity, one can expect that short range ordering in the distribution of nitrogen atoms eases the trapping of hydrogen atoms and retards the transport of hydrogen to the dilated zones of the crystal lattice.

The strengthening effect of nitrogen in martensitic steels also arises from the nitrogen-caused short range atomic ordering. Quench-and-temper martensitic steels are strengthened by precipitates. The size of nitride particles precipitated during the decomposition of as-quenched martensite is smaller and consequently the strengthening effect is higher as compared to carbides in martensitic carbon steels. The dualism of ordering-clustering in the behaviour of nitrogen and carbon atoms takes place in α exactly as in γ solid solutions (Sects. 1.2.2 and 2.5). Furthermore, nitrogen-assisted short range ordering (or delay of clustering) in austenite is inherited by the as-quenched martensite and retained during tempering, shifting the precipitation of nitrides to higher temperatures and retarding the growth of precipitates. That is why the secondary hardening of martensitic nitrogen steels is so effective.

In duplex ferritic-austenitic steels the partitioning of nitrogen between two phases is of particular significance. As an austenite-former, nitrogen is preferentially located in the γ phase strengthening it and thereby equalising the properties of the two phases, which is important e.g. for prolonged fatigue life. Moreover, affecting the chemical activity of substitutional alloying elements, nitrogen changes their partitioning between the phases increasing the content of chromium in austenite and making the latter chemically more resistant. At the same time, because of a strong interaction between nitrogen and chromium atoms, an increased nitrogen solubility in the chromium-rich ferrite occurs, which allows to predict the probability of the following important effect. As nitrogen delays short range decomposition (clustering) of iron-chromium solid solutions, the 475°C embrittlement of chromium-rich ferrite in duplex steels may be prevented at some sufficient nitrogen contents. This effect would allow to increase the service temperature of duplex nitrogen steels. Some data obtained for duplex high-nitrogen CrMn steels seem to confirm this idea (see Sect. 5.3.3).

***Planar dislocation slip*** caused by nitrogen in austenitic steels is an important consequence of short range atomic ordering. As mentioned above, it arises from the preference of localised slip in once activated planes where further gliding of dislocations does not need an expenditure of energy to break the atomic ordering.

Nitrogen-assisted planar slip affects the creep behaviour because it retards the climb of dislocations between active slip planes as well as it subdues the formation of subgrains that are usually formed during creep of nitrogen-free steels (Sects. 3.1.6 and 5.2.8).

Planar slip is also responsible for the increased low cycle fatigue life of austenitic nitrogen steels because it diminishes the tendency to cross slip providing reversible plastic deformation (see Sect. 3.1.5).

***Surface effects.*** A more homogeneous nitrogen-caused distribution of substitutional atoms in the solid solution appears to be superior to a clustered one in relation to corrosion properties as corrosion attacks the areas of lowest concentration. A higher resistance to general corrosion is one of the consequences of the nitrogen-assisted homogeneity in the distribution of alloying elements in the solid solution.

Another effect is concerned with the state of grain boundaries. Preventing $M_{23}C_6$ precipitation which would be accompanied by a decreased content of chromium in the grain boundary zones, nitrogen improves the resistance to intercrystalline corrosion. As clustering of chromium atoms is retarded by nitrogen, no acceleration and even a delay is observed in the precipitation of chromium nitrides in spite of a higher affinity of nitrogen to chromium (see Sects. 1.1.3.1 and 2.4.3) and a smaller chromium content in the nitrides that would require a smaller preliminary clustering of chromium atoms for the formation of nitrides.

The nitrogen-increased resistance to pitting and crevice corrosion can be also partly attributed to the increased homogeneity of the solid solution because pits are nucleated in chromium-depleted areas. Another important effect arises from the interaction of nitrogen cations with $Cl^-$ anions and the formation of $NH_4^+$ species resulting in a desorption of the former and a local raise of pH.

It has to be stressed again that the high resistance of nitrogen steels to local corrosion is their remarkable advantage over respective carbon steels. It allows to realise the unique possibility of strengthening stainless steels by interstitials and simultaneously improve their corrosion properties.

***Nitrides.*** The key benefits of HNS stem from nitrogen in solid solution as summarized in the previous paragraphs. However, in some applications nitrogen is used in form of nitrides precipitated during ageing of austenite and tempering of martensite. We have to discern this intentional precipitation from the unwanted one which is generally avoided in steels of the solid solution type. Short range atomic ordering induced by nitrogen is responsible for the fine precipitation structure in nitrogen steels affecting both stages of precipitation: nucleation and growth of chromium nitrides.

Lamellar chromium nitrides as well as globular ones were precipitated by appropriate alloying and ageing of austenitic steels to increase the creep resistance. The results are contradictory and lack long time confirmation. Alloying by vanadium or niobium allows to avoid the lamellar structure of nitrides (see Sect. 2.5) and to increase the yield strength of austenitic steels without deteriorating their toughness.

Because of higher supersaturation and faster diffusion martensite is more prone to develop nitrides. In hard martensitic stainless steels, though, a unique nitride phase enables secondary hardening with hardly any loss of corrosion resistance. This constitutes a major improvement over stainless carbon grades which in addition are less resistant to rolling contact fatigue because of some coarse carbides. The transition nitrides responsible for secondary hardening at about 450 °C are transformed to equilibrium nitrides as the tempering temperature exceeds 600 °C. Compared to respective carbides the nitrides consume less chromium and are fine and more homogeneously distributed, which improves the mechanical and chemical properties of HNS. Last not least some 10 vol % of nitrides are embedded in a martensitic matrix to enhance the wear resistance of stainless tool steels manufactured from nitrided powder by hot isostatic pressing. At equal wear resistance the corrosion resistance of HNS is distinctly superior to respective carbon grades.

## 6.3 Technical relevance

In general interstitial alloying elements have a stronger impact on the constitution and mechanical properties of steels than substitutional ones. Carbon has always been used because it is a remainder of the reduction process. The introduction of nitrogen started much later and made good progress during the last decade. The development of HNS is based on a superior combination of properties: improved mechanical properties in the bulk and enhanced chemical properties at the surface.

***Manufacturing.*** Compared to C the solubility of N is lower in the melt but higher in the austenite which entails differences in manufacturing between HNS and respective carbon grades. The high solubility of nitrogen in austenite provides the basis of HNS. Austenite is used in its stable form or as a parent phase of martensite which emerges during cooling or cold working. Ferrite is hardly suitable because of inferior N solubility unless in combination with austenite like in duplex steels. The volatility of N requires special measures to achieve a sufficient solubility in the melt (Sect. 4.1). The most common one is alloying, especially by Cr, leading to high alloy steels, as e.g. stainless grades. We encounter an N content of approximately (mass%) $\leq 0.2$ in martensitic HNS, $\leq 0.4$ in austenitic CrNi HNS and $\leq 1$ in austenitic CrMn HNS. A further rise of N calls for pressure or powder metallurgy which add to the costs. This holds also true for subsequent costs by impaired workability and weldability (Sects. 4.2 and 4.3). Pre-shaping by casting under pressure is excluded for costs as well. Pressurized electroslag remelting is e.g. applied to martensitic HNS for stainless bearings, tools and valves to gain a high hardenability and also to austenitic HNS for high strength and body friendly applications. Powder metallurgy may e.g. provide wear resistant stainless grades with up to about 3 mass% of N. As always, only a convincing performance justifies higher costs. Therefore the major part of HNS is produced by melting under normal pressure employing $N_2$ blowing or nitrided ferro-chromium to induce nitrogen. The manufacturing of certain products is addressed in Sect. 4.6.

***Advantages of HNS.*** The following aspects are particularly important for the application of HNS. (1) The solubility in austenite is higher for N than for C which jointly improves strength and toughness of HNS compared to respective high alloy carbon grades. (2) Nitrogen stabilizes the austenite without being as close to embrittling precipitations as carbon grades. This again promotes toughness, especially in thicker cross sections. (3) After aging or tempering carbides tend to decorate grain boundaries, while nitrides are more evenly distributed and smaller. As a result the mechanical and chemical properties of HNS are enhanced. (4) Quite in contrast to C, dissolved N suppresses localized corrosion. The combination of strength, toughness and corrosion resistance of HNS distinctly surpasses that of respective carbon grades and is the reason for the development and growing application of HNS. As shown in the previous sections these advantages are traced back to the influence of N on the electronic structure and the atomic distribution. It is demonstrated that N supports the metallic character of interatomic bonding and leads to short range atomic ordering which in turn assists planar slip. The differences between the effects of N and C on the atomic level reflect the diverging properties of the respective steels. This underlines that the advantages of HNS are not accidental but well-founded. One advantage is easily overlooked, namely the possibility to exploit the benefits of N in so many different groups of steel, of which more than twenty are listed in Chap. 5. We see N in various austenitic and martensitic grades as well as in steels of mixed microstructure like dual-phase and duplex steels. Moreover, solution nitriding offers a means to enrich the surface of near-net shape parts with dissolved N leading to a hard martensitic or a high strength, yet ductile austenitic case. However, ferritic steels are not among HNS because of their low N solubility. But this is more than compensated by completely new grades in other groups of steel like e.g. nickel-free body friendly austenitic steels. It is evident that N is not only used to enhance the properties of existing steels but also to design new ones for advanced applications.

***Austenitic steels.*** As mentioned in Sect. 6.1 the metallic component of interatomic bonding is enhanced up to about 0.5 mass% N. This corresponds well to the upper limit of high strength HNS (0.6 mass% N) or corrosion resistant HNS (0.5 mass%

N) melted under normal pressure. At higher N contents, e.g. by pressure melting or high CrMn contents an embrittlement starts at cryogenic temperatures. The transition temperature of the notch impact toughness is raised as the N content grows until it reaches room temperature at about 1 mass% N which is considered an upper limit for austenitic steels. A smooth shape and monotonous loading tend to lower the transition temperature of work pieces. But in case of cyclic loading, fatigue cracks support the embrittling effect of too high a nitrogen content. (Sect. 3.1.4)

Between 15 and 30 mass% Cr are alloyed to impart N solubility and corrosion resistance to the austenite. A high CrMo content stabilizes ferrite and promotes intermetallic precipitation, which are both counteracted by N. Carbon stabilizes austenite as well but impairs toughness and corrosion resistance because of covalent bonding and resulting clustering. Therefore the development centers on extra-low carbon HNS. Of the substitutional austenite stabilizers, Ni enhances the metallic bond, i.e. toughness, and the corrosion resistance but lowers the N solubility. Manganese strengthens the embrittling covalent bond, lowers the corrosion resistance but raises the N solubility.

This condensed set of rules and relations is helpful to catch the essentials of HNS. In standard stainless austenitic steels about 0.15 mass% N are added to improve strength and corrosion resistance (Sect. 5.2.1). Replacing expensive Ni by Mn and N yields low cost CrMnN steels of moderate corrosion resistance but high strength and wear resistance due to 0.5 to 1 mass%N introduced by melting under normal pressure (Sect. 5.2.2). Pressure melting allows to add Mo which lifts the resistance to localized corrosion. After cold working high strength strip, wire or expanded rings emerge (Sect. 5.2.3). Large retaining rings for electric power generators are a successful product of this kind (Sect. 5.2.3). Ni-free CrMnMoN grades are also designed to avoid Ni allergy and serve as body friendly alloys (Sect. 5.2.4).

The toughness of Ni-free grades at subzero temperatures is limited and therefore non-magnetic HNS for cryogenic service rely more on Ni than on Mn (Sect. 5.2.6). Because of ferromagnetic ordering the Ni content is however limited to about 20 at%. Therefore 0.3 to 0.4 mass% N assist the stabilization of austenite down to the temperatures of superconducting equipment. Manganese acts in the same direction and indirectly so by increasing the N solubility. The applied N level promises a large concentration of free electrons providing a high degree of metallic bonding in these CrNiMnN steels. At higher N contents a ductile-to-brittle transition is impending, especially at the abnormally intense solid solution strengthening of N at low temperatures.

If it comes to HNS of high corrosion resistance, Ni is again preferred to Mn which is added to lift the soluble N content (Sect. 5.2.5). Steels with up to 0.5 mass% N are commercially melted under normal pressure and pressure melted steels of up to 0.9 mass% N have been developed. Nitrogen suppresses localized corrosion, assisted by $\leq$ 7 mass% of Mo. The high CrMo level is only feasible with the help of N procuring the necessary stability of austenite to the formation of ferrite and to deteriorating intermetallic phases. These superaustenitic HNS are meeting new demands in chemical and process engineering.

Austenitic HNS for service at elevated temperatures contain up to about 0.25 mass% N in solid solution. In addition up to 0.7 mass% N are precipitated as nitrides in pressure melted steels. The short time creep resistance is raised by dissolved N. The influence of amount and morphology of nitrides needs further study and so does the long time behaviour (Sects. 5.2.7, 5.2.8). In a Ni alloy with 30 mass% Cr the precipitation of $\pi$ nitrides supports the creep strength up to 1000 °C considerably (Sect. 5.2.9).

To conclude we can state, that the aforementioned development of austenitic HNS for various applications did not merely generate better properties but broke new ground in industry.

***Martensitic steels.*** Compared to austenitic HNS the martensitic ones are of low Ni and Mn content to facilitate a transformation from austenite to martensite during cooling. The nitrogen solubility depends on the Cr content which is limited to about 17 mass% because of retained austenite. Therefore pressure melting is required to achieve N contents above 0.2 mass%. This applies to hard martensitic steels which are tempered at < 500 °C for decomposition of retained austenite and secondary hardening. The latter is brought about by a nitride phase which is not enriched in Cr and in consequence hardly affects the corrosion resistance of stainless CrMo HNS. In contrast respective carbon grades loose their corrosion resistance during secondary hardening because of a carbide precipitation rich in Cr. A combination of about 0.35 mass% N and 0.3 mass% C in a stainless CrMo steel offers superior corrosion resistance combined with hot hardness due to secondary hardening and with improved fatigue life in rolling contact due to the absence of coarse precipitates. This unique combination of properties was not available before and has e.g. extended the life of some advanced bearings by an order of magnitude.

Melting under normal pressure provides just the N content required for steels tempered above 600 °C, i.e. 0.1 to 0.2 mass%. Among these are the creep resistant grades and the stainless quench-and-temper steels. The former thrive on the smaller size and more homogeneous distribution of nitrides as compared to carbides which enhances the creep resistance. In the latter the refinement of microstructure by N entails a 40 % increase of toughness over respective carbon grades. In addition these HNS show a better corrosion resistance because the precipitated nitrides consume less Cr than respective carbides.

Powder metallurgy allows to introduce a few percent of nitrogen and bind it to V,Nb or Ti. This leads to a dispersion of hard nitrides embedded in a hardenable stainless matrix. The microstructure of these wear resistant HNS matches that of conventional PM tool steels but the corrosion resistance of the former is quite superior. This points to tooling for the processing of food and aggressive polymers.

In spite of better properties martensitic HNS will have a hard time to conquer those markets that rely mostly on stock. This holds true for quench-and-temper grades and for tooling materials. Creep resistant HNS will take time for a long test duration. Today the application of martensitic HNS has progressed most in the bearing industry where nitrogen is seen as a milestone for some advanced solutions.

**Steels of mixed microstructure.** Ferritic-martensitic dual-phases steels contain from 10 to 25 vol% martensite in which the interstitial contents is accumulated. In a stainless grade the use of N instead of C improves the corrosion resistance of martensite, which is inferior to ferrite in this respect. Although the N content of such a CrMo steel amounts to only 0.1 mass% a high nitrogen austenite is formed during intercritical annealing and transformed to martensite during quenching. Applications are seen in high strength stainless sheet for deep drawing but an industrial turnover is still lacking.

A similar partitioning of N between two phases is observed in ferritic-austenitic duplex steels. At a low interstitial content the austenite is weaker in respect to strength and corrosion resistance. Alloying with nitrogen leads to its accumulation in austenite which is improved in both respects. This favourable influence of nitrogen is widely used in industry today. On top nitrogen impedes the precipitation of detrimental intermetallic phases at high CrMo levels and helps to save Ni even down to body friendly CrMnMoN duplex grades. As superaustenitic steels the superduplex CrNi(Mn)MoN grades of highest corrosion resistance owe their development and industrial success to about 0.4 mass% N.

Martensite is formed in austenitic-martensitic steels either by cold working of instable austenite or by tempering of nickel-martensitic steels. Only the latter is accompained by partitioning of nitrogen between the two phases. Nitrogen enhances the corrosion resistance and strength of cold worked strip and wire as well as the toughness and creep resistance of reaustenitized steel. Applications are in progress.

The above HNS of mixed microstructure do not require pressure melting. The N content of dual-phase steels is low anyway. The high Cr content of duplex steels provides a sufficient N solubility taking into account that the N content in the austenite is almost twice as high as in the steel. Austenitic-martensitic grades are of moderate N level to retain toughness in the martensite.

# References

Aaronson HI, Domian HA, Pound GM (1966) Thermodynamics of the austenite→proeutectoid ferrite transformation. I, Fe-C alloys. TMS AIME 236(5): 753-767

Abrikosov IA, Ruban AV, Johansson B, Skriver HL (1998) Total energy calculations of random alloys: supercell, Connolly-Williams and CPA methods. Comput Mater Sci 10: 302-305

Adachi T (1991) Effect of alloying elements on SCC resistance and crevice corrosion of austenitic stainless steel in NaCl solution. In: Stainless Steels'91, Chiba. The Iron and Steel Institute of Japan, pp 189-195

Agren J (1979) A thermodynamic analysis of the Fe-C and Fe-N phase diagrams. Metall Trans 10A(12): 1847-1852

Akdut N, Foct J (1996) Microstructure and deformation behaviour of high nitrogen alloyed duplex stainless steels. ISIJ Intern 36(7): 883-892

Akdut N, Keichel J, Foct J (1997) The influence of nitrogen and orientation on the rolling deformation mechanisms of austenitic single crystals. steel research 68(11): 495-500

Amigood GG, Litvinov VS (1983) Short range order and stability of austenite in iron-manganese alloys of Fe-20Mn type. Metal Physics and Metallogr (in Russian) 56(6): 1132-1137

Andersson J-O (1985) Thermodynamic properties of chromium. Intern JournThermophys 6(4): 411-419

Andersson J-O, Sundman B (1987) Thermodynamic properties of the Cr-Fe system. CALPHAD, 11: 83-92

Andersson J-O, Fernández Guillermet A, Hillert M, Hansson B, Sundman B (1986) A compound-energy model of ordering in a phase with sites of different coordination numbers. Acta Metall 34(3): 437-445

Andrew J (1912) Iron and Nitrogen. In: Carnegie Scholarship Memoirs, The Iron and Steel Institute, London, vol ii, pp 210-226

Andrews KW, Brookes PE (1951) $\chi$ phase in alloy steels – its relationship to $\sigma$ phase. Metall Treat Drop Forg 18: 301-311

Anthamatten BR, Cui ML, Uggowitzer PJ (1987) Neue warmfeste ferritische Stähle mit Stickstoff. In: Speidel MO, Uggowitzer PJ (eds) Stickstofflegierte Stähle. Verlag Thubal-Kain, Institut Metallforschung und Metallurgie, ETH-Zürich, pp 205-217

Anthamatten BR, Uggowitzer PJ, Cui ML, Speidel MO, Stein G (1989) New high nitrogen ferritic steels. In: Foct J, Hendry A (eds) High Nitrogen Steels, HNS 88. Institute of Metals, London, pp 58-62

Anthamatten BR, Uggowitzer PJ, Solenthaler Ch, Speidel MO (1990) High-nitrogen 9-12% chromium steels for high-temperature applications. In: Stein G, Witulski H (eds) High Nitrogen Steels, HNS 90. Stahl&Eisen, Düsseldorf, pp 436-441

Antonov VN, Nemoshkalenko VV, Gavriljuk VG, Vakhnej AG, Foct J (1996) Electronic structure of some complex chromium carbides. Metal Physics Advanced Technologies 15: 697-709

Archard J (1953) Contact and rubbing of flat surfaces. J Appl. Phys 24: 981-988

Armstrong RW, Codd I, Douthwaite RM, Petch NJ (1962) The plastic deformation of polycrystalline aggregates. Phil Mag 7: 45-58

Armstrong RW (1995) Hall-Petch analysis of yield, flow and fracturing. In: Otooni MA et al. (eds) Grain size and mechanical properties – fundamentals and applications. Materials Research Society, Pittsburgh, Pennsylvania, 362, pp 9-17

Ashby M (1970) The deformation of plastically non-homogeneous materials. Phil Mag 21: 399-424

Atkinson D, Bodsworth C (1970) Thermodynamic properties of nitrogen in austenitic iron and iron-nickel alloys. JISI 208(6): 587-593

Azuma S, Miyuki H, Kudo T (1996) Effect of nitrogen on crevice corrosion of austenitic stainless steels. ISIJ Intern 36(7): 793-798

Avery H (1974) Work hardening in relation to abrasion resistance. Proc of Symposium on Materials for Mining Industry. Cimax Molybdenum Company, Vail, Colorado, pp 43-77

Bachelet GB, Hamann DR, Schlüter M (1982) Pseudopotentials that work: from H to Pu. Phys Rev B26(8): 4199-4228

Bain EC, Aborn RH, Rutherford JJB (1933) The nature and prevention of intergranular corrosion in austenitic stainless steels. Trans Amer Steel Treating Soc 21: 481

Balanyuk AG, Sozinov AL, Gavriljuk VG (1998) C-C interaction in iron-based austenite by using Monte Carlo computer simulation. Metal Physics Advanced Technologies 20(7): 11-14

Balanyuk AG, Bugaev VN, Nadutov VM, Sozinov AL (1998) Estimation of the energies of N-N and C-C interactions in fcc Fe-N and Fe-C alloys on the basis of Mössbauer spectroscopy data. Phys. Stat. Sol. (b) 207(1): 3-12

Bannykh OA, Blinov VM (1991) On the effect of discontinuous decomposition on the structure and properties of high-nitrogen steels and on methods for suppression thereof. steel research 62(1): 38-45

Banov RM, Parshorov IM, Kamenova CZ (1978) Relaxation phenomena in austenitic steels alloyed by nitrogen. Izvestia AN SSSR (in Russian) 1: 126-129; A model of relaxation in nitrogen austenitic steels. Izvestia AN SSSR (in Russian) 3: 178-182

Ban-ya S, Elliott JF, Chipman J (1969): Activity of carbon in Fe-C alloys at 1150 °C. TMS AIME 245(6), 1199-1206

Ban-ya S, Elliott JF, Chipman J (1970) Thermodynamics of austenitic Fe-C alloys. Metall Trans 1A(5): 1313-1320

Baran NP, Gavriljuk VG, Maximenko VM, Smouk EE, Smouk SYu, Shanina BD (1992) Spin resonance of conduction electrons in carbon and nitrogen austenites. Solid State Commn. 81(1): 55-58

Barcik J (1986) The process of σ-phase solution in 25 pct Cr-20 pct Ni austenitic steels. Metall Trans 18A(7): 1171-1177

Bashchenko AP, Izotov VI, Omelchenko OV, Soshnikov PS, Shcherbedinskij GV (1985) Alloying of iron and ist alloys by nitrogen under high pressure. Izvestia AN SSSR (in Russian) 4(4): 173-178

Bauer Ph, Uwakweh ON, Genin JMR (1988): CEMS study of the carbon distribution in austenite. Hyperfine Interactions 41: 555-558

Bauer Ph, Uwakweh ONC, Genin JMR (1990) Mössbauer study of the distribution of carbon interstitials in iron alloys and the isochronal kinetics of the aging of martensite: the clustering-ordering synergy. Metall Trans 21A: 589-602

Beachem CD (1972) A new model for hydrogen-assisted cracking (hydrogen „embrittlement"). Metall Trans A3(2): 437-451

Becker H, Brandis H, Küppers W (1986 I) Zur Verfestigung instabil austenitischer nichtrostender Stähle und ihre Auswirkung auf das Umformverhalten von Feinblechen (Schrifttumsübersicht). Thyssen Edelst. Techn.Ber. 12 (1): 35-54

Becker H, Brandis H, Küppers W (1986 II) Untersuchungen zur Auswirkung unterschiedlicher Walzbedingungen auf die mechanischen Eigenschaften von nichtrostendem Federbandstahl X12CrNi177 (Remanit 4310). Thyssen Edelst. Techn.Ber. 12 (1): 55-65

Beer SZ (1961) Solubility of nitrogen in molten iron manganese alloys. Trans TMS AIME 221: 2-8

Bechtoldt CJ, Vacher HC (1953) Phase-diagram study of alloys in the iron-chromium-molybdenum-nickel system. J Research National Bureau Standards 58(1): 7-19

Berg H-J, Serchen B, Spies H-J (1995) Randaufsticken hochlegierter Stähle zur Verbesserung der Korrosionsbeständigkeit. In: Erzeugung und Charakterisierung von Randschichten auf Bauteilen und Werkzeugen, Bergakademie Freiberg, pp 169-184

Bernauer J., Lichtenegger G, Hochörtler G, Lenger H (1998) Development of high nitrogen steels at Böhler Edelstahl Gmbh Kapfenberg. In: Hänninen H, Hertzman S (eds) High Nitrogen Steels, HNS 98, Proc of the 5th Intern Conf at Espoo/Stockholm, to be published

Bernhardsson S (1991) The corrosion resistance of duplex stainless steels. In: Charles J, Bernhardsson S (eds) Duplex Stainless Steels '91, Les Editions de Physique, Les Ulis, France, pp 185-210

Berns H, Wendl F (1986) Effect of Carbon Content in CrMoV Hot Working Tool Steel. steel research 57 (12): 671-676

Berns H, Krafft F (1987) Influence of nitrogen on the microstructure and properties of 12% chromium steels. Proc. Intern Conf. Advance in Material Technology for Fossil Power Plants ASME Chicago, pp 173-180

Berns H, Lueg J, Trojahn W, Wähling R, Wisell H (1987) The fatigue behaviour of conventional and powder metallurgical high speed steels. Powder Metallurgy International 19 (4): 22-26

Berns H, Krafft F (1989) Exchange of carbon by nitrogen in a 12%Cr steel. In: Foct J, Hendry A (eds) High Nitrogen Steels, HNS 88. Institute of Metals, London, pp 169-173

Berns H, Krafft F (1990) Influence of a thermomechanic treatment on the mechanical properties of creep resistant steel alloyed with nitrogen under pressure. In: Stein G, Witulski H (eds) High Nitrogen Steels, HNS 90. Stahl&Eisen, Düsseldorf, pp 338-342

Berns H, Wang G (1989) Influence of nitrogen on the mechanical properties of creep resistant steels of tubing. In: Foct J, Hendry A (eds) High Nitrogen Steels, HNS 88. Institute of Metals, London, 375-379

Berns H, Lueg J, Trojan W, Zoch HW (1990) High nitrogen steels for corrosion resistant rolling contact bearings. In: Stein G, Witulski H (eds) High Nitrogen Steels, HNS 90. Stahl&Eisen, Düsseldorf, pp 425-429

Berns H, Wang G (1990) High nitrogen hard alloys by powder metallurgy. In: Stein G, Witulski H (eds) High Nitrogen Steels, HNS 90. Verlag Stahleisen, Düsseldorf, pp 332-337

Berns H, Xie X (1992) Sekundarhärten weichmartensitischer nichtrostender Stähle durch Vanadin, Niob und Stickstoff. Radex-Rundschau 1: 71-78

Berns H, Siebert S (1993) Effect of temperature and nitrogen pressure on the case hardening with nitrogen. In: Gavriljuk VG, Nadutov VM (eds) High nitrogen steels, HNS 93. Institute for Metal Physics, Kiev, pp 566-571

Berns H, Trojahn W (1993) High nitrogen Cr-Mo steels for corrosion resistant bearings. In: Hoo JC (ed.) Creative Use of Bearing Steels. ASTM STP 1195, Philadelphia, pp 149-155

Berns H, Wang G (1993) Stainless martensitic PM – HNS. In: Gavriljuk VG and Nadutov VM (eds) High Nitrogen Steels, HNS 93. Institute for Metal Physics, Kiev, pp 415-419

Berns H, Bugajchuk SN, Duz' VA, Ehrhardt R, Gavriljuk VG, Petrov YuN, Yakubtsov IA (1994) Phase transformations during tempering of the Fe-15Cr-1Mo martensites containing nitrogen or carbon. steel research 65(10): 444-450

Berns H, Ehrhardt R, Duz' VA, Gavriljuk VG, Tarasenko AV (1995) Mössbauer study of tempering of the stainless martensitic steel alloyed with nitrogen and/or carbon. Metal Physics and Advanced Technologies 15: 561-570

Berns H, Ehrhardt R (1996) Carbon or nitrogen alloyed quenched and tempered stainless steels – a comparative study. steel research 67 (8): 343-349

Berns H, Kleff J, Krauss G, Foley RP (1996) Microstructure and tensile behavior of nitrogen-alloyed, dual-phase stainless steels. Metallurgical and Materials Trans. A. 27A (7): 1845-1859

Berns H, Siebert S (1996) High nitrogen austenitic cases in stainless steels. ISIJ Intern 36(7):927-931

Berns H, Liu J, Theisen W (1996) A new experimental approach to metal cutting. Z. Metallkde 87: 418-423

Berns H (1997) Entwicklung hochstickstoffhaltiger nichtrostender Stähle mit Mischgefüge. Abschlussbericht Nr. AZ 322-88-63, Minist. Wirtsch., Mittelst., Technol. des Landes NRW, Düsseldorf

Berns H, Duz' VA, Ehrhardt R, Gavriljuk VG, Petrov YuN, Tarasenko VA (1997) Precipitation during tempering of chromium-rich iron-based martensite alloyed with carbon and nitrogen. Z Metallkd 88(2): 109-116

Berns H, Duz VA, Gavriljuk VG, Petrov YuN, Tarasenko AV (1997) Structure and properties of cold-worked stainless steels alloyed with nitrogen. Met Phys Adv Tech 2: 37-48

Berns H, Gavriljuk VG (1997) Tempering of martensitic stainless steel with 0.6 w/o nitrogen and/or carbon. J.Physique IV, Coll. C5, Suppl.J.Phys. III: 263-268

Berns H, Kleff J (1997) Stickstofflegierte nichtrostende Dualphasenstähle. HTM 52 (5): 281-290

Berns H (1998) Verschleiß bei Raumtemperatur. In: Berns H (ed) Hartlegierungen und Hartverbundwerkstoffe, Springer Verlag Berlin/Heidelberg, pp 99-100

Berns H (1998) Stickstofflegierte Schnellarbeitsstähle, experimentelle Verifizierung. Final Report DFG - Be 1022/7-1, Bad Godesberg

Berns H, Escher Ch, Streich WD (1998 I) Druckaufgestickte Stähle für Verbrennungsmotoren. Ingenieur Werkstoffe, 3:36-39

Berns H, Escher Ch, Streich WD (1998 II) Martensitic high nitrogen steel for application at elevated temperatures. In: Hänninen H, Hertzman S (eds) High Nitrogen Steels, HNS 98. Proc of the 5th Intern Conf at Espoo/Stockholm, to be published

Berns H, Eul U, Heitz E, Juse RL (1998 III) Corrosion behaviour of solution nitrided stainless steels. In: Hänninen H, Hertzman S (eds) High Nitrogen Steels, HNS 98, Proc of the 5th Intern Conf at Espoo/Stockholm, to be published

Berns H, Juse RL, Bowman JW, Edenhofer B (1999) Verfahrenstechnik und Stähle für das Randaufsticken. HTM 54 (3): 128-135

Bernstein (1996) The role of hydrogen: is the story any clear? In: Thompson AW, Moody NR (eds) Hydrogen Effects in Materials. TMS AIME, Warendale PA, pp 3-11

Betrabet HS, Nishimoto K, Wilde BE, Clark WAT (1987) Electrochemical and microstructural investigation of grain boundary precipitation in AISI 304 stainless steels containing nitrogen. Corrosion 43(2): 77-84

Beyer E, Kalla G, Petschke U, Schmidt B, Menzel J, Troesken F (1990) Laser welding of high nitrogen steel. In: Stein G, Witulski H (eds) High Nitrogen Steels, HNS 90. Verlag Stahleisen, Düsseldorf, pp 309-313

Bezobrazov SV, Ponomarenko AG, Inozemtseva EN (1960) Nitrogen solubility in high-chromium melts of Fe-Cr system. News Acad of Sci of USSR, Metals (in Russian) 3: 53-59

Binder K (ed) (1987) Applications of the Monte Carlo Methods in Statistical Physics. Topics in Current Physics. Vol. 36. Springer, Berlin-Heidelberg-New York-Tokyo-London

Binder K (ed) (1992) The Monte Carlo Methods in Condensed Matter Physics. Topics in Applied Physics. Vol. 7. Springer, New York-Berlin-Heidelberg

Birnbaum HK (1990) Mechanisms of hydrogen related fracture of metals. In: Moody NR, Thompson AW (eds) Hydrogen Effects on Material Behaviour. TMS AIME, Warrendale, PA, pp 639-660

Blenkinsop PA and Nutting J (1967) Precipitation of the sigma phase in an austenitic steel. J Iron and Steel Inst 205(9): 953-958

Boas M, Rosen A (1977) Effect of load on the adhesive wear of steels. Wear 44: 213-222

Bondar VI, Danilchenko VY, Okhrimenko VA (1988) Crystal structure of martensite in the alloy Fe-31Ni. Phys Metals Metallogr (in Russian) 66(1): 157-161.

Bodsworth C (1970) Comment on the paper «The thermodynamics of nitrogen austenite». Scripta Metall 4(12): 971-972

Bott AH, Pickering FB, Butterworth GJ (1986) Development of high manganese high nitrogen low activation austenitic stainless steels. J Nuclear Mat 141-143: 1088-1096

Böttger A, Liu Cheng, Frikkee E, De Keiser ThE, Mittemeijer EJ (1990) Occurrence of incoherent $\alpha''$- $Fe_{16}N_2$ at room temperature in iron-nitrogen martensite observed by neutron and X-ray diffraction. J. Phys: Condensed Matter 2: 9237-9245

Böttger A, Van Cenderen MJ, Sijbrandij SJ, Mittemejier EJ, Smith GDW (1996) Atom-probe and X-Ray diffraction analysis of the composition and structure of precipitates formed on tempering of ternary iron-carbon-nitrogen martensites. ISIJ Intern 36(7): 764-767

Böttger A, Mittemejier EJ (1998) Phase transformations in high nitrogen martensites. In: Hänninen H, Hertzman S (eds) High Nitrogen Steels, HNS 98. Proc of the 5th Intern Conf at Espoo/Stockholm, to be published

Böttger A, Cerezo A, Sijbrandij SJ, Van Genderen MJ, Smith GDV, Mittemeijer EJ (1998) Ageing and tempering of iron-carbon-nitrogen martensites: (re)distribution of interstitial atoms. In: Hänninen H, Hertzman S (eds) High Nitrogen Steels, HNS 98. Proc of the 5th Intern Conf at Espoo/Stockholm, to be published

Brandis H, Heimann W, Schmidtmann E (1976) Einfluß von Stickstoff auf das Ausscheidungsverhalten des Stahles X3CrNiMoNbN 23 17 (Amagnit 3974) TEW-Technische Berichte 2(2): 150-166

Brass AM, Chanfreau A, Chene J (1990) The role of grain boundaries and cold work on H permeation in nickel. In: Moody NR, Thompson AW (eds) Hydrogen Effects on Material Behaviour. TMS AIME, Warrendale, PA, pp 19-31

Brezina P (1983) Martensitische Cr-Ni-Stähle mit niedrigem C-Gehalt. HTM 38 (5): 197-250 and (6): 251-262

Briant CL (1982) Effects of nitrogen and cold work on the sensitization of austenitic stainless steels. Report NP-2457 on the research project 1574-1, General Electric Company, New York

Briant CL (1987) Nitrogen segregation to grain boundaries in austenitic steels. Scripta Metall 21(1): 71-74

Bridelle R (1955) Sur les nitrides et carbonitrides de fer. Annales de Chimie, Delepine M et Lebeau P (eds), Paris, ser 12, tome 10: 824-870

Brill U (1998) Hochtemperaturwerkstoffe. Dr. habil. thesis, Technische Hochschule, Aachen

Bugaev VN and Chepulski RV (1995) The symmetry of interatomic lattice potentials in general crystal structures. Acta Crystallogr 51A(4): 456-462

Bugaev VN, Gavriljuk VG, Nadutov VM, Tatarenko VA (1983) Mössbauer study of carbon distribution in Fe-Ni-C austenite. Acta Metall 31(3): 407-418

Butler BD, Cohen JB, Zschack P (1991) An investigation of Fe-Ni order in a steel. Metall Trans 22A(11): 2807-2809

Byrnes MLG, Grujicic M, Owen WS (1987) Nitrogen strengthening of a stable austenitic stainless steel. Acta Metall 35(7): 1853-1862

Campillo Illanes BF, Sarkar AD (1986) Wear of thermochemically produced nitrogen stainless steel. J Tribology, Trans ASME 108: 334-339

Carpenter (1998) Preliminary data sheet on BioDur 108 alloy. Carpenter Technology Corporation, Reading PA

Carreker RP, Hibbard WR (1953) Tensile deformation of high-purity copper as a function of temperature, strain rate and grain size. Acta Metall 1(11): 654-663

Carreker RP (1957) Tensile deformation of silver as a function of temperature, strain rate and grain size. Trans AIME 209: 112-115

Carreker RP, Hibbard WR (1957) Tensile deformation of aluminium as a function of temperature, strain rate and grain size.Trans AIME 209: 1157-1163

Cavalleri A, Guzman L, Ossi PM, Rossi I (1986) On the wear resistance of nitrogen implanted 304 stainless steel. Scripta Metall 20(1): 37-42

Chabanel M, Janot C, Motto J (1968) Étude par effet Mössbauer des nitrures de fer $\varepsilon$ et $\zeta$ au voisinage de la composition $Fe_2N$. C R Acad Sci Paris, Serie B, 266: 419-422

Chen SR, Tang D (1990) Effect of interstitial atom concentration on lattice parameters of martensite and retained austenite in iron-carbon-nitrogen alloys. Materials Science Forum, 56-58: 201-206.

Chin HA, Bursey RW, Ehlert DD, Biroscak R, Streit E, Trojahn W (1994) CRONIDUR 30 – An advanced nitrogen alloyed stainless steel for advanced corrosion resistant fracture tough cryogenic bearings. In: Proc. NASA Earth to Orbit Propulsion Technology Conference, NASA Marshall Space Flight Center, Huntsville

Charles J. (1991 I) Duplex stainless steels: Materials to meet your needs. In: Charles J. Bernhardsson S (eds) Duplex Stainless Steels. Les Edition de Physique, Les Ulis, France, pp 3-48

Charles J. (1991 II) Super duplex stainless steels: structure and properties. In: Charles J. Bernhardsson S (eds) Duplex Stainless Steels. Les Edition de Physique, Les Ulis, France, pp 151-176

Chipman J (1955) Atomic interaction in molten alloy steels. JISI: 180: 97-106

Chipman J, Corrigan D.A. (1965) Prediction of solubility of nitrogen in molten steel. Trans AIME 233: 1249-1252

Chipman J (1972) Thermodynamics and phase diagram of the Fe-C system. Metall Trans 3A(1): 55-64.

Chumlyakov YuI, Kireyeva IV, Korotayev AD, Aparova LS (1993) Plastic deformation in austenitic stainless steel single crystals with high nitrogen content. In: Gavriljuk VG, Nadutov VM (eds) High Nitrogen Steels, HNS 93. Institute of Metal Physics, Kiev, pp 215-220

Chyou SD, Shih HC (1990) Structure and electrochemical properties of plasma-nitrided low alloy steel. Mater Sci Engng A 129: 109-117

Cihal V (1968) On some questions of intercrystalline corrosion of corrosion-resistant steels. Protect Met (in Russian) 4(6): 637-655

Clayton CR, Martin KG (1989) Evidence of anodic segregation of nitrogen in high nitrogen stainless steels and its influence on passivity. In: J.Foct J, Hendry A (eds) High Nitrogen Steels, HNS 88. Institute of Metals, London, pp 256-260

Cochardt A, Schoeck G, Wiedersich H (1955) Interaction between dislocations and interstitial atoms in body-centered cubic crystals, Acta Metall 3(11): 533-537

Combrade P, Audouard JP (1991) Duplex stainless steels and localized corrosion resistance. In: Charles J. Bernhardsson S (eds) Duplex Stainless Steels. Les Edition de Physique, Les Ulis, France, pp 257-281

Conrad H (1963) Effect of grain size on the lower yield and flow stress of iron and steel. Acta Metall 11(1): 75-77

Conrad H, Feuerstein S, Rice L (1967) Effects of grain size on the dislocation density and flow stress of niobium. Mater Sci Engng 2: 157-168

Copley SM, Kear BH (1968) The dependence of the width of a dissociated dislocation on dislocation velocity. Acta Metall 16(2): 227-231

Copson RP (1959) Physical metallurgy of stress corrosion fracture. Interscience, New York, p 247

Cordier-Robert C., Kliauga AM, Foct J (1998) Mössbauer surface analysis of nitrogen ion implanted austenitic and ferritic steels. In: Hänninen H, Hertzman S (eds) High Nitrogen Steels, HNS 98, Proc of the 5th Intern Conf at Espoo/Stockholm, to be published

Corney NS, Turkdogan ET (1955) The effect of alloying elements on the solubility of nitrogen in iron. J Iron Steel Institute 180:344-348

Cotton JA, Knutsen RD, Sundman B (1998 I) Modification of the stainless steel database for high manganese, chromium and nitrogen contents. In: Hänninen H, Hertzman S (eds) High Nitrogen Steels, HNS 98, Proc of the 5th Intern Conf at Espoo/Stockholm, to be published

Cotton JA, Knutsen RD, Lang CI (1998 II) The influence of niobium and vanadium on the microstructure and the mechanical properties of a high nitrogen steel. In: Hänninen H, Hertzman S (eds) High Nitrogen Steels, HNS 98, Proc of the 5th Intern Conf at Espoo/Stockholm, to be published

Cottrell AH, Bilby BA (1949) Dislocation theory of yielding and strain ageing of iron. Proc Phys. Soc A 62: 49-62

Cottrell AH (1953) Dislocations and plastic flow in crystals. Clarendon Press, Oxford

Cottrell AH (1958) Theory of brittle fracture in steel and similar metals. Trans TMS-AIME 212: 192-203

Cowley JM (1950) X-ray measurement of order in single crystals of $Cu_3Au$. J Appl Phys 21(1): 24-30

Darken LS and Smith RP (1946) Appendix to the article by Smith RP: Equilibrium of iron-carbon alloys with mixtures of $CO$-$CO_2$ and $CH_4$-$H_2$. J Amer Chem Soc 68(7):1163-1175

Darken LS, Smith RP, Filer EW (1951) Solubility of gaseous nitrogen in gamma iron and the effect of alloying constituents – aluminum nitride precipitation Trans AIME 191:1174-1179

Decker RF, Floreen S (1965) Am Inst Min Metall Pet Eng Conf Proc vol. 28, p 69

DeCristofaro N, Kaplow R (1977) Interstitial atom configurations in stable and metastable Fe-N and Fe-C solid solutions. Metall Trans 8A(1): 35-44

Degallaix S, Foct J, Hendry A (1986) Mechanical behaviour of high-nitrogen stainless steels. Mater Sci Technol 2(9): 946-950

Degallaix S, Taillard R, Foct J (1988) Role of nitrogen interstitials in plastic fatigue of austenitic stainless steels. In: Beevers et al. (eds) Fatigue. EMAS, 1, pp 45-49

Degallaix S, Dickson JI, Foct J (1989) Effect of nitrogen on fatique and creep-fatigue behaviour of austenitic stainless steels. In: Foct J, Hendry A (eds) High Nitrogen Steels, HNS 88. Institute of Metals, London, pp 380-386

Degallaix S, Seddouki A, Nilsson JO, Polak J (1993) Influence of nitrogen on monotonic and cyclic mechanical properties of duplex stainless steels. In: Gavriljuk VG and Nadutov VM (eds) High Nitrogen Steels, HNS 93. Institute for Metal Physics, Kiev, pp 420-425

Delong WT (1974) Ferrite in austenitic stainless steel weld metal. Weld J Research Suppl.: 273s-286s

Dennert R (1995) Einfluß der Elemente Schwefel, Stickstoff und Kohlenstoff auf die Anfangsstadien der Oxidation von Fe-20Cr und Ni-20Cr. Doctoral thesis, Universität Dortmund

Dhers J, Foct J, Vogt JB (1989) Influence of nitrogen content on fatigue crack growth rate at 77 K and 293 K of a 316L steel. In: J.Foct J, Hendry A (eds) High Nitrogen Steels, HNS 88. Institute of Metals, London, pp 199-203

Dietrich H (1964) Eigenschaften der nichtmagnetisierbaren Stähle und ihre metallkundliche Deutung. DEW Techn. Ber. 4 (3): 111-133, 4 (4): 163-181

Dijkstra LJ (1949) Precipitation phenomena in the solid solutions of nitrogen and carbon in alpha iron below the eutectoid temperature. Trans Amer Inst Min Met Eng 145: 252-260

Dingley DJ, McLean D (1967) Components of the flow stress of iron. Acta Metall 15(5): 885-901

Dinsdale AT (1991) SGTE data for pure elements. CALPHAD 15(4): 317-425

Driver JH, Sinclair R, Jack KH (1979) Modulated substitutional-interstitial solute-atom clustering in nitrided austenitic Fe-34Ni-V-alloys. Proc Roy Soc A 367: 99-115

Duhaj P, Ivan J, Makovicky E (1968) Sigma-phase precipitation in austenitic steels. J Iron and Steel Inst 206(12): 1245-1251

Dulieu D, Nutting J (1964) Effect of alloying elements on stacking fault energy in austenitic iron-nickel-chromium alloys. In: Metallurgical developments in high-alloy steels. Special report No. 86, The Iron and Steel Institute, pp 140-145

Dunn WW, McLellan RB (1970) The application of a quasichemical solid solution model to carbon austenite. Metall Trans 1A(5): 1263-1265

Dunn WW, McLellan RB (1971) The thermodynamic properties of carbon in body-centered cubic iron. Metall Trans 2A(4):1079-1086

Eckardt T (1996) Reaktives Plasmaspritzen in kontrollierter Stickstoffatmosphäre zur Herstellung nitridhaltiger Beschichtung. Doctoral thesis, Technische Universität Aachen

Eckel JF, Clevinger GS (1970) Ageing of AISI type 304 austenitic stainless steel containing nitrogen and its influence on stress corrosion cracking. Corrosion, 26(8): 251-255

Eckstein CB, Guimaraes JRC (1984) Microstructure-property correlation in martensite-austenite mixtures. J Mat Sc 19: 3043-3048

Ehrhardt R. (1995)Stickstofflegierte nichtrostende Einsatz- und Vergütungsstähle. Doctoral thesis, Ruhr-Universität, Bochum, also Fortschr.Ber. VDI Reihe 5 Nr. 418, VDI Verlag, Düsseldorf

Eliezer D, Chakrapani DG, Altstetter CJ, Pugh EN (1979) The influence of austenite stability on the hydrogen embrittlement and stress-corrosion cracking of stainless steel. Metall Trans 10A(7) 935-941

Elliot JF, Gleiser M, Ramakrishna V (1963) Thermochemistry for Steelmaking. Addison-Wesley, Reading, MA

Eltester B, Uebing Ch (1996) Surface segregation phenomena on Fe-3.5%Mo-N(100): the formation of a MoN surface compound. Surface Sci 347: 39-45

Ericsson T, Cohen JB (1971) Clustering in Fe-3.9 at.% Mo. Acta Crystall A27(1): 97-109

Ernst C, Rasche K (1992) Nitrogen alloyed tool steels. In: Berns H, Hofmann M, Norström LA, Rasche K, Schindler AM (eds) New materials, processes, experiences for tooling. MAT SEARCH Andelfingen, pp 481-497

Ernst P, Hasegawa Y, Tokomitsu N, Kawauchi Y, Masuyama F (1990) Microstructure and mechanical properties of a new creep resistant steel with a high-nitrogen content. In: Stein G, Witulski H (eds) High Nitrogen Steels, HNS 90. Stahl&Eisen, Düsseldorf, pp 132-137

Escher Ch (1999) Druckaufgestickte Stähle für Verbrennungsmotoren. Doctoral thesis, Ruhr-Universität, Bochum

Ezaki H, Morinaga M, Yukawa N, Adachi H (1986) Prediction of the occurrence of the σ phase in Fe-Cr-Ni alloys. Phil Mag A 53(5): 709-716

FAG (1999) Data sheets on CRONIDUR bearings. FAG Aircraft and Super Precision Bearings GmbH, Schweinfurt ,

Fast JD, Verrijp MB (1955) Solubility of nitrogen in alpha-iron. JISI, 180: 337-343

Fawley R, Quader MA, Dodd RA (1968) Compositional effects on the deformation modes, annealing twin frequencies, and stacking fault energies of austenitic stainless steels. TMS AIME 242: 771-778

Fedotov VP, Samarin AM (1958) Nitrogen solubility in liquid iron and melts of iron and silicon. Reports of Acad Sci of USSR (in Russian) 122(4): 597-599

Feichtinger H, Satir-Kolorz A, Zheng X-O (1989) Solubility of nitrogen in solid and liquid iron alloys with special regard to the melting range. In: Foct J, Hendry A (eds) High Nitrogen Steels, HNS 88. Institute of Metals, London, pp 75-80

Feichtinger H (1990) Alternative routes to the production of high-nitrogen steels. In: Stein G, Witulski H (eds) High Nitrogen Steels, HNS 90. Stahl&Eisen, Düsseldorf, pp 298-302

Feichtinger HK, Zheng X-O, Rennhard Ch (1990) Measurements of nitrogen solubility in iron and iron-nickel alloys, using a new temperature gradient method. steel research 61(1): 26-29

Feichtinger HK (1991) Alternative methods for the production of high nitrogen steels. In: Stainless Steels '91, Chiba. The Iron and Steel Institute of Japan, pp 762-770

Feichtinger HK, Zheng XH (1991) Thermodynamik und Verfahrenstechnik der Herstellung von Stickstoff-Stählen. In: Speidel MO, Uggowitzer PJ (eds) Stickstofflegierte Stähle. Thubal-Kain Verlag, Institute Metallforschung und Metallurgie, ETH-Zürich, pp 19-32

Feichtinger HK (1993) Concepts of nitrogen solubility. In: Gavriljuk VG and Nadutov VM (eds) High Nitrogen Steels, HNS 93. Institute for Metal Physics, Kiev, pp 45-54

Feichtinger HK, Stein G. (1998) Melting of high nitrogen steels. In: Hänninen H, Hertzman S (eds) High Nitrogen Steels, HNS 98, Proc of the 5$^{th}$ Intern Conf at Espoo/Stockholm, to be published

Feltner CE, Laird C (1967) Cyclic stress-strain response of fcc metals and alloys – I Phenomenological experiments. Acta Metall 15(10): 1621-1632

Fernández Guillermet A, Gustafsson P (1985) An assesment of the thermodynamic properties and the (p,T) phase diagram of iron. High Temp - High Pres. 16: 591-610

Fernández Guillermet A, Frisk K (1991) Thermodynamic properties of Ni nitrides and phase stability in the NiN system. Int Journ Thermophys 12(2): 417-431

Finnie I, Mayville RA (1990) Historical aspects in our understanding of the ductile-brittle transition in steels. Trans ASME 112: 56-60

Fisher JC (1954) On the strength of solid solution alloys. Acta Metall 2(1): 9-10

Fischmeister H, Karlsson B (1977) Plastizitätseigenschaften grob-zweiphasiger Werkstoffe. Z Metallkd 68(5): 311-327

Fleischer RL, Hibbard WR (1963) In: The relation between structure and mechanical properties. Proc of the Conference at N.P.L., London, H.H.Stationary Office, 1, p262

Flemings MC (1974) Solidification Processing. McGraw Hill Inc., New York

Flowers JW, Beck FH, Fontana MG (1963) Corrosion and age hardening studies of some stainless alloys containing ferrite. Corosion 19(5): 186-198

Foct J (1973) Conditions imposées aux configurations d´interstitiels dans l´austénite fer-azote par les résultats de spectrométrie Mössbauer. C R Acad Sci Paris Serie C 276(14): 1159-1163

Foct J, Dubois JM, LeCaer G (1977) Etude par spectrométrie des distributions d´interstitiels et de leurs évolutions dans les solutions Fe-C et Fe-N. J Physique 38: C7 231-234

Foct J, Magnin T, Perrot P, Vogt JB (1991) Nitrogen alloying of duplex stainless steels. In: Charles J, Bernhardsson S (eds) Duplex Stainless Steels. Les Edition de Physique, Les Ulis, France, pp 49-65

Foct J, Akdut N (1993) Cleavage-like fracture of austenite in duplex stainless steel. Scripta Metall 29(2): 153-158

Foct J, Akdut N, Magnin T, Taillard R, Vogt JB (1993) Fracture in duplex steels. J Phys IV Colloq C7, suppl J Phys III 3: 597-604

Forchhammer P, Engell HJ (1969) Untersuchungen über den Lochfraß an passiven austenitischen Chrom-Nickel-Stählen in neutralen Chloridlösungen. Werkstoffe und Korrosion 20: 1-11

Frehser J, Kubisch Ch (1963) Metallurgie und Eigenschaften unter hohem Druck erschmolzenerstickstoffhaltiger Stähle. Berg und Hüttenmännische Monatshefte 108(11): 369-380

Friedel J (1963) In: Electron Microscopy and Strength of Crystals. Proc. of 1st Berkeley Intern Mater Conf, Interscience Publ, New York, p 634

Friedel J (1964) Dislocations, Pergamon Press, Oxford-London-Edinburgh-New York-Paris-Frankfurt

Friedel J (1974) Electron theoretical fundamentals in order disorder transformations of metallic alloys. In: H.Warlimont (ed.) Order-disorder transformations in alloys. Springer Verlag, Berlin-Heidelberg-New York, pp 1-26

Frisk K, Hillert M (1989) Thermodynamics of the Fe-Cr-Ni-N system. In: Foct J, Hendry A (eds) High Nitrogen Steels, HNS 88. Institute of Metals, London, pp 1-9

Frisk K (1990) A thermodynamic evaluation of the Cr-Fe-N system. Metall Trans 21A(9): 2477-2488

Frisk K (1991) A thermodynamic evaluation of the Cr-N, Fe-N, Mo-N and Cr-Mo-N systems. CALPHAD 15(1): 79-106

Frisk K (1991) A thermodynamic evaluation of the Cr-Fe-Ni-N system. Z Metallkd 82(2): 108-117

Frisk K (1993) A thermodynamic evaluation of the Cr-Mn-N system. CALPHAD 17(3): 335-349

Frisk K, Qiu C (1994) A thermodynamic evaluation of the solubility of N in solid and liquid Cr-Fe-Mn alloys. Z Metallkd 85(1): 60-69

Fromm E, Gebhardt E (1976) Gase und Kohlenstof in Metallen. Springer-Verlag, Berlin

Fujikura M, Takada K, Ishida K (1975) Effect of manganese and nitrogen on the mechanical properties of Fe-18%Cr-10%Ni stainless steels. Trans Iron Steel Inst Jap 15(9): 464-469

Gallacher PT, Lambert JA, Oates WA (1969) Carbon activity in austenite by Monte Carlo computations. TMS AIME 245(4): 887-889

Gammal TE, Abdel-Karim R, Walter MT, Wosch E, Feldhaus S (1996) High nitrogen steel powder for the production of near net shape parts. .ISIJ Intern 36 (7): 915-921

Gavriljuk VG, Herzricken DS, Polushkin YuA, Falchenko VM (1981) Mechanism of decomposition of cementite during cold work of steel. Physics of Metals and Metallogr 51(1): 147-152

Gavriljuk VG, Nadutov VM (1983) Effect of carbon on the quadruple interaction in austenite. Physics Metals and Metallogr 55(3): 520-527

Gavriljuk VG, Yagodsinsky YuN (1986) Model of Snoek-Köster relaxation in bcc metals. Physics Metals Metallogr 62(2): 38-50

GavriljukVG, Duz' VA, Yephimenko SP, Kvasnevski OG (1987) Interaction between carbon/nitrogen atoms and dislocations in austenite. Physics Metals and Metallogr 64(6): 1132-1135

Gavriljuk VG (1987) Carbon distribution in steel (in Russian). Naukova Dumka, Kiev

Gavriljuk VG (1987 I) Private communication. Institute for Metal Physics, Kiev

GavriljukVG, Duz' VA, Yephimenko SP (1989) Dislocations in austenite and mechanical properties of high nitrogen steels. In: Foct J, Hendry A (eds) High Nitrogen Steels, HNS 88. The Institute of Metals, London, pp 447-451

Gavriljuk VG, Yephimenko SP (1990) Distribution of nitrogen atoms, their interaction with dislocations and properties of high-nitrogen austenite. In: Stein G, Witulski H (eds) High Nitrogen Steels, HNS 90. Stahl&Eisen, Düsseldorf, pp 11-21

Gavriljuk VG, Duz' VA, Yephimenko SP (1990) The structure and mechanical properties of cold worked nitrogen austenite. In: Stein G, Witulski H (eds) High Nitrogen Steels, HNS 90. Stahl&Eisen, Düsseldorf, pp 100-103

Gavriljuk VG, Nadutov VM, Gladun ON (1990 I) Nitrogen distribution in Fe-N austenite. Physics Metals and Metallogr 3: 128-34

Gerdes A, Redecker R (1990) N-alloyed 12% Cr-steel for welded steam turbine rotors. In: Stein G, Witulski H (eds) High Nitrogen Steels, HNS 90. Stahl&Eisen, Düsseldorf, pp 338-342

Gavriljuk VG, Nadutov VM, Neklyudov IM, Voyevodin VN, Platonov PV (1991) Structural change and redistribution of alloying elements under irradiation of nitrogen steels Cr19Ni6Mn24 and Cr22Ni6Mn10. In: Yephimenko SP, Gavriljuk VG (eds) High Nitrogen Steels (in Russian), Institute for Metal Physics, Kiev, pp 298-307

Gavriljuk VG, Jephimenko SP, Shanina BD (1993) Electron-spin-resonance study of electron properties in nitrogen and carbon austenites. Phys Rev B 48(5): 3224-3231

Gavriljuk VG, Hänninen H, Tereshchenko AS, Ullakko K (1993 I) Effect of nitrogen on hydrogen-induced phase transformations in stable austenitic steel. Scripta Metall Mater 28(2): 247-252

Gavriljuk VG, Hänninen H, Tereshchenko AS, Ullakko K (1994) Hydrogen-induced phases in AISI 310 type steel. Scripta Metall Mater 31(6): 781-785

Gavriljuk VG, Hänninen H, Tarasenko AV, Tereshchenko AS, Ullakko K (1995) Phase transformations and relaxation phenomena caused by hydrogen in stable austenitic stainless steels. Acta Metall Mater 43(2): 559-568

Gavriljuk VG, Bugaev VN, Petrov YuN, Tarasenko AV, Yanchitski BZ (1996) Hydrogen-induced equilibrium vacancies in fcc iron-base alloys. Scripta Mater 34(6): 903-907

Gavriljuk VG, Hänninen H, Smouk SY, Tarasenko AV, Ullakko K (1996 I) Internal friction in hydrogen-charged CrNi and CrNiMn austenitic stainless steels. Metall Mater Trans 27A(7): 1815-1821

Gavrilyuk VG, Sozinov AL, Balanyuk AG, Grigoriev SV, Gubin OA, Kopitsa GP, Okorokov AI, Runov VV (1997) Effect of carbon and nitrogen on chemical homogeneity of fcc iron-based alloys. Metall Mater Trans 28A(11):2195-2199

Gavriljuk VG, Berns H (1998) Precipitates in tempered stainless martensitic steels alloyed with nitrogen, carbon or both. In: Hänninen H, Hertzman S (eds) High Nitrogen Steels, HNS 98, Proc of the 5$^{th}$ Intern Conf at Espoo/Stockholm, to be published

Gavriljuk VG, Foct J, Bugajchuk SN, Sozinov AL (1998) Relaxation phenomena in nitrogen austenitic steels. Scripta Mater 37(12): 1889-1894

Gavriljuk VG, Sozinov AL, Foct J, Petrov YuN, Polushkin YuA (1998 I) Effect of nitrogen on the temperature dependence of the yield strength of austenitic steels. Acta Mater 46(4): 1157-1163

Gavriljuk VG, Shanina BD, Berns H (1999) On a correlation between electron structure and short range atomic order in iron-based alloys. Submitted to Acta Materialia

Gavriljuk VG, Berns H, Escher Ch, Glavatskaya NI, Sozinov AL, Petrov YuN (1999 I) Grain boundary strengthening in nitrogen austenitic steels. Accepted for publication in Mater Sci Engng A

Gavriljuk VG, Shivanyuk VN (1999) Effect of hydrogen on the interatomic bonds in austenitic steels. Private communication. Kiev, Institute for Metal Physics

Gavriljuk VG, Shivanyuk VN (1999 I) An evidence for hydrogen-enhanced localised plasticity in austenitic steel obtained by means of internal friction. To be published in Metal Physics and Advanced Technologies

Genin JMR, Flinn PA (1968) Mössbauer effect study of clustering of carbon atoms during the room-temperature aging of iron-carbon martensite. TMS AIME 242(7): 1419-1430

Gerlach H, Schmidtmann E (1968) Einfluß von Kohlenstoff, Stickstoff und Bor auf das Ausscheidungsverhalten eines austenitischen Stahles mit rd. 16% Cr, 2% Mo, 16% Ni und Niob. Arch Eisenhüttenw 39(2): 139-149

Gerold V, Karnthaler HP (1989) On the origin of planar slip in fcc alloys. Acta Metall 37(8): 2177-2183

Gimenez S, Aguirre I, Talacchia S. Iturriza I, Berasategui X (1998) Simultaneous sintering and quenching of T 42 HSS nitrided compacts in vacuum heat treatment furnace. Proc. Powder Metallurgy World Congress, Granada, pp 296-301

Göcmen A, Steins R, Solenthaler Ch, Ugowitzer PJ, Speidel MO (1996) Precipitation behaviour and stability of nitrides in high nitrogen martensitic 9% and 12% chromium steels. ISIJ International, 36(7): 768-776

Göcmen A (1997) Grundrisse der Gefügeausbildung und der Zeitstandeigenschaften martensitischer 9-12% Chromstähle. Doctoral thesis 12020, ETH Zürich

Göcmen A, Ernst P, Holmes P (1998) Principles of alloy design in high nitrogen 12% chromium steels. In: Hänninen H, Hertzman S (eds) High Nitrogen Steels, HNS 98, Proc of the 5$^{th}$ Intern Conf at Espoo/Stockholm, to be published

Golczewski J, Fischmeister HF (1996) Stickstofflegierte Schnellarbeitsstähle, thermodynamische Modellierung. Final Report DFG – Fi 340/5-1, Bad Godesberg

Gielen PM, Kaplow R (1967) Mössbauer effect in iron-carbon and iron-nitrogen alloys. Acta Metall 15(1): 49-63

Glavatska (1998) Effect of structure-texture evolution on cold work hardening caused by rolling of nitrogen Cr-Ni-Mn steels. In: Hänninen H, Hertzman S (eds) High Nitrogen Steels, HNS 98. Proc of the 5$^{th}$ Intern Conf at Espoo/Stockholm, to be published

Goldschmidt H (1967) Interstitial Alloys. Butterworths, London

Gooch T (1987) In: Proc Intern Conf "Stainless Steels '87", The Institute of Metals, York

Gow JT, Harder OE (1942) Balancing the composition of cast 25 per cent chromium-12 per cent nickel type alloys. Trans Am Soc Met 30: 855-935

Graat PCJ (1998) The initial oxidation of iron and iron nitride. Doctoral thesis, Technische Universiteit Delft

Grabke HJ, Uebing C, Viefhaus H (1989) Precipitation and cosegregation on the surface of Fe-Cr-N single crystals. In: Foct J, Hendry A (eds) High Nitrogen Steels, HNS 88. Institute of Metals, London, pp 261-265

Grabke HJ, Hirsch S, Reynders B, Stratmann M, Tomac S (1993) Role of nitrogen in the corrosion of high nitrogen alloys and steels. In: Gavriljuk VG and Nadutov VM (eds) High Nitrogen Steels, HNS 93. Institute for Metal Physics, Kiev, pp 513-518

Grabke HJ (1996) The role of nitrogen in the corrosion of iron and steels. ISIJ Intern 36(7): 777-786

Greenside HS, Schlüter M (1983) Pseudopotentials for the 3d transition-metal elements. Phys Rev B 28(2): 535-543

Gridnev VN, Gavriljuk VG, Meshkow JN (1974) Strength and plasticity of cold worked steel (in Russian) Naukova dumka, Institute of Metal Physics, Kiev

Gridnev VN, Gavriljuk VG, Nemoshkalenko VV, Polushkin YuA, Razumov ON (1977) A study of the structure of the iron-carbon martensite using Mössbauer effect. Phys Metal Metallogr (In Russian) 43(3): 582-590.

Grigorenko GM, Torkhov GF, Lakomskij VI (1970) On temperature dependence of solubility of nitrogen in liquid iron. Reports of Academy of sciences of USSR (in Russian) 194(4): 881-882

Grigorenko GM, Pomarin YuM (1989) Hydrogen and nitrogen in metals at plazma melting (in Russian). Naukova Dumka, Kiev. See also Grigorenko GM, Pomarin YuM (1990). In: Sov Tech Rev Weld and Surf 1, Harvard Acad Publ GmbH, GB

Grobner PJ (1981) Effect of molybdenum on creep properties of a ferritic 18Cr-Nb-Ti steel for catalytic converters. Climax Molybdenum Report pp 19-25

Groß V, Heuser H, Ladwein T (1998) Welding of high nitrogen superaustenitic and superduplex stainless steels. In: Hänninen H, Hertzman S (eds) High Nitrogen Steels, HNS 98, Proc of the 5$^{th}$ Intern Conf at Espoo/Stockholm, to be published

Grosskreutz JC (1972) Strengthening and fracture in fatigue (approaches for achieving high fracture strength). Metall Trans 3(5): 1255-1262

Grujicic M, Nilsson J-O, Owen WS, Thorwaldsson T (1989) Basic deformation mechanisms in nitrogen strengthened stable austenitic stainless steels. In: Foct J, Hendry A (eds) High Nitrogen Steels, HNS 88. Institute of Metals, London, pp 151-158

Grujicic M (1994) The core structure of (a/2)<110> screw dislocations in Fe-Ni-Cr-N austenite. Mater Sci Engng 183A: 223-232

Grujicic M (1995) The effect of nitrogen on the structure and mobility of dislocations in Fe-Ni-Cr austenite. J Mater Sci 30(22): 5799-5807

Grujicic M, Owen WS (1995) Models of short range order in a face-centered cubic Fe-Ni-Cr alloy with a high concentration of nitrogen. Acta Metall Mater 43(11): 4201-4211

Grützner G, Schüller HJ (1967) Die Entwicklung korrosionsbeständiger austenitischer Stähle mit erhöhtem Stickstoffgehalt. Stahl und Eisen 87: 466-469, 542-548

Grützner G (1973) Kornzerfallsanfälligkeit stickstofflegierter austenitischer Chrom-Nickel-Stähle durch Chromnitridausscheidungen. Stahl und Eisen 93(1): 9-18

Gümpel P, Michel E (1986) Über den Einfluß der Elemente Silicium und Stickstoff auf einige Eigenschaften von austenitischen chemisch beständigen Stählen. Thyssen Edelst. Techn. Ber. 12 (2): 181-189

Gümpel P, Ladwein T, Michel E, Strom FH (1988) Entwicklungen bei austenitischen Stählen mit erhöhten Festigkeitseigenschaften für den Einsatz im chemischen Apparatebau. Thyssen Edelst. Techn. Ber. 14 (1): 12-25

Gümpel P, Schaffrath W, Schmidt W (1988) Einfluß von Beanspruchungsdauer und Beanspruchungsgeschwindigkeit auf die mechanischen Eigenschaften einiger nichtrostender Stähle. Thyssen Edelst. Techn. Ber. 14 (1): 99-107

Gümpel P, Strom FH (1988) Über die Eigenschaften von kaltverformten Drähten aus stickstoffhaltigen nichtrostenden Stählen für die Seilherstellung. Thyssen Edelst. Techn. Ber. 14 (1): 108-117

Gurevich II, Tarasov TV (1965) Physics of low energy neutrons. Nauka, Moscow

Haasen P (1979) Solution hardening in fcc metals. In: Nabarro FRN (ed) Dislocations in Solids, North-Holland Publishing Co, Ch 15, pp 155-189

Haddick GT, Thompson LD, Parker ER, Zackey VF (1978) New nickel-free austenitic stainless steels for ambient and cryogenic service. Metal Progress 11: 37-40

Häfele J, Heine B, Kirchheim R (1992) On the kinetics of passive film formation of iron and iron-chromium alloys. Z Metallkd 83(6): 395-404

Hägg G (1929) X-Ray studies on the binary systems of iron with nitrogen, phosphorus, arsenic, antimony and bismuth. Nova Acta Regia, Soc Sci Upsaliensis, Ser 4, 7(1)

Hales R, Hill AC (1977) The diffusion of nitrogen in an austenitic stainless steel. Metal Sci. 7: 241-244

Hall EO (1951) The deformation slip and aging of mild steel: III Discussion of results. Proc Phys Soc B64: 747-753

Hall EO and Algie SH (1966) The sigma phase. Metallurgical Reviews 11: 61-88

Hansen M, Anderko K (1958) Constitution of Binary Alloys. McGraw-Hill Inc, New York – Toronto – London, pp 670-675

Hartline AG III (1974) The effect of nitrogen additions upon the pitting resistance of 18 pct Cr, 18 pct Mn stainless steel. Metall Trans 5A(11): 2271-2276

Harzenmoser MA, Uggowitzer PJ (1987) Neue aufgestickte austenitisch-rostfreie Stähle und Duplexstähle. In: Speidel MO (ed) Moderne Stähle. Verlag der Schweizerischen Akademie der Wissenschaften, Zürich, pp 219-246

Harzenmoser MAE (1990) Massive aufgestickte austenitisch-rostfreie Stähle und Duplexstähle. Doctoral thesis, Eidgenössische Technische Hochschule, Zürich

Harzenmoser MAE, Reed RP, Uggowitzer PJ, Speidel MO (1990) The influence of nickel and nitrogen on the mechanical properties of high-nitrogen austenitic steels at cryogenic temperatures. In: Stein G, Witulski H (eds) High Nitrogen Steels, HNS 90. Stahl&Eisen, Düsseldorf, pp 197-202

Hede A, Aronsson B (1969) Microstructure and creep properties of some 12 % Cr steels. J. Iron Steel Inst. (9): 1241-1251

Heimann W, Bischoff I, Buckstegge J (1979) Magnetisches Verhalten von X3CrNiMoNbN 1916 (Amagnit 3964) und X5NiCrTi2615 (Thermon 4980) bei tiefen Temperaturen. Thyssen Edelst. Techn. Ber. 5 (3): 194-198

Herblseb G (1982) Der Einfluß von Schwefeloxid, Schwefelwasserstoff und Kohlenmonoxid auf die Lochkorrosion von austenitischen Chrom-Nickel-Stählen mit bis zu 4 Massen-% Molybdän in 1M Natriumchlorid-Lösung. Werkstoffe und Korrosion 33: 334-340

Hertzman S, Sundman B (1982) A thermodynamic analysis of the Fe-Cr system. CALPHAD 6(1): 67-80

Hertzman S, Sundman B (1985) A thermodynamic analysis of the Fe-Cr-Ni system. Scand J Metallurgy 14: 94-102

Hertzman S, Jarl M (1987) A thermodynamic analysis of the Fe-Cr-N system. Metall Trans 18A(10): 1745-1752

Hertzman S (1987) A study of equilibria in the Fe-Cr-Ni-Mo-C-N system at 1273 K. Metall Trans 18A(10): 1767-1778

Hertzman S (1995) Highly alloyed stainless steels. A swedish perspective. Scand. J. Metallurgy 24: 140-146

Hertzman S, Jargelius Pettersson R, Blom R, Kivineva E, Eriksson J (1996) Influence of shielding gas composition and welding parameters on the N-content and corrosion properties of welds in N-alloyed stainless steel grades. ISIJ Int. 36 (7): 968-976

Hertzman S, Wessman S. (1998) An experimental and theoretical study of nitrogen flux in stainless steel TIG Welds. In: Hänninen H, Hertzman S (eds) High Nitrogen Steels, HNS 98, Proc of the 5$^{th}$ Intern Conf at Espoo/Stockholm, to be published

Heubner U, Rockel M, Wallis E (1989) Das Ausscheidungsverhalten von hochlegierten austenitischen Stählen mit 6% Molybdän und sein Einfluß auf die Korrosionsbeständigkeit. Werkstoffe und Korrosion 40: 459-466

Hill TL (1960) An Introduction to statistical thermodynamics. Addison-Wesley, Reading, MA

Hillert M, Jarl M (1975) A thermodynamic analysis of the iron-nitrogen system. Metall Trans 6A(3): 553-559

Hillert M, Jarl M (1978) A model for alloying effects in ferromagnetic metals. CALPHAD 2(3): 227-238

Hillert M, Qui C (1990) A reassessment of the Cr-Fe-Ni system. Metall Trans 21A(6): 1673-1679

Hirth JP, Lothe J (1970) Theory of dislocations. McGraw Hill Book Company, New York-St Louis-San Francisco-Toronto-Lndon-Sydney

Holmberg H, Nilsson J-O, Liu P (1990) Development of low cost non-magnetic stainless spring steels. ISIJ Intern 30(8): 594-599

Holzworth ML, Loutan MR (1968) Hydrogen-induced phase transformations in type 304 stainless steels Corrosion 24(4): 110-124

Hooli P, Kupari P (1998) Nitrogen control in AOD-converter. In: Hänninen H, Hertzman S (eds) High Nitrogen Steels, HNS 98, Proc of the 5$^{th}$ Intern Conf at Espoo/Stockholm, to be published

Hornbogen E, Rittner K, Becker J (1984) Abrasiver Verschleiß von Stählen mit Dualphasen-Gefüge. Sonderb. prakt. Metallogr. 15: 342-352

Horovitz MB, Beneduce Neto F, Garbogini A, Tschiptschin AP (1996) Nitrogen bearing martensitic stainless steels: microstructure and properties. ISIJ Intern 36(7): 840-845

Horvath W, Tabernig B, Werner E, Uggowitzer P (1997) Microstructures and yield strength of nitrogen alloyed super duplex steels. Acta mater 45(4): 1645-1654

Houska CR, Averbach BL (1962) Atom arrangements in some iron-aluminum solid solutions. J Phys Chem Solids 23(12): 1763-1769

Hsiao C-M, Dulis EJ (1957) Precipitation reactions in austenitic Cr-Mn-C-N stainless steels. TASM 49: 655-685

Hu Z, Hiong K, Yan X, Cui K (1990) Effect of nitrogen in high wear resistance cold work die steel. In: Stein G, Witulski H (eds) High Nitrogen Steels, HNS 90. Stahl&Eisen, Düsseldorf, pp 188-190

Huang C-C, Tsai W-T, Lee J-T (1996) Surface modification of carbon steel with laser treated nitrogen-containing stainless steel layers. Surface and Coatings Technology 79: 67-70

Hucklenbroich I, Stein G, Chin H, Trojahn W, Streit E (1998) High nitrogen martensitic steels for critical components in aviation. In: Hänninen H, Hertzman S (eds) High Nitrogen Steels, HNS 98, Proc of the 5$^{th}$ Intern Conf at Espoo/Stockholm, to be published

Hull FC (1973) Effects of composition on embrittlement of austenitic stainless steels. Weld J (London) 52(3): 104s-113s

Ieki Yu, Asano S (1994) Hydrogen-induced hardening and embrittlement in fcc Fe-Ni-Mn alloys subjected to cathodic charging. J Japan Inst Metals, 58(9): 1008-1014

Igata N, Kohno Yu, Ohno K, Tsunakawa H (1981) A study of vacansion swelling of stainless steel under electron irradiation. Ann Rept Eng Res Inst Fac Eng Univ Tokyo 40: 169-171

Ikeda S, Tone S, Tokashima S, Kaji H (1990) Effect of thermo-mechanical control process on strengthening of a 22Mn-13Cr-5Ni austenitic stainless steel plate for cryogenic use. ISIJ Intern 30(8): 600-607

Ikegami Y, Nemoto R (1996) Effect of thermo-mechanical treatment on mechanical properties of high-nitrogen containing Cr-Mn-Ni austenitic stainless steels. ISIJ Intern 36(7): 855-861

Il'insky AG, Korobov VA, Slyusarenko SI (1995) An atomic structure of four-component alloys in Bi-Ga-Sn-Ge system. Physics of Metals and Advanced Technologies 17(8): 29-34

Ioffe A, Kirpitscheva MW, Lewitzky MA (1924) Deformation und Festigkeitder Kristalle. Z für Physik, 22: 286-302

Irvine KJ, Murray JD, Pickering FB (1960) The effect of heat-treatment and microstructure on the high-temperature ductility of 18%Cr-12%Ni-1%Nb steels. JISI 196(1): 166-179

Irvine KJ, Llewellyn T, Pickering FB (1961) High-strength austenitic stainless steels. JISI 199(2): 153-175

Irvine KJ, Gladman T and Pickering FB (1969) The strength of austenitic stainless steels. JISI 199: 1017-1028

Ishizaki J, Orita K, Terao K (1992) The influence of chemical composition and pre-strain on impact toughness transition behaviour of 18%MN-18%Cr-N austenitic steels. J ISIJ (Tetsu to Hagane) 78(12): 1846-1853

Isomoto T, Ikeda H, Kouda T, Ichii K, Oishi T (1998) Mechanical properties and corrosion resistance of PM high nitrogen stainless steel consolidated by hot extrusion. In: Hänninen H, Hertzman S (eds) High Nitrogen Steels, HNS 98, Proc of the 5$^{th}$ Intern Conf at Espoo/Stockholm, to be published

Jack KH (1951) The occurrence and the crystal structure of $\alpha''$-iron nitride; a new type of interstitial alloy formed during the tempering of nitrogen martensite. Proc. Roy A, Ser A, No 1093, 208: 216-224

Jack KH (1989) Nitrogen precipitation - retrospect and prospect. In: Foct J, Hendry A (eds) High Nitrogen Steels, HNS 88. The Institute of Metals, London, pp 117-135

Jargelius R (1986) The influence of nitrogen alloying on the corrosion resistance of 20Cr-25Ni and 20Cr-25Ni-4.5Mo stainless steels. Report of Swedish Institute for Metal Research, IM 2179, Stockholm

Jargelius R, Johannson A, Gustafson P, Sjöberg A, Hertzman S, Arnberg L, Lagneborg R (1990) A nitrogen alloyed tool steel produced by solid state nitridation of powder. In: Stein G, Witulski H (eds) High Nitrogen Steels, HNS 90. Stahl&Eisen, Düsseldorf, pp 314-319

Jargelius-Pettersson RFA (1992) Sensitisation in nitrogen-alloyed stainless steels and its evaluation using intergranular corrosion testing methods. In: Proc 12$^{th}$ Scandinavian Corrosion Congress Eurocorr '92, pp 119-128

Jargelius-Pettersson RFA (1994) Phase transformations in a manganese-alloyed austenitic stainless steel. Scripta Metall Mater 30(9): 1233-1238

Jargelius-Pettersson RFA (1995) Localised corrosion of stainless steels: ranking, alloying, and microstructure effects. Scand J Metallurgy 24: 188-193

Jargelius-Pettersson RFA (1996) Sensitization behaviour and corrosion resistance of austenitic stainless steels alloyed with nitrogen and manganese. ISIJ Intern 36(7): 818-824

Jargelius-Pettersson RFA (1998) Precipitation trends in highly alloyed austenitic stainless steels. Z Metallkd 89(3): 177-183

Jargelius-Pettersson RFA (1998 I) The influence of N, Mo and Mn on the microstructure and corrosion resistance of austenitic stainless steels. Doctoral thesis, Swedish Institute for Metal Research, Stockholm

Jargelius-Pettersson RFA (1998 II) Application of the pitting resistance equivalent concept to some highly alloyed austenitic stainless steels. Corrosion 54(2): 162-168

Jarl M (1977) A thermodynamic analysis of the Cr-N system. CALPHAD 1(1): 91-95

Jarl M (1978) A thermodynamic analysis of the interaction between nitrogen and other alloying elements in ferrite and austenite. Scand J Metallurgy 7: 93-101

Jesper H, Wessling W, Achtelik K (1966) Festigkeitseigenschaften nichtrostender austenitischer Stähle mit höherem Stickstoffgehalt und deren Anwendungsmöglichkeiten. Stahl und Eisen 86: 1408-1418

Johnson AA (1962) The ductile-brittle transition in body-centered cubic transition metals. Phil Mag 7(73-74): 177-196

Jomard F, Perdereau, M (1991) SIMS study of nitrogen distribution in duplex stainless steels. In: Charles J, Bernhardsson S (eds) Duplex Stainless Steels. Les Edition de Physique, Les Ulis, France, pp: 719-725

Josefsson, B, Nilsson JO, Wilson A (1991) Phase transformations in duplex steels and the relation between continuous cooling and isothermal heat treatment. In: Charles J, Bernhardsson S (eds) Duplex Stainless Steels. Les Edition de Physique, Les Ulis, France, pp 67-78

Jones ER, Datta T, Almasan C, Edwards D, Ledbetter HM (1987) Low temperature magnetic properties of fcc Fe-Cr-Ni alloys: Effects of manganese and interstitial carbon and nitrogen. Mat Sci Engng 91: 181-188

Juse R (1999) Aspekte des Randaufstickens nichtrostender Stähle - Prozess, Gefüge, Eigenschaften – Doctoral thesis, Ruhr-Universität, Bochum

Juse RL, Berns H, (1997) Solution nitriding of stainless steels. Proc 6$^{th}$ Int Seminar of IFHT: 388-395

Kajihara M, Choi SK, Kikuchi M, Tanaka R, Seo Y, Okumura T, Kondoh Y (1986) Evidence of long range diffusion of nitrogen in cellular precipitation of $Cr_2N$ in Cr-Ni austenitic steel. Z Metallkd 77(8): 515-518

Kajihara M, Ono N, Kikuchi M (1993) Formation and stability of a nitride with the structure of β manganese in the Cr-Ni-N ternary and Cr-Fe-Ni-N quarternary systems. In: Gavriljuk VG and Nadutov VM (eds) High Nitrogen Steels, HNS 93. Institute for Metal Physics, Kiev, pp 63-72.

Kamachi Mudali U, Dayal RK, Gnanamoorthy JB, Rodrigez P (1996) Relationship between pitting and intergranular corrosion of nitrogen-bearing austenitic stainless steels. ISIJ Intern 36(7): 799-806

Kaputkina LM, Efimenko S.P., Dobatkin SV, Leschinskaya EM, Panovko VM, Kazakov MA (1998) Diagrams of dynamic recrystallization of austenitic stainless steels with Nitrogen. In: Hänninen H, Hertzman S (eds) High Nitrogen Steels, HNS 98, Proc of the 5$^{th}$ Intern Conf at Espoo/Stockholm, to be published

Kashyap VC, Parlee N (1958) Solubility of nitrogen in liquid iron and iron alloys.Trans TMS AIME 212 : 86-91

Kashyap BP, Tangri K (1995) On the Hall-Petch relationship and substructureal evolution in type 316L stainless steel. Acta Metall Mater 43(11): 3971-3981

Kasper JS (1954) The ordering of atoms in the chi-phase of the iron-chromium-molybdenum system. Acta Metall 2(5): 456-461

Katz JD, King TB (1989) Kinetics of nitrogen absorption and desorption from a plasma arc by molten iron. Met Trans B (20B): 175-185

Kautz HR, Gerlach H (1968) Eigenschaften nichtstabilisierter vollaustenitischer Stähle für die Verwendung im Reaktorbau und Dampfkesselbetrieb. Archiv Eisenhüttenw 39(2): 151-158

Ke TS, Tsien CT (1957) On the mechanism of the internal friction peaks associated with the stress-induced diffusion of carbon in face-centered cubic alloy-steels. Physics Metals Metallogr 4(2): 291-305

Kearns JR (1985) The effect of nitrogen on the corrosion resistance of austenitic stainless alloys containing molybdenum. In: Lula RA (ed) New developments in stainless steel technology, ASM, Detroit, pp 117-127

Kestenbach HJ (1977) The effect of applied stress on partial dislocation separation and dislocation substructure in austenitic stainless steel. Phil Mag 36(6): 1509-1515

Kikuchi M, Wakita S, Tanaka R (1973) Trans. Iron Steel Inst. Jpn., vol. 13, p226 (1973).

Kikuchi M, Tanaka T, Tanaka R (1974) Quenching defects in a nitrogen-containing austenitic stainless steel. Metall Trans 5A(6): 1520-1521

Kikuchi M, Tanaka R, Hamagami K, Ogura Y, Tanaka R (1976) Lattice dilation of 25Cr-28Ni and 25Cr-28Ni-2Mo austenitic steels by dissolved nitrogen. Metall Trans 7A(6): 906-908

Kikuchi M, Nishimura T, Tanaka T, Tanaka R (1978) Matrix precipitation of $Cr_2N$ in a phosphorus-containing austenitic stainless steel. Metall Trans 9A(9): 1337-1339

Kikuchi M, Sekita T, Wakita S, Tanaka R (1981) J. Iron Steel Inst Jpn 67: 1981

Kikuchi M (1985) Precipitation processes in austenitic stainless steels. In: Igata N, Lei TC (eds) The University of Tokyo-Harbin Institute of Technology Symposium on Materials Science. Department of Materials Science, The University of Tokyo, Tokyo, pp 22-30

Kikuchi M, Kajihara M, Frisk K (1989) Solubility of nitrogen in austenitic stainless steels. In: Foct J, Hendry A (eds) High Nitrogen Steels, HNS 88. Institute of Metals, London, pp 63-74.

Kikuchi Y, Matsuda F, Okabe T, Ohta M (1996) Nitrogen content of 316L weld metal and its fine particle by means of high-pressure MIG Arc Welding. ISIJ 36 Intern (7): 977-982

Kleff J (1996) Gefüge und mechanische Eigenschaften stickstofflegierter nichtrostender Dualphasenstähle. Abschlussbericht KL 951/1-1, Deutsche Forschungs Gemeinschaft (DFG) Bad Godesberg

Kliauga, AM (1997) Randschichtbeeinflussung von ferritisch-austenitischen Chrom-Nickel-Stählen durch Stickstoffeinsatz. Doctoral thesis, Ruhr-Universität, Bochum

Klotz UE, Uggowitzer PJ, Speidel M (1998) Nitrogen alloyed 9-12% Cr-steels with martensitic-austenitic microstructure. In: Hänninen H, Hertzman S (eds) High Nitrogen Steels, HNS 98, Proc of the $5^{th}$ Intern Conf at Espoo/Stockholm, to be published

Knott JF (1996) Fracture toughness and hydrogen-assisted crack growth in engineering alloys. In: Thompson AW, Moody NR (eds) Hydrogen Effects in Materials. TMS AIME, Warendale PA, pp 387-408

Ko C, McLellan RB (1983) Thermodynamics of ternary nitrogen austenite. Acta Metall 31(11): 1821-1827

Koch W (1965) Metallkundliche Analyse. Stahl und Eisen, Düsseldorf

Kocis F, Matlock WM (1978) Some characteristics of automotive exhaust valve alloys. Z. Werkstofftech 9: 132-140

Koehler JS (1952) The production of large tensile stresses by dislocations. Phys Rev 85(3): 480-481

Köhler M, Heubner U, Eichenhofer KW, Renner M (1995) Alloy 33, a new corrosion resistant austenitic material for the refincry industry and related applications. Corrosion 95, NACE Int Ann Conf and Corr Show, Paper 338: 1-14

Köhler M, Heubner U, Eichenhofer KW, Renner M (1996) Progress with alloy 33 (UNS R20033), a new corrosion resistant chromium-based austenitic material. Corrosion 96, NACE Int Ann Conf and Corr Show, Paper 428: 1-18

Koneva NA, Lychagin DV, Teplyakova LA, Trishkina LI, Kozlov EV (1988) In: Vladimirov VI (ed) Disclinations and rotation deformation of solids (in Rusian). Ioffe Institute of Physical Engineering, Leningrad, pp 103-113

Koneva NA, Kozlov EV (1990) Physical nature of stages in plastic deformation. In: Panin VE (ed) Structural levels of plastic deformation and fracture (in Russian), Nauka, Novosibirsk, pp 123-186

Kopp R, Holl A (1990) Metal forming strategy for newly-developed materials. In: Stein G, Witulski H (eds) High Nitrogen Steels, HNS 90. Stahl&Eisen, Düsseldorf, pp 288-292

Kornyushin YuV (1970) Electrostatic interaction of a dislocation with a charged point defect. Physics of Metals and Metallogr (in Russian) 29(3): 659-661

Korshunov LG, Mints RI (1969) A study of wear-resistant unstable austenitic steels at dry sliding. Phys Chem Mech Mater (in Russian) 5(5): 569-572

Köstler HJ, Sidan H (1977) Einfluß der Korngößе auf die Kaltverfestigung des nichtmagnetisierbaren Stahles X40MnCrN19. Z Wirtsch Fert 72(10): 785-788

Kotlar A, Achour M, Dode M (1973) Determination des pressions partielles d'azote en equilibre avec les nitrures non-stochiometriques du systeme Cr-N entre 900 et 1100 °C pour les compositions 0.37<N/Cr<0.98. Revue Chimie Minerale 10(4): 651-659

Kozlov EV, Koneva NA, Olevkov AV, Cherkasova TV (1998) A study of solid solution hardening nature of austenitic steels with various nitrogen content. In: Hänninen H, Hertzman S (eds) High Nitrogen Steels, HNS 98. Proc of the $5^{th}$ Intern Conf at Espoo/Stockholm, to be published

Krafft F (1991) Druckaufgestickte warmfeste Chromstähle. Doctoral thesis Ruhr-Universität, Bochum, also Fortschr.Ber. VDI Reihe 5 Nr. 192, VDI Verlag, Düsseldorf

Krauss G (1990), Steels: Heat Treatment and Processing Principles, ASM International, Materials Park, Ohio

Krivoglaz MA, Tikhonova EA (1960) Theory of X-ray scattering by multicomponent ordered solutions. J Ukr Physic Soc (in Russian) 5(2): 158-173

Kühl A, Berger D (1991) Chemische Diffusion vonStickstoff in hochlegierten austenitischen CrNi(Mo)-Stählen. Materialwissenschaft und Werkstofftechnik 22: 462-467

Kühl A (1999) Solution nitriding of stainless steel for applications in mechanical and chemical engineering. Final technical report of project BE-95-1209, European Comission, Brussels

Kulmburg A, Wagner J, Kubisch Ch (1972) The metallography of steels melted under pressure with high N-contents. Z Prakt Metallogr 9: 256-269

Kumar D, King AD, Bell T (1983) Mass transfer of nitrogen from $N_2$ – $H_2$ atmospheres into Fe-18Cr-Ni-Mn alloys. Metal Sci 17 (1): 32-40

Küntzler KR (1973) Low temperature specific heat of ordered and disordered FeCo. Phys Stat Sol (b) 58(2): 519-522

Küppers W (1982 I) Zur Auswirkung der Korngröße auf die Bildung von $\alpha$'-Martensit und die Verfestigung nichtrostender austenitischer Stähle. Thyssen Edelst. Techn. Ber. 8 (1): 38-46

Küppers W (1982 II) Untersuchung zur Abhängigkeit des optimalen Streckziehverhaltens von der Austenitstabilität. Thyssen Edelst. Techn. Ber. 8 (2): 153-161

Küppers W (1984) Zum Fließverhalten instabil austenitischer nichtrostender Feinbleche. Thyssen Edelst. Techn. Ber. 10 (1): 44-53

Kuwana T, Kokawa H, Naitoh K (1989) Effect of chromium and nickel on nitrogen absorption of arc-melted iron. Trans Jpn Weld Soc 20: 10-16

Labusch R (1972) Statistische Theorien der Mischkristallhärtung. Acta Metall 20(7): 917-927

Lagerberg G, Josefsson A (1955) Influence of grain boundaries on the behaviour of carbon and nitrogen in α-iron. Acta Metall 3(5): 236-244

Lardon JM, Charles J, Dupoiron F, Bavay JC (1989) Duplex-austenitic-ferritic stainless steels, mechanical properties and corrosion resistance. In: Foct J, Hendry A (eds) High Nitrogen Steels, HNS 88. Institute of Metals, London, pp 280-287

Larikov LN, Falchenko VM, Mazanko VF, Gurevich SM, Kharchenko GI, Ignatenko AI (1975) Abnormal acceleration of diffusion during impulse loading of metals. Reports of Academy of Sciences of USSR (in Russian) 221(5): 1073-1075

Ledbetter HM, Austin MW (1987) Dilation of an fcc Fe-Cr-Ni alloy by interstitial carbon and nitrogen. Mater Sci Technology 3: 101-104

Lee HM (1974) Carbon activity in austenite. Metall Trans 5A(5): 787-789

Lenel UR, Knott BR (1987) Structure and properties of corrosion and wear resistant Cr-Mn-N steels. Metall Trans 18A(5): 847-855

Levis MH, Hattersley B (1965) Precipitation of $M_{23}C_6$ in austenitic steels. Acta Metall 13(11): 1159-1168

Li JCM (1963) Petch relation and grain boundary sources. Trans TMS-AIME 227(2): 239-247

Li JCM, Chou YT (1970) The role of dislocations in the flow stress grain size relatioships. Metall Trans 1(5): 1145-1159

Lismer RE, Pryce L, Andrews KW (1952) Corrosion by retained treatment chemicals on phosphated steel surfaces. J Iron and Steel Inst 171: 49-58

Litvinov VS, Poptsov ME, Ivtchenko VA (1986) An analysis of atomic distribution of the components in chromium-manganese steel by means of Mössbauer spectroscopy and ion field microscopy. Metal Physics and Metallogr (in Russian) 61(2): 361-364

Liu Ch (1990) Phase transformations in iron-based interstitial martensites. Doctor thesis, Delft Technical University

Liu Ch, Böttger A, Mittemeijer EJ (1992) Tempering of iron-carbon-nitrogen martensites. Metall Trans 23A(4): 1129-1145

Liu Cheng, Böttger A, Keijser ThH, Mittemeijer EJ (1990) Lattice parameters of iron-carbon and iron-nitrogen martensites and austenites. Scripta Metall et Mater 24(3): 509-514

Liu J (1996) Modellversuche zur Drehbearbeitung metallischer Werkstoffe. Doctoral thesis, Ruhr-Universität, Bochum, also Fortschr. Ber. VDI Reihe 2 Nr. 376, VDI Verlag, Düsseldorf

Liu R, Narita N, Altstetter C, Birnbaum H, Pugh EN (1980) Studies of the orientations of fracture surfaces produced in austenitic stainless steels by stress-corrosion cracking and hydrogen embrittlement. Metall Trans 11A(9): 1563-1574

Liu SC, Hashida T, Takahashi H, Kuwano H, Hamaguchi Y (1998) A study of fractography in the low-temperature brittle fracture of an 18Cr-18Mn-0.7N austenitic steel. Metall Mater Trans 29A(3): 791-798

Longinow AW, Bates JF (1969) Influence of alloying elements on the stress corrosion behaviour of austenitic stainless steel. Corrosion, 25(1): 15-22

Lorenz K, Medawar G (1969) Über das Korrosionsverhalten austenitischer Chrom-Nickel-(Molybdän-)Stähle mit und ohne Stickstoffzusatz unter besonderer Berücksichtigung ihrer Beanspruchbarkeit in chloridhaltigen Lösungen.Thyssen Forschung 1(3): 97-108

Lösche T, Reichel J, Werries H (1997) Langlebige Wälzlager auch unter schwierigsten Einsatzbedingungen. In: Gold PW (ed.) Antriebstechnisches Kolloquium des IME der RWTH, Verl. Stercken Aachen

Lu YC, Luo JL, Ives MB (1991) Effect of nitriding on the anodic behaviour of iron and its significance in pitting corrosion of iron-based alloys. Corrosion 47(11): 835-839

Lu YC, Luo JL, Ives MB (1991 I) NACE Western Region Conference Sasakatoon, February

Lu YC, Ives MB, Clayton CR (1993) Synergism of alloying elements and pitting corrosion resistance of stainless steels. Corrosion Science, 35(1-4): 89-96

Lueg, J (1990) Stickstofflegierte Werkzeugstähle. Doctoral thesis Ruhr Universität, Bochum, also Fortschr.Ber. VDI Reihe 5 Nr. 188, VDI Verlag, Düsseldorf

Lueg J (1993) High nitrogen tool steels for various applications. In: Gavriljuk VG, Nadutov VM (eds) High nitrogen steels, HNS 93. Institute for Metal Physics, Kiev, pp 580-585

Lueg J, Berns H (1991) Stickstofflegierte nichtrostene Stähle für Wälzlager und Werkzeuge. In: Speidel MO, Uggowitzer PJ (eds) Stickstofflegierte Stähle. Verlag Thubal-Kain, Institut Metallforschung und Metallurgie, ETH-Zürich, pp 73-85

Lupis CHP and Elliott JF (1967) Prediction of enthalpy and entropy interaction coefficients by the „central atoms" theory. Acta Metall 15(2): 265-276

Lupis CHP (1983) Chemical Thermodynamics of Materials. Elsevier Science Publishing Co., North-Holland, N.-Y.-Amsterdam-Oxford

Lysak LI, Vovk YaN (1965) A nature of phase transformations during quenching of manganese steel. Phys Metals Metallogr (In Russian) 20(4): 540-545.

Lysak LI, Nikolin BI (1966) About positions of carbon atoms in the crystal lattices of ε-, ε′- and κ′-martensites, Phys Metals Metallogr (In Russian) 22(5): 730-736.

Lysak LI, Vovk YaN (1971) Formation of κ′-martensite in carbon steels. Phys Metals Metallogr (In Russian) 31(3): 646-648.

Lysak LI, Danilchenko VY (1971) Formation of κ′-martensite in nickel steel. Phys Metals Metallogr (In Russian) 32(3): 639-641.

Machado IF, Padilha AF (1996) Precipitation behaviour of 25% Cr – 5.5% Ni austenitic stainless steel containing 0.87% nitrogen. steel research 67(7): 285-290

Magdowski RM, Speidel MO (1989) Stress corrosion cracking of high nitrogen steels. In: Foct J, Hendry A (eds) High Nitrogen Steels, HNS 88. Institute of Metals, London, pp 251-255.

Marcinkovski MJ, Fisher RM (1965) Atomic order and Petch relation in an Fe-Co alloy. Trans TMS-AIME 233(2): 293-298

Margolin H, Mahajan Y, Saleh Y (1976) Grain boundaries, stress gradients and fatigue crack initiation. Scripta Metall 10(12): 1115-1118

Masuyama F, Hiromatsu K, Hasegawa Y (1996) High temperature oxidation behavior of high nitrogen ferritic steels. ISIJ Intern 36(7): 825-833

Mateo A, Llanes L, Iturgoyen L, Anglada M (1996) Cyclic stress strain response and dislocation substructure evolution of a ferrite-austenite stainless steel. Acta Mater 44(3): 1143-1153

Matsuo T, Morioka N, Kaise S, Kikuchi M, Tanaka R (1989) Effect of nitrogen on creep deformation of 25Cr-28Ni austenitic steels – solid solution strengthening due to nitrogen. In: Foct J, Hendry A (eds) High Nitrogen Steels, HNS 88. Institute of Metals, London, pp 213-217

Matsuo T, Fujita N, Kikuchi M (1990) Effect of nitrogen on creep resistance and dislocation substructures of a 25Cr-28Ni austenitic steel – the stress dependence of solid solution strengthening due to nitrogen. In: Stein G, Witulski H (eds) High Nitrogen Steels, HNS 90. Stahl&Eisen, Düsseldorf, pp 182-187

McLellan RB (1964) The thermodynamics of dilute silver-oxygen and iron-nitrogen interstitial solid solutions. Trans TMS AIME 230(6): 1468-1475

McLellan RB (1965) The thermodynamics of dilute Fe-C solid solutions. TMS AIME 233(9): 1664-1670

McLellan RB, Garrard TL, Horowitz SJ, Spraque JA (1967) A model for concentrated interstitial solid solutions; its application to solutions of carbon in gamma iron. TMS AIME 239(4): 528-535

McLellan RB, Dunn WW (1969) A quasi-chemical treatment of interstitial solid solutions: its application to carbon austenite. J Phys Chem Sol 30(11): 2631-2637

McLellan RB, Alex K (1970) The thermodynamics of nitrogen austenite. Scripta Metall 4(12): 967-970

McLellan RB, Dunn WW (1970) The thermodynamics of austenite. Scripta Metall 4(5): 321-326

McLellan RB, Chraska P (1971) Thermodynamics of iron-carbon solid solutions. Mat Sci Engng 7(6): 305-317

McLellan RB (1972) Thermodynamics of solid solutions. Mat Sci Engng 9(3): 121-140

McLellan RB, Farraro RJ (1980) Thermodynamics of the iron-nitrogen system. Acta Metall 28(3): 417-422

McLellan RB (1982) Cell models for interstitial solid solutions. Acta Metall 30(1): 317-322

McLellan RB (1988) The thermodynamics of intertitial-vacancy interactions in solid solutions. J Phys Chem Sol 49(10): 1213-1217. See also McLellan RB (1988) The diffusivity of lattice atoms in dilute interstitial solid solutions. Acta Metall 36(8): 1923-1928

Medovar BI, Saenko VYa, Grigorenko GM, Pomarin YuM, Kumysh VI (1996) Arc-Slag Remelting of Steel and Alloys. Cambridge International Science Publishing

Mekata M, Yoshimura H, Takaki H (1972) Magnetic study of hexagonal nitrides of 3d transition metals. J Phys Soc Japan 33(1): 62-69

Menshikov AZ, Yurchikov YeV (1984) Mössbauer effect in fcc Fe-Ni alloys. J Exper Theor Phys (in Russian) 63(1): 190-198

Menshikov AZ, Sidorov SK, Arkhipov VE (1971) Magnetic structure of fcc iron-nickel alloys. J Exper Theor Phys (in Russian) 61(1): 311-319

Menzel J, Stein G (1993) The PESR process, a way of producing HNS on an industrial scale. In: Gavriljuk VG, Nadutov VM (eds) High Nitrogen Steels, HNS 93. Institute of Metal Physics, Kiev, pp 572-579

Menzel J, Kirschner W, Stein G (1996) High nitrogen containing Ni-free austenitic steels for medical applications. ISIJ Intern 36(7): 893-900

Mielitynen-Tiitto K (1979) Precipitation of $Cr_2N$ in some nitrogen-alloyed austenitic stainless steels. Acta Polytechn Scand Ch 141: 3-69

Mihalisin JR, Bieber CC, Grant RT (1968) Sigma – its occurence, effect, and control in nickel-base superalloys. Trans Am Inst Min Metall Pet Eng 242: 2399-2414

Mills T (1970) Pressure-temperature relations in the chromium-nitrogen system. J. Less Common Met 22(4): 373-381

Mirzayev DA, Rushitz SV, Ustinov AI, Goykhenberg YuN (1982) Diffraction effects caused by twins {011}<011> in tetragonal crystals. I. $\kappa'$-martensite and twins of system {011}<011>. Metal Physics 4(4): 43-48; II. Methods of analysis of experimental data on diffraction from single and polycrystals containing twins {011}<011>. Metal Physics 4(5): 26-30

Misawa T, Tanabe H (1996) In-situ observation of dynamic reacting species at pit precursors of nitrogen-bearing austenitic stainless steels. ISIJ Intern 36(7): 787-792

Mittemeijer EJ (1983) Tempering of Iron-Nitrogen-Martensite. Z.Metallkd. 74 (7): 473-483

Miura H, Omuro K, Ogawa H (1996) Preparation of amorphous high nitrogen iron alloys with ternary additions by mechanical alloying. ISIJ Intern 36 (7): 951-957

Miyahara K, Sugihara R, Satoh T, Hosoi Y (1991) Effects of alloying elements of manganese, carbon and nitrogen and aging treatments on corrosion resistance of Fe-Cr-Mn alloys. In: Stainless Steels'91, Chiba. The Iron and Steel Institute of Japan, pp 139-145

Miyahara K, Bae DS, Kimura T, Shimoide Y, Hosoi Y (1996) Strength properties and microstructure of high Mn-Cr austenite steels as potential high temperature materials. ISIJ Intern 36(7): 878-882

Morinaga M, Yukawa N, Adachi H (1984) Alloying effect on the electronic structure of $Ni_3Al$ ($\gamma'$). J Phys Soc Jap 53(2): 653-663

Morinaga M, Yukawa N, Ezaki H, Adachi H (1985) Solid solubilities in transition-metal-based fcc alloys. Phil Mag A 51(2): 223-246

Morley JI, Kirkby HW (1952) Sigma-phase embrittlement in 25Cr-20Ni heat-resisting steels. JISI 172: 129-142

Moss SC (1967) Static atomic displacements in iron-carbon martensite. Acta Metall 15(12): 1815-1826

Mott NF (1950) Imperfections in Nearly Perfect Crystals. Shockley W (ed), Wiley

Mozhi TA, Clark WAT, Nishimoto K, Johnson WB, Macdonald DD (1985) The effect of nitrogen on the sensitization of AISI 304 stainless steels. Corrosion 41(10): 555-559

Mozhi TA, Betrabet HS, Jagannatan V, Wilde BE, Clark WAT (1986) Thermodynamic modeling of sensitization of AISI 304 stainless steels containing nitrogen. Scripta Metall 20(5): 723-728

Mozhi TA, Nishimoto K, Wilde BE, Clark WAT (1986 I) The effect of nitrogen on stress corrosion cracking of AISI 304 stainless steel in high-temperature sulfate solution. Corrosion 42(4): 197-203

Müller R, Weintz R (1998) Ventilwerkstoffe für Verbrennungsmotoren. Mat-wiss und Werkstofftech 29: 97-130

Müllner P, Solenthaler C, Uggowitzer P, Speidel MO (1993) On the effect of nitrogen on the dislocation structure of austenitic steels. Mater Sci Engng A 164: 164-169

Müllner P, Solenthaler C, Speidel MO (1994) Second order twinning in austenitic steel. Acta Metall Mater 42(7): 2211-2217

Müllner P, Sollenthaler C, Uggowitzer PJ, Speidel MO (1994 I) Brittle fracture in austenitic steel. Acta Metall Mater 42(7): 2211-2217

Müllner P (1997) On the ductile to brittle transition in austenitic steel. Mater Sci Engng A 234-236: 94-97

Mummert K, Engelmann HJ, Schwartz S, Uhleman M (1996) Influence of the Ni-content on the cathodic and corrosive hydrogen induced cracking behaviour of austenitic alloys. In: Thompson AW, Moody NR (eds) Hydrogen Effects in Materials. TMS AIME, Warendale, PA, pp 679-688

Murphy S, Whiteman J (1970) The precipitation of ε-carbide in twinned martensite. Metall Trans 1A(4): 843-848

Nabarro FRN (1946) Proc Phys Soc 58: 669

Nabarro FRN (1977) The theory of solid solution hardening. Phil Mag 35(3): 613-622

Nagakura S, Hirotsu Y, Kusunoki M, Nakamura Y (1983) Crystallographic study of the tempering of martensitic carbon steel by electron microscopy and diffraction. Metall Trans 14A(6): 1025-1031

Nakajima H, Hirano K (1978) Electromigration of carbon in Fe-0.52 wt.%C alloy. Trans Japan Inst Metals 19(7): 400-409

Nakajima H, Yoshida K, Shimamoto S (1990) Development of new cryogenic steels for the superconducting magnets of the fusion experimental reactor. ISIJ Intern 30(8): 567-578

Nakamura N, Takaki (1996) Structural control of stainless steel by nitrogen absorption in solid state. ISIJ Intern 36 (7): 922-926

Nakamura N, Tsuchiyama T Takaki S (1998) Effect of structural factors on the mechanical properties of high nitrogen austenitic steels. In: Hänninen H, Hertzman S (eds) High Nitrogen Steels, HNS 98, Proc of the 5th Intern Conf at Espoo/Stockholm, to be published

Nakazawa T, Abo H, Tanino M, Komatsu H, Nishida T, Tashimo M (1989) Effect of nitrogen and carbon on creep properties of type 316 stainless steels. In: Foct J, Hendry A (eds) High Nitrogen Steels, HNS 88. Institute of Metals, London, pp 218-224

Narita N, Altstetter CJ, Birnbaum HK (1982) Hydrogen-related phase transformations in austenitic stainless steels. Metall Trans 13A(8): 1355-1365

Nassau van L, Meelker H, Hilkes J (1991) Welding duplex and super-duplex stainless steels. In: Charles J. Bernhardsson S (eds) Duplex Stainless Steels. Les Edition de Physique, Les Ulis, France, pp 303-324

Nemchenko VP, Maldin IP, Popel SI, (1968) On thermodynamics of solution of nitrogen in Fe-Cr, Fe-Mn and Fe-Cr-Mn melts (in Russian). Izvestia Vuzov, Chern. Metall., 12: 5-8

Newman RC, Sharabi T (1987) The effect of alloying nitrogen or dissolved nitrate ions on the anodic behaviour of austenitic stainless steel in hydrochloric acid. Corrosion Sci 27(8): 827-838

Newman RC, Lu YC, Bandy R, Clayton CR (1984) In: Proc 9th Intern Congr Metallic Corrosion, National Research Council, Canada 3, p 394

Nevitt MV (1963) Alloy chemistry of transition metals. In: Electronic Structure and Alloy Chemistry of the Transition Elements, Interscience Publishers, New York, pp101-169

Nilsson JO (1983) The effect of slip behaviour on the low cycle fatigue behaviour of two austenitic stainless steels. Scripta Metall 17(5): 593-596

Nilsson JO, Thorvaldsson T (1985) The influence of nitrogen on microstructure and strength of a high-alloy austenitic stainless steel. Scand J Metallurgy 15: 83-89

Nilsson JO (1992) Super duplex stainless steels. Mater Sci Techn 8(8): 685-700

Nilsson JO, Kangas P, Karlsson T, Wilson A (1998) Microstructural stability and mechanical properties of a high nitrogen super duplex stainless steel. In: Hänninen H, Hertzman S (eds) High Nitrogen Steels, HNS 98. Proc of the 5th Intern Conf at Espoo/Stockholm, to be published

Norström LA (1977)The influence of nitrogen and grain size on yield strength in type AISI 316L austenitic stainless steel. Metal Sci 11(6): 208-212

Noskova NI, Pavlov VA, Nemnonov SA (1965) A correlation between the stacking fault energy and structure of metals (in Russian). Physics Metals Metallogr 20(6): 920-924

Novak CJ (1977) Structure and Constitution of Wrought Austenitic Stainless Steels. In: Handbook of Stainless Steels, McGraw-Hill, New York, pp 4-1 to 4-78

Novick AS, Berry BS (1972) Anelastic Relaxation in Crystalline Solids. Academic Press, New York - London

Nyilas A, Obst B (1989) Tensile and fracture properties of high nitrogen bearing austenitic steels at cryogenic temperatures. In: Foct J, Hendry A (eds) High Nitrogen Steels, HNS 88. Institute of Metals, London, pp 194-198

Nyilas A, Obst B, Nakajima H (1993) Tensile properties, fracture and crack growth of a nitrogen strengthened new stainless steel (Fe-25Cr-15Ni-0.35N) for cryogenic use. In: Gavriljuk VG and Nadutov VM (eds) High Nitrogen Steels, HNS 93. Institute for Metal Physics, Kiev, pp 339-344

Nyström M, Lindstedt U, Karlsson B, Nilsson JO (1997) Influence of nitrogen and grain size on deformation behaviour of austenitic stainless steels. Mater Sci Technol 13(7): 560-567

Obst B, Nyilas A (1991) Experimental eveidence on the dislocation mechanism of serrated yielding in f.c.c. metals and alloys at low temperatures. Mater. Sci. and Engng A 137: 141-150

Obst B (1998) Basic aspects of tensile properties. In: Seeber B (ed) Handbook of Applied Superconductivity, Institute of Physics Publishing, Bristol-Philadelphia, 2, F1.1, pp 969-993.

Oda K, Kondo N, Shibata K (1990) X-ray absorption fine structure analysis of interstitial (C,N)-substitutional (Cr) complexes in austenitic stainless steels. ISIJ Intern 30(8): 625-631

Oda K, Umezu K, Ino H (1990) Interaction and arrangement of nitrogen atoms in fcc γ-iron. J Phys: Condensed Matter 2: 10147-10158

Oda K, Fujimura H, Ino H (1994) Local interactions in carbon-carbon and carbon-M (M: Al, Mn, Ni) atomic pairs in fcc γ-iron. J Phys: Condensed Matter 6: 679-692

Ogawa H, Omuro K, Miura H (1998) Amorphization and nanoscale refinement of high nitrogen containing Fe-based crystalline materials by mechanical alloying. In: Hänninen H, Hertzman S (eds) High Nitrogen Steels, HNS 98, Proc of the 5$^{th}$ Intern Conf at Espoo/Stockholm, to be published

Okamoto M, Tanaka R, Naito T, Fujimoto, Tetsu to Hagané (1962). On manufacture of high-chromium steels in high-pressure nitrogen atmosphere and heat-resisting properties of 316L type steels . Overseas 2: 25-37

Okayama S, Tsujikawa S, Kikuchi K (1987) Effect of alloying elements on the repassivation potentials for crevice corrosion of stainless steels in 3% NaCl solution. Corrosion Eng (Jpn) 36: pp 157-167

Olsson COA (1995) The influence of nitrogen and molybdenum on passive films formed on the austenitic-ferritic stainless steel 2205 studied by AES and XPS. Corrosion Sci 37(3): 467

Ono N (1986) Master of Engineering Thesis. Tokyo Institute of Technology, Tokyo

Oriani RA, Josephic PH (1974) Equilibrium aspects of hydrogen-induced cracking of steels. Acta Metall 22(9): 1065-1074

Orita K, Ikeda Y, Iwadate T, Ishizaka J (1990) Development and production of 18Mn-18Cr non-magnetic retaining rings with high yield strength. ISIS Intern 30 (8): 587-593

Orlic J, Rose A, Wiest P (1973) Atlas zur Wärmebehandlung der Stähle, Band 3. Stahleisen, Düsseldorf

Owen WS (1990) Nitrogen strengthening of austenitic stainless steels at temperatures above 500 °K. In: Stein G, Witulski H (eds) High Nitrogen Steels, HNS 90. Stahl&Eisen, Düsseldorf, pp 42-46

Palma RH, Urrutibeaskoa I Martinez V, Urcola JJ (1992) Metallographic changes in the sintering of grade T high speed steels in an industrial atmosphere and vacuum. J Mat Sci 27: 2026-2034

Pant P, Dahlmann P, Schlump W, Stein G (1985) Eine neue Technologie der Massivaufstickung – ein Weg zur deutlichen Eigenschaftsverbesserung austenitischer Stähle. Techn.Mitt. Krupp-Forsch.-Ber. 43 (3): 67-82

Paranjpe VG, Cohen M, Bever MB, Floe CF (1950) The iron-nitrogen system. TMS AIME 188(2): 261-267

Paul G, Charman-Kpodo H, Hendry A (1996) Effect of enhancement in surface nitrogen concentration on the corrosion and fatigue properties of austenitic steel wire rod. ISIJ Intern 36(7): 867-872

Pauling L (1960) The Nature of the Chemical Bond. Cornell University Press, Ithaca, New York

Paulus N, Magdowski R.,Speidel, M.O. (1993) Cold worked high nitrogen superaustenitics. In: Gavriljuk V, Nadutov VM (eds) High nitrogen steels. Institute for Metal Physics, Kiev, pp 394-400

Pedrazzoli RM, Speidel MO (1990) Stress corrosion cracking of nitrogen containing steels. In: Stein G, Witulski H (eds) High Nitrogen Steels, HNS 90. Stahl&Eisen, Düsseldorf, pp 208-213

Pepperhoff W (1992) Physical properties. In: Verein Deutscher Eisenhüttenleute (ed) Steel Vol 1. Springer Verlag Berlin, pp 397-402

Petch NJ (1953) The cleavage strength of polycrystals. JISI 174: 25-28

Petrov YuN (1978) Defects and diffusionless transformation in steel (in Russian). Naukova dumka, Kiev, 262 pp

Petrov YuN (1993) On the carbon distribution at structural imperfections in manganese austenite. Scripta Metall et Mater 29(11): 1471-1476

Petrov YuN, Trophimova LN (1998) Effect of nickel on segregation of carbon at grain boundaries in alloyed austenite. Metal Physics and Advanced Technologies (in Russian) 20(9): 11-14

Petrov YuN, GavriljukVG, Berns H, Escher Ch (1999) Nitrogen partitioning between matrix, grain boundaries and precipitates in high-alloyed austenitic steels. Scripta Materialia 40(6): 669-674

Pomarin YuM, Grigorenko GM, Lakomski VI, Torkhov GF, Sherevera AV (1972) About solubility of nitrogen in the iron-nickel melts. Izvestiya AN USSR, Metalli (in Russian) 4: 32-36

Pichard I, Girodin D, Dudragne G, Moraux JY (1998) Metallurgical and tribological evaluation of 32CrMoV13 deep nitrided steel and XD15N high nitrogen martensitic steel for aerospace applications. In: Hoo JJC, Green WB (eds) Bearings steels: Into the 21st century. ASTM STP 1327, West Conshohocken, pp 391-405

Pinedo CE, Vatavuk J, de Oliveira SD, Tschiptschin AP (1998) Solid state alloying by plasma nitriding and diffusion annealing treatment for austenitic stainless steel. In: Hänninen H, Hertzman S (eds) High Nitrogen Steels, HNS 98, Proc of the 5th Intern Conf at Espoo/Stockholm, to be published

Pohl M, Wischnowski F, Ibach A (1998) Gefüge und Verschleißverhalten eines nickelfreien Duplex-Stahls. Prakt. Metallogr., Sonderband 27: 363-366

Pomarin YuM, Grigorenko GM, Latash YuV, Kanibolotskij SA (1983) A study of nitrogen solubility in multicomponent iron alloys at nitrogen gaseous pressure up to 1000 KPa. Izvestiya AN USSR, Metalli (in Russian) 2: 27-33

Pomarin YuM, Grigorenko GM, Latash YuV, Kanibolotskij SA (1984) A study of nitrogen solubility in Fe-Cr melts under pressure. Izvestiya AN USSR, Metalli (in Russian) 6: 10-14

Presser R, Silcock JM (1983) Aging behaviour of 18Mn-18Cr high nitrogen austenitic steel for end rings. Met Sci 17(5): 241-247

Pridantsev MV, Talov NP, Levin FL (1969) High-strength austenitic steels (in Russian). Metallurgia, Moscow

Qiu C, Fernández Guillermet A (1993) Predicative approach to the entropy of manganese nitrides and calculation of the Mn-N phase diagram. Z Metallkd 84(1): 11-22

Qiu C (1993) A thermodynamic evaluation of the Fe-Mn-N system. Metall Trans 24A(3): 629-645

Quick NR, Johnson HH (1979) Permeation and diffusion of hydrogen and deuterium in 310 stainless steel, 472 to 779 K. Metall Trans 10A(1): 67-70

Rabinovich AV, Zaslavsky YuB, Tregubenko GN, Milova UM (1993) Development of technology of an experimental batch of thin–wall tubes made of high-nitrogen corrosion-resistant steel using solid-phase decarburization and alloying by nitrogen method. In: Gavriljuk VG, Nadutov VM (eds) High nitrogen steels, HNS 93. Institute for Metal Physics, Kiev, pp 602-604

Rapatz F (1941) Verwendungsmöglichkeiten von nichtrostenden und hitzebeständigen Stählen mit Stickstoffzusatz. Stahl und Eisen 61 (48): 1073-1078

Rashev Ts, Ivanov R (1979) Die Stickstofflöslichkeit in Eisen-Mangan-Schmelzen bei 1600 °C und Gasdrücken bis zu 25 bar. Arch Eisenhüttenw 50(9): 369-371

Rashev Ts, Dzhambalova LP, Kovacheva RS, Andreev CA (1981) Processes of precipitation and intercrystalline corrosion in high-nitrogen Cr-Mn steels after isothermal annealing. Met Sci Heat Treatm 23: 310-313

Rashev Ts, Rasheva J, Jonov Ts, Angelov I (1993) Some conclusion on the investigation of nitrogen rapid steel mechanical characteristics. In: Gavriljuk VG, Nadutov VM (eds) High nitrogen steels, HNS 93. Institute for Metal Physics, Kiev, pp 432-437

Rashev T (1995) High nitrogen steels, metallurgy under pressure. Publ. House Bulgarian Acad. of Sci, Sofia

Rasheva J, Rashev Ts (1990) High nitrogen tool steels. In: Stein G, Witulski H (eds) High Nitrogen Steels, HNS 90. Stahl&Eisen, Düsseldorf, pp 381-386

Rasheva J, Rashev Ts, Grigorova N, Kamenova Ts, Zlateva G, Dimitrova M (1993 I) Nitrogen effect on the phase composition of high nitrogen rapid steels. In: Gavriljuk VG, Nadutov VM (eds) High nitrogen steels, HNS 93. Institute for Metal Physics, Kiev, pp 300-305

Rawers JC, Dunning JS; Assai J, Reed RP (1992) Characterization of stainless steels melted under high nitrogen pressure. Metall Trans 23A(7): 2061-2068

Rawers JC, Gokcen NA (1993) High-temperature, high-pressure nitrogen concentration in Fe-Cr-Mn-Ni alloys. steel research 64 (2): 110-113

Rawers JC, White H, Doan R (1996) Nitrogen addition to bcc-Fe. ISIJ Intern 36(7): 746-749

Rawers J, Govier D, Korth G (1996) Consolidation, mechanical properties, and phase stability of mechanically alloyed Fe-N powder composites. ISIJ Intern 36 (7): 947-950

Rawers J, Krabbe R, Cook D (1998) Nitrogen addition to bcc-Fe by attrition milling. In: Hänninen H, Hertzman S (eds) High Nitrogen Steels, HNS 98, Proc of the 5$^{th}$ Intern Conf at Espoo/Stockholm, to be published

Rawers JC, Ugowitzer P (1998) Characterization of Fe-C/N steel. In: Hänninen H, Hertzman S (eds) High Nitrogen Steels, HNS 98. Proc of the 5$^{th}$ Intern Conf at Espoo/Stockholm, to be published

Rayaprolu DB, Hendry A (1988) High nitrogen stainless steel wire. Materials Science and Technology (4): 136-145

Rayaprolu DB, Hendry A (1989) Cellular precipitation in a nitrogen alloyed stainless steel. Mater Sci Technology 5(4): 328-332

Reed RP, Purtscher RT, Yushchenko KA (1986) Nickel and nitrogen alloying effects on the strength of austenitic stainless steels at 4K. Adv Cryog Eng Mater 32: 43-50

Reed RP, Simon NJ (1989) Nitrogen strengthening of austenitic stainless steels at low temperatures. In: Foct J, Hendry A (eds) High Nitrogen Steels, HNS 88. Institute of Metals, London, pp 180-188

Reichel J (1993) Hydraulische Antriebe unter Tage mit umweltverträglichen Druckflüssigkeiten. Final report of project 65-11-287-2, DMT Gesellschaft, Essen

Rennhard CF (1998) New industrial applications of HNS. In: Hänninen H, Hertzman S (eds) High Nitrogen Steels, HNS 98, Proc of the 5th Intern Conf at Espoo/Stockholm, to be published

Reppich B (1982) Ein auf Mikromechanismen abgestütztes Modell der Hochtemperaturfestigkeit und Lebensdauer für teilchengehärtete Legierungen. Z. Metallkde. 73 (11): 697-705

Richter E, Günzel R (1997) Härten von Edelstahl durch Stickstoff-Plasma-Immersions-Ionenimplantation. Ing. Werkst. 6 (3, 4): 44-46

Ridley N, Stuart H and Zwell L (1969) Lattice parameters of Fe-C austenites at room temperature. Trans TMS AIME 245: 1834-1836

Ridley N, Stuart H (1970) Met Sci J 4: 219-222

Ritter AM, Henry MF (1985) Phase transformations during aging of a nitrogen-strengthened austenitic stainless steel. Metall Trans 16A(10): 1759-1771

Rochegude P, Foct J (1986) Interstitial atom ordering in binary Fe-N solid solutions studied by Mössbauer spectrometry. Phys Stat Sol (a) 98: 51-62

Rochegude P, Cordier-Robert C, Kouam J (1993) On thermal expansion of the iron-nickel-nitrogen invar alloys. In: Gavriljuk VG, Nadutov VM (eds) High nitrogen steels, HNS 93. Institute for Metal Physics, Kiev, pp 388-393

Rockel M, Jasner M, Kirchheiner R (1990) Application of an austenitic nitrogen alloyed 6-Mo stainless steel in the chemical process and other industries. In: Stein G, Witulski H (eds) High Nitrogen Steels, HNS 90. Stahl&Eisen, Düsseldorf, pp 442-447

Roitburd AL, Khachaturyan AG (1970) Interstitial atoms and a crystallographic mechanism of martensitic transformations in steels, Phys Metals Metallogr (In Russian) 30(6): 1189-1199

Romu J, Terwo J, Hänninen H, Liimatainen J (1993) Wear resistance of high nitrogen austenitic stainless steels manufactured by molten and powder metallurgy routes. In: Gavriljuk VG and Nadutov VM (eds) High Nitrogen Steels, HNS 93. Institute for Metal Physics, Kiev, pp 372-378

Romu J, Terwo J, Hänninen H, Liimatainen J (1996) Development of properties of P/M austenitic stainless steels by nitrogen infusion. ISIJ Intern 36(7): 938-946

Roth M, Chamberod A, Billard L (1978) Short range order in a 70-30 FeNi alloy. J. Magn. Magn. Mater., 7: 104-106

Rotman F, Gilbon D, Dimitrov O (1990) Periodic decomposition of electron-irradiated pure austenitic Fe-Cr-Ni alloys. In: Russel KC and Smith DF (eds) Physical Metallurgy of Controlled Expansion Invar-Type Alloys. TMS, pp 145-158

Rozenak P (1990) Effects of nitrogen on hydrogen embrittlement in AISI type 316, 321 and 347 austenitic stainless steels. J Mater Sci 25(5): 2532-2538

Rozenak P, Robertson IM, Birnbaum HK (1990) HVEM studies of the effects of hydrogen on the deformation and fracture of AISI type 316 austenitic stainless steel. Acta Metall Mater 38(11): 2031-2040

Rozin RM, Finkelstein NV (1953) A study of phase transformations using the method of internal friction. Reports of Acad of Sci of USSR (in Russian) 91(4): 811-812

Rudy ML, Huggins RA (1966) Grain boundary segregation and the cold work peak in iron containing carbon and nitrogen. TMS AIME 236(12): 1662-1666

Rybin V (1986) Large plastic deformations and fracture of metals (in Russian). Metallurgia, Moscow, 224 pp

Sakamoto T, Abo H, Okazaki T, Ogawa T, Zaizen T (1980) Alloys for the eighties, Climax Molybdenum Company, Ann Arbor, p 269

Sakamoto T, Nakagawa Y, Yamauchi I, Zaizen T, Nakajima H and Shimamoto S (1984) Nitrogen-containing 25Cr-13Ni stainless steel as a cryogenic structural material. Adv Cryog Eng Mater 30: 137-144

Sandström R, Bergqvist H (1977) Temperature dependence of tensile properties and strengthening of nitrogen alloyed austenitic stainless steels. Scand J Metallurgy 6: 156-169

Sassen J, Garrat-Reed AJ, Owen WS (1989) Electron microscopy of austenitic Fe-Ni-Cr alloys containing nitrogen. In: Foct J, Hendry A (eds) High Nitrogen Steels, HNS 88. Institute of Metals, London, pp 159-162

Satir-Kolorz A, Feichtinger HK, Speidel MO (1989) Über die Löslichkeit von Stickstoff in Eisen- und Stahlgußlegierungen unter erhöhtem Druck. Gießereiforschung 41(4): 149-165

Satir-Kolorz A, Feichtinger HK, Speidel MO (1990) Literaturstudie und theoretische Betrachtungen zum Lösungsverhalten von Stickstoff in Eisen-, Stahl- und Stahlgußschmelzen. Gießereiforschung 42(1): 36-49

Satir-Kolorz A, Feichtinger HK (1991) On the solubility of nitrogen in liquid iron and steel alloys using elevated pressure. Z Metallkd 82(9): 689-697

Schaeffler AL (1949) Constitutional diagram for stainless steel weld metal. Metal Progress 56 (5): 680

Schellewald M (1999) Zerspanbarkeit hochstickstofflegierter nichtrostender Stähle. Diploma thesis, Ruhr-Universität Bochum

Schenck H, Frohberg MG, Graf H (1958) Untersuchung über Beeinflussung der Gleichgewichte von Stickstoff mit flüssigen Eisenlösungen durch den Zusatz weiterer Elemente (I). Arch Eisenhüttenw 29(11): 673-676

Schenck H, Frohberg MG, Graf H (1959) Untersuchungen über die Beeinflussung der Gleichgewichte von Stickstoff mit flüssigen Eisenlösungen durch den Zusatz weiterer Elemente (II). Arch Eisenhüttenw 30(9): 533-537

Schenck H, Frohberg MG, Heinemann H (1962) Untersuchungen zur Stickstoffaufnahme in flüssigen Eisenlegierungen im Druckbereich bis zu vier Atmosphären. Arch Eisenhüttenw 33(9): 593-600

Schenck H, Frohberg MG and Reinders F (1963) Beitrag zur Kentniss der Löslichkeit des Stickstoffs in Eisenlegierungen im Temperaturbereich von 700 bis 1200°C. Stahl und Eisen 83(2): 93-99

Schlump W (1975) Mechanische und magnetische Eigenschaften austenitischer Chrom-Nickel-Stähle bei tiefen Temperaturen. Techn. Mitt. Krupp Forsch.Ber. 33 (3): 93-99

Schmidt W, Domalski HH, Schaffrath W (1986) Zur Verfestigung chemisch beständiger austenitischer Drähte durch Kaltverformung. Thyssen Edelst. Techn. Ber. 12 (1): 101-112

Schmidt W, Küppers W (1986) Der Einfluß der Austenit-Stabilität auf mechanische Eigenschaften und Umformverhalten von Chrom-Nickel-Stählen. Thyssen Edelst. Techn. Ber. 12 (1): 80-100

Schoeck G (1963) Friccion interna debido a la interaccion entre dislocaciones y atomos solutos. Acta Metall 11(6): 617-622

Schön CG, Rechenberg HR, Goldenstein H (1993) Mössbauer study of iron-chromium-carbon austenite. Scripta Metall et Mater 29(11): 1483-1488

Schramm RE, Reed RP (1975) Stacking fault energies of seven commercial austenitic stainless steels. Metall Trans 11A(6): 1345-1351

Schug R (1993) Entwicklung eines hochstiffstoffhalten Duplex-Stahles. Diploma thesis, Ruhr-Universität Bochum

Schumann H, Fircks v HJ (1969) Die martensitischen Umwandlungen in kohlenstoffarmen austenitischen Chrom-Nickel-Stählen. Archiv für das Eisenhüttenwesen 40 (7): 561-568

Schürmann E, Kunze HD (1967) Gießereiforschung 19: 101-108

Schürmann E, Kättlitz W (1981) Untersuchungen über Konzentrations- und Temperaturabhängigkeit der Stickstofflöslichkeit in eisenreichen Dreistoffschmelzen Fe-N-X. Arch Eisenhüttenw 52: 219-224

Seeger A (1954) The temperature dependence of the critical shear stress and of work-hardening of metal crystals. Phil Mag 45(366): 771-773

Seeger A (1955) The generation of lattice defects by moving dislocations, and its application to the temperature dependence of the flow stress of fcc crystals. Phil. Mag 46(382): 1194-1217

Seeger A (1979) A theory of the Snoek-Köster relaxation (cold-work peak) in metals. Phys Stat Sol (a) 55(2): 457-468

Segtrop K (1995) Einfluß einer Drehbearbeitung auf die Randzone von Hartlegierungen. Doctoral thesis, Ruhr-Universität, Bochum, also Fortschr.Ber. VDI Reihe 5 Nr. 402, VDI Verlag, Düsseldorf

Seith W (1955) Diffusion in Metallen, Platzwechselreaktionen. Springer-Verlag, Berlin-Göttingen-Heidelberg

Sekiguchi T, Haga M, Arai H (1991) 17%Cr-15%Mn austenitic stainless steel for elevated temperature spring use. In: Stainless Steels '91, Chiba. The Iron and Steel Institute of Japan, pp 1256-1262

Sewell PB, Mitchel DF, Cohen M (1972) Early stages of oxidation of Fe(001) observed by electron diffraction and X-Ray emission. Surface Sci 33(3): 535-552

Shanina BD, Kolesnik SP, Konchitz AA, GavriljukVG, Smouk SYu, Tarasenko AV (1994) The influence of nitrogen on the paramagnetic properties of the multicomponent d-element iron-based alloy. Solid State Commn 90(2):109-113

Shanina BD, Gavriljuk VG, Konchitz AA, Kolesnik SP, Tarasenko AV (1995) Exchange interaction between electron subsystems in iron-based fcc alloy doped by nitrogen or carbon. Phys Status Solidi (a) 149: 711-722

Shanina BD, Gavriljuk VG, Konchitz AA, Kolesnik SP (1998) The influence of substitutional atoms upon the electron structure of the iron-based transition metal alloys. J Physics: Condensed Matter 10: 1825-1838

Shanina BD, Gavriljuk VG, Kolesnik SP, Shivanyuk VN (1999 I) Paramagnetic spin resonance in hydrogen-charged stainless austenitic steel. J Phys D: Appl Phys 32: 298-304

Shanina BD, GavriljukVG, Konchitz AA, Glavatskij IN (1999 II) Electron spin resonance in CrMn nitrogen austenitic steels. To be published

Shiflet GJ, Bradley JR, Aaronson HI (1978) A re-examination of the thermodynamics of the proeutectoid ferrite transformation in Fe-C alloys. Metall Trans 9A(7): 999-1008

Shimada M (1990) Effect of phosphorus and boron on cryogenic mechanical properties of a sensitized 17Cr-12.5Ni-2Mo-0.05Nb-0.2N steel. ISIJ Intern 30(8): 579-586

Shimada T, Abe S, Yamamoto A (1994) A method to obtain fine Cr carbides in high carbon martensitic stainless steel. Metallurg Sci and Technol 12 (2): 55-63

Shimamoto S, Nakajima H, Yoshida K, Tada E (1986) Requirements for structural alloys for superconducting magnet cases. Adv Cryog Eng Mater 32: 23-32

Shin KK (1985) Effect of nitrogen ion implantation on impact wear. Wear 105: 341-347

Siebert S (1994) Randaufsticken nichtrostender Stähle. Doctoral thesis, Ruhr-Universität, Bochum, also Fortschr.Ber. VDI Reihe 5 Nr. 383, VDI Verlag, Düsseldorf

Siegel RW, Fougere GE (1995) Grain size dependent mechanical properties in nanophase materials. In: Otooni MA et al. (eds): Grain size and mechanical properties – fundamentals and applications, Materials Research Society, Pittsburgh, Pennsylvania, 362, pp 219-229

Sieverts A (1931) Die Absorbtion von Stickstoff durch Eisen. Z Phys Chem A 155: 299-313

Sieverts A, Zapf G (1935) Eisen und Stickstoff. Z Phys Chemie A 172: 314-315
Sieverts A, Zapf G, Moritz H (1939) Die Löslichkeit von Wasserstoff, Deutcrium und Stickstoff in Eisen. Zs Phys Chemie A 183: 19-37
Simmons JW (1996) Overview: high-nitrogen alloying of stainless steels. Mater Sci Engng 207A: 159-169
Simmons JW, Covino BS Jr., Hawk JA, Dunning JS (1996) Effect of nitride ($Cr_2N$) precipitation on the mechanical, corrosion and wear properties of austenitic stainless steel. ISIJ Intern 36(7): 846-854
Singhal LK, Martin JW (1968) The formation of ferrite and sigma-phase in some austenitic stainless steels. Acta Metall 16(12): 1441-1451
Singhal LK, Martin JW (1969) Sigma-phase precipitation in austenitic steels. JISI 207(10): 1382
Smith TR, Armstrong RW, Hazzledine PM, Masumura R, Pande CS (1995) Pile-up based Hall-Petch considerations at ultra-fine grain sizes. In Otooni MA et al. (eds): Grain size and mechanical properties – fundamentals and applications, Materials Research Society, Pittsburgh, Pennsylvania, 362, pp 31-37
Solomon HD, Levinson LM (1978) Mössbauer effect study of 475 °C embrittlement of duplex and ferritic stainless steels. Acta Metall 26(3): 429-442
Sozinov AL, Balanyuk AG, Gavriljuk VG (1997) C-C interaction in iron-base austenite and interpretation of Mössbauer spectra. Acta Mater 45(1): 225-232
Sozinov AL, Balanyuk AG, Gavriljuk VG (1999) N-N interaction and nitrogen activity in iron base austenite. Acta Mater 47(3): 927-935
Sozinov AL, Gavriljuk VG (1999) Estimation of interaction energies Me-(C, N) in fcc iron-based alloys using Thermo-Calc thermodynamic database. Accepted for publication in Scripta Mater
Speidel MO (1981) Nichtmagnetisierbare Stähle für Generator-Kappenringe gegen Korrosion und Versprödung. VGB Kraftwerkstechnik 61 (5): 417-427
Speidel MO (1981) Stress corrosion cracking of stainless steels in NaCl solutions. Metall Trans 12A(5): 779-789
Speidel MO, Uggowitzer PJ (1985) Erste experimentelle Hinweise auf Struktur und Eigenschaften massiv aufgestickter 9-12% Chromstähle. Interner Forschungsbericht 38, IMM ETH Zürich
Speidel MO (1987) In: Proc Intern Conf "Stainless Steels '87", The Institute of Metals, York
Speidel MO (1989) Properties and applications of high nitrogen steels: austenites and duplex. In: Foct J., Hendry A (eds) High Nitrogen Steels, HNS 88. Institue of Metals, London, pp 92-96
Speidel MO (1991) Corrosion science of stainless steels. In: Stainless Steels'91, Chiba. The Iron and Steel Institute of Japan, pp 25-35
Speidel MO, Pedrazzolli RM (1992) High nitrogen stainless steels in chloride solutions. Mater Perform 31(9): 59
Speidel MO, Uggowitzer PJ (1993) In: Lula RA (ed) Proc. Materials Week 92, Chicago, ASM Int: 135
Speidel MO (1998) High nitrogen nickel-free stainless steels . In: Hänninen H, Hertzman S (eds) High Nitrogen Steels, HNS 98, Proc of the 5th Intern Conf at Espoo/Stockholm, to be published
Speiser R, Spretnak JW (1955) Thermodynamics of binary Fe-C austenite and cementite. TMS AIME 47(3): 493-507
Spraque JA, McLellan RB (1968) Discussion of «A model for concentrated interstitial solid solutions: the application to solutions of carbon in gamma iron». TMS AIME 242(4): 733-734

Srinivasan A, Reynders B, Grabke HJ (1995) Localised corrosion behaviour of high and low nitrogen Cr-Mn steels. steel research 66(10): 439-443

Steigerwald EA, Schaller FW, Troiano AR (1960) Role of stress in hydrogen induced delayed failure. TMS AIME 218: 832-841

Stein G (1987) P900 – Der Grundwerkstoff für Kappenringe zur Erfüllung gegenwärtiger und zukünftiger Anforderungen. EPRI Generator Retaining Ring Workshop, Charlotte, USA

Stein G, Menzel J, Dörr H (1989) Industrial manufacture of massively nitrogen-alloyed steels. In: Foct J, Hendry A (eds) High Nitrogen Steels, HNS 88. Institute of Metals, London, pp 32-38

Stein G, Trösken F (1990) Arc-welding steels nitrogen-alloyed beyond the solubility limit. In: Stein G, Witulski H (eds) High Nitrogen Steels, HNS 90. Stahl&Eisen, Düsseldorf, pp 303-308

Stein G, Menzel J (1991) Herstellung, Weiterverarbeitung und technische Anwendungsmöglichkeiten massiv aufgestickter Stähle. In: Speidel MO, Uggowitzer PJ (eds) Stickstofflegierte Stähle. Verlag Thubal-Kain, Institut Metallforschung und Metallurgie, ETH-Zürich, pp 33-45

Stein G, Hucklenbroich I, Feichtinger H (1998) Current and future application of high nitrogen steels. In: Hänninen H, Hertzman S (eds) High Nitrogen Steels, HNS 98, Proc of the 5$^{th}$ Intern Conf at Espoo/Stockholm, to be published

Stein G, Hucklenbroich I, Wagner M (1998) P 2000 – A new austenitic high nitrogen steel for power generating equipment. In: Hänninen H, Hertzman S (eds) High Nitrogen Steels, HNS 98, Proc of the 5$^{th}$ Intern Conf at Espoo/Stockholm, to be published

Stoichev T, Kamenova Tz, Ivanov K (1993) Heat treatment effect on abrasive wear resistance of GW-5Mo-0.2N high speed steel. In: Gavriljuk VG, Nadutov VM (eds) High nitrogen steels, HNS 93. Institute for Metal Physics, Kiev, pp 438-442

Stoltz RE, Vander Sande JB (1980) The effect of nitrogen on stacking fault energy of Fe-Ni-Cr-Mn steels. Metall Trans 11A(6): 1033-1037

Strehblow HH (1976) Nucleation and repassivation of corrosion pits for pitting of iron and nickel. Werkstoffe und Korrosion 27: 792-799

Sumitomo H, Nakatuka J (1991) Development of high-strength nonmagnetic stainless steel for cryogenic services. In: Stainless Steels '91, Chiba. The Iron and Steel Institute of Japan, pp 480-486

Sun H, Diener M, Uggowitzer PJ, Speidel MO (1990) Low cycle fatigue behaviour of high nitrogen steels. In: Stein G, Witulski H (eds) High Nitrogen Steels, HNS 90. Stahl&Eisen, Düsseldorf, pp 220-223

Sundman B, Jansson B, Andersson J-O (1985) The Thermo-Calc databank system. CALPHAD 9(2): 153-190

Sundman B (ed) (1995) Thermo-Calc software and databases for thermodynamic calculations. Users' Guide, Royal Institute of Technology, Stockholm.

Sundman B (1997) ThermoCalc users' guide, version L, Royal Institute of Technology, Stockholm

Suyazov AV, Usikov MP, Mogutnov BM (1976) A study of structural transformations in iron-nitrogen alloys. Physics Metals Metallogr (in Russian) 42(4): 755-763

Svyazhin AG, Siwka J, Skuza Z, Hutny A (1998) The gas blow-holes in high nitrogen iron alloys and steels during their crystallization. In: Hänninen H, Hertzman S (eds) High Nitrogen Steels, HNS 98, Proc of the 5$^{th}$ Intern Conf at Espoo/Stockholm, to be published

Suzuki J, Kitamura T (1972) Critical potential for growth of localized corrosion of stainless steels in chloride media. Corrosion 28(1): 1-6

Suzuki H ([1979) In: Haasen P *et al.* (eds) Strength of Metals and Alloys, Pergamon Press, Oxford, 3, p 1595

Svejcar J, Hubackova J, Cihal V, Mazanec K (1987) Austenite stabilization in Cr 13% Ni 4-6% steels. steel research 58 (1): pp 13-17

Swann PR (1963) Dislocation substructure vs transgranular stress corrosion susceptibility of single phase alloys. Corrosion 19(3): 102-112

Swartz JC (1969) The solubility of graphite and cementite in (alpha, delta) iron. TMS AIME 245(5): 1083-1092

Szumer A, Janko A (1979) Hydride phases in austenitic stainless steels. Corrosion 35(10): 461-464

Taillard R, Foct J (1989) Mechanisms of the action of nitrogen interstitials upon low cicle fatigue behaviour of 316 stainless steels. In: Foct J, Hendry A (eds) High Nitrogen Steels, HNS 88. The Institute of Metals, London, pp 387-391

Takaki S, Nakamura N, Goto H. (1998) Alloy design for suppressing eutectoid reaction in high nitrogen austenitic steels. In: Hänninen H, Hertzman S (eds) High Nitrogen Steels, HNS 98, Proc of the 5$^{th}$ Intern Conf at Espoo/Stockholm, to be published

Takahashi Y, Yoshida K, Shimada M, Tada E, Miura R, Shimamoto S (1982) Mechanical evaluation of nitrogen-strengthened stainless steels at 4 K. Advan Cryog Eng 28: 73-82

Takemoto T, Igawa T, Murata Y, Uematsu Y (1990) Effects of nitrogen and themomechanical treatment on mechanical and magnetic properties of metastable austenitic stainless steel. In: Stein G, Witulski H (eds) High Nitrogen Steels, HNS 90. Stahl&Eisen, Düsseldorf, pp 161-166

Tanaka M (1994) Grain refinement by thermal cycling in high-nitrogen austenitic heat-resistant steels. Z Metallkd 85(6): 446-452

Tanaka H, Nagakura S, Nakamura Y, Hirotsu Y (1997) Electron crystallography study of tempered iron-nitrogen martensite and structure refinement of precipitated $Fe_{16}N_2$. Acta Mater 45(4): 1401-1410

Tanaka T, Ito K, Hoshino K (1983) Effect of alloying elements, cold-rolling reduction and deformation-induced martensite on mechanical properties of low carbon type 301 hard stainless steel (Development of low carbon high strength steinless steels–II) Transactions ISIJ 23: B 139, 140

Tendo M, Takeshita T, Nakazawa T, Abo H (1991) Room temperature creep behaviour of austenitic stainless steels. In: Stainless Steels'91, Chiba. The Iron and Steel Institute of Japan, pp 487-493

Tervo J (1997) Wear properties of high nitrogen austenitic stainless steels. Doctor thesis. Luukkala M (ed) Acta Polytechnica Scandinavica, Mechanical engineering series No. 128, The Finnish Academy of Technology, Helsinki

Tervo J (1998) Wear properties of HNS. In: Hänninen H, Hertzman S (eds) High Nitrogen Steels, HNS 98. Proc of the 5$^{th}$ Intern Conf at Espoo/Stockholm, to be published

Tervo J, Tarasenko A, Hänninen H (1998) Effect of nitrogen on elastic coefficients of austenitic stainless steels. In: Hänninen H, Hertzman S (eds) High Nitrogen Steels, HNS 98. Proc of the 5$^{th}$ Intern Conf at Espoo/Stockholm, to be published

Theisen W (1997) Bearbeiten verschleißbeständiger Legierungen aus werkstofftechnischer Sicht. Dr. habil. thesis, Ruhr-Universität, Bochum, also Fortschr. Ber. VDI Reihe 2 Nr. 428, VDI Verlag, Düsseldorf

Thien V (1979) Supraleiter Generatorläufer. Thyssen Edelst. Techn. Ber. 5 (3): 198-200

Thier H (1967) Der Einfluss von Stickstoff auf das Ausscheidungsverhalten des austenitischen Chrom-Nickel-Stahles X 5 CrNiMo 1713. Doctoral Thesis, Rheinisch-Westfälische Technische Hochschule, Aachen

Thier H, Bäumel A, Schmidtmann E (1969) Einfluß von Stickstoff auf das Ausscheidungsverhalten des Stahles X 5CrNiMo 1713. Arch Eisenhüttenw 40(4): 333-339

Thomas G (1963) The effect of short-range order on stacking fault energy and dislocation arrangements in fcc solid solutions. Acta Metall 11(12): 1369-1371

Thomas B (1976) Microscopie electronique et deformation plastique des aciers inoxydables austentiques industriels. J Microsc Spectrosc Electron 1: 623-626

Thompson AW, Bernstein IM (1980) In: Stress Corrosion Cracking. Freund Publishing House, Tel-Aviv, pp 193-230

Tobler RL, Reed RP (1980) Interstitial carbon and nitrogen effects on the tensile and fracture parameters of AISI 304 stainless steels. In: Reed RP (ed) Materials Studies for Magnetic Fusion Energy Applications at Low Temperatures-III, NBSIR 80-1627, Fracture and Deformation Division, National Bureau of Standards, Boulder, Colorado, pp 17-48

Tobler RL, Reed RP (1984) Interstitial carbon and nitrogen effects on the cryogenic fatigue crack growth of AISI 304 type stainless steels. J Testing Evaluation 12(6): 364-370

Tobler RL, Meyn D (1988) Cleavage-like fracture along slip planes in Fe-18Cr-3Ni-13Mn-0.37N austenitic stainless steel at liquid helium temperature. Metall Trans 19A(6): 1626-1631

Tomasello CM, Maloney JL, Ward PC, Materkowski JP (1998) A new corrosion resistant martensitic stainless steel for improved performance in miniature bearings. In: Hoo JJC, Green WB (eds) Bearings steels: Into the 21st century. ASTM STP 1327, West Conshohocken, pp 437-446

Tomota Y, Endo S (1990) Cleavage-like fracture at low temperatures in an 18Mn-18Cr-0.5N austenitic steel. ISIJ Intern 30(8): 656-662

Tomota Y, Xia Y, Inoue K (1998) Mechanism of low temperature brittle fracture in high nitrogen bearing austenitic steels. Acta Mater 46(5): 1577-1587

Tomota Y, Nakano J, Xia Y, Inoue K (1998 I) Unusual strain rate dependence of low temperature fracture behaviour in high nitrogen bearing austenitic steels. Acta Mater 46(9): 3099-3108

Torkhov GF, Grigorenko GM, Lakomski VI, Pomarin YuM (1971) Nitrogen behaviour in liquid iron-chromium alloys. Avtomatical Welding (in Russian) 10: 16-20

Torkhov GF, Latash YuV, Fessler RR, Clauer AH, Fletcher EE, Hoffmanner AL (1978) Development of melting and thermomechanical-processing parameters for a high-nitrogen stainless steel prepared by plasma-arc melting. Journal of Metals: 20-27

Trojahn W (1992) High nitrogen martensitic steels – A new family of martensitic corrosion resistant steels for improved aerospace bearing performance. Proc. Gas Turbine and Aeroengine Congress and Exposition, Cologne, ASME 92-GT-338, New York

Trojahn W (1998) Die Entwicklung des hochstickstofflegierten Stahles Cronidur 30 von der Forschung zum Einsatz. Oral Presentation Werkstofftechn. Kolloquium, Ruhr-Universität Bochum

Truman JE (1989) Effects of nitrogen on corrosion behaviour of high alloy steels. In: J.Foct and A.Hendry (eds) High Nitrogen Steels, HNS 88. Institute of Metals, London, pp 225-239

Truman JE, Lomax KB (1998) Effect of section size on the properties of a nitrogen-bearing austenitic steel of high corrorions resistance. In: Hänninen H, Hertzman S (eds) High Nitrogen Steels, HNS 98, Proc of the 5th Intern Conf at Espoo/Stockholm, to be published

Tsai WT, Reynders B, Stratmann M, Grabke HJ (1993) The effect of applied potential on the stress corrosion cracking behaviour of high nitrogen steels. Corrosion Sci 34(19): 1647

Tseng CC, Shen Y, Thompson SW, Mataya MC, Krauss G (1994) Fracture and the formation of sigma phase, $M_{23}C_6$, and austenite from delta-ferrite in an AISI 304L stainless steel. Metall Mater Trans 25A(6): 1147-1158

Tuma H, Landa V, Löbl K (1981) Changements de phasespar maintien isotherme de longue durée dans les aciers austénitiques du type 16Cr-14Ni-2,5Mo en fonction de la teneur en azote jusqu'à 0.2%. Mémoires et Etudes Scientifiques Revue de Métallurgie 5: 2 55-259

Uetz H, Sommer K (1986) Abrasiv-Gleitverschleiß. In: Uetz H (ed) Abrasion and erosion, Carl Hanser Verlag, München, pp 144-148

Uggowitzer PJ, Holm Ch, Speidel MO, Harzenmoser M (1985) Das Ausscheidungsverhalten stickstofflegierter austenitischer Stähle. Interner Forschungsbericht Nr XLVI am Institut für Metallforschung und Metallurgie, ETH, Zürich

Uggowitzer PJ, Harzenmoser M (1989) Strengthening of austenitic steels by nitrogen. In: J.Foct and A.Hendry (eds) High Nitrogen Steels, HNS 88. Institute of Metals, London, pp 174-179

Uggowitzer PJ, Speidel MO (1990) Ultrahigh-strength austenitic steels. In: Stein G, Witulsky H (eds) High Nitrogen Steels, HNS 90, Stahl&Eisen, Düsseldorf, pp156-160

Uggowitzer PJ, Speidel MO (1991) Ultrahigh-strength C r-Mn-N-steels. In: Stainless Steels '91, Chiba. The Iron and Steel Institute of Japan, pp 762-770

Uggowitzer PJ, Paulus N, Speidel MO (1992) Ductile-to-brittle transition in nitrogen alloyed austenitic stainles steels. In: Nordberg H, Börklund J (eds), Application of Stainless Steels'92, The Institute of Metals ASM Intern, Stockholm, pp62-72

Uggowitzer PJ (1993) Room temperature creep of high nitrogen steels. In: Gavriljuk VG, Nadutov VM (eds) High Nitrogen Steels, HNS 93. Institute for Metal Physics, Kiev, pp 345-352

Uggowitzer PJ, Magdowski R, Speidel MO (1993) High nitrogen austenitic stainless steels – properties and new developments. In: Assozione Italiana di Metallurgia, "Innovation Stainless Steel", Florence, Italy, pp 2.359-2372

Uggowitzer PJ, Magdowski R, Speidel MO (1994) High nitrogen austenitic stainless steels - Properties and new developments. La metallurgia italiana 86 (6,7) 347-353

Uggowitzer PJ, Magdowski R, Speidel MO (1996) Nickel free high nitrogen austenitic steels. ISIJ Intern 36(7): 901-908

Uggowitzer PJ, Bähre W-F, Wohlfromm H, Speidel MO (1998) Nickel-free high nitrogen austenitic stainless steels produced by metal injection moulding. In: Hänninen H, Hertzman S (eds) High Nitrogen Steels, HNS 98, Proc of the 5$^{th}$ Intern Conf at Espoo/Stockholm, to be published

Ulitchny MG Gibala R (1973) Internal friction and strain aging of ferrous austenite. Metall Trans 4A(2): 497-506

Ullakko K, Gavriljuk VG (1992) Effects of coherent interfaces in the freshly formed iron-nickel-carbon martensites, Acta Metall Mater 40(10): 2471-2482

Ullakko K, Tarkiainen R, Levonmaa R, Gavriljuk VG (1993) Hot and cold rolling of high-nitrogen Fe-Mn-Cr-N and Fe-Ni-Cr-N austenitic stainless steels. In: Gavriljuk VG, Nadutov VM (eds) High nitrogen steels, HNS 93. Institute for Metal Physics, Kiev, pp 401-403

Ullakko K, Gavriljuk VG, Nadutov VM (1994) Aging of freshly formed Fe-based martensites at low temperatures. Metall Mater Trans 25A(5): 889-909

Ulmer DG, Altstetter CJ (1993) Phase relations in the hydrogen-austenite system. Acta Metall Mater 41(7): 2235-2241

Urrutibeaskoa I, Jauregi S, Fernández F, Talacchia S, Palma R, Martinez V, Urcola J (1993) Improved sintering response of vanadium-rich high speed steels. Int. J. Powder Metall. 29 (4): 367-378

Ustinovshikov Y, Ruts A, Bannykh O, Blinov V (1996) The microstructure of the Fe-18%Cr alloys with high N contents. Acta Mater 44(3): 1119-1125

Van Aswegen JST, Honeycombe RWK, Warrington DH (1964) Precipitation on stacking faults in Cr-Ni austenitic steels. Acta Metall 12(1): 1-14

Van Cenderen MJ, Böttger A, Mittemeijer EJ (1992) First stage of precipitation in iron-carbon-nitrogen martensites; diffraction analysis using synchrotron diffraction. Scripta Metall et Mater 26(6): 883-888

Van Genderen MJ, Böttger A, Mittemeijer EJ (1997) Formation of $\alpha''$ iron nitride in FeN martensite: nitrogen vacancies, iron-atom displacements, and misfit-strain energy. Metall Mater Trans 28A(1): 63-77

Vanderschaeve F, Taillard R, Foct J (1993) Effect of heat treatment on the microstructure of a high nitrogen 12% chromium martensitic steel. steel research 64(4): 221-227

Vanderschaeve F, Taillard R, Foct J (1995) Discontinuous precipitation of $Cr_2N$ in high nitrogen chromium-manganese austenitic stainless steel. J Mater Sci 30(23): 6035-6046

Van Gent A, Van Doorn FC, Mittemejier EJ (1985) Crystallography and tempering behaviour of iron-nitrogen martensite. Metall Trans 16A(8): 1371-1384

Varin RA, Kurjidlowski KJ (1988) The effects of nitrogen content and twin boundaries on the yield strength of various commercial heats of type 316 austenitic stainless steel. Mater Sci Engng A 101: 221-226

Vehanen A, Hautojärvi P, Johansson J, Yli-Kauppila J (1982) Vacancies and carbon impurities in $\alpha$-iron: electron irradiation. Phys Rev B 25(2): 762-780

Vergnol JFM, Grilhe JR (1984) Relatioship between extrinsic stacking faults and mechanical twinning in fcc solid solutions with low stacking fault energy. J Physique 45: 1479-1490

Verner VD, Finkelstein BN, Shalimova AV (1961) Internal friction study of nitrogen behaviour in fcc iron alloys. Solid State Physics (Soviet Physics) 3(11): 3363-3366

Verner VD (1965) To the nature of internal friction peak in solid solutions with fcc lattice. Solid State Physics (Soviet Physics) 3(11): 2318-2326

Vincze I, Campbell IA (1973) Mössbauer measurements in iron based alloys with transition metals. J Phys F: Metal Phys 3: 647-662

Vladimirov VI, Romanov AE (1986) Disclinations in crystals (in Russian). Nauka, Leningrad, 223 pp

Vladimirov VI (ed) (1988) Disclinations and rotation deformation of solids. Ioffe Institute of Physical Engineering, Leningrad, 226 pp

Vlasova EN (1963) Diffusion scattering of X-rays and the fine structure of iron-aluminum alloys. Phys Met Metallogr (in Russian) 16(3): 355-360

Vlasova EN (1972) Ordering in iron-silicium alloys. Phys Metal Metallogr (in Russian) 33(1): 130-136

Vogt JB, Foct J, Regnard C, Robert G, Dhers J (1991) Low-temperature fatique of 316L and 316LN austenitic stainless steels. Metall Trans 22A(10): 2385-2392

Vogt JB, Magnin T, Foct J (1993) Effective stresses and microstructure in cyclically deformed 316L austenitic stainless steel: effect of temperature and nitrogen content. Fatique Fract Engng Mater Struct 16(5): 555-564

Vogt JB, Messai A, Foct J (1994) Cleavage fracture of austenite induced by nitrogen supersaturation. Scripta Metall Mater 31(5): 549-554

Vogt JB, Bigeon C, Foct J (1994 I) Combined effect of nitrogen and silicon on low cycle fatigue of 12%Cr martensitic stainless steels. Z Metallkd 85(2): 92-99

Vogt JB, Messai A, Foct J (1996) Sensitivity of a high nitrogen austenitic stainless steel to fatigue crack initiation. ISIJ Intern 36(7): 862-866

Vogt JB, Massol K, Foct J (1998) Low cycle fatigue of high nitrogen duplex stainless steels. In: Hänninen, H, Hertzman S (eds) High nitrogen Steels, HNS 98, Proc. of the 5$^{th}$ Intern Conf at Espoo/Stockholm, to be published

Vogt JB, Saadi BA, Foct J (1999) Analysis of the fatigue dislocation structures in a duplex stainless steel alloyed with nitrogen. Z Metallkd 90(5): 323-328

Vonsovsky SV (1971) Magnetism (in Russian). Nauka, Moscow

VSG (1995) Data sheets on CRONIDUR, Vereinigte Schmiedewerke, Essen

VSG (1996) Data sheet on P 900, Vereinigte Schmiedewerke, Essen

Wada H and Pehlke RD (1977) Solubility of nitrogen in liquid Fe-Cr-Ni alloys containing manganese and molybdenum. Metall Trans 8B(12): 675-682

Wada H and Pehlke RD (1977 I) Nitrogen solution and titanium nitride precipitation in liquid Fe-Cr-Ni alloys. Metall Trans 8B(9): 443-450

Wagemann B (1995) Der Einfluß der Nichtmetalle Schwefel, Stickstoff und Kohlenstoff auf die Oxidation der Legierungen Fe-20Cr und Ni-20Cr. Doctoral thesis, Universität Dortmund

Wagner C (1962) Thermodynamics of Alloys. Addison-Wesley, Reading, MA

Wahlberg G, Dunlop GL (1987) Nitrogen strengthening of duplex stainless steels. In: Proc. Stainless Steel '87, York. The Institute for Metals, pp 291-299

Wahlberg G, Rolander U and Andrén HO (1989) Interactions between nitrogen and substitutional elements in the austenitic phase of duplex austenitic-ferritic stainless steels. In: Foct J, Hendry A (eds) High Nitrogen Steels, HNS 88. The Institute of Metals, London, pp 163-168

Walker LR, Wertheim GK, Jaccarino V (1961) Interpretation of the $^{57}$Fe isomer shift. Phys Rev Letters 6(3): 98-101

Wang G (1992) Härtbare nichtrostende PM-Stähle und Stahlverbunde mit hohem Stickstoff-gehalt. Doctoral thesis, Ruhr-Universität, Bochum, also Fortschr.Ber. VDI, Reihe 5 Nr 277, VDI-Verlag, Düsseldorf

Wang J (1998) High strength nickel-free duplex stainless steels for structural engineering applications. Doctoral thesis 12649, Swiss Federal Institute of Technology, Zürich

Wang J, Uggowitzer PJ, Magdowski R, Speidel MO (1999) Nickel-free Duplex Stainless steels. Scripta Materialia 40(1): 123-129

Warlimont H (1968) Elektronenmikroskopische Untersuchung der Gleichgewichte und Umwandlungen der $\alpha$-Eisen-Silizium-Phasen. Z Metallkd 59(8): 595-602

Watanabe M, Wayman CM (1971) Highly tetragonal martensite in Fe-Al-C steels. Scripta Metall 5(2): 109-116

Watanabe K, Okuma N, Fukai Y, Sakamoto Y, Hayashi Y (1996) Superabundant vacancies and enhanced diffusion in Pd-Rh alloys under high hydrogen pressures. Scripta Mater 34(4): 551-557

Weiss RJ (1963) The origin of the invar effect. Proc Phys Soc 82(525): 281-288

Weiss B and Stickler R (1972) Phase instabilities during high temperature exposure of 316 austenitic stainless steel. Metall Trans 3A(4): 851-866

Weller M, Diehl J (1976) Internal friction studies on reaction of carbon and nitrogen with lattice defects in neutron irradiated iron. Scripta Metall 10(2): 101-105

Wendl F (1985) Einfluß der Fertigung auf Gefüge und Zähigkeit von Warmarbeitsstählen mit 5% Chrom. Doctoral thesis, Ruhr-Universität, Bochum, also Forschr.Ber. VDI Reihe 18 Nr. 29, VDI Verlag, Düsseldorf

Wentrup H, Reif O (1949) Über die Löslichkeit von Stickstoff in Eisenschmelzen mit Chrom-,Mangan- und Nickelzusätzen. Arch Eisenhüttenw 20(11/12): 359-362

Werner E (1988) Solid solution and grain size hardening of nitrogen-alloyed austenitic steels. Mater Sci Engng 101A(1): 93-98

Werner H, Voigt C, Riedel G, Günzel M, Simmchen R (1990) Korrosion, Dresden 21: 3

Werries H (1995) Korrosionsbeständige Wälzlager. Vorhab. Nr. 129, AIF Nr. 7771, Forschungsheft 201, FKM-FVA, Frankfurt

Wessling W, Heimann W (1993) Nonmagnetizable steels. In: Verein Deutscher Eisenhüttenleute (eds.) Steel Vol 2. Springer Verlag, Berlin, pp 547-555

Wessling W, Ulm F (1993) Steels vor valves in internal combustion engines. In: Verein Deutscher Eisenhüttenleute (ed.) Steel Vol 2. Springer Verlag, Berlin, pp 452-467

Whitefield DJ, Van Bennekom A (1996) Abrasive wear properties of experimental metastable duplex stainless steels. Wear 196: 92-99

Wiegand H and Doruk M (1962) Einfluß von Kohlenstoff und Molybdän auf die Ausscheidungsvorgänge, besonders auf die Bildung intermetallischer Phasen in austenitischen Chrom-Nickel-Stählen. Archiv Eisenhüttenw 33(8): 559-566

Wright CS, Mascarenhas JMG, Oliveira MM (1998) Sintering mechanism and microstructures of high speed steels containing 6-11% vanadium sintered in a $N_2$-7%$H_2$ atmosphere. Proc. Powder Metallurgy World Congress, Granada, pp 377-382

Wronski AS, Wright CS, Iturriza I (1998) Some recent developments in direct sintering of water-atomised – high speed steel powders to full density. Proc. Powder Metallurgy World Congress, Granada

Wu TL, Wang CM (1958) Mechanism of carbon diffusion peak in fcc iron-nickel alloys. Acta Physica Sinica 14(4): 354-368

Wood JH (1962) Energy bands in iron via the augmented plane wave method. Phys Rev, 126(2), pp 517-522

Woodyatt LRC, Sims CT, Beattie HJ (1966) Prediction of sigma-type phase occurrence from compositions in austenitic superalloys. Trans TMS-AIME 236(4): 519-527

Zapp (1997) Data sheet on CROMANITE, Robert Zapp Werkstofftechnik, Düsseldorf

Zheng X-H (1991) Nitrogen solubility in iron-base alloys and powder metallurgy of high nitrogen stainless steels. Doctoral thesis 9488, Swiss Federal Institute of Technology, Zürich Zheng X-O (1991) PhD Thesis, Diss. ETH No. 9488, Swiss Federal Institute of Technology, Zürich

Zitter H, Habel L (1973) zur Löslichkeit des Stickstoffs in Reineisen und austenitischen Chrom-Nickel-Stählen. Arch Eisenhüttenw 44(3): 181-188

# Index

Printing: Mercedes-Druck, Berlin
Binding: Stürtz AG, Würzburg